The FORGES du Saint-Maurice

Beginnings of the Iron and
Steel Industry in Canada

1 7 3 0 - 1 8 8 3

The FORGES du Saint-Maurice

Beginnings of the Iron and Steel Industry in Canada

1 7 3 0 - 1 8 8 3

ROCH SAMSON

 Canadian Heritage Patrimoine canadien

Parks Canada Parcs Canada

LES PRESSES DE L'UNIVERSITÉ LAVAL

The publishers gratefully acknowledge the financial
assistance of the Government of Canada through
the Book Publishing Industry Development Program.

Canadian Cataloguing in Publication Data

Samson, Roch

The Forges du Saint-Maurice: beginnings of the iron and steel industry in Canada,
1730–1883

Issued also in French under title: Les Forges du Saint-Maurice. Includes bibliogr-
aphical references and index.

ISBN 2-7637-7549-7

1. Forges du Saint-Maurice (Trois-Rivières, Quebec) – History. 2. Foundries –
Quebec (Province) – Trois-Rivières Region – History. 3. Iron industry and trade –
Quebec (Province) – Trois-Rivières Region – History. I. Title.

TS229.5.C2S2513 1998   338.4'76691'0971445   C98-940379-3

Book design: Norman Dupuis

Page layout: Marc Brazeau
              Norman Dupuis

Cover illustration
Eighteenth-century blast furnace. In the centre, two men skim
the slag from the surface of molten iron. On the right, one man collects
the cast iron with a ladle while another makes a sand casting.
Léonard Defrance, *Inside the Foundry*, oil on wood,
Musées royaux des beaux-arts de Belgique, Brussels

Published by Les Presses de l'Université Laval and the Department
of Canadian Heritage—Parks Canada in co-operation with the Department
of Public Works and Government Services Canada.

Catalogue No.: R64-201-1998E

ISBN 2-7637-7549-7

SEP 1 8 1998

Note to readers

Throughout the book, the units of measure current in 18th-
century France have been retained for easier understanding
of certain technical discussions. The French foot (*pied*)
(equal to 12 inches or *pouces*) is equivalent to 32.484 cm
and the French pound (*avoirdupois*), 489.41 g. The *millier* or
thousandweight (1,000 French pounds) is equivalent to
489.41 kg. The tonne (t) is the metric ton, equivalent to
2043.3 French pounds and the imperial ton is the long ton,
equal to 2,240 lbs. Units of imperial, or English, measure
have been converted to the metric system except where
explicitly stated.

The French pound currency (*livre*) is denoted as such and
represented in tables in italics as *l s d*, whereas the British
pound sterling is shown in roman as £ s d . The *livre* is sub-
divided into *sols* and *deniers*, with 12 *deniers* making a *sol*,
and 20 *sols*, a *livre*. Similarly, the pound sterling (£) is divided
into shillings (s) and pence (d), with 12 pence making a
shilling and 20 shillings, £1. At the time of the Conquest of
New France in 1760, the £ was equivalent to 24 *livres*.

For historical accuracy, Quebec City as it is now known is
referred to by its old name "Quebec," which was the established usage in English throughout the lifetime of the Forges
(pronounced *à la française*).

*Words shown in bold are defined
in the glossary.*

The Forges du Saint-Maurice National Historic Site is a source of great pride to me as a Canadian and a native of La Mauricie. When I presided over its creation in 1973 as Minister responsible for Parks Canada, there was no site commemorating the origins of Canadian industry in the National Historic Sites System. The Forges du Saint-Maurice, which takes us back to the beginnings of the iron and steel industry in Canada, is a landmark site, a jewel of our industrial heritage.

Already in its 25th year, this National Historic Site has lost none of its appeal thanks to the originality of its development concept, which is without doubt one of Parks Canada's most brilliant successes. The publication of this prestigious book marks this anniversary by presenting, for the first time, a comprehensive history of the Forges incorporating the fruits of a remarkable research effort on the part of Parks Canada.

As a salute to Canada's industrial roots and in celebration of its own 100th anniversary, the Canadian Institute of Mining, Metallurgy and Petroleum graciously accepted to sponsor this publication and contribute to its funding. As honorary president of this centennial, I can appreciate the extent of the commitment of the Institute's Centennial Corporation, which has made it possible to acknowledge both the birthplace of early industry and the origins of modern industry in Canada.

Such an initiative required, as well, the participation of a major representative of contemporary iron and steel producers, in the form of financial support for this book on the industry's origins. I thank Ispat Sidbec for paying this fitting tribute to Canada's industrial history as a successor to the 18th-century pioneers of the Forges du Saint-Maurice.

I am pleased at the excellent quality and the achievements of the industrial partnership formed to commemorate the rich history of the Forges du Saint-Maurice. This collaboration reflects the deep roots of an industry that has grown with Canada since its earliest days and has played an important role in building our national identity.

I wish you many hours of interesting reading as you learn about the entrepreneurs and workers who were responsible for establishing Canada's first industrial community.

Jean Chrétien

OTTAWA
1998

*To Gladys
and Jean-Christophe*

Preface

This book stems from the massive multidisciplinary research program undertaken by Parks Canada to develop the Forges du Saint-Maurice National Historic Site created in 1973. I have been personally involved in this project since 1982 as a researcher and as a contributor to the extensive interpretive program for the historic site. This monograph is thus rooted in both the findings of the archaeological excavations and the historical research launched in the 1960s by the Quebec Department of Cultural Affairs and continued by Parks Canada. If I have been able to revisit and distil the fruits of over 20 years of research it is thanks to the working conditions conducive to the pursuit of research and professional endeavour at Parks Canada. My colleagues on the Forges project, be they historians, archaeologists, ethnologists, engineers, architects, designers, illustrators, museologists, curators, communications officers, historic site superintendents or interpretors, not to mention administrative support personnel, have all been of inestimable help throughout the process. I extend to them all my deepest gratitude.

I would especially like to thank the management of Parks Canada, Department of Canadian Heritage, particularly Quebec Executive Director Laurent Tremblay and his predecessor Gilles Desaulniers, for their support in allowing me to immerse myself so deeply in such a research project. I am deeply grateful to Louis Richer, Head, Cultural Resource Management, for the confidence he has shown in me over all these years. Thanks also to my colleague André Bérubé, my long-time supervisor and frequent source of inspiration, and co-workers Réal Boissonnault, Michel Bédard and Pierre Drouin for their unfailing assistance and great generosity. Sincere thanks to Marie Lavoie, Director, Mauricie District Office and to Carmen Desfossés Lepage, Superintendent of the St Maurice Forges, for their support and interest in my research, and especially for their patience in the long wait for it to bear fruit. I am particularly grateful to my colleagues and friends Michel Barry and Pierre Lessard, and to Albert Nollet, Richard Hébert, Martine Bugeaud and Pierre Demers for their professional complicity and for their constant encouragement and motivation in keeping the flame burning brightly. I owe a profound debt of gratitude to my former colleague, engineer Achille Fontaine, for instilling in me an appreciation of the special features of the Forges site and the hydromechanical works designed by the ironmasters.

I am immensely grateful to Professor Marc Vallières of Laval University for his enthusiastic endorsement of the idea of making the book my doctoral thesis and for his subsequent guidance. Thanks also to Professors Jacques Mathieu and Marcel Moussette of the History Department of Laval University for their sage advice in the formulation, refining and elaboration of my thesis project.

Various organizations and archives were of great assistance in helping me track down and access collections of manuscripts, maps and illustrations. I wish to thank the archivists of the National Archives of Canada, National Archives of Quebec in Quebec City, Trois-Rivières and Montreal, Archives of Ontario, Archives du Séminaire de Trois-Rivières, Archives de l'Évêché de Trois-Rivières, the archives of Laval and McGill universities and the University of Montreal, of the Parish of L'Immaculée Conception des Trois-Rivières and St James Anglican Church of Three Rivers. Various museums and depositories of picture collections have also been of invaluable help: the Bibliothèque nationale de France, Musée des beaux-arts de Montbard, la Documentation française, Les Musées royaux des beaux-arts de Belgique, the Quebec Department of Culture and Communications, National Library of Quebec, McCord Museum, Stewart Museum, Regional Museum of Vaudreuil-Soulanges, and the Hôpital général de Québec.

I am also indebted to private collectors, some of whom are descendants of Forges masters and workers, for their contribution to the historical record of the Forges: John McGreevy, Lawrence and David McDougall, Eric Sprenger, Armour Landry and Raoul Rathier.

I extend my warm thanks to Joan Dyer, Translation Bureau, Public Works and Government Services Canada, co-ordinator of the English version of this book and her team of translators for their exemplary professionalism in this mammoth task.

I sincerely thank the following professionals who contributed in various ways to the making of this book: Bernard Duchesne and Yolande Larochelle (historical reconstructions), Jean Audet, Jean Jolin and Jacques Beardsell (photographs), François Pellerin, Louis Lavoie, Jacques Laplante and Nicole Delisle (drawings and illustrations), Diane Lebrun (artifacts), Andrée Heroux (maps and diagrams), Andrée Raîche-Dussault (glossary) and Noëlla Gauthier (bibliography).

My sincerest thanks to Louis Richard and Suzanne Adam-Filion, of the Parks Canada Publications Unit, Ottawa and to Joanne Joanisse, Public Works and Government Services Canada, who co-ordinated the editing and co-publishing of the book, not to mention editor André Larose who painstakingly edited the manuscript.

I am also deeply grateful to the Presses de l'Université Laval, former director Denis Vaugeois, Léo Jacques, Director of Development and his production team Geneviève Saladin, Jocelyne Naud and Peter Frost and especially art director Norman Dupuis for the high quality of the graphic design and layout.

My special thanks go to the engineers and industrialists who contributed to the production and distribution of this book which traces their roots back to the St Maurice Forges. I owe a particular debt to Professor Fathi Habashi, Chairman of the Historical Metallurgy Committee of the Metallurgical Society of the CIM, for introducing me to the world of industry today.

Roch Samson

Table of Contents

List of Tables and Figures

The site of the lower forge as it was some time before the St Maurice Forges closed.
The chimney of the old, disused *renardière* chafery can be seen, and on the plateau in the background,
the Grande Maison, or Master's House.
LUCIUS RICHARD O'BRIEN, *OLD CHIMNEY AND CHATEAU*, 1882, IN GEORGE MUNRO GRANT, ED.,
PICTURESQUE CANADA: THE COUNTRY AS IT WAS AND IS (TORONTO: BELDEN BROS., 1882), VOL. 1, P. 96.

Introduction

As the expense of a forge and its furnaces

is not inconsiderable, every precaution must be taken

not to embark lightly on such a venture.

Intendant Jean Talon to Colbert, 1670

IRONMAKING IN NEW FRANCE

On his third voyage, in 1541, Jacques Cartier discovered, near Quebec, "a fine **mine** of the best iron ore in the world [...] ready for the **furnace**."[1] Cartier was searching for greater prizes—gold and diamonds—and would be disappointed, but the tenor of his remark reveals that a good **vein** of pure iron ore was nevertheless worth its weight in gold. Next to nobler metals such as gold, copper and lead, iron ore ranked high on the list of minerals sought by the early explorers. That is why four smiths doubling as mine prospectors were part of Cartier's crew. It would be another 200 years before Canada's iron ore mines were actually worked. In the meantime, mapping their location kept interest in these deposits alive. In 1604, Samuel de Champlain also had a miner with him, Simon Le Maistre, who would discover ore deposits in Acadia. More than 10 years later, in his inventory of resources, the founder of Quebec painted an enticing picture of the million-or-so *livres* in revenue that mining the different metals of Canada, including iron ore, could generate. Mine prospecting went no further until administration of the colony reverted to the Crown in 1663.

The chimney of the lower forge chafery photographed in 1921. Close to 15 m high, it is, even today, the most imposing part of the remains of the Forges, which closed for good in 1883.
BIBLIOTHÈQUE NATIONALE DU QUÉBEC, MONTREAL, MASSICOTTE COLLECTION, TROIS-RIVIÈRES.

In the interim, the French Crown established its right to govern the mining of underground resources through provisions included in land-grant deeds under the seigneurial system. In some cases, the Crown imposed the royal tithe, as in its concession to the Compagnie des Cent-Associés (1627–63); in other cases it waived that right, as in the charter of the Compagnie des Indes occidentales (1663–74). Throughout the French regime, except in a few rare cases, all seigneurial grants required the seigneur to "advise the King of any mines" with a view to possible mining.[2] That is what the seigneur of St Maurice, François Poulin de Francheville, did in 1729 to obtain a mining privilege from the King.

Iron mining began to attract interest again in 1663. That year, a special commissioner, Sieur Gaudais, was sent to New France to examine the possibility of opening an iron mine, but it was only with the arrival of the first intendant of New France, Jean Talon, that a willingness to proceed with actual mining manifested itself for the first time. An initial expedition to the mines of Baie St Paul by Sieur de la Tesserie, founder of the Compagnie des Indes occidentales, led nowhere. In 1670, Talon sent an ironmaster, Sieur de La Potardière, to inspect the seigneuries of Champlain and Cap de la Madeleine near Trois-Rivières. The ironmaster confirmed to Talon the quality of the ore there, but the Intendant prudently waited for confirmation from tests on the ore sample that La Potardière took back to France. By 1671, Talon had 1,500 *pipes* of ore collected in anticipation, but La Potardière never returned, and in 1672, Talon himself was recalled to France for good.

Ruins of the blast furnace at the St Maurice Forges in 1903, 20 years after they closed for good. COLLECTION OF ARMOUR LANDRY.

Although it took another 60 years before any action was taken, Talon's efforts had shown that the Trois-Rivières region offered real mining potential. When the time was finally ripe, it would be the region designated for mining. After the Talon episode, the issue of iron ore mining resurfaced almost every five years, except in the 1690s, until Francheville made his proposal in 1729 (see Appendix 1).

It was Jean-Baptiste Colbert, during his time as Minister of Marine (1669–83), who urged Governor Louis de Buade, Comte de Frontenac and Intendant Jacques de Meulles to propose mining projects.[3] Indeed, Colbert himself owned an **ironworks** in France's Berry region and in Mazarin's time as prime minister he took care of the many iron forges belonging to the Cardinal.[4] Given Colbert's aim of "making France at once a State and a Manufactory," the iron industry played a strategic role in his plans to turn the country into an industrial power.[5] Despite the protectionist climate that flowed from his policy of mercantilism, Colbert was not against the idea of iron ore mining in New France, since he saw it as a way of replacing the iron imported from Sweden and other northern countries. In the last quarter of the 17th century, the French market was flooded with iron from Sweden and Spain, which explains the slump in the French iron industry at the beginning of the 18th century.[6] It is therefore hardly surprising that there was not more support for plans to establish an ironworks in the colony, at a time when competition was driving **blast furnaces** and **forges** in France out of business. Seignelay, Colbert's son and successor at the Ministry of Marine, put a stop to colonial mining exploration in 1684.[7] In Berry, one of France's major iron-producing regions, half of the blast furnaces were idle in 1716, and everywhere the industry was reported to be in decline. Ironworks owners demanded that France's borders be closed to Spanish and Swedish imports.[8] It is no wonder, therefore, that in 1717 the Duc d'Orléans, regent at the

time, dismissed another proposed iron mining project in the colony with the terse comment: "There is enough in France to supply all of Canada."[9] This was the only time that a clearly mercantilist argument was used, but it was sparked by the sorry state of the French iron industry.

The colonial authorities nevertheless continued to submit proposals, stressing all of the benefits that New France would derive from the establishment of an ironworks: land clearing would be advanced by the attendant woodcutting; it could manufacture cannons and anchors and provide the iron needed for shipbuilding; it could make stoves and necessities for the colony; moreover, surplus iron would be shipped to France and thus help redress the colony's trade balance. When Jean-Frédéric Phélypeaux, Comte de Maurepas became Minister of **Marine** in 1723, everything became possible once again. His attitude was favourable, and so the Intendant of New France, Gilles Hocquart, devised a strategy tying ironmaking to shipbuilding in the colony, which resulted finally in Francheville's proposal.

Beginning of the first archaeological excavations in 1966, conducted by the Quebec Department of Cultural Affairs, under the direction of archaeologist Michel Gaumond. The remains of the Grande Maison can be seen on the plateau overlooking the St Maurice.
QUEBEC, MINISTÈRE DE LA CULTURE ET DES COMMUNICATIONS, MICHEL GAUMOND COLLECTION, 1966.

When the Forges du Saint-Maurice National Historic Site was established
in 1973, the most extensive archaeological digs in Canada were undertaken.
The remains unearthed by Parks Canada archaeologists helped improve
knowledge of the first ironworks in Canada.
PHOTOGRAPH BY PARKS CANADA/JEAN AUDET, 130/PA/PR-6/S-16-6, 1977.

The time lag in setting up an ironworks in New France was thus due largely to the state of the iron industry and market in France. We also know what it cost in France to set up an ironworks. In a great many cases, the government propped up the industry by providing the necessary capital in the form of loans and advances or by taking over bankrupt establishments.[10] The first cost estimate for an ironworks in the colony, drawn up by Breton **ironmaster** Pierre Hameau, in 1689, was over 200,000 *livres*, entailing investment of 100,000 *livres* by the Crown.[11] This was 40 years before the St Maurice Forges began their long history, yet Hameau was not far off in his estimate. An ironworks would also require skilled manpower not available in the colony; in his project, Hameau had estimated that such a work force would be "one of the largest expenses." Until 1730 the French government was clearly not prepared to invest heavily in such an undertaking.

Since Colbert's day, the preference had been for privately financed ventures, but that predicated a thriving economy. In France, the recession that began in the last quarter of the 17th century effectively blocked the initiatives of French entrepreneurs. In New France, there was no private individual with deep enough pockets to make the capital outlay required. In addition, the wars in which France was almost continually engaged between 1667 and 1713 made conditions even less favourable and swallowed up the capital that the government might otherwise have invested in the colony.[12] With peace restored in 1713, capital did not immediately become available, at least in part, as we have seen, because of the poor state of the French iron industry. But Maurepas's colonial policy, which favoured increased sea trade, justified ironmaking as a means of supplying shipbuilding and made government capital available for colonial projects. Maurepas was of course in favour of free enterprise, which was why Francheville's proposal was well received. But it was not the modest **bloomery** forge proposal put forward by the seigneur of St Maurice, nor his personal fortune, that would finally allow the project to see the light of day. The scope of the desired establishment, and the role it was intended to play in supplying the shipbuilding industry, were beyond Francheville's means. The cost of such an endeavour was well known, and this time the French government was ready to invest.

THE EARLY IRON INDUSTRY

Establishing an ironworks in New France was a considerable undertaking. Clearly, funding from the royal exchequer was inevitable in view of the inherent costs and constraints. The early ironmaking industry was a world unto itself, with its own internal constraints. The technology used and the seasonal rhythm of work it imposed, as well as its voracious appetite for raw materials, were served by a multifarious work force operating to a precise schedule. Ironworkers were a breed apart with their own ways, technical and social, forged by over two centuries of tradition. None of this existed in New France in the early 18th century. Everything had to be imported from France: capital, technology and labour. This in-bred world had to be transplanted to the New World, a process that would prove to be dramatic and difficult, but lasting. Indeed, the history of the first industrial community in Canada would have been completely different had the industry in question been a different sort of industry. Thus, recognizing the characteristics of the iron industry helps us to grasp more fully the different dimensions and constraints involved in setting up an ironworks in New France.

Ironworks of the type that the Forges would be patterned on were legion in 18th-century Europe. There were hundreds of such plants in France,[13] most commonly in the form of an integrated complex comprising a blast furnace and forges. In the Champagne region of northeastern France, birthplace of the ironmaster Pierre-François Olivier de Vézin, the most common model was that of a furnace and two forges. Vézin followed that model in establishing the St Maurice Forges.

The ironworks originated in the 14th century when the invention of the blast furnace introduced the process of **indirect** ore reduction;[14] in the furnace, the ore was first smelted into **cast iron**, which was then refined into **wrought iron** in a forge hearth called a **finery** or **chafery**. This technological innovation created a two-step process divided between **founders**, **fillers** and **moulders** toiling at the blast furnace, and **forgemen** labouring at the forges. This division of labour would not be altered substantially until the introduction of the coke-fired blast furnace and puddling furnace in 18th-century England and 19th-century France.

In addition, the invention of the blast furnace greatly increased requirements in water power and especially in fuel, namely **charcoal**. Forests were razed to the point that ironworks became known as forest devourers. The needs of the new iron industry also created employment for many peasants as woodcutters, **colliers**, miners and carters.

Remains of the last blast furnace at the site of the original furnace
of 1736. Alterations over the Forges' long lifespan resulted in a complex
pattern of walls and masonry structures.
PHOTOGRAPH BY PARKS CANADA/JEAN AUDET, 130/21.1/PR-6/A-58, 1976.

The remains unearthed at the site of the upper forge provide evidence
of several major changes, the most significant of which was the building, in 1881,
of a second blast furnace. Its massive foundation can be seen in the foreground.
PHOTOGRAPH BY PARKS CANADA/JEAN-PIERRE ÉLIE, 130/22.1/PR-6/A-553, 1977.

What was thus a common industry in Europe, France in particular, would be unique in New France. Until 1800, the St Maurice ironworks was the only such establishment in Canada. Importing this industry to New France brought with it the organization of work typical of the French iron industry. Most of the workers were originally from eastern France, particularly the Franche-Comté region. They were the most sought-after ironworkers in France, since Franche-Comté was, along with the Ariège and Normandy regions, one of the three ironmaking centres that exported skilled labour.[15] The success of the *méthode comtoise*, the **fining** process developed in Franche-Comté, stemmed largely from its economical use of charcoal: fining and chafering were done in a single hearth, whereas the **Walloon process**, also widely used, split the two operations and used two hearths.[16]

Skilled workers were also rare in France.[17] The shortage was attributable to the growing number of ironworks and the practice of keeping trades in the family, passing them down from father to son. The historian H. C. Pentland ascribed the generous working conditions of ironworkers at the St Maurice Forges to the dearth of such skilled manpower and to the non-existence of a labour market in New France in the 18th century.[18] The phenomenon was not unique to the remote colony. Even in France, where there was a labour market for ironworkers, workers received benefits comparable to those at St Maurice. There were also complaints in France that ironworkers were arrogant and self-willed.[19] Another typical feature of the early iron industry was the restrictions imposed on the mobility of workers. In some ironmaking centres in France in the 18th century, ironmasters obtained royal privileges for their workers, such as exemption from the *corvée*, the *taille* and other forms of tax, and from service in the militia.[20] Furthermore, as was the case at St Maurice, workers were given room and board along with other benefits such as the right to own property. An ironmaster at the time estimated fringe benefits to be worth a week's wages.[21] What Pentland calls "untypical overhead costs" for workers at the St Maurice Forges were on the contrary perfectly common in the early iron industry and cannot be singled out as a particular feature of production in a pre-industrial colonial economy. It all depended on the type of industry introduced to the colony.

Set against the workers' benefits, coercive measures to prevent desertion were necessary both in France and at St Maurice, but in the old country those measures were mitigated by competition among ironmasters for workers. The Forges had to compensate a French ironmaster for enticing a founder away, and the Minister of Marine warned Vézin "not to deprive French ironworks of their workers."[22] At St Maurice there was also the fear that workers might defect to the British colonies, and such desertions were subject to heavy fines and corporal punishment.[23] With no regulations to secure the manpower needed by ironmasters in New France, specific ordinances had to be issued to govern employment at the St Maurice Forges.

It was thus a microcosm of the Old World iron industry that would be reproduced at the St Maurice Forges for close to 150 years. Five generations of workers succeeded one another in the service of an industry whose technological base remained unchanged until the latter half of the 19th century.

THE FORGES OVER TIME

[...] something to ponder other than a sterile litany of administrative facts; tiresome examples of internal quarrels; [...]

Édouard Montpetit, 1925[24]

The primary aim of this work is to describe an early ironworks that was one of the most significant legacies of French colonial rule in Canada. Previous treatments have all taken a chronological approach, thanks no doubt to the very longevity of the Forges and the historical method. It is true that an enterprise that endured over 150 years, through all the political changes in Quebec and Canada, deserves to have this feat recognized. However, despite successive administrations, the Forges always operated substantially along the same lines. The process of ore reduction using charcoal was never superseded by the more modern coke-fuelled method that was the practice in England from the end of the 18th century onwards. The Forges plant remained as originally built by the French, although changes were made to improve performance. Similarly, the organization of work was based almost to the very end on a dynastic tradition whereby families handed down skills from generation to generation.

The story of the Forges over time is thus set against a backdrop of industrial continuity. We will examine the historical conditions that allowed the Forges to survive so long—in the context of Canadian history, of course. By concentrating on the enduring conditions that contributed to the ironworks' survival, we hope to shed light on how the early iron industry operated. The conditions in question reflect a particular combination of factors that governed the production of cast and wrought iron. They are found in the ways that the ironworks was financed and its raw materials were managed, how it figured in the market, organized work and controlled the work force. This approach to the history of the Forges should also lead to a better understanding of how an enterprise created in the 18th century—outside any industrial context in a colonial economy—was able to continue operating right up to the dawn of the 20th century, without being part of the Industrial Revolution.

With a view to synthesis, this work takes the form of an industrial monograph. Each aspect—the management of the Forges; the key element of mineral, timber and water resources; the industrial plant and processes; ironworking; the Forges wares; the working population; and finally the industrial village— will be discussed in turn in order to capture as faithfully as possible the social and technical world of the St Maurice Forges, Canada's first industrial community.

GILLES HOCQVART

Plate 1.1a
Intendant Gilles Hocquart (1695–1783). Intendant of New France from 1729 to 1748,
he was the most enthusiastic promoter of the St Maurice Forges.

I

Masters of the St Maurice Forges, 1730–1883[1]

The objection that arises is to know why this establishment which is now come to perfection has not turned a profit: I have touched on the reasons in this dispatch, but the solidest reason, and the real one, in my estimation, is the lack of money.

Intendant Hocquart, 1741

Maladministration, creditors going bankrupt, treasurers going missing, such was, in short, the financial life of the French iron industry in the 18th century: and all of these facts prove a lack of financial education on the part of industrialists and a distinct shortage of capital.

Bertrand Gille, Les origines de la grande industrie métallurgique en France, 1947

No modern improvements in iron Manufactures have been introduced at the Forges of St. Maurice [...] [they] are daily expecting, from Scotland, an Engineer [...] fully competent to [...] conduct such a Manufacture.

Andrew Stuart and John Porter, 1852

THE FORGES POST, 1730–41

[...] the conditions he makes are costly neither to the King nor to individuals [...] The King runs no risk.[2]

Francheville and the Compagnie des Forges de Saint-Maurice

In the fall of 1729, François Poulin de Francheville, a Montreal merchant and seigneur of St Maurice, petitioned the King of France for the right to invest his own capital in iron mining on his seigneury and the adjoining seigneuries of Yamachiche and Cap de la Madeleine, conditional on his being granted the exclusive right, for 20 years, to work the mines. He also asked the King to grant him the monopoly on the manufacture and trade in iron from his mining operations and authorization to harness the necessary streams. Francheville undertook to open the mines within two years of the date of concession.

In support of his petition, Francheville stressed the "considerable advantage" of such a venture, which could supply the colony with iron, particularly for shipbuilding, and replace iron imported from France at high cost.[3] Francheville's proposal was forwarded to the Comte de Maurepas, Minister of Marine, by the Governor of New France, the Marquis de Beauharnois, and the Intendant, Gilles Hocquart (Plate 1.1a, 1.1b, 1.1c), who supported the project, seeing it as entailing "no risk" to the King. They reiterated the advantages pointed out by Francheville, and vouched for "this merchant who is well-to-do and will find associates to join him in this enterprise."[4]

Five months later, on 25 March 1730, the King granted Francheville a royal warrant on terms even more generous than those he had requested. The exclusive 20-year privilege would apply not from the date of the warrant, but from the start-up date of operations, two years hence. The King also exempted Francheville and his heirs from payment of any indemnity or tithe. He gave the Intendant the judicial authority to settle, free of charge, any claims or legal disputes that might arise from working mines outside Francheville's seigneury.[5]

Upon learning of the royal decision, Francheville had two ironworkers recruited in France. They arrived in late 1731 with chests of specially ordered tools (Table 1.1) and began ore prospecting and testing the following spring. That summer, Jean-Baptiste Labrèche, Francheville's overseer, travelled to New England to study the small American bloomery forges that Francheville planned to use as a model for his operation at St Maurice. On his return, Labrèche tested the **direct ore reduction process** in a regular smithy at Quebec and produced a few iron bars which were sent to Maurepas, in France. Labrèche returned to New England in the spring of 1733, accompanied by edge-tool maker Christophe Jamson, *dit* Lapalme, and carpenter Louis Bellisle.[6] Francheville had begun building his establishment and had a cart road laid to connect it with Trois-Rivières. By the end of 1732 he had already invested over 9,000 *livres* in the first phase of setting up his operation.[7]

Before damming the stream and starting up the operation, Francheville asked the King, on the recommendation of the colonial authorities, for an advance of 10,000 *livres*. Having realized how costly such a venture was for a single man to bear, on 16 January 1773 he also formed a company with four partners, though he retained executive authority and majority control (Table 1.2). "Francheville et Compagnie" thus gave birth to the "Compagnie des Forges de Saint-Maurice."

Table 1.1

WORKMEN AT FRANCHEVILLE'S FORGE[8]

Worker's name	Trade	Date	Origin
François Trébuchet	founder	1731	Bardonnière, Brittany
Jean Godard	keeper	1731	Ferrière-sur-Eure, Normandy
Jean-Baptiste Labrèche (overseer)		1732	Montreal
Christophe Jamson, *dit* Lapalme ["masonry and carpentry smith"]	edge-tool maker	1733	Montreal
Louis Bellisle, *dit* Chèvrefils ["to work [...] in the ironworks and on building the frame"]	carpenter	1733	Montreal
Nicolas Grand'Maître*	unspecified	1733	Combeau-Fontaine, Franche-Comté
Jean Chassé*	unspecified	1733	Combeau-Fontaine, Franche-Comté
Nicolas Camiré*	unspecified	1733	Persé-le-Grand, Franche-Comté
Mathieu Lussau*	unspecified	1733	La Tessoir-Aunis, La Rochelle
Maurice Herbet*	unspecified	1733	Mont-Jean, diocese of Angers
Jean Pommier	unspecified	1734	
"Benoits"	unspecified	1734	
Créqui			Quebec

* Job: "[...] to work [...] at the iron forge [...] and elsewhere regardless of the work, whether in labouring of otherwise [...]," with board and lodging at 120 *livres* a year.

Table 1.2

PARTNERS IN FRANCHEVILLE ET COMPAGNIE[9]

Name	Social status	Role in company	Shares
François Poulin de Francheville	Montreal fur merchant	treasurer, Forges agent, Montreal	10
François-Étienne Cugnet	director of the Western Domain, senior councillor on the Superior Council of New France	keeper of the minute book, Forges agent, Quebec	4
Louis-Frédéric Bricault de Valmur	secretary to Intendant of New France	approval of financial records	2
Pierre Poulin de Francheville	Quebec merchant, brother of François Poulin de Francheville	shareholder	2
Ignace Gamelin	Montreal fur merchant	shareholder	2

In the fall of 1733, Francheville received the requested royal advance of 10,000 *livres*. The establishment was completed and ready to go into operation (Table 1.3 and Plate 1.2) when suddenly, on 28 November in Montreal, Francheville died.[10] Hocquart immediately safeguarded the King's interests by having Francheville's widow, Thérèse de Couagne, sign an undertaking to honour her husband's commitments.

Table 1.3

THE FORGE OF FRANCHEVILLE ET COMPAGNIE, 1733–35[11]

Location	At the mouth of the St Maurice Creek on the west bank of the St Maurice River	
Area	1.18 acres (90 *toises* by 14 *toises*)	
Buildings	**Dimensions***	**Materials**
forge	45 x 25 *pieds* (14.6 m x 8.1 m)	lumber, no foundation, except wheelrace, of masonry
shop	12 x 12 *pieds* (3.9 m x 3.9 m)	*pieux en terre* palisading, roofed with boards
house	38 x 24 *pieds* (12.3 m x 7.8 m)	*pièce sur pièce* log construction
stables	28 x 20 *pieds* (9.1 m x 6.5 m)	*poteaux debout et pieux de travers* (post and beam)
Equipment	1 flume 2 waterwheels 1 chafery 2 bellows 1 hammer 1 anvil[12]	"a hearth that is properly a French chafery […]"
Ore reduction process	direct reduction	
Ore	bog ore	Goethite (HfeO₂) Limonite (2Fe₂O₃)
Fuel	charcoal	
Chafery capacity	400 pounds of bar iron (195.8 kg)	every 24 hours
Chafery yield	33.3%	50 pounds of iron for every 150 pounds of ore (24.5 kg for 73.4 kg)
Forge workers	4 workers	
Operating period	2 months	January–February 1734
Total production	1,600–2,000 pounds** (783–979 kg) bar iron	

* Dimensions in French feet (*pieds*) according to Vézin in 1735 except for the shop. Different dimensions are given in other documents. A French foot or *pied* = 32.482 cm

** French pound = 489.41 g

Plate 1.1b
Minister of Marine, Jean-Frédérick Phélypeaux, Comte de Maurepas (1701–81).
FRANCE, BIBLIOTHÈQUE NATIONALE, PARIS.

The company decided nevertheless to start up the ironworks during the winter of 1734, but in two months, and at the cost of high charcoal and ore consumption, the four workers managed to produce only about 2,000 pounds of iron of uneven quality. The inexperienced workers were unable to master the process of direct ore reduction; moreover, the spring thaw revealed how unstable the forge's machinery was. On the recommendation of the overseer, Labrèche, who admitted his lack of experience, it was decided to suspend operations.

The outcome was that Francheville's forge was abandoned, though this did not put an end to plans to make iron at St Maurice.

So far, 21,583 *livres*, 1 *sol*, 6 *deniers* had been invested in the forge.[13] Since the experiment had at least succeeded in proving the abundance and quality of the ore, Hocquart decided to shelve the project pending the opinion of French ironmasters, which he requested from Maurepas in the fall of 1734. Hocquart then sent to France three iron bars and a model of Francheville's forge so that Maurepas could obtain expert views on the quality of its iron and on its location and set-up.

Plate 1.1c
Governor Charles de La Boische, Marquis de Beauharnois (1671–1749), Governor of New France from 1726 to 1746.
STEWART MUSEUM AT THE FORT ON ST HELEN'S ISLAND, MONTREAL.

The experts came to the conclusion that the entrepreneur and his workers lacked the necessary know-how, and Maurepas decided to have the project re-evaluated on site.[14] In the fall of 1735, therefore, he sent ironmaster Pierre-François Olivier de Vézin, a native of Champagne, who, after spending five weeks at the site, recommended that operations resume, though on a new basis, using the process of indirect ore reduction. Vézin's proposal led to the dissolution of the Compagnie des Forges de Saint-Maurice and the immediate pursuit of a new venture with the young Vézin in charge of erecting a full-scale ironworks.

Plate 1.2
View of Francheville's forge, built in 1733.
RECONSTRUCTED BY ILLUSTRATOR BERNARD DUCHESNE.

Cugnet et Compagnie

The Ironworks Project

Vézin's project led to the definitive establishment of the St Maurice Forges. The works built between the summer of 1736 and the fall of 1739, modelled on the *grosses forges* furnace and forge pattern of Europe, were not substantially altered until 1854. And even then, after the complete rebuilding of the blast furnace, the plant remained essentially that of the French era, since the object of the overhaul was merely to increase the capacity of the existing works, without altering the process of indirect ore reduction using charcoal.

Vézin submitted his "observations" and his plans to the colonial authorities on 17 October 1735. He estimated that his project would cost close to 100,000 *livres*: 36,000 *livres* for the construction of the works and dwellings, and 61,250 *livres* for annual operating expenses. He also projected annual revenue of 116,000 *livres*, which would more than cover operating costs and leave a profit of close to 55,000 *livres*.

Financial Structure

Less than a week later, on 23 October, the partners in Francheville et Compagnie dissolved the company and made a proposal to the King, the same day, that they relaunch the venture along the lines proposed by Vézin. As a condition, however, they asked for a royal advance of 100,000 *livres*, spread over three years, to cover the expenses estimated by Vézin, to be used in the first year to erect the works, in the second year to collect and lay in raw materials, and in the third year to pay the wages of workers during the start-up phase.[15]

The colonial authorities, adding their vigorous endorsement, forwarded the proposal along with Vézin's project to the Minister. Maurepas cautiously sought expert advice before reaching a decision. An investment of 100,000 *livres* would represent 10 times the sum advanced to Francheville in 1733. Furthermore, Maurepas wanted to ensure that colonial production, estimated at 600,000 pounds of iron (294 t) annually, would not harm the industry in France. After receiving a favourable opinion from the Council of Commerce, which wanted to substitute iron from New France for some of the imports from Sweden and Spain and to stimulate shipbuilding in Canada, Maurepas agreed to invest the 100,000 *livres* requested.[16] The decision was made in mid-March, but word did not reach New France until the end of June. Construction of the Forges thus only began in July 1736, but the colonial authorities, anticipating a favourable response from the Minister, kept Vézin in the colony and made a start on cutting wood and building roads.

Articles of Association

Vézin and two of Francheville's former partners, Cugnet and Gamelin, took on other partners and established what became the new Compagnie des Forges de Saint-Maurice,[17] for which Cugnet, director and treasurer of the company, was authorized to sign "Cugnet et Compagnie" (Table 1.4). Two newcomers, Burgundian ironmaster Jacques Simonet, and Thomas-Jacques Taschereau, agent of the treasurers general of the Marine in New France, were imposed by Maurepas. Simonet, sent by the Minister to assist Vézin, was paid an annual salary, while Taschereau acted as the Minister's representative; advances to the Company were sent through him.[18] In "recognition of the kindness" of Hocquart for "his protection" of those involved in the Forges venture, the partners made provision in their agreement to offer him one-fifth of the interest of the four main partners, equal to 3.6 shares (1/5 of 18). The following year, Hocquart was authorized by Maurepas to accept the offer.[19]

Table 1.4

**PARTNERS IN CUGNET
ET COMPAGNIE, 1736–41**[20]

Name	Social status	Role in company	Shares (sols)	Salary (livres)
François-Étienne Cugnet	director of the Western Domain, senior councillor on the Superior Council of New France	director and treasurer of the company, Forges agent at Quebec, keeper of the minute book	3.6	nil
Pierre-François Olivier de Vézin	ironmaster	management of the Forges	3.6	3,000/year
Jacques Simonet	ironmaster	management of the Forges	3.6	1,500/year
Ignace Gamelin	Montreal fur merchant	account auditing, procurement of goods and other victuals, Forges agent at Montreal	3.6	nil
Thomas-Jacques Taschereau	Marine treasurer, member of the Superior Council of New France	approval of accounts	2	nil
Gilles Hocquart	Intendant of New France	partner	3.6	nil

Privileges

The fledgling company, solidly backed by the colonial authorities, also enjoyed certain privileges. In 1737, the King officially transferred to the company the 20-year monopoly on mining rights granted to Francheville in 1730. That same year, the Minister advanced 14,000 *livres* to Simonet to purchase tools in France and to recruit workers to whom the King granted passage to New France aboard the royal vessel. In addition, Hocquart persuaded the Minister to allow the King's advances to be repaid in two instalments, one in 1740, the other in 1742. To ensure that the company had sufficient timber reserves and to avoid causing forest fires, the fief of St Étienne and adjoining land were annexed to the seigneury of St Maurice.[21] The two fiefs together, plus the lands to the rear, formed a concession of 19,200 ha. Combined with the tract of adjoining land (113,300 ha) from which raw materials could be collected, the Forges lands, or reserve, totalled some 132,500 ha (see Appendix 2).

Implementation

Considering its structure and backing, financing and privileges, the new company had every chance of succeeding. It managed to build the Forges within three years, as planned, but at a cost and under conditions such that it would be forced to declare bankruptcy after six years of existence, only three of which it was in production. The brief years of Cugnet et Compagnie were so turbulent that they continue to stand out as the most remarkable in the long history of the St Maurice Forges.

The construction of the Forges and the various problems encountered by the ironmasters are discussed in detail in Chapter 4. Because of the atmosphere of mistrust and dissension that soon developed between the ironmasters and the other company officials, the difficulties encountered during those six years were greatly exacerbated. The long accusatory **memorials** by Cugnet and Vézin, in which each gives chapter and verse on his side of the story, have been a rich source of historical documentation and inspiration for many Forges historiographers. Technical details of the sort that Vézin felt compelled to provide to explain his actions and justify his decisions would probably never have come down to us had the Forges come into being without incident.

No attempt will be made here to reconstruct in detail the saga that marked the establishment of the Forges. The main difficulties that arose between 1736 and 1741 are summarized in Appendix 3.

The chief problems surrounding the construction of the Forges stemmed from the fact that haste was the order of the day, coupled with lax control over the way in which the project was carried out. By underestimating the constraints inherent in such a venture on colonial soil, the builders made one technical mistake after another and demonstrated a complete lack of planning. Vézin would suggest that his knowledge of the Canadian terrain and climate had been inadequate and that the rigours of winter had caused him problems and additional expense;[22] the effects of freezing and thawing, in particular, put foundations and structures to a severe test. Vézin also misjudged the nature of the ground on which he erected his shops and in particular the high water table, which caused serious problems when the blast furnace was fired up, or **blown in**. And even though he had seen the unfortunate consequences at Francheville's small forge, he underestimated the extent of the spring flooding of the St Maurice River, which forced him to modify the wheels of his forge, erected on the very site of Francheville's bloomery. The river's swollen waters in the spring and during the rainy season delayed the important job of quarrying limestone on its banks.

Table 1.5

DEFICIT OF THE BANKRUPT COMPAGNIE DES FORGES DE SAINT-MAURICE, OCTOBER 1741 [23] (*livres*)

Expenditures 1735–41	Revenues 1735–41	Deficit
527,072*l* 15s 1d	177,769*l* 2s 0d	349,303*l* 13s 1d

Table 1.6

DISCREPANCY BETWEEN ESTIMATED AND ACTUAL EXPENDITURES, 1735–41 (*livres*) [24]

Estimates according to Vézin's calculations, 1735		Expenditures according to Cugnet and Hocquart's calculations, 1741	
Construction	Operations 1738–41 (3 years)	Construction	Operations 1738–41 (3 years)
36,003*l* 6s 8d	183,750*l*	(146,000*l*)[25]	unspecified
Total: 219,753*l* 6s 8d		**Total**: 527,072*l* 15s 1d	

Discrepancy between estimated and actual expenditures
307,319*l* 8s 5d
240%

Vézin also faced a shortage of skilled workers. All the extensive masonry work, for instance, involved a good many of the masons available in New France, and building a large stone house on top of it was more than could reasonably be handled. But the main mistakes lay in the timing and conditions under which these scarce craftsmen were hired. The precipitate hiring of masons and carpenters when work began in the summer of 1736 forced the ironmaster to use them as woodcutters or labourers. Similarly, the premature arrival of ironworkers not only led to unnecessary expense, but also caused serious disciplinary problems that permeated the atmosphere of the **post** from the outset, even after the plant went into operation. This working environment did nothing to diminish the arrogance of the workers, who were already felt to be too independent and overpaid.

The promptness with which this establishment was begun greatly increased its cost and the rigours of the Canadian climate made it necessary to have buildings that were much more solid, better sealed and consequently of much greater expense than those made for ironworks in France. It is essential to shelter the movements and wheelraces from the excessive cold of the country.

Bankruptcy

The fledgling Compagnie des Forges de Saint-Maurice had accumulated a deficit of close to 350,000 *livres* when it finally relinquished its rights to the King in 1741 (Table 1.5).

The difference between estimated and actual expenditures was considerable, although cost overruns were common in major Crown projects at that time (Table 1.6). The cost of royal shipbuilding and of the fortifications of Quebec also far exceeded their initial estimates.[26] Such situations were also common in the French iron industry.[27] After all, this was a first attempt to build an ironworks in a fairly underdeveloped colony, and that fact alone would have a profound impact on its creation.

It should not be forgotten, however, that in contrast to public works projects the St Maurice ironworks was a production facility capable, according to Vézin, of generating revenue of 116,000 *livres* a year, amply covering its operating costs (61,250 *livres*) and turning a profit of close to 55,000 *livres*.[28] In light of those projections, it is easier to understand Maurepas's willingness to increase the advances needed to set up the Forges, and to put off their repayment; it is easier also to understand Hocquart's tolerance in allowing Cugnet to use funds from the **Western Domain**'s coffers to complete the work. In reality, Vézin's figures were too optimistic. When bankruptcy was declared in October 1741, Hocquart admitted to Maurepas that he doubted whether the value of production could ever have reached 110,000 *livres* and that he had never hoped for more than 12,000 to 15,000 *livres* in net profit (11–14%).[29]

The fact that the Forges failed to generate a profit in their initial years of operation drove the Compagnie des Forges straight into bankruptcy. The King finally recovered his advances by seizing the Forges, and Cugnet was the only person really hurt financially by the poor performance of the company; he ended up over 140,000 *livres* in debt because of the Forges.[30]

The enterprise set up at great expense and in an atmosphere of tension and haste was nevertheless turned over "in its perfection," as Hocquart reported to Maurepas in the fall of 1741. The establishment of an industrial facility like the Forges would have major repercussions on the future of New France, and the players of the time should not be blamed for trying to provide the colony with an industrial base essential to its development, even if the enterprise was born in controversy.

Were we to assess the performance of the fledgling company through all the twists and turns of the creation of the Forges, we would have to acknowledge that two main types of mistake were made: technical ones and management ones. These mistakes were attributable to the unco-ordinated actions of two men, one responsible for construction work and the other for the budget. Vézin was largely to blame for the technical mistakes, though he rejected responsibility for some of them, and Cugnet was largely responsible for the management mistakes, though Vézin may deserve his share of the blame for his own management. There were indeed shortcomings in their respective spheres of responsibility. In their individual reports on the bankruptcy, each accused the other of the errors directly related to his particular sphere. Vézin referred to "excessive haste" in the conduct of the project and in particular in engaging the workers, while Cugnet attacked Vézin directly, questioning his competence as ironmaster.

Cugnet was not sufficiently prudent in the financial management of operations and he paid the price by having to declare personal bankruptcy. It should be said in his defence that Hocquart was largely responsible for Cugnet's lack of caution, and the Intendant himself admitted as much to the Minister: "it is I who persuaded him through my incitements to make new efforts to support an enterprise that all regard as a good thing." Hocquart also exonerated Cugnet from having used funds belonging to the Western Domain, admitting to the Minister that Cugnet's dual duties as receiver general for the Domain and treasurer of the company "would have been easily reconciled if the Forges had had the success that was hoped. The opposite occurred."[31]

Cugnet, for his part, said that he had been "forced" to take financial risks, though he did not explicitly name Hocquart as the source of what the latter called "incitements."

Despite the delays and wrangling that followed the bankruptcy of 1741, the debts of Cugnet et Compagnie to the Crown (192,642 *livres*) were finally written off in 1749 against the estimated value of the Forges in 1744 (174,849 *livres*).[32] The other debts to the Marine exchequer and private creditors were assumed by Cugnet. However, he was granted certain concessions by Maurepas and Hocquart to help him meet his obligations. Immediately after bankruptcy was declared, his property was inventoried and sequestered. At the same time he lost his position as receiver general for the Western Domain. The Minister could not officially endorse Cugnet's practice of diverting 70,006 *livres* from the Domain to the Forges, even though he had been encouraged to do so by Hocquart himself. To help Cugnet pay off his debts, Maurepas allowed him to keep the right to farm the taxes of the King's posts at Tadoussac and extended the lease by six years from 1747 so that he could repay 10,000 *livres* a year to the exchequer. Moreover, to help him pay his creditors, the Minister granted him the right for nine years to farm the taxes at three other trading posts. These concessions were not a problem as long as Hocquart was Intendant, but when François Bigot replaced him in 1748, the new Intendant, not favourably disposed towards Cugnet, whom he considered a profiteer, withdrew the Tadoussac lease and gave it to the widow Fornel. Cugnet protested to the new Minister, Antoine-Louis Rouillé, Comte de Jouy, who had replaced Maurepas in 1749. He finally learned the following year, from an unofficial source, that the Minister would "place his debts on the King's account, by making the necessary arrangements with creditors"; there was talk of a satisfactory arrangement to replace the Tadoussac lease. Pending Rouillé's official decision, Cugnet anticipated it "as the end of the anxieties that have assailed me for over ten years, the beginning of my peace of mind, the preservation of my family and the assurance that the creditors of the forges will be paid."[33]

On 19 August 1751, Cugnet died, apparently leaving his widow comfortably off and on her way to being free of her husband's debts.[34]

The risks taken by Cugnet would have appeared justified had Vézin succeeded in getting the Forges running sooner and producing as planned. The establishment was eventually completed, but too late for the company that gave birth to it.[35] To absolve himself, Vézin wrote after the fact that he would have needed more time to adapt to conditions in the colony. That would be his only admission. Relieved of his duties, he left New France.

That was the end of Cugnet et Compagnie, but not of the St Maurice Forges. Despite the discord and frequent interruptions that marked the initial years of operation, the Forges still succeeded in supplying iron suitable for the Rochefort Arsenal and the royal shipyard at Quebec. It was inconceivable that an enterprise "now come to perfection" and in which the government had invested so much, be abandoned. Indeed, Maurepas had sensed that the company was in trouble in 1739,[36] but had allowed Hocquart to continue operations.

The state of mind of the authorities at the time of bankruptcy was similar to their attitude after Francheville's experiment. Relatively speaking, the King's investment in the small bloomery had been similar to his subsequent investment in Vézin's works. Even though Francheville's project had failed, it had proved the colony's ability to support an ironworks, and the high grade of its ore. After the Francheville episode, the French authorities were certain of the value of the ore at St Maurice and, after the experience of Cugnet et Compagnie, they remained convinced of the value of the St Maurice Forges.

Immediately after accepting the resignations of the partners of Cugnet et Compagnie, Hocquart, not wishing to interrupt operations, placed a Quebec merchant, Guillaume Estèbe, in charge of the Forges as trustee for one year.[37] However, the enterprise was never managed efficiently under the French, although Estèbe's efforts in the aftermath of the bankruptcy did help to remedy the situation somewhat.

CROWN CONTROL, 1741–64

French Crown

For all the difficulties encountered in the six years of Cugnet et Compagnie, it proved relatively easy to raise capital in France and to bring over the necessary equipment and skilled workers. When the Crown took control in 1741, the outbreak of war placed constraints on the movement of capital and workers and on the management of the Forges. The war also meant that a good part of the ironworks' output would be war materiel, mainly munitions. In 1741, France became involved in the War of the Austrian Succession, which lasted until 1748, and later in the Seven Years' War, which broke out in the colony in 1754.

France's involvement in those conflicts prevented a new private company from being formed to take charge of the Forges. It also blocked the recruitment in France and the free movement of skilled workers between France and the colony. The workers who had arrived before 1741 perforce had to stay, yet they were already being described in 1743 as needing to be replaced owing to poor health, lack of discipline and low productivity.[38] Repeated requests for new manpower were made throughout this period, but the only known new workers were two moulders who arrived in 1745. Locally, there were problems in recruiting woodcutters and miners because of the small population in the region. The Trois-Rivières garrison had to be reinforced to make soldiers available for labour at the Forges. In addition, able-bodied *habitants* were pressed into service through *corvées*. Food shortages in 1742–44 caused supply problems, which did not help operations. Moreover, wood for charcoal had to be sought farther and farther afield, boosting transportation costs and hence, the cost of production. Competition from the French iron industry was constant, periodically forcing the ironworks to lower its prices.

Both Hocquart and Bigot tried to have the Minister put a stop to French exports, but without success.[39]

Initially put under trusteeship as a temporary measure, the Forges remained under Crown control until 1760. Yet that was not the goal after the bankruptcy of Cugnet et Compagnie. The colonial authorities and Maurepas wanted rather to attract private entrepreneurs by offering them various incentives, such as a monopoly on the sale of metals in the colony, a monopoly on the sale of goods at the Forges and even the trading rights over the King's posts at Tadoussac, to bolster a profit margin that was not considered attractive enough. But with France at war, Maurepas postponed his plans to bring in a new private entrepreneur until after the war in Europe had ended. In the interim, control would stay with the Crown, a situation that would continue even after the War of the Austrian Succession ended in 1748. The Minister enjoined Hocquart to manage the Forges "economically," keeping the establishment running but spending as little as possible. In 1745, after three years of operation that turned a profit of 42,846 *livres*, Hocquart was able to announce that the Forges were no longer a financial burden on the Crown, and when he was replaced by Bigot in 1748, Hocquart was able to produce a positive balance sheet. But the problems of administration were not completely resolved, since they reappeared immediately after Hocquart's departure when it was found that the ironworks was in rather bad shape. In fact, 1748 and 1749 were not very good years, producing a deficit again of close to 25,000 *livres*.[40]

Estèbe's trusteeship was above reproach: for the first time, the enterprise seemed close to breaking even (Table 1.7), even though production was down below the previous year's level.[41]

Figures available for the next three years show that the value of output gradually rose to close to 480,000 *livres* in 1744–45. Overall, the four years of production immediately following the bankruptcy generated, according to Hocquart, a profit of over 50,000 *livres*.[43] The Forges did not reach the anticipated production level, but the profit margin, though also well below initial estimates, showed that the business was quite viable (Table 1.8).

| Table 1.7 | ESTÈBE'S ACCOUNTS FOR 1741–42[42] |

Reference	Expenditures	Revenue	Deficit
Accounts for 8 October 1742	65,208*l* 9s 0d	65,185*l* 16s 8d	22*l* 13s 8d
Accounts for 14 October 1746	61,911*l* 16s 8d	61,911*l* 16s 8d	nil
Accounts for 8 August 1748*	61,911*l* 16s 8d	53,733*l* 17s 9d	8,177*l* 18s 11d

* Figures taken from statements of account for 21 September 1750 and 8 July 1752

| Table 1.8 | IRON PRODUCTION IN THE FIRST SEVEN YEARS |

Year	Pounds of iron	Tonnes*
1738–39	140,427	70
1739–40	355,071	178
1740–41	394,533	197
1741–42	337,345	169
1742–43	376,545	188
1743–44	400,000	200
1744–45	479,333	240
Total	**2,483,254**	**1,242**
Average	**354,751**	**177**

* Figures are rounded off

Transferred officially to the **King's domain** in 1743, the Forges would remain the property of the King until the end of French rule. After Estèbe, the Forges were managed by government officials: Jean-Urbain Martel de Belleville (1742–50)[44] and Jean Latuilière (1750–5?).[45] In 1749, following a very unproductive year, Bigot appointed an "inspector," René-Ovide Hertel de Rouville, to oversee operations; he held his position until 1760.[46] Jacques Simonet's son, Jean-Baptiste, was Estèbe's ironmaster until he was charged with theft and sent back to France in the summer of 1742.[47] It was Claude Courval-Cressé, Francheville's cousin and former clerk under Vézin, who served as ironmaster from 1742, probably until 1760. Vézin and the two Simonets, father and son, were in fact the only ironmasters worthy of the name who ever ran the Forges.[48] Until the end of French rule in the colony, the Forges were managed by merchants, clerks, overseers and inspectors, each with his own area of jurisdiction. This diffuse management structure did little to improve the atmosphere that had reigned in the time of Cugnet et Compagnie. The Governor of Trois-Rivières, Rigaud de Vaudreuil, severely criticized the poor management of the Forges in 1749.[49] That same year, the naturalist Pehr Kalm visited the Forges and noted the presence of "several officers and overseers" who "appear to be in very affluent circumstances" despite the fact

that the revenues of the ironwork do not pay the expences which the king must every year be at in maintaining it [...][50]

Three years later, the engineer Louis Franquet noted the presence of a director, a treasurer, a clerk, a merchant and a chaplain, and commented:

> one has to feel that this system must lead to many abuses, especially since full authority is not vested in the Director, but shared with the Treasurer, and each reports to the Intendant directly or to his subdelegate, on the part that is assigned him, and the clerk in charge of the property maintained at the King's expense believes himself to be independent.[51]

Franquet recommended that a director be appointed fully accountable to the Intendant and that "a master craftsman with skill and expertise in all sorts of works" be brought from France. In 1754, two years after Franquet's visit and 13 years after the bankruptcy of Cugnet et Compagnie, an ironmaster was still being sought. To follow up on his recommendation of 1752, Franquet gave the Minister, Rouillé, the name of Mathieu Molérac, a man belonging to a line of ironmasters "spread throughout the Kingdom," who probably did not come to New France.[52] In 1755, Vézin, passing through New France, expressed concern over the management of the Forges and offered to take over the ironworks, provided the terms were favourable. But his offer was not taken up.[53]

The final four years of the French regime were marked by war raging again in New France, which is probably why information becomes sparse until 1760. The last available information concerns operations in the year 1756, showing that the Forges were involved in the war effort, making large quantities of munitions for the artillery.[54]

British Crown

In September 1760, the surrender of French troops to the British was assured with the fall of Montreal. That same month, as he was having the people of Trois-Rivières pledge allegiance and submission to His Britannic Majesty, the military governor, Ralph Burton, received orders from the Governor General, Jeffery Amherst, on the management of the St Maurice Forges. Hertel de Rouville, the outgoing director, conducted an inventory of the establishment on 8 September, and on 1 October, instructions were given to Claude-Joseph Courval, the new inspector of the Forges, to keep them running.[55] According to the terms of those orders and the contents of the inventory, it appears that operations had been suspended for some time. The order to convert the cast iron in stock into wrought iron would indicate that the furnace was not in blast at that point. Also, the stocks of supplies of raw materials recorded in the inventory show that reserves, especially of charcoal, were very low. The 250 *pipes* of charcoal recorded were far below the 35,000 to 40,000 *pipes* required annually for the blast furnace and the two forges.[56] Combined with the low stocks of cast iron (35,000 pounds or 17 t) and wrought iron (3,100 pounds or 1.5 t), those figures suggest that the workers had been idle the previous year (1759–60). The advance of the British troops probably explains this "idleness." The workers were still on the post, however, since Courval was ordered, on 1 October 1760, to keep seven French ironworkers—the founder and his furnace **keeper**, and five forgemen (Table 1.9).[57] This order would have a decisive impact on the subsequent history of the Forges because those workers settled in the colony for good and their descendants would man the Forges for another hundred years.

after the Conquest of this country, they all wanted to cross to France, since they did not expect to find any more work in this province; but his Excellency General Amherst [...] ordered them all to remain and continue working at the forges just as they had always done under the French.

Keeping the key workers on the job made it possible to continue operations for four years, though at a slower pace. The aim was merely to keep things running, mainly by converting the cast iron in stock, scrap iron and old ordnance. From September 1760 to September 1764, the two successive governors of Trois-Rivières, Ralph Burton and Sir Frederick Haldimand, were charged with monitoring operations and managing the business. The revenue generated covered the expenditures of the Government of Trois-Rivières.[59] The results were deemed most satisfactory since close to 500,000 pounds of **bar iron** and 280 cast iron stoves were produced during that period (Table 1.10). When Haldimand

handed over power to Cramahé, the civil secretary, in September 1764, there were still close to 400,000 pounds of iron (245 t) and 148 stoves in stock, the equivalent in iron of a good year's output under the French.

In addition to forcing the ironworkers to stay on the job, the British had to institute statute labour, like the *corvées* in the final years of French rule, pressing the *habitants* to cut wood for charcoal. The military authorities also made repairs to the plant and undertook road maintenance.

Immediately after the transfer of the Forges to civil authority in September 1764, operations came to a halt. The workers remained idle at the Forges for a year, for which they were paid two years later.[61] On 1 August 1765, General James Murray ordered them off the post, which they left the following October, when Courval turned the Forges over to Conrad Gugy, the governor's secretary. The Forges were then placed under military guard of a dozen troops of the 27th Regiment and would remain closed until 1767.[62]

Table 1.9

WORKERS ORDERED TO REMAIN AT THE FORGES IN 1760[58]

Department	Name	Craft
Blast furnace	Delorme	founder
	Belu	keeper
Forges	Marchand	hammerman
	Robichon	hammerman
	Terreau	finer
	Michelin	finer
	Imbleau	finer

Note: There is also mention in the 1765 accounts of the following workers: Robichon *fils*, helper; Marchand *fils*, helper; F. Grenier, carter.

Table 1.10

IRON PRODUCTION UNDER THE MILITARY, 1760–64[60]

Year	No. of months	Pounds of iron	Period
1760–61	14	127,784	October 1760 to end of December 1761 (with interruptions)
1762	6	150,476	10 May to 30 July; 23 August to 16 November
1763	9	136,982	March (or May) to November
1764	4	67,659	June to 30 September
Total		**482,901**	

The military administration demonstrated the Forges' production capacity and the importance of keeping the ironworks in operation. The specific instructions that London gave Cramahé, to protect and even expand the Forges lands, show quite clearly the government's desire to keep the Forges running.[63] However, the civil authorities took their time deciding on how the Forges should be managed in future; they had to choose between government control and private control. We know that in 1764 at least two private entrepreneurs expressed interest in the Forges. In June, John Marteilhe, a Quebec merchant, submitted a memorial to the Board of Trade, seeking preferment should the government decide to lease the Forges. In support of his request he produced a balance sheet on the Forges for the year 1756, along with budget estimates. He also asked for advances from the King to get the enterprise back on its feet, but his request was not approved. In November, a Mr McKenzie, partner of a Mr Ocks, spent a month at the post making observations which he then took to London. In the meantime, the authorities consulted Courval, acting inspector of the Forges, and Voligny, the overseer, son of Rouville, the former inspector under the French. Both recommended government control over private control, with Courval expressing concern that greedy private entrepreneurs would soon exhaust the mineral and timber resources by overexploiting them. He also pointed out the problem of recruiting seasonal labourers, given the region's small population, which could be addressed only "by authority" of the government.[64] In January 1765, the Governor, Thomas Gage, decided instead to lease the Forges to private entrepreneurs, but not to McKenzie, whom he considered insufficiently wealthy.[65] The first lease was not granted, however, until 1767.

THE FORGES UNDER LEASE, 1767–1846

When at last Christophe Pélissier signed the first lease on the Forges in June 1767, the ironworks had been at a standstill for almost three years; it is surprising still to find at the Forges the same workers who had been dismissed in August 1765.[66] There was no official explanation for the delay in resuming production, but it is possible that during the period

Plate 1.3
Mathew Bell (1769–1849), master of the St Maurice Forges from 1793 to 1846.
COLLECTION OF JOHN MCGREEVY.

of indecision various options were under consideration and that other parties expressed an interest in the Forges. The last known one was Peter Hasenclever, a German-born entrepreneur. Beginning in 1764, Hasenclever established ironworks in the Province of New York, to which he brought ironworkers from Europe. In a memorial addressed to Lord Shelburne, in London, in January 1767, Hasenclever made it plain that he wished to expand his business; he sought the St Maurice Forges to keep the workers recruited in his native Germany fully employed. He seemed well informed about the Forges, since he also specifically asked to be granted the rights to the stream that rose about two miles above the facilities. Hasenclever's request never bore fruit since on 9 June 1767, the Pélissier syndicate obtained a 16-year lease to the Forges.[67]

The government set terms that imposed minimal obligations on the lessees. Until the sale of the Forges in 1846, the terms of the lease were adapted to the changing situation, mainly with respect to the Forges lands, but they would remain very favourable to the tenants. For this first lease, the only return the government expected was the £25 currency in annual rental, payable in four instalments. In exchange, the lessees could run the Forges for their profit. They were also granted the land of the seigneury of St Maurice (19,200 ha)[68] to collect the ore, stone and wood needed for operations; they would therefore not have to pay for their raw materials. The only stipulation was that they keep the plant and equipment in good working order and make the necessary repairs; upon expiry of the lease, they had to return the establishment in the state in which it had been leased to them, as specified in advance in a detailed inventory.[69]

The masters of the destiny of the Forges for close to 80 years were mainly merchants primarily interested in profiting from a business that would long be the only one of its kind in Canada and that drew on a solid tradition of skilled ironworkers. Their chief concern centred on maintaining the material conditions—in the form of ore and timber resources—needed to operate the ironworks, which were freely granted them by the government. The land question would figure large on their agenda to secure the long-term supply of raw materials. They were thus no captains of industry interested in growing the business, by increasing its production capacity, for instance, developing other, subsidiary, businesses (edge-tool or nail manufacturing), or diversifying production. They were simply interested in running a business whose overhead was low thanks to the generosity of the government, and they had no desire to develop the iron industry. In bringing moulders from Scotland at the beginning of their tenure and opening a **foundry** in Trois-Rivières, Mathew Bell and his associates would diversify and boost production, but this was purely in response to competition from foreign products on the Canadian market.

The Pélissier Syndicate, 1767–78

Christophe Pélissier was a merchant who headed a syndicate of partners made up of other merchants and prominent people, which, in 1767, obtained the first lease to the Forges, for a period of 16 years. Pélissier was a Frenchman from Lyons who had arrived in the colony in 1752. Married to a Canadian woman in 1758, he was one of the French *bourgeois* who had decided to stay in Canada after the Conquest. Pélissier's partners, Benjamin Price, Colin Drummond, Thomas Dunn and George Allsopp, were not industrialists but influential members of the English merchant class at Quebec. They were, among other things, members of the Council of Quebec created by Governor James Murray in 1764. Dunn and Allsopp would hold important positions in the civil administration, and Dunn was known to enjoy government patronage.[70] Their social standing, wealth and influence must have been signifcant factors in their obtaining the Forges lease for a token £25 a year. Nevertheless, they invested over £6,000, by their own account, in putting the ironworks back on its feet. The value of their shares on disposition seems to support their claim (Table 1.11). Most of the partners would sell their shares to Pélissier before very long, however.

Table 1.11

THE PÉLISSIER SYNDICATE, DISTRIBUTION OF SHARES, 1767[71]

Partner	Social status in 1767	Share value*	Shares bought by Pélissier
Christophe Pélissier#	Trois-Rivières merchant	£615	
George Allsopp	merchant, assistant clerk of the Council of Quebec, deputy provincial secretary	£615	1771
Colin Drummond	merchant, member of the Council of Quebec (1768)	£615	1771
Alexandre Dumas#	Quebec merchant	£615	1772
Jean Dumas Saint-Martin#	Montreal merchant, justice of the peace	£615	
Thomas Dunn	member of the Council of Quebec, receiver general for the province in 1770	£615	1771
James Johnston	Quebec merchant	£615	1770
Benjamin Price†	merchant, member of the Council of Quebec	£615	
Brook Watson	merchant, London	£615	1771
Total value of shares		**£5,535**	

* Value in current £ at the time of their purchase by Pélissier between 1770 and 1772.

† Died in 1768; his share was bought from his heirs in 1777 by Pierre de Sales Laterrière for £900 currency

Forges agents. Dumas and Saint-Martin were cousins.

Despite the size of the land reserve granted to the Forges lessees, problems of ore availability were not long in arising. In 1768 and 1769, documents show, the Pélissier administration was forced to mine ore beyond the allotted reserve. In fact, the lessees asked for the right, as under the French regime, to mine ore in an area beyond the St Maurice seigneury. We do not know what became of this request, which was probably granted in part, but the issue was never fully resolved, since the problem would resurface periodically. Timber would continue to be cut on the seigneury of Cap de la Madeleine thanks to a land concession that Pélissier obtained from the Jesuits in 1767, of a strip measuring 20 arpents by 2 leagues (1,148 ha) on the east bank of the St Maurice River, opposite the Forges.[72]

Production does not appear to have resumed right away in 1767, at least not at the blast furnace,[73] because the following October a notice advertising the availability in the near future of bar iron and ploughshares appeared in the *Quebec Gazette*. Major repairs, including an overhaul of the blast furnace, were undertaken that year and possibly the following year, so that by the end of 1768, over £6,000 had been invested.[74] It was apparently only in the spring of 1769 that the Forges really got going again. Pélissier reported hiring "new workers," of British origin,[75] to complement those of French origin returning to their jobs at the Forges. An announcement in the *Quebec Gazette* on 29 June 1769 confirms that the Forges were in production again. Information on operations in the years that followed is scant, apart from the cannonballs produced for the Americans during the invasion of 1775 (see Chapter 6). Pélissier's purchase of the shares of six of his partners between 1770 and 1772, to be paid for in **pig iron**, is an indicator of the healthy state of the company at the time.[76]

From inspector of the Forges at the beginning of the lease, Pélissier succeeded quite rapidly in taking control of the company and, by the end of 1773, he held seven of the nine original shareholdings. But his claim to fame rests on his association with the American invaders in 1775–76. Notes on the invasion of Canada in the journal of notary J.-B. Badeaux are highly incriminating for Pélissier. Sympathizing with the Americans, he supplied their troops with materiel (cannonballs, bombs, stoves, shovels, pickaxes), dined with General Benedict Arnold and even provided information on how best to take Quebec. Badeaux's suspicions would be confirmed when Pélissier joined the American troops on their retreat south, taking with him some £2,000 of Forges proceeds. He spent some time in the United States and later resurfaced in Lyons, France. In 1778, he returned to Canada on the same ship that was carrying the new governor, Sir Frederick Haldimand, to Quebec! Pélissier stayed only long enough to wind up his affairs and returned immediately to France.

The person who suffered the worst consequences of Pélissier's collaboration with the Americans was his inspector, Pierre de Sales Laterrière, who was imprisoned for a month in the spring of 1776. The authorities had found it suspicious that he had passed so easily through the lines of the retreating American troops on his way to Quebec. In June, Laterrière replaced the exiled Pélissier as director of the Forges, and became his partner by buying out Benjamin Price in 1777.

Alexandre Dumas, 1778–83

After Pélissier's departure for France in the fall of 1778, a merchant and former shareholder, Alexandre Dumas, took over the lease of the Forges. In January 1779, he brought in Laterrière as equal partner and inspector of the ironworks,[77] but Laterrière was arrested again in February for having allegedly given comfort and assistance to an American prisoner, and would be imprisoned until the fall of 1782.[78] Dumas therefore oversaw operations until the expiry of the lease in June 1783.

Following a visit by Captain William Twiss of the Royal Engineers, in the spring of 1779, the Forges filled government orders for cannonballs and wrought iron, as well as manufacturing different types of iron and castings mainly for the local market.

Two years before the expiry of Dumas's lease, discussions on the granting of the next lease to the Forges had already been initiated by the future lessee, Conrad Gugy. He had come to New France as a Dutch officer with one of General Wolfe's regiments. It was probably his position as secretary to Governor Haldimand in Trois-Rivières, under the military government, that had acquainted him with the business of the Forges. In fact, he moved to the region not long after as seigneur of Grosbois and Grandpré, parts of the seigneury of Yamachiche.[79] In April 1781, Gugy, as a member of the Legislative Council, gave Governor Haldimand his opinion on the type of company that should take over the Forges. He recommended that the number of shareholders be limited to five, each of whom, in his view, would have to invest at least £500, for a total investment of £2,500 to run the Forges. In December that same year, Laterrière, still in prison, offered to manage the Forges on behalf of the government, and even volunteered to work there as a prisoner. It would appear that the option of government control had resurfaced, but it was not viewed favourably by the Governor.

Conrad Gugy, 1783–87

Following a request made in November 1782, it was finally Gugy who obtained a 16-year lease on the same terms as the previous lease (£25 a year), on the recommendation of the Executive Council, of which he was actually a member.[80] The lease was seen as a mark of gratitude by the Council, which felt that since his arrival in Canada Gugy had "served the Crown well for no consideration."[81] Although the terms of the previous lease required the tenants to return the establishment in the state in which they had found it, the facilities had deteriorated severely under Dumas, who asked for a grace period in which to make repairs. An inspection conducted in June 1783, on the date of expiry of the lease, confirmed the sorry state of the Forges. Despite an explicit agreement that he had signed with Gugy to repair the facilities, Dumas apparently did not do all the necessary work. The matter was still not settled in 1785 and Gugy submitted an estimate of £3,500 to restore the Forges to the condition in which they should have been handed over to him. As is so often the case for that period, there is no documentary record of what followed. However, things were apparently going well, as borne out by an announcement in the *Quebec Gazette*, in 1784, advertising a variety of Forges wares, in terms suggestive of a readiness to compete against European products.[82]

Gugy's 16-year lease was due to run until 1799, but in 1786 his fortunes took a turn for the worse. Following a lawsuit brought against him in 1782 by François Lemaître Duhaime, on 5 April 1786, a court ordered Gugy to pay £7,000 in damages. Gugy died five days later. As fate would have it, the ruling, seen as unjust, was later reversed. Having probably sensed that he would lose, Gugy had made a gift of his property, including the Forges lease, in January of that year, to his companion, Elizabeth Wilkinson. She therefore resisted when the authorities sought to sequester Gugy's property at the Forges. This was finally done in April 1786 and cancelled a month later. On 10 March 1787, following various proceedings, Elizabeth Wilkinson sold the remainder of the lease, covering 12 years (1787–99), to Alexander Davison and John Lees for £2,300. The purchase was handled by their representative, François Lemaître, the very man who had brought down Conrad Gugy.

Alexander Davison and John Lees, 1787–93

Alexander Davison and John Lees were major Quebec import-export merchants. They held the lease to the King's domain on the North Shore and were the official purveyors to the British troops in North America.[83] Their partnership was dissolved in 1792. Davison, who had moved to Britain a year earlier, acquired Lees's share. In 1793, Davison sold the Forges lease to his brother George, Mathew Bell and David Monro for the sum of £1,500.

In a petition dated 1788,[84] Davison and Lees requested, in vain, a 10-year extension of their lease. They stated that they had made improvements to the establishment and intended to invest more money. We know little about their tenure, but an inventory from 1807 confirms their outlays (see Appendix 4). Details of the work they had done on a dozen buildings show that it was under them that long multi-unit **tenement houses** were first erected, anticipating the population increase that would occur under Monro and Bell.

The Bell Years, 1793–1846

David Monro and Mathew Bell, partners since 1790, were already managing the affairs of the Davison brothers at Quebec when they went into partnership with George to manage the Forges. They administered the King's domain on behalf of Alexander Davison and were also in charge of supplying British troops. The Davisons, John Lees and Mathew Bell were all originally from Northumberland, in England.[85] It was at Quebec, as a clerk working for John Lees, that Bell had begun his business apprenticeship at around age 15. This new group of entrepreneurs was to be in charge of the Forges for 53 years, thanks mainly to one of them, Mathew Bell, who was in sole charge for 30 years (Plate 1.3). The documentary record on this lengthy tenure is scant considering how long it endured. Comparatively speaking, the six years of Cugnet et Compagnie are far better documented. No account books or correspondence from the Bell years have been found to this day.[86] Other sources, however, shed light on this period and on the stability and continuity of the business. The first nominal censuses of Lower Canada date from that time (1825, 1831) and the employee rolls drawn up by the company (1829, 1835, 1842) attest to the workers' long service and their deep roots at the Forges. In addition, detailed accounts by travellers and reports by inspecting military officers bear witness to the fact that the ironworks was running smoothly. This was also the period in which the first views of the village were painted by artists visiting the Forges, and some plans of the site, together with inventories of buildings, give us a more concrete idea of the layout of the Forges. Though not extensive, the documentation is of good quality and allows us to reconstruct the various stages of the Bell years at the St Maurice Forges.

Table 1.12

LEASES OVER THE BELL YEARS, 1793–1846 [87]

Date issued	Term	Rent	Lessee
6 June 1793 (continuation of lease of 1783)	6 years 1793–10 June 1799	£25	G. Davison, D. Monro, M. Bell
30 March 1799	2 years 10 June 1799–1 April 1801	£25	G. Davison, D. Monro, M. Bell (1799–1800) D. Monro, M. Bell (1800–1801)
6 June 1800	5 years 1 April 1801–1 April 1806	£850	D. Monro, M. Bell
15 July 1805	1 year 1 April 1806–1 April 1807	£850	D. Monro, M. Bell
1 October 1806 (sale of lease not ratified by the government)	20 years 1 April 1807– (previous lease extended to 1810)	£60	D. Monro, M. Bell
7 June 1810	21 years 1 January 1810–31 March 1831	£500	D. Monro, M. Bell (1810–16) M. Bell (1816–31)
27 May 1830	1 year 31 March 1831–31 March 1832	£500	M. Bell
8 March 1831	1 year 31 March 1832–31 March 1833	£500	M. Bell
14 March 1832	1 year 31 March 1833–31 March 1834	£500	M. Bell
25 November 1834	10 years 1 January 1834–1 January 1844	£500	M. Bell
18 April 1844	1 year 1 January 1844–1 January 1845	£500	M. Bell
19 September 1844	5 months 1 January 1845–1 June 1845	£500	M. Bell
11 April 1845	4 months 1 June 1845–1 October 1845	£500	M. Bell
25 June 1845	1 year 1 October 1845–1 October 1846	£500	M. Bell

As Table 1.12 shows, the lease on the Forges, granted first to Davison, Monro and Bell, then to Monro and Bell, and finally to Mathew Bell alone, was renewed 13 times in 53 years. Each renewal gave rise to petitions and discussions, and also led to opposing bids and public auctions, from which some combination of the trio always emerged as the winner. Since the first lease granted to Pélissier in 1767, the tenants of the Forges had never paid more than £25 a year in rent for the beneficial use of the establishment. Davison, Monro and Bell continued to enjoy those terms during their first lease and for two more years when their lease was renewed in 1799. The terms changed in 1801, however, at the second renewal. For the first and only time, another entrepreneur, Thomas Coffin, who headed the sole competitor of the St Maurice Forges in Canada, the Batiscan Iron Works Company,[88] drove up the bidding by making a formal offer to pay £500 in rent for the Forges. This triggered a series of bids and counterbids by the two competitors, resulting ultimately in the acceptance, by Governor Shore Milnes, of Monro and Bell's bid of £850. They had offered to top the highest bid by £50, a practice that was apparently deemed acceptable in those days. The new terms, 34 times the previous rent, lasted only nine years.[89] In 1806, with the auction of a 20-year lease, Monro and Bell, owing this time to a lack of serious competition, were once again granted bargain terms: rent of £60 a year along with an expanded land reserve.[90] The sharp decrease in rent did not please the Executive Council, which referred ratification of the sale to London. This inevitably raised doubts about the legality of 1806 auction, which were dispelled after investigation. An annual rent of £60 was still unacceptable, though, and it was finally the new governor, Sir James Craig, who reached an amicable agreement with Monro and Bell that led to the signing, on 7 June 1810, of a 21-year lease with rent set at £500 a year. These terms prevailed on subsequent renewals of the lease right up to the sale of the Forges in 1846.

The terms under which the Forges lease and its land reserve were granted are particularly well documented, owing to the fact that these matters were by law under civil authority and were the subject of petitions, investigations and reports submitted to the Executive Council and debated in the House of Assembly. It was the land question that basically brought down Mathew Bell, whose fame rested more on the control he exerted over the vast tracts reserved for the use of the Forges than on his admittedly efficient management. Under Bell, the importance that the land issue took on stemmed largely from the fact that the land in question was in the public domain and began to be coveted by the people of Trois-Rivières after 1800.

The government had always reserved to the Forges the land and resources of the combined seigneuries of St Maurice and St Étienne and other adjacent land, but those resources had been considerably depleted by the beginning of the 19th century and more land was required for raw materials. Since supplies of wood and ore were vital to the Forges, it was not surprising that Bell devoted considerable energy and exerted all his influence as a member of the House of Assembly and of the Legislative Council to ensure the continued operation of the ironworks.[91] The many land extensions he obtained when renewing his lease brought the Forges reserve to upwards of 40,000 ha.[92] It is not surprising that his detractors, many of whom had designs on some of that land for settlement and the timber trade,[93] complained of an outright monopoly of public land in the hands of one man.

In the successive petitions by Monro and Bell for access to the resources of new Crown lands, the reasons they gave changed over time. By 1796, they were anticipating the depletion of wood and ore reserves in the near future on the land already granted. In 1799, they claimed they had been forced to buy the right to cut over 3,000 cords of wood from a neighbouring seigneury.[94] That same year they also obtained the concession, from the Jesuits, of a plot of land measuring 2 leagues by 20 arpents in depth (1,148 ha) on the seigneury of Cap de la Madeleine, on the east bank of the St Maurice River, opposite the Forges. After the Jesuit estates reverted to the Crown in 1800,[95] Bell requested and eventually obtained the right to cut wood and collect ore on a portion of that seigneury as well.[96] The depletion of resources, timber in particular, does appear to have been a real problem; Monro and Bell obtained the land extensions they asked for.[97] In 1805, they began to bring up the need to add a strip of 2 to 3 leagues (10–15 km) on the northern boundary of the Forges reserve as a protective buffer against forest fires; they claimed to have lost 1,500 cords of wood to fire the previous year.[98] In 1808, they specified that this new land was to keep settlers away and reduce the risk of forest fires.[99]

With such requests, the lessees, already generously treated by the government, went beyond the vital needs of the business and crossed a line that suggested they had monopolistic aims. The 21-year lease signed in 1810 met most of their demands, and it was probably when Bell became the sole lessee of the Forges, beginning in 1816, that his land monopoly became most apparent and also most contested. Later, in 1819, Bell obtained a further extension of his rights to the resources of the neighbouring seigneury of Cap de la Madeleine, this time justifying his request by the rising cost of carting materials from farther and farther away. It was in 1825, when he began wanting to extend his hold over the resources of Cap de la Madeleine and coveting those of the Champlain seigneury, with the aim of acquiring the former Jesuit estates, that resistance to his monopoly really began to take shape, leading to debate in the House of Assembly in 1829.[100]

On a motion by the member for Trois-Rivières, Pierre-Benjamin Dumoulin,[101] a committee of the House was set up, which concluded that the vast lands under Bell's control constituted an obstacle to settlement in the Trois-Rivières district. There followed government inquiries into the quality of the soil on the land in question and the needs of the people of the region. Their claims were finally expressed in a petition presented by Dumoulin, to which the House of Assembly gave its assent in 1831, calling for the concession of lots within the reserve that accompanied the Forges lease, which was due to expire that year. Lafontaine tabled a similar petition before the House in 1832. In the meantime, the Governor, Lord Aylmer, referred the matter to London, and the House tried in vain to obtain a copy of the instructions he received. Pending settlement of the terms of a new lease, the Governor extended from year to year, until 1834, the 21-year lease signed in 1810. Called before the House to argue his case, Bell underlined the contribution of the St Maurice ironworks to the local economy, pointing out that the business employed over 400 people directly and provided work for the surrounding population as well. He claimed that the land so coveted by prospective settlers was poor and that it was actually the timber reserves of the Forges that the people of Trois-Rivières were after. He also continued to stress the need to maintain a buffer zone around his reserve to protect against fire.[102]

In 1832, Bell had asked for a 10- to 15-year lease, in order to preserve the stability of the business. On 25 May 1833, ignoring the representations of the citizens of Trois-Rivières and the House of Assembly, London authorized Aylmer to grant Mathew Bell a 10-year lease. The Governor's arrogance earned Bell the label of "a grantee of the Crown, who has been unduly and illegally favoured by the Executive" in one of the 92 Resolutions passed by the House of Assembly in February 1834 expressing its grievances with regard to the executive level of government.[103] These protests did not prevent Governor Aylmer from granting Mathew Bell, on 24 April 1834, a 10-year lease that also gave him the right to exploit the resources on 25,940 additional arpents (8,868.63 ha) of the Cap de la Madeleine seigneury.[104] That was to be the last tract of land added to Mathew Bell's reserve until the expiry of his lease, which was extended for a further two years before the Forges were put up for sale in 1846.

The additional land made available to Bell only served to confirm, in the eyes of the people of Trois-Rivières, Bell's relentless desire to expand his holdings, which were blocking settlement. The granting of the 1834 lease aggravated the mounting wave of disenchantment and continued to fuel bitter arguments between Bell and the citizens backed by the House of Assembly. In 1841, *Le Canadien* published the view of a Trois-Rivières man that the task of the next member of the House would be to work to eliminate the land monopoly of the Forges. In 1842, another petition concerning the release of land behind Trois-Rivières for settlement was referred to a committee of the House of Assembly. In 1843, a report of the Executive Council, now sensitive to public pressure, written by Étienne Parent, concluded that the concession of farm lots in the vicinity would not hurt the Forges, since the settlers would be encouraged to sell their wood to the ironworks. Parent suggested an approach that was finally adopted by the government, questioning "whether a sale or concession of the Forges and Lands would not be more advantageous to the public, than a lease."[105] In 1845, a survey of the Forges lands (fiefs of St Maurice and St Étienne) was carried out in order to plan the division of lots, while reserving land in the immediate vicinity of the Forges for the creation of a future village. The Forges were finally sold at auction on 4 August 1846. Bell was present and lost out by only £125.

The tenacity of Mathew Bell and the length of his tenure are proof that the ironworks was profitable; Bell himself confirmed this on several occasions, although we lack detailed information. In 1793, George Davison, Mathew Bell and David Monro had acquired the balance of Alexander Davison's lease for the sum of £1,500, plus £2,934 11s 8d for property and tools left on the premises. Their initial investment therefore totalled £4,434 11s 8d. The respective shares of the three partners were as follows: George Davison, half; Mathew Bell and David Monro, the other half. Though we lack detailed figures regarding operations, the appreciation in the value of the shares is indicative of the health of the business (Table 1.13). The year after Davison's death in 1799, Mathew Bell and David Monro jointly acquired Davison's share, valued at £10,523 18s 7d, which means that the total equity was £21,047 17s 2d, or almost five times the initial investment, after only seven years of operation. A few years later, in 1804, Lord Selkirk reported that the annual produce, or value of production, was £10,000 or £12,000.[106] In 1816, David Monro in turn sold his share to Mathew Bell for the sum of £13,123 10s 2d, indicating that their shares were now worth £26,247 0s 4d. This sum did not represent the value of the Forges as such, which remained Crown property, but of Bell's property on the post, mainly ironwares in stock and stores of raw materials. An inventory conducted in 1833 put the value of Bell's assets at £48,072 10s 6d, or more than 10 times the capital outlay of 1793 (Table 1.13). At the time (1833), Bell stated that the Forges pumped £10,000 to £12,000 a year into the Trois-Rivières region, and estimated the annual value of production at £30,000, or three times the figure reported by Lord Selkirk 30 years earlier.[107]

Table 1.13 SHARE VALUE APPRECIATION, 1793–1833

Lessee	Year	Value of shares
Davison, Monro, Bell	1793	£4,434 11s 8d
Monro, Bell	1800	£21,047 17s 2d
Bell	1816	£26,247 00s 4d
Bell	1833	£48,072 10s 6d

Insofar as these figures indicate the money-making potential of the Forges, one can understand the vigilance of Bell and his partners when it came to renewing their lease and protecting the land reserve with its free resources; collecting, "dressing" and carting them were the only costs attached. This Crown privilege contributed significantly to their bottom line, keeping down overhead and increasing their profit margin accordingly.

Under Bell and his partners, the Forges operated at full capacity manufacturing castings. The foundry side of the business was emphasized over the finery side, in order to meet local demand for the goods needed by an expanding colony. The premises were reorganized to make more space available for **moulding** work. The increase in output of **castings** also led Bell to operate a foundry in Trois-Rivières at which he employed, in 1829, at least seven workers (five moulders and two forgemen) from the St Maurice Forges.[108] The foundry produced mainly large hollow ware, such as the sugar and potash kettles in great demand at the time, and also steam-engine parts.[109] It was at that time that a real tradition of moulding was established by the British moulders hired by Mathew Bell.

Probably to reduce the cost of carting raw materials the long distance to the Forges, carters and day labourers began to be hired on an annual rather than seasonal basis. The arrival of these new workers and their families swelled the population on the post from 200 in 1805 to 425 in 1842.[110] By the end of Mathew Bell's tenure at the Forges, although the production capacity of the blast furnace had not been increased, diversification in the product line had created a bustling establishment that had every appearance of a true industrial village.

Bell left, but his legend lived on in oral history. Albert Tessier and Dollard Dubé recount the story of how he took his leave:

> When he blew out the furnace in 1846, the old folk say that he came out, in front of the big house, where the store was at the time, and set fire to the books, saying to the little crowd present, some of whom owed him money: "My friends, you have worked hard for me. Things didn't always go the way we would have wished, but I am happy all the same; now then [...] no one can ever tell you again that you owe anything to Mr Bell."[111]

If there is any truth to the story, Mathew Bell thus deprived historians of invaluable archival material. The list of account books kept subsequently under John Porter & Company (see Appendix 5) gives us an idea of the wealth of information that such documents contained on every facet of the company's operations. No account book of Bell's or of the administrations that preceded and followed him has been found to this day. While Bell's ledgers may have been burned, there is no indication that the others were destroyed. The discovery of those ledgers would provide some of the most important documents on Canada's industrial history.

THE FORGES UNDER PRIVATE OWNERSHIP, 1846–83

Managing the Forges became more complicated once they were taken over by private enterprise. The basis on which they operated became a problem. The new proprietors needed the same ore and timber resources as their predecessors, but now they would have to pay for them. The government had decided, under pressure from public opinion, to revoke the privilege of the land reserve that had always gone hand in hand with the Forges. The cost of buying Crown land or raw materials on private land, on top of the cost of acquiring the works themselves, would place the proprietors in debt from the start. These problems, unknown since the days of Cugnet et Compagnie under the French, would plague the first 15 years of private ownership. In contrast to the privileged tenants who had preceded them, the new owners had to face the real cost of operating the Forges. Their money problems would lead to trouble with lenders, who more than once turned to the courts to press their claims.

Although the first private owners of the Forges were not industrialists either, they made the beginnings of an attempt to rationalize operations. Henry Stuart (1841–46) had the **blower** of the blast furnace modified to improve its performance. This was the work of the first engineer reported at the Forges, in 1848, the Frenchman Nicolas-Edmond Lacroix.[112] A few years later, Andrew Stuart and John Porter, operating as John Porter & Company (1851–61), called in a British engineer, William Hunter, to upgrade the plant. He altered the blast furnace and the blowing engine to double ore reduction capacity. It was also under John Porter & Company that the Forges branched out into the manufacture of railcar wheels, at a time when the Canadian railway network was expanding.

But it was not until the McDougalls took over (1863–83) that real industrial entrepreneurship was seen. They streamlined operations to concentrate on the Forges' basic product in what had become the competitive world of the iron and steel industry. By discontinuing the manufacture of consumer products and specializing in the production of pig iron, they earned the industrial village a 20-year reprieve. A few years after the takeover by the McDougalls, it was reported in *Le Constitutionnel* that the Forges were "managed with rare intelligence."[113]

The shift to private enterprise, with the accompanying changes in production, altered the working environment that had characterized the ironworks since its inception. Master craftsmen, who had handed down their skills from father to son for over a century, left the post and were replaced by labourers whose work required no special skills. The workers who witnessed the final shutdown in 1883 were relative newcomers with no strong ties to the Forges.

Henry Stuart, 1846–51

It was with a bid of £5,575 that Henry Stuart acquired the Forges at the public auction held on 4 August 1846. Starting from an upset price of £3,000 set by the government, Stuart outbid four others, including Mathew Bell, who stopped at £5,450![114] Henry Stuart was a 38-year-old lawyer, born at Quebec and living in Montreal. He was the son of Andrew Stuart (1785–1840), who had been solicitor general of Lower Canada.[115] Three months later, on 6 November, at another auction, Stuart also acquired the fiefs of St Maurice and St Étienne for the bargain price of £5,900.[116] The terms of sale, however, required Stuart to sell 100 acres of land (40 ha) at no more than 6 shillings an acre to any interested party, but he kept the right to mine ore on unconceded land for five years.[117] At the time of these two successive sales, the Crown Lands Department required the purchaser to pay only an initial instalment, roughly a quarter (£1,404 13s 2d) for the Forges and a little over one-tenth (£595 18s 8d) for the two fiefs.

Stuart thus started out with a debt to the government of £9,474 8s 10d. During the entire time that he owned the Forges, until 1851, he repaid only a mere £515, in June 1847, thus remaining in debt to the Crown for £8,959 8s 10d, even though the terms of sale called for final payment to be made by 4 August 1849.[118] He also spent a "considerable sum" (about £8,000) to buy the personal property left on the post by Mathew Bell.[119] Stuart personally ran the Forges for only one season, that of 1847. He was forced to borrow a little over £1,700 from a Montreal merchant, James Ferrier, who also loaned him another £5,000, in October of the same year. The terms of repayment were then set down in an agreement between the two men. Stuart mortgaged the Forges and the two fiefs to Ferrier and agreed to lease the Forges to Ferrier for five years for him to recover his loans out of operating revenues. The agreement gave Ferrier full authority over the running of the Forges without his having to assume the full cost. In addition to reimbursing himself out of the profits, Ferrier paid himself an annual stipend of £500. Under the second clause of the agreement, Ferrier had to settle all debts incurred by Stuart that were related to the operation of the Forges, although Stuart remained ultimately liable. Later documents show that Stuart went deeper into debt to Ferrier in 1847–48 to the tune of £15,234 9s 6d,[120] probably to pay for property at the Forges purchased from Mathew Bell and also for improvements made to the blast-furnace blower.[121]

Ferrier ran the Forges for four years. The records of the legal proceedings instituted in 1852 are the first indication that both parties were in breach of their agreement. The agreement was not a lease as such, but provided for the signing of one. Ferrier had therefore been operating the Forges without a lease, while Stuart had also failed to respect the provision calling for him to grant such a lease. On 8 November 1851, Henry Stuart sold the Forges a few days after receiving (he claimed) verbal notice from James Ferrier that he intended to cease operations.[122] The 1847 agreement between the two had provided that Ferrier could cease operating the Forges at any time on two months' written notice. Since Stuart had not waited for the written notice, Ferrier would later take advantage of this technicality to claim that Stuart had ended the lease prematurely by selling the Forges without Ferrier's consent. Since both men were in breach of their agreement of 1847, they finally settled out of court in 1853.

Ferrier ran the Forges at a profit, according to two of the earliest historians of the Forges, F. C. Wurtele (1887) and Napoléon Caron (1889), who talk of success and making a profit, without mentioning sources, however. Yet during the legal proceedings in 1852, Ferrier maintained that he had not made sufficient returns to recoup his loans to Henry Stuart, and claimed a further £5,000 from him. All the same, Ferrier had recovered close to £17,000 advanced to Stuart, not counting his annual stipend of £500.[123] However, he left the Forges in a sorry state, according to Étienne Parent, who reported in 1852 that "it was plain that the hand and eye of the interested owner had not been there for years."[124] The new owners of the Forges, John Porter & Company, accused Ferrier of costing them an entire year's production because he had failed to lay in supplies for the following year before giving up the Forges in the fall of 1851. It was also reported that workers had begun to leave under James Ferrier.[125]

John Porter & Company, 1851–61

On 8 November 1851, Andrew Stuart,[126] Henry's brother, and John Porter, operating under the name of John Porter & Company, acquired the Forges and the two fiefs, along with appraised movable property, for a total of £16,559. The bulk of that sum, £11,659 (70.4%), was owed to the government, and the remainder, £4,900 (29.6%), to Henry Stuart and the heirs of William Conolly. The purchase price was reduced twice, in 1852 and 1853, as the result of reassessments of the new company's debt to the government. As Table 1.14 shows, the debt was adjusted downwards from £11,659 to £7,526 12s 8d, thus reducing the purchase price to £12,426 12s 8d in 1852, because of an overestimation by the government of the land area of the fiefs of St Maurice and St Étienne.

Leaving aside the sums owed directly to Henry Stuart and the Conolly heirs, the £7,526 12s 8d owed to the government was due in five years. Stuart and Porter requested a further readjustment following their urgent efforts to obtain land concessions. Indeed, a major condition imposed by the government at the time of the transaction forced the purchasers of the two fiefs sold with the Forges to concede 100-acre (40 ha) lots to anyone making such a request, at a price that was not to exceed 6 shillings an acre. From December 1851 to May 1852, Stuart and Porter sold 130 lots (13,499.5 acres or 5,463 ha) in the fief St Étienne for a total of £4,400 7s 4d. The government did not see that money, of course, because the buyers had 20 years to pay off the principal, with a mere 6% interest payable annually until final settlement. Stuart and Porter discovered, as others had before them, that the value of the Forges was diminished by the sale of the adjoining land. For the first time in the history of the Forges, they had to pay for their raw materials. Terms for collecting ore

Table 1.14

SALE OF THE FORGES TO JOHN PORTER & COMPANY IN 1851 AND ADJUSTMENT OF THE PURCHASE PRICE IN 1852

Creditor	Value of Forges and fiefs	Value of movables	Balance owed by John Porter & Company	
			initial amount	adjusted amount
Government	£7,459	£4,200	£11,659	(£10,062 7s 11d) (£7,526 12s 8d) #
Henry Stuart		£3,400	£3,400	£3,400
William Conolly*	£1,500		£1,500	£1,500
TOTAL	£8,959	£7,600	£16,559	(£14,962 7s 11d) (£12,426 12s 8d)

* The heirs of William Conolly[127] held a mortgage on the Forges for a loan he had made to Henry Stuart in 1846.

\# More realistic amount calculated by John Porter & Company.[128] The sum of £10,062 7s 11d calculated by Félix Fortier of the Crown Lands Department was in fact a re-evaluation of the price that Henry Stuart should have been charged in 1846 based on the corrected land area of the fiefs. In imputing this sum to John Porter & Company, Fortier neglected to subtract the payments already made by Henry Stuart.[129]

from settlers' land were not reached easily in the first few years, and in the first year of operation the new owners even had to pay £500 for wood. Stuart and Porter also discovered that the proximity of settlers' homes to the woodlots increased the risk of fire. In fact they claimed, in their memorial of 23 June 1852, to have lost close to 2,000 cords of wood to fire that spring.

Pleading that "the value of the Forges [was] very materially diminished by the Concessions of Land around them," and faced with the considerable sums they planned to invest in repairing and improving the plant, Stuart and Porter, in their memorial to the government, proposed a reassessment of their debt and of the terms of payment. They proposed paying the balance of £7,526 12s 8d in three instalments: half in the form of an assignment of the £3,763 6s 4d owed by the settlers to whom they had sold lots on the St Maurice and St Étienne fiefs, and the other half in two instalments, payable in two years. The two partners were thus requesting "a rebate of the arrears of interest," or forgiveness of the interest owed by Henry Stuart on the outstanding principal of £7,526 12s 8d by their calculations, saving them £2,384 15s 4d.[130] Their request was turned down, but they succeeded in having a

"competent person" visit the Forges to report on the state of the establishment and the effect of the sale of land on their business.

The Crown Lands Commissioner appointed Étienne Parent, assistant provincial secretary, to conduct an inquiry. Parent's report, dated 20 September 1852, is a mine of information, particularly with regard to the dilapidated state of the Forges at that time. Parent had to report specifically on three points relating to the requests of John Porter & Company: rebate of outstanding interest, approval of an assignment of the conceded lots, and the reserving of 150 unconceded lots to the Forges. Parent's recommendations generally supported the company's requests, except for some details. Having received a petition from squatters living on some of the 150 lots claimed by John Porter & Company, however, the Crown Lands Department had to commission another inquiry, this time from Crown Lands officer Oliver Wells, whose report, dated 1 March 1853, was less favourable to the company. Some sixty squatters had settled without title on land that Parent had deemed of poor quality and therefore more useful to the Forges. After considering the two reports, the Executive Council finally settled the question on 23 May 1853 and agreed to issue letters patent to John Porter & Company under certain terms (Table 1.15.)

This decision to reduce the company's debt to £6,200 brought the total price of the 1851 transaction down to £11,100 (£6,200 + £4,900)—all in all, a very modest price, considering that it covered the Forges, the two fiefs and the movable property. This was very close to what Henry Stuart had paid (£11,475) for the Forges and fiefs, without the movables (for which, it was said, he had paid Mathew Bell a considerable sum). In short, Henry Stuart was the big loser in this transaction (Table 1.16). While John Porter & Company did well in its dealings with the government, it would have other headaches with the merchants who became joint partners in the business.

In 1851, less than two weeks after the purchase, Stuart and Porter formed John Porter & Company and went into joint partnership with Weston Hunt & Company[131] to operate the Forges for a period of 10 years. Under the arrangement, Weston Hunt & Company managed the business through their representative on the spot, Jeffrey Brock.

In 1853–54, engineer William Hunter was commissioned to supervise major repairs and improvements to the Forges (Table 1.18). His biggest job was to completely overhaul the blast furnace to double its production capacity. The overhaul was botched and led to tragedy. The new furnace exploded, killing two workers and burning two others. The explosion also set fire to the **casting house** and caused a three-month halt in operations in the summer of 1854. A few months after this unfortunate accident, on 5 September 1854, the joint partnership between the two companies was dissolved.[133]

Table 1.15

TERMS OF SETTLEMENT OF THE SALE OF THE FORGES AND THE TWO FIEFS TO JOHN PORTER & COMPANY

OBLIGATIONS
- repayment of the principal within five years of the issuance of letters patent
- annual payment of interest on the balance
- mortgage on the Forges and the fief of St Étienne in favour of Her Majesty

INVESTMENT
- £4,000 over two years to improve the Forges

RIGHTS AND PRIVILEGES
- exclusive use by the Forges of roughly half the 150 lots on the first range of the fief of St Étienne, conditional on agreement being reached with the squatters
- price per acre for the lands of the fief of St Étienne reduced by 1s 6d, conditional on the same rebate being given to the settlers who had already bought lots
- remission of interest on the price of the Forges, but not on the price of the fief of St Étienne
- ADJUSTED BALANCE: £6,200

Table 1.16

PRICE OF THE ST MAURICE FORGES, 1846–63

Buyer	Year	Price of Forges and two fiefs	Price of movables	TOTAL
Henry Stuart #	1846	£11,475 (Forges £5,575, fiefs £5,900)	between £7,600 and £20,000	between £19,075 and £31,475
John Porter & Company	1851	£8,959	£7,600	£16,559
(adjusted price)*	1852			£12,426 12s 8d
(adjusted price)†	1853			£11,000
Government seizure	1861	for debt of £8,948 13s 4d	unspecified	seizure £8,948 13s 4d
Onésime Héroux	1862	Forges, farm and 1,195 acres (484 ha)**	unspecified	£1,750
John McDougall	1862 1863	Forges, including 69 acres (28 ha) £1,075	£250	£1,325

\# It is not known how much Henry Stuart paid Mathew Bell for the movables, but according to Stuart and Porter, it was a "large sum of money," which they estimated at £20,000 or higher. JLAPC, 16 Victoriae, 1852, Appendix CCC, p. 25, Stuart and Porter to Étienne Parent, 6 September 1852, and ANQ, Superior Court, docket no. 614, John Porter et al. v. James Ferrier, 1853, affidavit, 3 May 1853.

* The price of the fiefs was reassessed by the government in 1852, owing to an overestimation of their area at the time of sale in 1846. See Table 1.14 for an explanation of the adjusted total.

† Further reduction of the sale price because of a downward reassessment of Stuart and Porter's debt to the government, from £7,526 12s 8d to £6,200.

** This area comprises the Forges post (408 acres) plus seven adjoining lots; ANQ-TR, Not. Rec. Petrus Hubert, No. 4575, 27 April 1863, "Vente des Forges St.Maurice," Onésime Héroux to John McDougall.

Table 1.17

JOINT PARTNERSHIP BETWEEN JOHN PORTER & COMPANY AND WESTON HUNT & COMPANY TO OPERATE THE ST MAURICE FORGES

	John Porter & Company	Weston Hunt & Company
Responsibilities	owners	managers agents at Quebec
Investment	repayment of investment by Weston Hunt & Company plus interest	up to £7,500/year[132]
Annual instalments*	£600	£400
Distribution of profits	2/3	1/3

* Before profits.

Table 1.18

WORK RECOMMENDED BY ENGINEER WILLIAM HUNTER,* 1852

Structure	Repairs to be done
Large (blast) furnace	complete repair (1 month)
Cupola furnace	to be rebuilt
Floors for the conveyance of water to the wheels (headraces)	to be renewed
Wall of moulding shop	fallen down, to be rebuilt
Watercourse	to be deepened
Dams	to be repaired and raised
Sawmill	new one to be erected
Finishing shop	to be fitted in the brick building
Grist mill	to be repaired
Barns (for charcoal)	fallen down, 6 new ones to be built
Large house	new windows and doors; plastering throughout
Estimate of repairs: £3,600 to £4,000	

Improvements needed	Comments
New blast furnace of twice the capacity	
Double hot-air furnace	to replace the present single furnace
Erection of a rolling mill	to manufacture round and square iron
Construction of lathes	for machinery purposes
Estimate of improvements: £7,000 to £8,000	

Other possible improvements	
Steelworks	after experimenting with steelmaking
Estimate of these improvements: £1,500 to £2,000	

* "The Forges are so old-fashioned and so much out of repair, that they will be to make nearly new. The Cottages for the workmen, and the Workshops, all require extensive repairs." William Hunter, 24 August 1852.[134]

The partnership was thus dissolved on 5 September 1854, but not without the two companies acknowledging their financial obligations to each other for the division of debts, profits and, in particular, costs (Tables 1.19 and 1.20).

John Porter & Company agreed to repay Weston Hunt & Company £22,574 18s 1d, although this did not constitute *per se* an acceptance of their partners' accounting. Weston Hunt & Company's figure was in fact the balance of the working capital advanced by them for close to three years; only part of those advances had been repaid by John Porter & Company (Table 1.19). That figure would later be contested in court. Table 1.20 shows that part of the debt was to be paid out of sales revenues, with the balance secured by a mortgage on the Forges and by promissory notes signed by William Henderson, due in three years.[135] When five of those notes, valued at £1,000 each, were put into circulation the following spring, the terms of the agreement were breached, triggering a series of legal proceedings. The five £1,000 notes were held by the Quebec Bank, which called them in.[136] Faced with John Porter & Company's refusal to honour them on the grounds that under the terms of the agreement the notes were not due for three years, the bank took legal action and won its case in Superior Court on 10 December 1855. Having obtained a ruling against William Henderson,[137] the Quebec Bank demanded that he be dismissed as manager of the Forges, a position he had held since 14 October 1854. Until the £5,000 was repaid, the Quebec Bank was entitled to manage the Forges to recover its money. That would doubtless explain William Henderson's dismissal and his replacement by a more competent person. But the bank, for which running the Forges was apparently a money-losing venture, withdrew from the business, bringing operations to a halt in 1858.[138]

The outcome of the various lawsuits launched by John Porter & Company against their former partners is not known. They would appear to have reached a settlement in 1867, the terms of which are also unknown to us.[139]

Other proceedings, initiated by the government in October 1860 to recover the balance (£6,200) plus outstanding interest (£2,748) on the sale of the Forges and the fief of St Étienne, were what officially ended the tenure of John Porter & Company in 1861. The ruling of 7 November 1860 called for the company to be discharged of its debts against the return of its letters patent and the income from the land conceded in the fief of St Étienne. On 3 May 1861, the bailiff seized all the land in the company's possession, and on 22 October the Crown Lands Commissioner bought in most of that land from the sheriff. Only in 1866 did John Porter & Company finally receive its discharge.[140]

All in all, the first two sets of Forges proprietors did not have much success with their partners. As soon as they acquired the Forges, they gave over management to third parties with whom they became embroiled in lengthy legal proceedings. They did try to upgrade the plant, however, with Henry Stuart setting the example, and John Porter & Company following his lead. Despite the two explosions that occurred after the rebuilding of the blast furnace, the increased production capacity made it possible to branch out into new products such as railcar wheels. It was also under John Porter & Company that charcoal began to be made in **kilns**, as recommended by employee Timothy Lamb in 1852.[141]

Table 1.19

STATEMENT OF ACCOUNT BETWEEN JOHN PORTER & COMPANY AND WESTON HUNT & COMPANY ON THE DISSOLUTION OF THEIR PARTNERSHIP, 1851–54

Working capital advanced by Weston Hunt & Company	Amount repaid by John Porter & Company	Balance owed by John Porter & Company to Weston Hunt & Company
£39,755 12s 11d	£19,804 15s 7d	£19,950 17s 4d

Note: From Bédard 1986, pp. 42–44. The balance owing was adjusted to £22,575 4s 11d by Weston Hunt & Company from the £22,574 18s 1d established in the agreement of 5 September 1854.

Table 1.20

SCHEDULE OF DEBT OWED BY JOHN PORTER & COMPANY TO WESTON HUNT & COMPANY, 1854*

Total debt	Staggered debt	Repayment	Terms
£22,574 18s 1d	£3,000	drafts drawn on Frothingham and Workman, Forges agents at Montreal	
	£3,000	proceeds of sale of Forges wares in the hands of Weston Hunt & Company at Quebec	
	£8,287 9s 0d	promissory notes endorsed by William Henderson, renewable quarterly and payable only after three years	secured by a mortgage on Henderson's land
	£8,287 9s 0d	payable in six years, interest paid semi-annually	secured by a mortgage on the Forges

* According to the agreement signed on 5 September 1854 concerning the dissolution of their partnership. Bédard 1986, p. 35.

These initiatives did not make up for the weakness and instability of the current management. A whole year's production was lost at the end of Henry Stuart and James Ferrier's administration, while John Porter & Company ceased production in an atmosphere of confusion in 1858. In both cases, the Forges were left in a state of ruin. The ironworks began to lose its old-stock workers, many of whom migrated to the foundries of Montreal and the new Radnor Forges, which set up near the St Maurice Forges in 1854.[142] The village must have missed Mathew Bell a little during those 12 years. After some fifteen years, the Forges were back in the hands of the Crown, which had never really been out of the picture. The end seemed to be at hand. But after five years at a standstill, the Forges were started up again in 1863.

Things were never the same again, however, in terms of either the Forges' output or working population. The census of 1861, noting that the Forges were **out of blast**, recorded only 210 people living there, compared with 395 ten years earlier. Of the 89 workers recorded in 1851, only 37 were left in 1861.[143] The exodus, which had already begun in the 1840s, continued and changed the make-up of the industrial community. Workers from the old-stock families (Michelin, Imbleau, Terreau, Robichon, Marchand) and the other families that had put down roots there in the late 18th century and early 19th century, left the Forges.

Practically abandoned by its skilled work force, the ironworks was run down, its value depressed. H. R. Symmes, an inspector who came with the sheriff to see what state it was in before it was put up for sale, reported:

> We found the place generally in a very dilapidated state. That several of the dwellings had been entirely taken down and used as firewood. That all of the unoccupied buildings had been forcibly entered and more or less injured, and that the fencing upon the farm had almost entirely disappeared.
>
> [...] That most of the buildings are in very bad repair—the majority of them not repairable.
>
> That with four or five exceptions the buildings are of no value except for extensive manufacturing purposes.
>
> That the farm is good and for purely agricultural purposes is probably worth $3000.[144]

In November 1861, a committee of the Executive Council recommended that the Forges be sold with a closing date of 15 January 1862, at an upset price of £1,900. The sale was postponed to 15 September when the bidding failed to reach the upset price. The Forges were finally sold for £1,750 to the merchant Onésime Héroux in November 1862, comprising the ironworks itself, along with a farm and seven adjoining lots valued at just over 40% (£750) of the total price, as appraised by Symmes ($3,000) in September 1861. Héroux therefore paid only £1,000 for the Forges proper. More interested in the farm and the land, on 27 April 1863, Héroux resold the ironworks alone, carefully setting its boundaries within a tract of 28 ha, for £1,075.[145]

The McDougall Era, 1863–83

The McDougalls were the only true industrialists in the history of the Forges. They did not run the Forges as simple iron merchants but sought their own particular niche in the changed world of the iron and steel industry. Their analysis of the situation and positioning of the Forges made it possible to carry on for a while yet, despite the ironworks' obsolete technology. By focusing on the special characteristics of **charcoal iron**, they turned the ironworks into a supplier of pig iron to heavy industry and thereby managed to take the Forges almost to the threshold of the 20th century. But the McDougalls, like their predecessors, were not immune from financial problems, which would precipitate the Forges' ultimate demise in 1883.

In their last 20 years of operation, the Forges were managed successively by John McDougall (1863–67), John McDougall & Sons (1867–76), George and Alexander Mills McDougall (1876–80) and George McDougall (1880–83). As their family tree shows,[146] they were all related, but some members of the family lived in Trois-Rivières (John McDougall and his sons, including Alexander) and the others in Montreal (George, son of James, nephew of John McDougall). A third, William, brother of John and James and living in Baltimore, Ontario, was also embroiled in the financial affairs of the Forges.

John McDougall, 1863–67

John McDougall was a Trois-Rivières businessman, an agent of the Quebec Bank, and well acquainted with the business and the iron industry in general (Plate 1.4). His entry into the world of ironmaking in the Trois-Rivières region appears to have been well planned. His contacts with his brother James, an industrialist in Montreal, and a cousin, named John as well, also a Montreal industrialist and owner of the Caledonia Foundry, no doubt smoothed his path. The sequence of John McDougall's transactions shows a methodical man at work. His first aim was to put himself on a solid financial footing. In 1860, he obtained a $4,000 (£1,000) bond from his brother James and John Paterson as surety for his financial soundness as a local agent.[147] The bank's confidence in him left him well placed to raise capital. In December 1862, four months before acquiring the Forges, John McDougall purchased the movable assets of John Porter & Company for the sum of $1,000 (£250). On 16 April 1863, he bought for $8,000 (£2,000), including movables, the L'Islet Forges, a plant founded in 1856 on the east bank of the St Maurice. On 27 April, he bought the St Maurice Forges from Onésime Héroux for $4,300 (£1,075). In the

Plate 1.4
John McDougall (1805–70), who purchased the Forges in 1863 and founded
John McDougall & Sons in 1867. He died at the Forges in 1870.
NOTMAN PHOTOGRAPHIC ARCHIVES, MCCORD MUSEUM OF CANADIAN HISTORY, 6359-I.

space of five months, John McDougall had therefore invested $13,300 (£3,325) in the two ironworks.[148]

No industrialist invests so much money in so short a time without the prospect, or even the guarantee, of lucrative contracts. McDougall wanted to have the production capacity to fulfil prospective major contracts. He operated the Forges only briefly as a foundry, as we will see in Chapter 6. A large contract for pig iron, signed in 1865 with John McDougall & Company of Montreal, would absorb almost the entire output of the two ironworks. Having given the business a new focus, McDougall set about reorganizing it and all his affairs.

John McDougall & Sons, 1867–76

John McDougall founded a family firm, bringing in his sons and obtaining the backing of his brother James, owner of a mill in Montreal. In addition to William, who held a one-third share in the two plants, and James, who acted as proxy in real estate transactions, John McDougall brought four other sons (Robert, George, David and Alexander) into his business and its operations. All four came to live at the St Maurice Forges. McDougall also hired Henry Symmes as superintendent, the same Henry Symmes who had inspected the establishment for the government in 1861. McDougall formalized the involvement of all his sons by creating John McDougall & Sons on 26 April 1867 (Table 1.21).[149]

In contrast to the Forges' previous masters, John McDougall & Sons did not have a ready-made reserve of Crown land to supply basic raw materials. As soon as he had purchased the two ironworks, John McDougall set out to assemble a land reserve in their immediate vicinity. Table 1.22 shows that at the time of its founding in 1867, John McDougall & Sons held 2,490 ha of land, and that by the time the company was dissolved, the reserve had grown to 4,250 ha. Most of the land was located on the east bank of the St Maurice River within the parishes of St Maurice and Mont Carmel, precisely where the L'Islet Forges were situated. The shift of woodcutting and mining to the east bank can also be explained by the gradual depletion of the reserves on the fiefs of St Étienne and St Maurice. There was nothing new in exploiting the resources of the east bank: since the Pélissier administration, the resources of the Cap de la Madeleine seigneury had been exploited, by Mathew Bell in particular. The reserve assembled in 1876 appears to have been sufficient to support both ironworks, since when the Forges finally closed, George McDougall still held roughly the same area of land (3,822.9 ha). The financial credibility of the company was backed by the value of the land, which served to offset debts at the time of closing.

In terms of production infrastructure, the McDougalls operated much the same plant as had their predecessors, John Porter & Company. According to Henry Symmes's description in 1861 of the sorry state of the Forges (quoted above), the establishment must have required major investment to be put back into shape, as it had not been in blast since 1858. After a single year of diversified production, similar to that of their predecessors, the new owners shifted to producing pig iron for the railcar-wheel industry, prompted by the five-year contract won in 1865, renewed for two years in 1871. It was probably at that time

Table 1.21

SHAREHOLDERS OF JOHN MCDOUGALL & SONS, 1867–76

Shareholder	Shares in 1867	Shares in 1868	Role
John Sr	1/2	5/18	principal shareholder—1870†
William	1/9	1/9	shareholder
Robert	1/18	3/18	company director until 1874*
George	1/18	3/18	company director
David	1/18	1/18	company director
Alexander	1/18	1/18	company director
John	1/18	1/18	shareholder
James	1/18	1/18	shareholder/proxy
Thomas	1/18	1/18	shareholder
Total	**9/9**	**9/9**	

† Died 21 February 1870.

* Left the St Maurice Forges in 1874 to take charge of the St Francis Forges, owned by John McDougall of Montreal.

Table 1.22

THE FORGES LANDS UNDER JOHN MCDOUGALL & SONS

Year	Total area (in ha)	Location on the St Maurice River (in ha and in %)			
		East bank		West bank	
1867	2,490.51	1,902.75	**76.4%**	587.76	**23.6%**
1876	4,250.90	3,488.67	**82.1%**	762.23	**17.9%**

that the **campaign** or working season was extended to 10 and even 13 months. At least that is what the geologist Dr B. J. Harrington observed in 1873, when John McDougall & Sons was still supplying John McDougall & Company of Montreal with pig iron. The new focus on pig iron had a direct impact on the make-up of the work force. Skilled moulders were no longer required, and the shift away from bar iron meant that fewer forgemen were needed. In the 1871 census, however, four workers still described themselves as moulders, and five as forgemen. In 1881, a single worker identified himself as a forgeman and 41 gave their occupation as day labourers (compared with 34 in 1871 and 17 in 1861). As will be seen in Chapter 7, the men employed at the Forges at that time were no longer related to the old-stock families, who had left for other parts.

The McDougalls made certain improvements to the works, to better accommodate their new product. To reduce the risk of forest fires, and no doubt mainly to improve **coaling** productivity in order to sustain iron production over 10 to 13 months at a time, they had six brick kilns built on the post itself. This charcoal-making process cut timber costs by roughly 40% in comparison with the pit-coaling process that had been in use since the French regime.[150]

In late 1871, the McDougalls set up an axe factory on the site of the former lower forge. The terms of the 1865 contract earmarked 150 tons of pig iron for conversion to bar iron and also to make castings such as stoves, which would be manufactured for a few more years. Some pig iron was thus used to make axes, for two years only (1872–73). To link the two ironworks (St Maurice and L'Islet), which stood about 5 km apart, in 1872 the McDougalls installed a tramway running on wooden rails along the east bank of the St Maurice River.

Both ironworks ran at a profit until 1876. According to estimates (discussed in detail in Chapter 6), revenues from the sale of pig iron from the company's two plants were between $25,000 and $50,000 annually.[151] In 1875, after 12 years of operation by the McDougall family, the Quebec Bank was still extending the business a $100,000 line of credit. Table 1.23 shows that despite substantial liabilities the company had the confidence of three banks.

The accumulated debts ($128,000) of John McDougall & Sons were secured by the company's stock in trade, by certain properties, and also by the assets of James McDougall, John's brother, the Montreal industrialist. We cannot say exactly why the company carried such a large debt load, but the family store in Trois-Rivières seems to have had something to do with it.[152] On the basis of a positive balance sheet in June 1875, the Quebec Bank loaned a further $80,500 (Table 1.23). Though it is not known precisely how that money was used, there is reason to believe that it was tied to the bankruptcy of the Trois-Rivières store that same year. Thus, from September 1875 onwards, the company's financial situation was closely monitored by the Quebec Bank and by James McDougall, the chief backer of John McDougall & Sons. To recover the debt, now reduced to $60,000, the Quebec Bank demanded more collateral from John McDougall & Sons and its backer, James McDougall. In the following months, John McDougall & Sons mortgaged to the bank, the St Maurice and the L'Islet Forges, including their land reserve, along with some buildings in Trois-Rivières. For his part, James McDougall assigned a $20,000 mortgage that he held on his brother William's mill in Baltimore, Ontario. Complications surrounding the validity of that mortgage were what led, technically, to the shutdown of the St Maurice Forges.

As chief backer of John McDougall & Sons, James McDougall (Plate 1.5a) became liable for the St Maurice Forges now that the Quebec Bank was demanding concrete guarantees of solvency. The dissolution of John McDougall & Sons in September 1876 and the transfer of its operations to a new company were probably the result of James McDougall's desire to exert closer control over the two ironworks he had been financing for several years. The new company was called G. & A. McDougall and was made up of James's son George and his nephew Alexander Mills McDougall.

G. & A. McDougall, 1876–80

The two cousins were equal owners of the company, with George, James's son, in charge of finance and sales, while Alexander was responsible mainly for managing the two plants. The new company only operated for one campaign: an economic downturn affected iron sales and led to the temporary suspension of operations at the St Maurice Forges in the fall of 1877 and the closing of the L'Islet Forges in 1878. Alexander explained the situation:

> In 1878 it did not pay to smelt iron ore; we ourselves closed our establishments known as the Vieilles Forges Saint-Maurice and the L'Islet Forges for that very reason. Iron was not selling at all.[153]

The new company had also inherited the debts of the former John McDougall & Sons, totalling $114,400 (Table 1.24).

Table 1.23

FINANCIAL STATEMENTS OF JOHN MCDOUGALL & SONS, 1875

Assets (in $)		Liabilities (in $)	
Value	Form	Value	Creditors
42,000	merchant iron	88,000	Quebec Bank
38,000	raw materials	25,000	Union Bank
*	fixed assets and bond	15,000	Molson Bank
		128,000	**Combined**

* In addition, the loans from the Quebec Bank were secured by fixed assets in Trois-Rivières and a bond from James McDougall in Montreal.

Plate 1.5a
James McDougall (1816–89), John's brother and George's father. He gave his brother and son financial backing to operate the St Maurice Forges.
NOTMAN PHOTOGRAPHIC ARCHIVES, McCORD MUSEUM OF CANADIAN HISTORY, 4394-I.

Table 1.24

DEBTS OF G. & A. MCDOUGALL, 1876 *

Debts	
Value (in $)	**Creditors**
80,500	Quebec Bank
9,000	Anne Paterson
2,400	Margaret Allen
2,000	Edward Armstrong
2,500	Janet Purvis
6,000	Geo. B. Houliston & Co.
12,000	heirs of John McDougall
114,000	**Combined**

* Debts inherited from John McDougall & Sons

The terms of repayment to the Quebec Bank and disagreements between the two cousins led to new upheavals. Acknowledging his personal liability of $80,615.22 ($78,500 + $2,115.22 in interest), James McDougall was forced to give the bank a $25,000 mortgage on his Montreal mill, and he agreed to pay back $78,500 within five years. The bank also demanded immediate payment of the $2,115.22 in interest from G. & A. McDougall. The company paid that amount and some other sums owed to private creditors from the proceeds of a real estate transaction. However, it was granted a deferment for the repayment of $34,723 advanced against the stock in trade. In 1879, another creditor, Anne Paterson, demanded that $9,000 lent to the former John McDougall & Sons be repaid within a year.

In the middle of this financial turmoil, James McDougall and his son George tried to dissolve G. & A. McDougall in order to gain control of the business. Alexander Mills McDougall first opposed the proposal, but finally agreed to withdraw from the company, although not empty-handed. He was granted final discharge as a partner and member of the former John McDougall & Sons and was appointed manager for two years of the St Maurice Forges at an annual salary of $1,400 guaranteed by his uncle James. George McDougall was now in charge of the business.

George McDougall, 1880–83

The agreement between the two cousins was signed in January 1880, and already by August of that year disagreement had arisen between them. Alexander had gone into partnership with Louis Dussault in a firm that subleased a competing ironworks, the Shawinigan Iron Mines Company (Grondin Forges).[154] Alexander was immediately dismissed by his cousin, whom he later took

Plate 1.5b
George McDougall (1844–1915), John's nephew and James's son, was the last master of the St Maurice Forges. Notman Photographic Archives, McCord Museum of Canadian History, 6258-I.

to court. George McDougall then moved to Trois-Rivières and appointed another cousin, Alexander's brother, also named George, as manager. The involvement of yet another son of the late John McDougall gives one to wonder at the remarkable strength of the ties between the two McDougall families, despite the financial problems that marked their association. James McDougall, George's father, even risked his own mill, which in fact he would end up losing; this commitment can elicit only astonishment. Unfortunately, we do not know what terms James received or whether he stood to gain from backing the business of his Trois-Rivières cousin and nephews. It is nonetheless interesting to note that his attitude does not, at first glance anyway, show a strictly capitalist logic. The explanation may well be that blood is thicker than water.

In 1880, George McDougall wanted to reopen the L'Islet Forges to supply pig iron to the railcar-wheel foundry in Trois-Rivières that he had taken over. To this end he obtained an additional $20,000 loan from the Quebec Bank, but instead of starting up the ironworks at L'Islet, he had a second blast furnace built at the St Maurice Forges to cut down on costs and save on labour.[155] Built for approximately $6,000, the new blast furnace, the base of which was located by archaeological digs on the site of the former upper forge, was a circular brick kiln with metal cladding. The air compressor coupled to a hot-air furnace was driven alternately by a hydraulic turbine[156] and by a steam engine from the L'Islet Forges. At the same time, George McDougall had the carts and sleighs for haulage modified to double their capacity, in response to a rise in transportation costs.[157] In increasing the production capacity of the St Maurice Forges, along with operating a wheel foundry in Trois-Rivières, George McDougall was capitalizing on good economic conditions and showed no signs of running out of steam and of the slumping fortunes that would force him to close down barely three years later. Had it not been for the crushing debt load borne by his backer, his father, he would probably have continued operating ironworks in the region.

Let us now turn to the circumstances that led to the ultimate demise of the Forges in 1883, which has never been properly elucidated in the historiography. Resource exhaustion has generally been adduced as the principal cause, though without firm evidence.[158] The historian Michel Bédard has re-examined the circumstances and has brought to light material in court documents and in the archives of the Quebec Bank that point to the immediate reasons for the closing of the Forges. Bédard shows that the shutdown was tied directly to the company's debt level, and in particular, that of its backer, James McDougall. In a situation where debt has reached a critical level, all it takes is for one loan to be called or an unexpected lawsuit to be launched to completely destroy a company's financial credibility.

We have seen that in the fall of 1875, to secure the debts of John McDougall & Sons, the Quebec Bank forced the company to mortgage the two ironworks, St Maurice and L'Islet, and obliged James McDougall to assign a $20,000 mortgage that he held on his brother William's mill in Baltimore, Ontario. The following spring, on 24 April 1876, the Quebec Bank had the assignment registered at the Northumberland West registry office in Ontario. It was then that it was discovered that another mortgage, worth $4,000, had already been registered 10 years earlier (15 February 1866) by a David Campbell, apparently without the knowledge of James McDougall.[159] The registration of the $20,000 mortgage assignment to the Quebec Bank effectively gave the bank precedence as a creditor over David Campbell. To recover his $4,000 plus interest, Campbell launched a series of lawsuits against the heirs of William McDougall, the Quebec Bank and James McDougall, in the Ontario Court of Chancery, the Ontario Court of Appeal, and finally the Supreme Court of Canada. On 7 December 1881, the Supreme Court upheld the Court of Appeal's ruling, that the plaintiff

Table 1.25

LIABILITY OF JAMES MCDOUGALL TO THE QUEBEC BANK FOR THE FORGES, 1883

Amount (in $)	Origin
76,800	balance of debt inherited from John McDougall & Sons
25,000	debts of G. McDougall between 30 January and 30 April 1883
7,408.51	accumulated interest
109,208.51	

Table 1.26

SETTLEMENT OF JAMES MCDOUGALL'S DEBTS TO THE QUEBEC BANK, 1883

Amount (in $)	Origin
43,520.98	proceeds of the sale of his mill
43,520.98	claims against his son George
39,112.89	claims against his son George for loans made between 1876 and 1879
126,154.85 *	

* The bank collected the $109,208.51 owed and remitted the balance to James McDougall.

was entitled to seek remedy, but only against James McDougall. On 26 December 1882, the representatives of the late David Campbell sought that remedy by suing James McDougall for the sum of $6,407.85 and won their case before the Superior Court of Montreal on 5 February 1883. The Forges closed one month later, on 11 March.

The Forges were brought down by a crushing debt load. James McDougall personally owed the Quebec Bank over $126,000 (Table 1.26). Since his liability for the Forges was over $100,000 (Table 1.25), the repercussions on the ironworks, headed by his son George, were immediate. In the months that followed, James McDougall obtained a discharge from the Quebec Bank against the value of his mill and his claims against his son George. The Quebec Bank then sued George, who turned over a whole series of buildings before obtaining, in April 1884, the bank's discharge of the sum of $156,982.55.

George McDougall (Plate 1.5b) was therefore unable to pursue his project to breathe new life into the old Forges, reputed to be the oldest active blast furnace in North America. This did not stop him from considering new ventures, for, some time later, in 1890, he was thinking of establishing blast furnaces in Trois-Rivières itself to take advantage of the transportation facilities offered by the harbour there. But this project came to naught.[160]

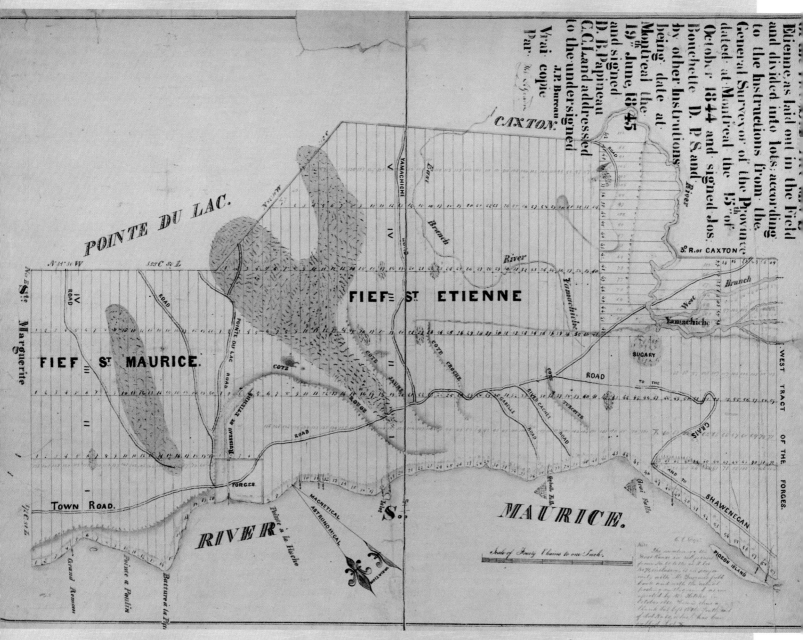

Figurative plan of the Fiefs St Maurice and St Étienne,
by Joseph-Pierre Bureau, 1845. The St Maurice Creek is marked
"Ruisseau St Étienne" [St Étienne Creek].

II

The Forges Lands

THE IRON PLANTATION AND ITS RESOURCES

Are you planning to build, purchase or lease a forge?
Combine your health, money, with knowledge of the land
and neighbouring property, the streams, forests, mines, iron quality
and trade, and there you have the first step.

Encyclopédie, 1757

The early iron industry depended on three types of raw material: iron ore, most often found on the surface in the form of iron oxide mixed with earthy **gangue**; charcoal, which acted as a combustible and reducing agent by releasing the carbon needed to reduce the ore; and **flux**, in the form of **limestone**, which was used to separate the gangue from the ore by clumping it into **slag**. The final ingredient needed was air to supply oxygen for the combustion and smelting of the materials. Water powered the wheel or turbine which activated a **bellows** or air compressor (see "The Chemistry of the Blast Furnace" in Chapter 4).

In this industry, material factors were critical, with resource availability being crucial to the siting of an ironworks. Transporting raw materials was a significant factor in the cost of production, and their estimated price delivered at the blast furnace was a criterion in selecting a site. The cost of getting products to market also accounted for a large share of the sales price of ironwares produced in a competitive situation.[1] The fact that all these elements came together at St Maurice was not lost on the colonial authorities as they set about launching a new ironworks after Francheville's venture failed in 1734:

> The mine is very rich and lies two leagues from the forge, stretching at least as far as Batiscan, seven or eight leagues below Trois-Rivières. The ore can be very conveniently carried in winter by sled, charcoal is made on the spot, and close to the forge the stream which operated it is never short of water and despite the current the Trois Rivières River is easily navigable by canoe up to one hundred or one hundred and fifty *toises* below the Establishment. It is only two leagues from Trois-Rivières by land and Sieur Francheville had a cart road built. *All these circumstances rarely come together in an Establishment of this type* and it is impossible, with proper management, that this would not be of great benefit for the colony and its trade with France.[2]

This careful planning is indicative of the industrial scale of early ironworks. They were distinguished from small-scale, craft operations not only by their scope and output, but also by the economic rationality that governed their creation and operation.[3] In the 18th century, however, this rationality was applied in a primitive fashion. It had a more empirical than scientific basis, so that French experts who had successfully set up ironworks in their own familiar territory failed lamentably elsewhere, especially when "elsewhere" was in the colonies.[4] Limited knowledge on the part of the ironmasters and poor entrepreneurship could sometimes be offset by a good supply of raw materials. Landes remarks on this in comparing the textile and metallurgical industries:

> Good entrepreneurship often seems to have been a decisive advantage in textiles [...] In metallurgy, however, cheap ore and charcoal could cover a multitude of sins and all the ingenuity in the world could not compensate for their absence.[5]

For over 100 years, the Forges masters did not have to pay for their resources! It is no exaggeration to say that the seeds of the demise of the Forges were sown after 1850, once the raw materials had to be paid for. The vast land reserve set aside by the Crown for the company's ore, timber and water power needs was an incalculable advantage. Government protection also obviated all the charges and bother of paying mining and stumpage dues and negotiating to buy raw materials and to obtain rights of way to build roads to mines and **charcoal pits** on private land.[6] All these problems were to surface once the government, under pressure from land-hungry settlers, rescinded the right to the land reserve. The idea of concentrating large tracts of land in the hands of a single user, albeit an industry, ran counter to the principle of land distribution and access implicit in the colonization movement. The colonial philosphy, based on agricultural development, did not recognize industrial activity as a factor of development. Yet the forest industry would be crucial to the viability of colonization, particularly in the 19th century.[7]

The need for a land reserve guaranteeing ready access to resources remained a constant and pressing concern for the Forges masters. Until 1846, each lease included advantageous provisions with regard to land.[8] Even Henry Stuart, the first private owner, in 1846, obtained the right to mine ore for five years on waste land in the fiefs of St Maurice and St Étienne. It was only in 1863, when John McDougall bought the business, that Crown land ceased to be part of the package. To be of benefit to the Forges, resources not only had to be abundant, but access to them had to be legally and economically advantageous.

Thus, the mix of natural resources was the prime factor in siting the Forges. The legal framework was the seigneurial system, making the local seigneur the original entrepreneur. After Seigneur Francheville had chosen the St Maurice Creek as the location for his establishment, and his choice had been endorsed by ironmaster Pierre-François Olivier de Vézin, it became apparent that the site was not in his fief of St Maurice but in St Étienne, which belonged to the heirs of Étienne Lafond. Francheville, as the third generation of his family to hold the St Maurice fief, was surely aware of this! Whatever the case, Francheville was already in possession of a 20-year royal warrant, and the King saw this as reason enough to annex the fief of St Étienne to that of St Maurice for the greater good of the Forges masters. The seigneurial system was thus manipulated for the convenience of the enterprise. Allowing the territorial logic of an industrial enterprise to override the area's original agricultural vocation presaged the conflict over land that would arise a century later.

THE FORGES LANDS

The Forges land reserve was therefore a highly strategic factor throughout the 150-year history. A significant factor in the longevity of the Forges was the abundance of mineral and timber resources on this land. The Crown almost always guaranteed the Forges sufficient land with more than enough resources. The St Maurice Forges, unlike many contemporaneous ironworks in Europe, never had problems with the supply of raw materials and was always self-sufficient in this regard. Despite considerable resource depletion throughout 150 years of operation, it was not resource exhaustion that caused the Forges' demise in 1883.[9]

From the time the Forges were established, this land reserve was so vast that it was long a byword for immensity and even excess. The King of France set the tone in 1730 by granting Francheville the right to work the mineral and timber resources in an area of over 120,000 ha. The Forges reserve encompassed seven seigneuries, including St Maurice and the enormous seigneury of Cap de la Madeleine, which in itself represented 80% of the accessible land (Plate 2.1a and Appendix 2).

Plate 2.1a

THE FORGES LANDS

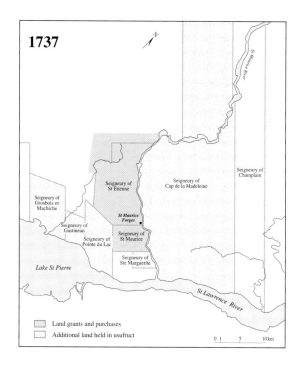

1737

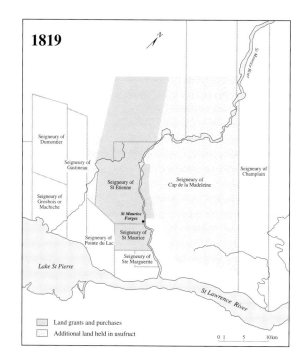

1819

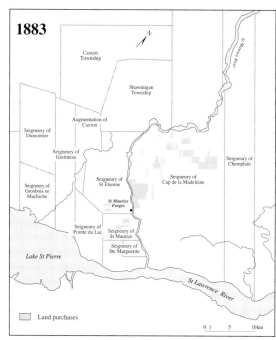

1883

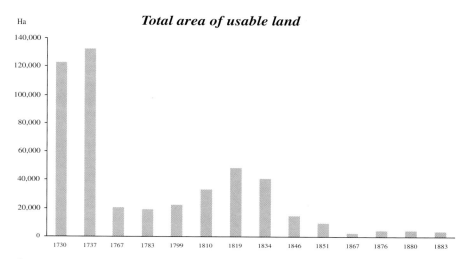

Total area of usable land

Sources: Allan Greer, *Le territoire des Forges du Saint-Maurice, 1730-1862;* Michel Bédard, "Le territoire des Forges du Saint-Maurice, 1863-1884." Manuscript Report No. 220 (Ottawa: Parks Canada, 1975–76). *Limites seigneuriales,* Quebec Department of Natural Resources, 1972, map 311, Trois-Rivières.

MAP BY ANDRÉE HÉROUX.

▼

Plate 2.1b
Map of the vicinity of the St Maurice Forges, 1815.
JOSEPH BOUCHETTE'S MAP OF 1815 (DETAIL), NATIONAL ARCHIVES
OF CANADA, MAP DIVISION, P/300-1815, PART 2.

This land reserve seems out of proportion to Francheville's modest project—a bloomery forge that could produce only 400 pounds of iron every 24 hours. In 1735, Vézin, sent to reconnoitre the lie of the land, merely hinted at the wealth of timber available in his "Observations," commenting simply that "hardwood is very common around St Maurice." In 1749, the Swedish naturalist Pehr Kalm visited the establishment and reported that "Charcoals are to be had in great abundance here, because all the country round this place is covered with woods, which have never been stirred."[10] In 1815, the Surveyor General Joseph Bouchette again remarked upon the "immense surrounding forests."[11] The land reserve always appeared excessive to the inhabitants of Trois-Rivières and neighbouring seigneuries, who began to covet parcels for settlement. At the height of their demands, their imaginations were still gripped by the idea of so vast an area being set aside for a single company, and doubt was even cast on the actual needs of Mathew Bell. The matter was debated in the House of Assembly at Quebec, with Mr Kimber, the member for Nicolet, declaring:

> The lands in the rear of Three-Rivers now lay waste, to the great detriment of that town, and the inhabitants along the river. Three-Rivers was thereby cribbed in and curtailed of the benefits it ought to derive from its advantageous possessions. [...] It was vexatious to find that this wide space was so kept in favour of one individual, who had no use for it, but to sport over it with a pack of hounds.[12]

The impression of vastness was no doubt partly fostered by the fact that the exact size and boundaries of the land reserve were unknown. It was not until 1806 that the first official survey was made of the land comprised in the Forges lease.[13] The Surveyor General at that time estimated the areas of the fiefs of St Maurice and St Étienne, considered to be the core area included in the lease, at 158 km².[14] A further tract roughly equivalent in area (172 km²) was also available for cutting wood and collecting ore. The land reserve thus totalled some 330 km² at that time (1810). The survey also confirmed that Francheville had indeed originally set up his forge on the neighbouring fief of St Étienne! In reuniting the two fiefs to the King's domain in 1737 for the benefit of the Forges,[15] the authorities were acting to protect their investment in the ironworks they were paying to set up. The real risk of forest fire was cited, and three days after the confiscation of the fief of St Étienne, another ordinance was issued, prohibiting the *habitants* from hunting and "lighting any fire in the woods bordering the seigneury of St Maurice for the purpose of clearing land."[16] After the Conquest, London also sent instructions (to Governor Cramahé of Trois-Rivières) that "no part of the lands upon which the said Iron Works were carried on be granted to any private person whatever" and even anticipated expanding the territory if need be.[17] In 1782, royal instructions to Governor Haldimand reaffirmed that "as large a district of land as conveniently may be adjacent to and lying around the said Iron Works over and above what may be necessary be reserved for our use."[18] At the height of the challenge to Mathew Bell's land monopoly in the House of Assembly in the 1820s and 1830s, the government did not hesitate to grant Bell additional land in an area now coveted by settlers and other interests. That is the reason why, until their sale in 1846 and even later, the Forges, and especially the land reserve, were considered a government venture, which Mathew Bell would dearly have wished to see acknowledged:

> The Iron Works of St Maurice I humbly conceive ought to be considered more a Provincial establishment, than that of a Local or District one; and I cannot but here remark the very erroneous statements that have been, and are still being circulated, respecting the great extent of the territory leased with the Iron Works, to refute which, reference need only be had to the *Plan** which accompanies the Lease; the Seigniories neighbouring have many of them a greater extent of unsettled and waste Lands.[19]

Today, the "Provincial establishment" that Bell would have liked to see acknowledged as such would have the status of a Crown corporation, like SOQUEM or SIDBEC. The Forges were set up from the public purse and the Crown retained ownership for over 100 years (1741–1846). It should thus be no surprise that the land grants to the lessees were so generous. In addition, the Forges were located in the heart of the vast St Maurice forest in an area that would never be threatened by encroaching settlement. At the time, the delay in settling the region was attributed to the land monopoly of the Forges, but even after this monopoly was abolished with the sale of the Forges to private ownership, the limited population of the former Forges lands never really threatened the company's ability to find supplies of raw materials. During the final 20 years of the Forges' life, the McDougalls, even though they had to pay a good price for much of their land,[20] never really had any problem obtaining supplies of raw materials from the Cap de la Madeleine land so coveted by settlers 30 years earlier.

The map in Plate 2.1a, showing the changes over time in the extent of the Forges lands and in their geographic distribution, illustrates the main features of their tenure, which was of two types: land that was granted or sold to the Forges, and additional land over which the Forges had **usufruct**.

Land Grants and Purchases

The land granted or sold to the Forges constituted the core area under the direct control of the Forges masters. It was granted "en fief et en seigneurie" as under the French regime, included in the lease from 1767 to 1845, or sold in whole or in part to the proprietors between 1846 and 1883. Until 1861, this land consisted essentially of the fiefs of St Maurice and St Étienne. In 1730, only the St Maurice fief was owned by Seigneur Francheville in his own right. The St Maurice and St Étienne fiefs were combined into one (St Maurice) and granted to Cugnet et Compagnie in 1737,[21] but, in fact, the two fiefs remained separate. From then until 1846, the Forges lands remained fairly stable at between 15,000 and 20,000 ha (150–200 km²). New survey data (1806 and 1845) and the addition or withdrawal of concessions changed their extent from time to time. In 1846 and 1851, the government sold the fiefs of St Maurice and St Étienne in their entirety to the private owners of the Forges. However, they were obliged to sell lots to settlers seeking to buy them, resulting in a decrease in usable area in the two fiefs.[22] From 1861 until the closure in 1883, the McDougalls bought, piece by piece, the land which made up their reserve, formed only in part from land in the fiefs of St Maurice and St Étienne, with the largest part concentrated on the east bank of the St Maurice River.

Additional Land Held in Usufruct

Government involvement in the land aspect of the business was sometimes generous, sometimes restrictive. In addition to granting a core area, in the form of concessions or leases, the government intervened directly to make further land available to the Forges. There were two periods when this liberal intervention was especially strong: under the French regime (1730–60), and during the Bell years (1799–1846). The two expansions occurred, however, in different contexts. Under the French regime, the vast territory was not granted because of strong pressure on resources, since the enterprise was just getting started, but rather to guarantee long-term operations. Since the ore was widely distributed in variable concentrations, it was preferable to make available the largest possible area. During the Bell years, expansions were granted as Mathew Bell felt that resources in the fiefs of St Maurice and St Étienne were becoming exhausted or inaccessible.[23] The all-powerful Bell also wanted to guard against forest fire by keeping other interests and potential settlers as far as possible from Forges territory. In 1825, he complained of the incursions of Yamachiche farmers who, since 1819, had been cutting wood on Crown land (between the Gatineau and St Étienne fiefs) reserved for the Forges. Between 1834 and 1845, 46 charges of illegal woodcutting on Forges land were brought against local farmers.[24]

After the Forges were sold in 1846, government intervention became restrictive in the sense that land that had, until then, been earmarked for the Forges, was removed from the land reserve, although some accommodation was made to allow continued use of a large part of it.

The fluctuations in the size of the Forges land reserve indicate, in fact, that government involvement in the matter made all the difference. As long as the government guaranteed land-use privileges, the Forges masters had access to a minimum of about 20,000 ha (200 km²), but once the government withdrew its guardianship, the Forges land reserve shrank to less than 5,000 ha (50 km²).

Land Expansion

The expansion of the Forges land reserve (Plate 2.1a) shows that, except for a small parcel in the seigneury of Cap de la Madeleine, the bulk of the resources was collected on the west bank of the St Maurice until 1819. To the original area of the combined fiefs of St Maurice and St Étienne was added the neighbouring Crown land to the southwest and northwest. Later, as a result of the impetus provided by Mathew Bell, the east bank of the St Maurice gradually began to be worked, and subsequently, the vast territory of the seigneury of Cap de la Madeleine.[25] However, exploitation of land in this seigneury was halted for 15 years after the Forges were sold in 1846. From 1846 to 1861, under Henry Stuart, and then under John Porter & Company, operations were concentrated once again on the fiefs of St Maurice and St Étienne, no doubt taking advantage of the regeneration of the timber cut down by Bell 20 years before. During the final 20 years of the McDougalls' tenure, the main area of operations was the former seigneury of Cap de la Madeleine, in the parishes of Mont Carmel and St Maurice.[26]

The McDougalls were not alone in owning land there, since Auguste Larue and Company and the Radnor Forges held an area five times the size (119 km²) of that owned by John McDougall & Sons (25 km²) in the late 1860s.[27]

Mainly to meet its need for charcoal, the ironworks was the first forestry operation in the St Maurice Valley. The enclave created by its operations could be tolerated as long as settlement of Trois-Rivières and its hinterland was not curbed by Forges land. Although marginal for agriculture, the land was of interest to settlers and timber merchants because of its timber. It was mainly the fact that they were forbidden to enter this forest to cut firewood and tap maple trees that created animosity among local inhabitants.[28] The pressure created by the march of settlement was felt mainly in the Yamachiche, where squatters along the part of the Yamachiche River that crossed the Forges land finally forced Mathew Bell to give up this territory.

During Mathew Bell's era, it was not so much the extent of the land reserve, as the fact that it blocked access to the interior, that raised opposition. Forges land was hampering the development and expansion of the town of Trois-Rivières, which found itself "surrounded by a ring of iron."[29] The proximity of a settled area finally became harmful to the enterprise. However, the sale of the Forges and the reduction of the land reserve in 1846 did not necessarily mean the end of large concentrations of timber concessions in the St Maurice region. On the contrary, it helped open up the St Maurice River to other interests. In 1852, the government granted William Price, George Baptist and G. Benson Hall areas of land four to 10 times as large (from 1,248 km² to 4,947 km²) as that held by Mathew Bell at the height of his monopoly. Like Bell, they were criticized for taking much more than they could use. There was radical intervention by the government,

which, in 1855, confiscated unused concessions, yet "compared to timber merchants in
the Ottawa Valley, those in the St Maurice
Valley held four times the area called for by
their production."[30] Nevertheless, the woodland
of the St Maurice Valley continued to be concentrated in the hands of a few concession
holders, a practice that was followed into the
20th century with the establishment of the
pulp and paper industry.[31]

Other St Maurice iron companies continued to benefit from government largesse.
In the wake of the federal government's
National Policy of 1879, the federal, provincial
and even municipal governments began to
institute incentive measures in the form of
tariff protection, grants, tax exemptions and
free access to Crown land. That is how the
Canada Iron Furnace Company, which bought
the Radnor Forges in 1889, was granted
30,000 acres (121 km²) of land in Radnor
Township by the Quebec government in
1895—by having itself recognized as a colonization society "desirous of establishing its
employees"![32]

THE LIE OF THE LAND

Implications of the Location of Resources

The mining of **bog ore** called for special
planning. The ore lay close to the surface,
below the soil, in the form of nodules of varying sizes distributed in swampy or lacustrine
areas. Since it was scattered over a large area,
special methods were required to work the land
containing it. Mining it involved setting up a
road system, using suitable means of transportation and employing a large seasonal and
far-flung work force of miners and carters.
These constraints would shape how those
hands, working several miles from the establishment, would be controlled. In addition,
distance, combined with climate, was a factor
in deciding on the best time to build up and
cart ore reserves. Since the ore was mined in
swampland, it was stockpiled at the mines over
the summer and hauled by sleigh over the
snow or frozen roads in winter. The conditions
under which the ore was mined thus had
considerable implications for planning how it
was collected and organizing how it was transported. The same was true of the limestone and
building stone quarried a few miles above the
Forges on the banks of the St Maurice, which
were transported by scow, barge and raft by
boatmen employed by the company during
the spring and summer.

The same constraints applied to wood-cutting and subsequent charcoal making. Large tracts of forest were involved, occupying over 200 woodcutters annually, in addition to road builders and the carters recruited seasonally from among the local *habitants*. The charcoal burners or colliers plied their trade at charcoal pits in the depths of the forest, where they spent the summer and fall setting hundreds of pits in which more than 10,000 cords of wood were charred annually. Charcoal burning frequently caused forest fires, despite the care of colliers and even though a *garde-feu* or fire warder was employed by the company. Sometimes all the Forges hands would be mobilized to fight a fire.[33]

The dispersal of resources and the times of year they were collected and transported imposed a strict schedule if production operations were not to be unduly delayed or endangered. Before the blast furnace was blown in at the start of a campaign, all the raw materials required for the whole campaign had to be stockpiled and prepared. It was even recommended that stocks be built up more than a year in advance. Most of the problems at the Forges involved failure to adhere to a strict schedule for collecting and transporting raw materials. Fairly often, especially during the early years of operation, but also at other times, a lack of charcoal reserves delayed or interrupted a campaign (see Chapter 5).

Another major constraint was the gradual need to go farther and farther afield for raw materials. From the start, in 1735, some of the ore beds were already quite distant from the Forges. The mines at Pointe du Lac and Cap de la Madeleine were 2 leagues or over 10 km from the works, and the quarries were 6 to 10 km away. A network of roads was created at the outset, as was a system for transportation on the St Maurice. Nearly 100 years later, in 1827, Lieutenant Baddeley reported that ore was being brought from up to 9 miles (14.5 km) away from the Forges. Baddeley also reported that the wood on the St Maurice seigneury was considered practically exhausted.[34] Exploitation of the various resources required to run the operation created real pressure on the environment that periodically forced the Forges masters to seek more land. Given its far-flung resources and the cost and problems of finding seasonal labour, the company had to adopt another approach, especially for charcoal making, and a different strategy for employing labourers, by replacing part of the seasonal work force with permanent jacks of all trades.

▼

Prospecting for Resources

In 1735, Vézin mapped and commented in his "Observations" on the raw materials, each under its own heading:[35] "Wood," "Ore," "Flux" and "Building stone." These, essentially, were the four types of raw material required to run and build the ironworks. Charcoal making, with its voracious appetite for wood, was the most sensitive operation because of the risk of fire. The mines containing the surface deposits of ore were very widespread in the swamp lands, but would be worked only where major concentrations made intensive mining worthwhile. Vézin inventoried the ore deposits that showed the most promise (Table 2.1).

Limestone and building stone were always quarried at the same place on both sides of the St Maurice River, a few kilometres upstream from the Forges, at a quarry known as La Gabelle near Les Grès (Plate 2.2) and on some of the small islands in the river.[36]

Prospecting was an ongong concern of the Forges masters, but few of them took the trouble to map or evaluate resources accurately. In 1735, Vézin and Grand Voyer Lanouiller de Boisclerc, the chief road commissioner, created the only two known maps of ore mines (Plates 3 and 4). These maps, despite their inaccuracies, pinpoint the main ore beds that would be mined throughout the history of the Forges. Some of them were in the fief of St Maurice, but most were located in the fief of Pointe du Lac or, on the east bank of the St Maurice, in the seigneury of Cap de la Madeleine. This makes it even clearer why Francheville sought a warrant encompassing these seigneuries, since the best supplies were not on his own land. Francheville's first mining operations were carried out on the Cap de la Madeleine side, and Cugnet et Compagnie continued for a time to exploit the resources of this seigneury, which were not really drawn upon until the 19th century.

Table 2.1

MINES PROSPECTED BY VÉZIN, 1735 *

Location	Distance from Forges	Area Prospected	Evaluation
Pointe du Lac	2 leagues **	1 league x ¼ league	cost price: 2 *livres/pipe*
Pointe du Lac (Dupont habitation)	2 leagues	10 arp. x 1 arp.	
Cap de la Madeleine	2 leagues	3 mines measuring 1 arp.² 10 in. to 1.5 ft. thick	
Cap de la Madeleine (north of the River)		1 arp.² 6 in. to 1 ft. thick	

   * These figures, taken from Vézin's "Observations" in 1735, correspond to those on the map in Plate 2.3, which was probably drawn by Vézin or based on his observations.

  ** 1 league = 84 arpents; 1 arpent = 180 French feet.

Table 2.2.

MINE WORKINGS INVENTORIED BY LANOUILLER DE BOISCLERC, 1740 [37]

Location	Quantity of ore mined per year	Area of veins mined			Miners
		Length	Width	Depth*	
Pointe du Lac	95 ore heaps (1,021 m³)				Langevin Dainevert Maurice Déry Girard
Pointe du Lac — in the Brulé and surrounding area		¼ league	12 arp.	1 in.	
Pointe du Lac — along Lafonderie stream		5 arp.	3 ft.	4 in.	
Pointe du Lac — mined by Girard		12 arp.	8 ft.	2 in.	Girard
Pointe du Lac — mined by Langevin		16.5 arp.	12–15 ft.	2.5 in.	Langevin

   * Depth remaining to be mined.

LOCATION OF BASIC RAW MATERIALS IN THE VICINITY OF THE ST MAURICE FORGES

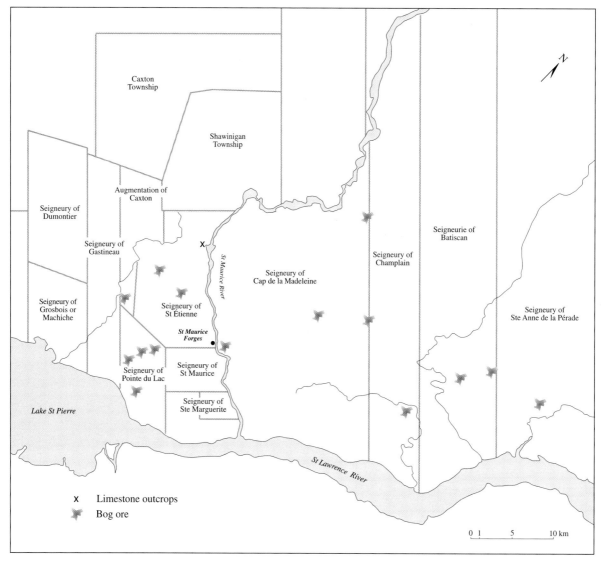

x   Limestone outcrops

   Bog ore

0 1   5   10 km

MAP BY ANDRÉE HÉROUX.

Quantities of ore were sometimes evaluated in terms of the area of a vein, sometimes in terms of a campaign; for example, a vein might be said to be good for two or three campaigns (Table 2.2).

We will now take a look at the resources, first examining their location, characteristics and what is known of their exploitation. We will then discuss the quantities required for the annual operations of the Forges.

Mines

The ore of the Trois-Rivières region had already been prospected in the mid-17th century, and since the Sieur de La Potardière had already taken 20 **barriques** of it to France, the quality of this ore, concentrated in surface veins most often located in bogs, was known. The quality and abundance of this bog ore was confirmed subsequently on a number of occasions, both before and after the Forges began operations. In 1734, experts from the Rochefort Arsenal judged the ore to be high quality, despite the poor yield obtained by Francheville's workers.[38] It was estimated that the ore would

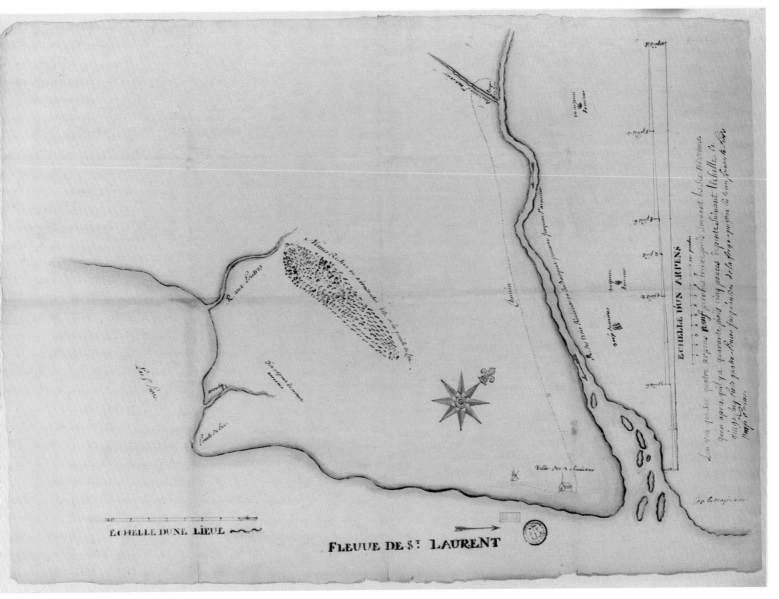

Plate 2.3
Plan of the mines at Trois-Rivières, by Olivier de Vézin, 1735.
On the right-hand side, Vézin has added his measurements of the degree of slope of the gully at the Forges.
FRANCE, BIBLIOTHÈQUE NATIONALE, PARIS, CARTES ET PLANS, PORTEFEUILLE 127, DIVISION 8, PIÈCE 50.

produce iron "at least as good as Berry iron" in France.[39] Contemporary analyses are silent on the composition of the ore, but comment on some of its attributes such as its purity, richness and weight or how easily and cheaply it could be mined "with a shovel." It was said at the time that two *pipes* of ore would make a **thousandweight** of pig iron, that is, about 2,200 pounds of ore would produce 1,000 pounds of pig iron (French measure), a yield of 45% at the blast furnace. This estimate was confirmed in 1828 by Lieutenant Baddeley, commenting on the ore's excellent quality.

The first laboratory analyses were carried out in 1852 by geologists[40] from the Geological Survey of Canada,[41] on the basis of ore samples from the Forges. This bog ore was characterized as "bright red limonite, with a brilliant black fracture," and its quality confirmed.[42] In 1855, the Survey Director, William Edmund Logan, received other samples of the ore from the Superintendent of the Forges, who also referred to it as "limonited bog ore." The samples, as well as examples of limestone and firestone, were shown by Logan at the Paris Exhibition (see Appendix 8).[43]

The ore samples analysed by geologists Logan and Hunt (Table 2.3) found iron oxide combined with small quantities of manganese, silica and phosphorous and a good proportion of water and organic matter. The high content of organic matter in the ore facilitated reduction, according to Dr B. J. Harrington, who visited the Forges in 1874.

Harrington, summarizing the work of the geologists who had preceded him, explained the origin of the ore, which was found in sandy soil in the form of porous nodules ready for the blast furnace:

> Bog ores are mainly of recent age, occurring at or near the surface, and generally in sandy regions, ferruginous sands often being the source of the iron [...] The variety employed for smelting occurs in concretionary lumps or masses often showing a curious cavernous structure [...] The concretions are either scattered through the soil, or else form continuous layers, generally only a few inches in thickness, though sometimes several feet [...].[44]

Harrington added that, based on the laboratory analysis of ore samples from different regions (Table 2.4), the proportion of metallic iron averaged 50%, but that, in general, yield at the blast furnace was only from 30 to 40% because of the presence of sand (silica), which was not easily removed even by washing. He commented that the manganese and phosphorous content gave a **white** or **mottled** iron, stating that the low phosphorous content of the ore used at the Forges resulted in a soft and malleable iron, unlike that normally expected from an ore of this type.[46]

Table 2.3

COMPOSITION OF IRON ORE FROM THE FORGES C. 1850 *

Constituents	Sample 1 (%)	Sample 2 (%)	Sample 3 (%)
Peroxide of iron	77.60	74.30	64.80
Sesquioxide of manganese	0.30	trace	5.50
Silica	5.40	3.60	4.80
Phosphoric acid	1.81	1.80	undetermined
Volatile matter (water and organic matter)	17.25	22.20	23.65
%	102.36	101.90	98.75
Metallic iron	**54.32**	**52.01**	**45.36**

* Analyses reproduced in "Notes on the Iron Ores of Canada and their Development" by Dr B.J. Harrington in 1874

Table 2.4

IRON CONTENT OF SAMPLES OF BOG ORE FROM THE PROVINCE OF QUEBEC IN THE 19TH CENTURY[45]

Origin of sample	Metallic iron content
Petite Côte, Vaudreuil	52.15
Côte St Charles, Vaudreuil	53.86
Ste Angélique, Vaudreuil	28.67
Upper Rocky Point, Eardley	54.56
Bastard, 20th lot, 2nd concession	40.00
L'Islet Forges	54.36
St Maurice Forges—a	54.32
St Maurice Forges—b	52.01
St Maurice Forges—c	45.36

In the early years of operation, the ore reserves on the Forges land seemed inexhaustible:

> at present, ore has been taken from this mine [Pointe du Lac] for use at the furnace for two years. Should any mine begin to diminish, there is iron ore all over the country, from Pointe du Lac to Batiscan, an infallible sign that veins of ore can be found in many places by looking for them, and new ones will assuredly be found while the old ones are being mined. Consequently, the only inconvenience to be feared is that fetching it from farther afield will cost more, but this would not be a sizable consideration [...].[47]

The annual needs of the blast furnace—over 800 t of ore during the French regime (Table 2.8)—were readily met by the abundant resources dispersed throughout an area of over 1,200 km². The distance of the ore beds from the blast furnace was the only drawback seen.[48] The author cited above felt that this would have an impact only on the cost of transport.

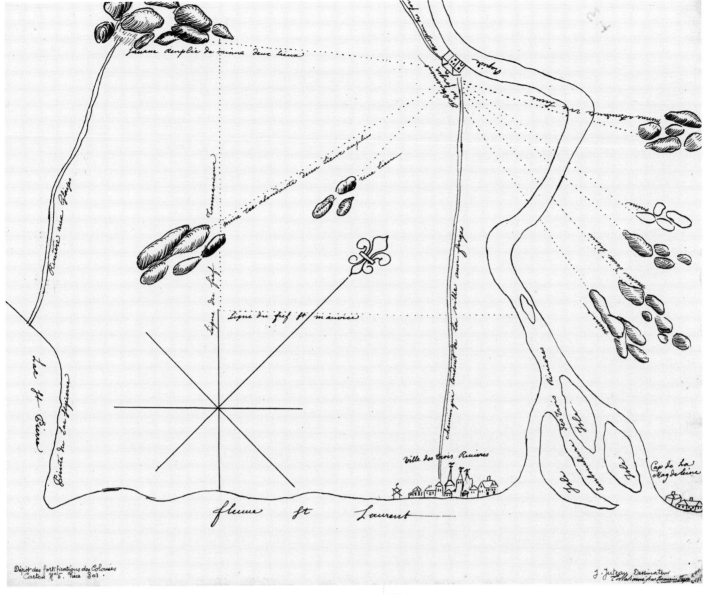

Plate 2.4
Plan of Trois-Rivières Forges, 1735. The unknown mapmaker has
marked the locations of the various iron-ore deposits near the Forges.
NATIONAL ARCHIVES OF CANADA, MAP DIVISION, H3/340,
TROIS-RIVIÈRES [1735].

After all, the Forges were still operating under
the royal warrant of 1730 granted to Francheville, authorizing the mining of ore in the
area between the seigneury of Yamachiche
and that of Cap de la Madeleine. The neighbouring seigneurs had to be compensated only
when the ore was mined on improved land
already under cultivation by the seigneurs or
their tenants.

After the Conquest, resources were just
as plentiful, but henceforth access to them was
limited and access conditions changed. With
the first lease, in 1767, the land reserve was
reduced to the combined fiefs of St Maurice
and St Étienne alone. This restriction did not
prevent the Pélissier syndicate from continuing
to exploit the rich ore bed of Pointe du Lac. But
Joseph Godefroy de Tonnancour, the local

seigneur, reacted quickly by contracting with
Pélissier:

> that the said S. Pellisier may collect and remove on
> his land for the number of years he wishes, the
> quantity of iron ore he deems appropriate. In
> return for which the said S. Pellisier agrees to pay
> the suppliant the sum of two hundred Spanish
> *piastres* for each campaign.[49]

This amount was the equivalent of £60 currency[50] and represented over 60% of the annual labour cost for miners at that time.[51] Pélissier's associates contested this contract on the grounds that mining rights belonged to the King alone. Tonnancour, through François-Joseph Cugnet, the son of François-Étienne, the former director of the Forges, asked Governor Carleton to enforce the contract by agreeing to pay the King a tithe of 20 Spanish *piastres* (£6).[52] This is the first mention of payment of this fee since the King had exempted Francheville in 1730. There is no documented sequel that we know of, but other sources reveal that attempts to regain free access to the resources of the vast territory of the French regime were unsuccessful. Mathew Bell was able to recover some of the privileges of the French era by obtaining the right to exploit the resources of waste Crown lands. After 1846, however, the new private owners were again obliged to pay for the ore that they mined or had mined on private land. Statements of account from 1856–58 indicate that John Porter & Company bought ore from the *habitants*.[53]

At this time, ore was obtained from a variety of sources. Samples of ore sent by Superintendent Henderson to Logan in 1855 came from Yamachiche, Pointe du Lac and Nicolet; ore from the seigneuries of Champlain and Cap de la Madeleine, used by the new Radnor Forges established the previous year, were also mentioned. Henderson still found abundant resources in the region:

> The supply of ore appears to be inexhaustable. It is found near the shores of the St. Lawrence and mere 10 to 15 miles back in allmost every part of this district & on both sides of the River.[54]

Table 2.5

MINE WORKINGS C. 1850[55]

Location	Area of veins worked (English measure)		
	Length	Width	Depth*
Caxton Township – near the Grande Rivière Yamachiche owned by P. Boivin	0.75 mi.		10–15 in.
St Étienne fief – in the 4th range	12 ft. 30 ft. 35 ft.	15 ft. 7 ft. 6 ft.	
St Étienne fief – in the 2nd range (7 veins)	30 yd.	2–3 yd.	6–9 in.
Pointe du Lac (2 veins)	1 mi. 1 mi.	60–100 yd. 10–220 yd.	6–20 in. 6–20 in.
Pointe du Lac – St Nicolas range, owned by C. Vincent	350 yd.	100 yd.	3–6 in.
Pointe du Lac – owned by Étienne Berthiaume	0.75 mi.	0.5–1 arp.	
Cap de la Madeleine – St Félix range (5 or 6 veins)			2–4 in.
Cap de la Madeleine – 1 mile northeast of St Félix range	0.5 mi.	2–6 yd.	6–10 in.
Champlain seigneury – near Richardson mill towards Batiscan	3 mi.	12–18 arp.	
Champlain seigneury – near Rivière à la Lime	30 arp.	3–6 arp.	3 in.–1 ft.
Batiscan – owned by Desaulniers	0.3 arp.		3–6 in.

During this same period, Logan located several ore deposits in the region (Table 2.5), including a very rich 15-km² deposit on the east bank of the St Maurice River, in the St Félix and Ste Marguerite ranges of the Cap de la Madeleine and Champlain seigneuries.

Several years later, the McDougalls acquired a number of parcels of land in this area and had a variety of arrangements with the *habitants* for reserving the ore on lots sold to them. At the same time, the McDougalls were ensuring a supply for their second establishment, the L'Islet Forges, located in this same area.

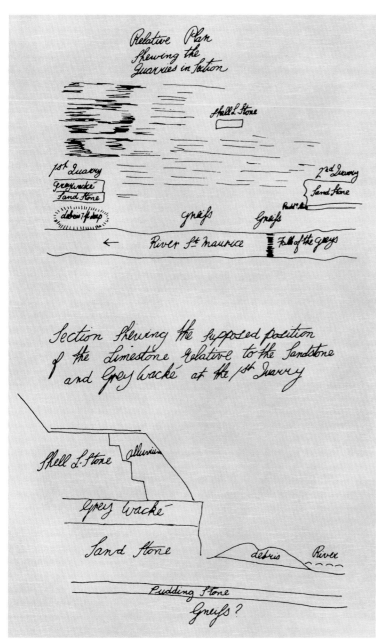

Plate 2.5
Map and sketch of the quarry at La Gabelle,
by Lieutenant Baddeley of the Royal Engineers, 1828.
NATIONAL ARCHIVES OF CANADA, MG12, B (W.O. 44),
VOL. 613, ENGINEERS 1826–41, MICROFILM B-3459.

Data from the 1871 census also indicate that the *habitants* of Yamachiche and Pointe du Lac had built up stockpiles of ore.[56] After over 100 years of mining, the ore deposits of Pointe du Lac, among the longest worked, were therefore still productive. These data, together with the fact that other ironworks were operating in the region from 1854 to 1910, contradict suggestions of a lack of ore, long adduced to explain the closing of the Forges in 1883. Ore depletion had already been mentioned in

the 1863 report of the Geological Survey of Canada to explain the shutdown of the Forges in 1858. Yet, Superintendent Henderson had stated in 1855 that the mineral resources were inexhaustible. The geologist, who obviously wrote his report before John McDougall bought the Forges that same year, specified that the closing in 1858 was caused by "the growing scarcity of ore and charcoal in the immediate vicinity." But the Forges had for a long time been mining ore over 9 km away in areas that geologists, in the same report, still considered to be rich in ore.

Quarries

> *Quarries of lime-stone, a good grey stone,*
> *and some other hard species fit for building,*
> *are opened on the banks of the St Maurice near*
> *the falls of Gros (sic), and those of Gabelle,*
> *a little below.*[57]

Sandstone deposits near the Forges on the banks of the St Maurice, where St Étienne des Grès is now located, were a determining factor in keeping the establishment on the site chosen by Francheville.

In his "Observations" in 1735, Vézin described his discovery, upstream from Francheville's establishment (Table 2.6), of a quarry that could supply limestone for flux and building stone and which would be easy and cheap to exploit, since the stone could be transported by flat-bottomed boat. He estimated the cost of the limestone at only 10 *sols* a *pipe* "delivered at the furnace."[58] In addition, it was the discovery of sandstone at the Gabelle quarry (Plate 2.5) and the promise that it could be quarried at low cost that motivated the building of a large stone house, the Grande Maison.[59] Proximity to a sandstone quarry which could meet annual needs for flux, fire-stone and building stone[60] was an inestimable benefit for an ironworks. According to Mathew

Bell, who ensured that this quarry was used exclusively for the Forges, annual requirements were about 600 to 800 tons. He was convinced that these quarries of high quality stone had been the deciding factor for the French in building the Forges at that location:

> The quarry in question contains a stone of a very superior quality to resist fire; no such other to my knowledge has yet been discovered in Lower Canada; and from the information I obtained nine and thirty years ago when I was much interested in the prosperity of these works, I have reasons to believe that the Ironworks were erected where they now stand by the French Government more on account of the very same quarry, than either of the wood or ore in that neighbourhood.[61]

At the same time, Lieutenant Baddeley, who visited the Forges, reconnoitred the St Maurice sandstone quarries. Accompanied by the quarryman, he inspected two quarries on the west bank of the river, which he mapped (Plate 2.5), sketching out their stratigraphy. The first, located at La Gabelle, was 9.6 km upstream from the Forges.[62] Its bottom layer consisted mainly of building stone used in building the Forges. He noted that the exploitable layer was 6 to 8 feet (2–2.5 m) above the level of the river. He remarked that during the spring runoff, the water level rose from 10 to 15 feet (3–4.5 m), thus submerging the quarry.[63] Between this first quarry and the Grès falls, he also saw a ledge of shell limestone 50 feet (15 m) high and about 200 yards (180 m) from the river, where a small quarry had long been in operation to supply the Forges with limestone for flux. Between the Forges and the first quarry he saw a number of smaller layers of shell limestone. The second large quarry was upstream from the Grès falls, 2.4 km above the first one. However, Baddeley found that the stone there was inferior in quality and a portage was required to bypass the falls. He also observed beds of sandstone on the east bank of the St Maurice, but these were low and covered with sand, making quarrying

difficult. We know, however, that these deposits were used by the Forges, as were the deposits on the small islands in the river.[64]

Baddeley also examined the walls of the Grande Maison, noting that stone of varying quality had been used in its construction 90 years earlier (1737). Some stone showed traces of deterioration—"marks of external decay or weathering"—while others had remained white. Most of the stonework showed iron discolouration. Baddeley also inspected the stone of the blast furnace and noted its remarkable quality and the rarity of the sandstone used as firestone:

> The whole or most part of this sandstone is fire proof—search might be made in vain a long time before a material could be found so well adapted for the interior coating of furnaces [...].

and he went on to say:

> The blowpipe does not appear to produce the slightest effect upon it.[65]

Table 2.6		QUARRY LOCATIONS	
Location	Type of stone	Date	Source
West bank of the St Maurice			
1.5 leagues* (7.4 km) upstream from the Forges	limestone	1735	Olivier de Vézin, "Observations"
2 leagues (9.8 km) upstream from the Forges	sandstone building stone	1735	Ibid.
9.6 km upstream from the Forges	flux limestone	1827	Lt Baddeley
9.25 km upstream from the Forges	fire stone building stone (lower beds)	1827	Ibid.
East bank of the St Maurice, at the foot of the Gabelle Rapids, downstream from the mouth of the Cachée River	refractory sandstone used in the construction of the hearths and other parts of the St Maurice forges	1853	Geological Survey of Canada, *Report of Progress for the Year 1852-53*, p. 63

* 1 league = 3.5 miles = 4.9 km.

Note: The various types of stone were found in the same quarries, but in different layers. Very dense sandstone, used as firestone and building stone, was found in the bottom layer. Less dense, more porous sandstone, also called limestone or flux after the purpose for which it was used, was found in the top layers.

Baddeley's description thus concurred with that of Bell, who understandably wanted to keep this stone exclusively for the needs of the Forges. The 1853 report of the Geological Survey of Canada, which identified the sandstone as of the Potsdam Formation,[66] was no less generous in praising its qualities:

> it has been found capable of resisting a very strong heat without injury, so much so, that the deposit had been resorted to for the material used in the construction of the hearths and other parts of the St. Maurice forges.[67]

During this same period, Jeffrey Brock spoke of an annual consumption of more than 500 tons of stone, taken from the Gabelle quarries, near the Grès falls. William Henderson, who succeeded Brock as superintendent, stressed the exceptional resistance of this stone in delivering two samples for the Paris Universal Exhibition of 1855:

> [...] two pieces of what we call here Fire stone, as it is used in building the furnace and stands the fire so well that a furnace usually lasts from 3 to 4 years. It is also an excellent building material, and is found only at one place viz the Gabelle above referred to. I am not aware of its existing elsewhere in this part of the Province. It forms the face of the steep & high bank of the River and in quantity appears inexhaustible.[68]

According to Harrington's 1874 figures, about a ton of flux was used per day, in addition to the stone required for annual repairs to the linings of the blast furnace.[69] For the smaller, pre-1854 furnace, the amount required for relining was estimated at about 8 *toises*; an interesting detail is that the types of stone used for this job were categorized separately:

> In order to renew the furnace lining, five *toises* of white stone, one and a half *toises* of bedded limestone for the boshes, two hearth bottoms of grey stone measuring four feet by three and a half feet and eight to ten inches thick, and a crucible of dressed limestone four feet by two to three feet and one foot thick. These two items are estimated at one *toise*, and together with the first two items make a total of seven and a half *toises*. This is more than eight *toises* which, at the furnace, including the boshes and crucible, for which limestone must be dressed, at fifty *livres* per *toise* amounts to ....400.[70]

Thus the French-designed blast furnace used until 1854 required some 157 t of sandstone[71] for repairs to the linings and internal parts. Sandstone was used as firestone at least until 1878, since the McDougalls showed sandstone intended for this purpose at the Paris Exhibition.[72] The most recent furnace discovered in an archaeological dig, as well as the second furnace built two years before the Forges closed, featured **boshes** made of refractory brick. The change therefore took place only a few years prior to closure.

Timber

*[...] we cannot estimate the length of time
this wood will last. So far, cutting has not been
regulated and the forest has been poorly exploited,
almost always at the wrong time for the regrowth;
once there is no more wood on these two seigneuries
(St Maurice and St Étienne), the neighbouring
seigneuries, which are heavily wooded,
will provide it. There is no fear that the
Establishment will fail for want of wood.*[73]

The St Maurice Valley boasts one of the best areas of mixed forest in Quebec.[74] This was even truer at the time when the St Maurice Forges were in operation. In 1736, ironmaster Vézin had to cut down "monstrous" trees to clear the site for the Forges. The St Maurice forest thus had nothing in common with the situation in France, where felling was already being rationalized[75] by the time Jacques Cartier was exploring Canada, two hundred years before the Forges were founded. The Forges masters made free with this resource, with no thought that it would one day be exhausted.[76] The surrounding forest satisfied practically all the company's needs for both hard and soft wood, which were used as fuel, firewood and building material. Only oak, used in building the wheels and hydraulic mechanisms, came from outside the region, from the Chambly area on the Richelieu River.

The forest cover consisted of a number of communities dominated by yellow birch:

> In general, this is the domain of the yellow birch.
> Small cedar groves and stands of red spruce are
> also found in some locations, with sugar maple-
> beech stands on the summits.[77]

Table 2.7

FOREST COMMUNITIES ON THE FORGES LANDS

Site	Community	Associates
Well drained	sugar maple	red maple
	sugar maple-beech	yellow birch and red maple
Wetter	sugar maple-elm	red maple
	elm-ash	red maple
Wet	cedar	elm, fir and black ash
Peat	cedar	spruce and fir
Elevations	sugar maple-beech	white birch and aspen
	sugar maple-red oak	white birch and aspen
	white pine	white birch and aspen

From Moussette 1978, pp. 60–61

These forest communities can be classified by site, as shown in Table 2.7.

Both hardwoods and softwoods were used in making charcoal; hardwood was used mainly for smelting the iron ore in the blast furnace, while softwood was used for refining in the forges. The naturalist Pehr Kalm described the beliefs of French ironmasters about the uses of different types of charcoal:

> The charcoals from ever-green trees, that is from
> the fir kind are best for the forge, but those of
> deciduous trees are best for the smelting oven.[78]

This remark, which is consistent with the opinion of 18th-century writers on this subject (Duhamel Dumonceau and Walter de Saint-Ange), is based on the fact that the higher heat of hardwood combustion was needed to smelt ore, and that softer and more malleable iron resulted from refining with charcoal from softwood. A century later, this had not changed. The clerk Hamilton Rickaby confirmed this: "Hard and soft wood are both used for charcoal. The hard wood charcoal is used for **smelting** and the soft wood charcoal for making forged iron."

▼

During this period (1852–58), accounts for hardwood and softwood were always kept in separate books[79] (see Appendix 5), and the two types of charcoal were used as long as both cast and wrought iron were produced. John Porter & Company was the last to carry on both these activities. With the arrival of the McDougalls, the large-scale production of wrought iron was very soon abandoned for the production of pig iron. This concentration on the manufacture of pig iron, accompanied by a change in the method of making charcoal, no doubt affected the timber being cut. Charcoal quality standards for the production of pig iron ingots for Montreal factories do not seem to have been the same as those previously applied. In 1874, Harrington remarked that, unlike the custom observed until 1860, mostly softwood charcoal was now being used to smelt the ore. He also noted that the charcoal was very inferior in quality, compared with the charcoal in other Canadian establishments; it weighed 11 to 12 lbs.(5–5.5 kg) a bushel, which was 5 to 10 lbs. (2.2–4.5 kg) less than charcoal produced from a mixture of soft and hard woods or from hardwood alone.[80] Harrington did not blame this drop in quality on a scarcity of wood. It is, however, plausible that repeated haphazard cutting for over 125 years could have resulted in wood of poorer quality. This was what Jeffrey Brock implied 20 years earlier in discussing the low productivity of third and fourth growth wood on Forges land.[81] The various sources of supply used by the McDougalls could also have had an impact on the quality of wood used for charcoal. Their wood was harvested mainly on the east bank of the St Maurice, in contrast with the practice of earlier administrations, which had almost always taken their wood from the west bank, from the fiefs of St Maurice and St Étienne and adjacent land. The sale of the former Forges land to settlers does not, in itself, explain this shift in the areas where wood was harvested.

Over a hundred years of repeated haphazard cutting, even in such a vast area, had truly depleted the surrounding forest, as Forges manager Jeffrey Brock implied in 1852. During the same period in the United States, it was said that the cumulative effects of repeated clearcutting for the needs of American ironworks had made the land unusable and unproductive, and that the forest would not regenerate until charcoal metallurgy had disappeared.[82] After travelling through the fief of St Étienne in 1852, inspector Étienne Parent pitied the settlers recently established on "poor sandy land despoiled of all that would have made it valuable" and that had been "all chopped over several times since the establishment of the Forges."[83]

RAW MATERIALS: SUPPLY AND CONSUMPTION

Ore

In 1735, ironmaster Vézin planned to produce a million pounds of cast iron in an 8-month campaign. Consequently, he estimated the annual ore requirement at 2,000 *pipes*, the weight of a *pipe* being 1,107 French pounds (541.6 kg),[84] or 1,084 t. Production never matched his estimate, however, at least not during the French regime. For eight months' output, the objective was revised to 800,000 pounds of cast iron, requiring from 1,500 to 1,600 *pipes* of ore (800–900 t), but this objective was not always met.

The figures available for the entire lifespan of the Forges (Table 2.8) indicate that, for a comparable 8-month campaign, ore consumption more than doubled after the blast furnace was rebuilt in 1854.[86] An equivalent increase in cast iron production can also be seen. Table 2.9 shows that the yield for the first blast furnace (1737–1853) was 2.2 t of ore per tonne of pig iron, while for the second furnace (1854–83), it increased to three for one. The yield reported by Harrington resulted from a **cold blast**; the yield noted by Hunt in 1868 was from a **hot blast**, comparable with the cold blast technology still in use at the time.

Table 2.8

ANNUAL CONSUMPTION OF RAW MATERIALS, ST MAURICE FORGES, 1740–1874[85]

Year	Production		Campaign (months)	Ore (t)	Charcoal (t)
	Cast iron	Wrought iron			
1740	391.5	343	8	867	2,498
1764	391.5	147	8	813	3,416
1828	520	81	8	1,152	1,344
1870	1,327	—	8	2,976	1,144
	1,990		12	4,464	1,716
1874	910	—	8	2,731	880
	1,365		12	4,096	1,320

Note: The tonnes of cast iron and tonnes of wrought iron do not add up, since the wrought iron was made from the cast iron produced.

Table 2.9

PIG IRON YIELD OF ORE[87]

Year*	Reference output	Campaign	Pipes of ore		Total weight of ore	Ore per tonne of pig iron	Ore yield
				Price	(t)	(t)	(%)
1735	1,000,000 French pounds	8 months	2,000	3 l	1,084	2.2/1	45
1740	800,000 French pounds	8 months	1,600	3 l	867	2.2/1	45
1743	810,000 French pounds	6 months	1,620	3 l	878	2.2/1	45
1764	800,000 French pounds	8 months	1,500	4 l 10 d	813	2.1/1	47
1828	16 long tons	7 days			36	2.2/1	45
1868	163.25 long tons	1 month			378	2.3/1	43
1874	4 long tons	24 hours			12	3/1	30

* The 1735 figures are based on Vézin's initial estimate; those from 1740 to 1764 are projections based on actual production figures; figures from 1828 to 1874 result from observation.

Logan's analyses, carried out in the 1850s, demonstrated that the ore used at the Forges contained 45–55% iron. Harrington, repeating these analyses in the laboratory, found the blast furnace yield was rather in the order of 30–40%.

Table 2.9 indicates that, before the blast furnace was rebuilt, ore yield was generally 45%. The 30% yield observed by Harrington in 1874 likely resulted from the use of poor quality ore. Ore was mined in a number of areas, where it was found in varying states of purity, depending upon its percentage content of other matter. This was indicated by the three samples analysed by Logan in 1853–54, which testify to the diversity of sources of supply confirmed by Superintendent Henderson in 1855. Henderson noted in writing to Logan that the ore from the Nicolet region contained less clay and required less washing than ore from the north bank of the St Lawrence River within a 16–24-km perimeter of the Forges.[88] Hamilton Rickaby stated that, at this time, about a third of the ore was being taken from the Nicolet mines, on the other side of the St Lawrence.[89]

Flux and Firestone

In addition to being a chemical compound (Fe_2O_3), iron ore is also mixed with muddy, sandy or clayey organic matter, which forms gangue that must be removed during the ore reduction process in the blast furnace.[90] When the requisite temperature ($1,200^\circ C$–$1,400^\circ C$) is reached, the flux helps to separate the gangue from the metallic iron by liquifying the vitrified organic matter, known as **scoria** or slag. In plain terms, it is waste from the smelting process, which floats on the surface of the molten iron accumulated in the **crucible** and is discarded just before the furnace is tapped. Limestone[91] was generally used as flux in the case of ironstones, while **clay marl** was used for calcareous iron ore.[92]

Limestone and Building Stone

Before the blast furnace was rebuilt in 1854, about 300 t of limestone and sandstone were required annually, about half of it used as flux and the other half as firestone and building stone for annual repair of the linings, boshes and the crucible of the blast furnace. With the new blast furnace, the volume of stone probably doubled, according to the limited information available.

The data are most precise for the French regime. In 1740 and 1743, the requirements for limestone and building stone were detailed separately. It was estimated that one *toise* per month was needed for flux, based on a projected campaign of six or eight months. This meant a cubic *toise*, which is equal to about 20 t.[93] For 1743, requirements were estimated for six full months of production, taking into account the many breakdowns and interruptions during a campaign which could last seven to eight calendar months. In 1740, for example, 7 *toises* of flux (140 t) were used for eight months of operation. The volume of firestone required to repair the linings was estimated very accurately at 8 *toises* (160 t). In 1764, Inspector Courval estimated the amount required "for the crucible, flux, linings and hearth bottom."[94] His overall estimate is on the same order as that of 1743, about 20 t. After that, we find no further figures on the volume of building stone required for annual repair of the blast furnace. The amount likely remained the same until construction of the new blast furnace, although there would have been variations, depending on how dilapidated the various parts that made up the **belly** of the blast furnace were after a campaign. Since the new blast furnace was double the height of the old, we can assume that the quantities would have doubled.

In 1827, contradictory figures appear for the volume of stone used. Based on Lieutenant Baddeley's report, which gives the amounts of raw materials required to reduce 3 tons of ore, we can estimate the volume used for flux alone at 130 t. This is similar to what was previously reported because the furnace was of the same type as that used during the French regime. The other figure, of 30 to 50 *toises* a year (600–1,000 t), comes from Mathew Bell, who held the lease of the Forges at that time. His figures are double or triple all previous estimates. However, Bell advanced these figures in support of his exclusive use of the Gabelle quarry, as we have already seen. He therefore had to inflate them to show that exclusive rights to the quarry were vital to his business.

After the Forges passed into private hands in 1846, the proprietors also tried to keep the quarries along the St Maurice for themselves in order to avoid an increase in costs following subdivision of the land. In 1852, Jeffrey Brock, who was concerned about the matter, estimated that over 500 tons (508 t) of stone was quarried annually for flux alone.[95] Here again, the figures were produced in the context of land claims and were greatly inflated. During this period, the small furnace was still being used, and it did not require more than 150 t of flux a year. Brock's estimate greatly exceeds that of Harrington in 1874 (335 t), which describes the consumption figures of a blast furnace with double the capacity of that of 1852! John Porter & Company finally succeeded in reserving the lots on which the quarries were located, in the first range of St Étienne,[96] but the McDougalls subsequently lost this privilege; they then had to pay for the right to mine ore and had to purchase land and flux from the local *habitants*.[97]

The other available data, on flux only, come from the later 1868 and 1874 reports of the Geological Survey of Canada. Based on the proportions given for the blast furnace charge, we can estimate the annual consumption of flux in the new blast furnace at 200–300 t, depending on whether a hot blast or a cold blast was used. The ratio for a hot blast was 1 t of flux to 20 t of ore, while a cold blast called for 1 t of flux to 13.3 t of ore.[98] The latter proportion is roughly double the volume used by the first cold blast furnace, which consumed 1 t of flux to 6 t of ore in 1740. Consumption doubled again with the installation of a second blast furnace in 1881.

Clay Marl

Clay marl was also used as a flux in the blast furnace. In his initial estimate of raw materials in 1735, Vézin allowed for 600 *pipes* of marl. The use of clay marl was mentioned again in 1740 and by the naturalist Pehr Kalm in 1749.[99] In 1828, Lieutenant Baddeley mentioned that clay was found in the immediate neighbourhood of the establishment and specified that it was used with limestone in precise proportions: "Ten bushels of limestone mixed with four of clay is a proportionate flux for three tons of ore."[100]

Clay, particularly ferruginous clay, was recommended by metallurgists of the time, who felt that it helped to strengthen the metal during the smelting process.[101]

Moulding Sand

The iron produced in the blast furnace was cast in sand. The floor of the casting house was completely covered with sand, in which the **sow**, **pigs** and other objects were cast. The carter at the blast furnace was responsible for seeing to the supply of sand.[102] A document from 1740 includes 100 *barriques* of sand at 3 *sols* per *barrique*.[103] A finer sand was also used for **box or flask moulding**. Passing through the Forges in 1752, the engineer Louis Franquet described **open sand moulding** in detail (see Chapter 5). The sand had to have enough body to receive the impression of the mould and resist the heat of the molten cast iron, and could not contain fusible fragments. To make the sand compact, it could be mixed with clay.[104] According to Lieutenant Baddeley, there were only two or three locations in England where fine sand, suitable for moulding, could be found.

Except for some references to its use and transportation, the documentary record from the French regime yields no information on the origin of the sand. In the early 19th century, the traveller John Lambert and the surveyor Joseph Bouchette reported that sand was imported from England. Later, in 1828, Lieutenant Baddeley specified that fine layers of imported sand were used on top of local sand in the manufacture of fine **castings**. The local sand came from the fief of St Maurice, along the border of the fief of Ste Marguerite.[105] In 1855, Superintendent Henderson wrote to Logan that moulding sand was found locally, but did not specify what type it was.[106] However, a statement of account for December 1857 mentions payment for the transportation of 100 tons of Belfast moulding sand.[107] It is possible that this sand was used for the manufacture of railcar wheels during these few years. The 1896 report of the Geological Survey of Canada implies that the sand requirements of the St Maurice and Batiscan ironworks were both met from local sources, and describes the sand used as "fine quartz ore, containing at the same time small quantities of argillaceous and ferruginous matters."[108]

Charcoal

Timber Acreage Required

According to the figures available, the Forges used between 10,000 and 20,000 cords of wood annually for charcoal. This meant felling on a large scale, over an area that we will try to estimate here.

In assessing the timber acreage required to meet annual charcoal requirements, a number of factors must be taken into account. The yield of a harvest of wood may vary according to the species and age of the trees; the felling method and timing are also very important, since clearcut forests take longer to regenerate than those where cutting is selective. Once the yield of the species being used is known, the time lapse between cuts is especially important. The interval required between cuts to produce trees of a suitable diameter (2–4 French inches) for charcoal production was estimated at 15–20 years. During the French regime, the principles and practices of silviculture were already known, since the fundamentals of a well-regulated cut were laid down during the early years of operation. The colonial authorities even issued ordinances to protect new growth in order to ensure the regeneration of the woodland used in the vicinity of the Forges. In fact, however, no great attention was paid to the principles of a well-regulated cut, and the vastness of the surrounding forest, which was perceived as inexhaustible, encouraged a careless rather than careful approach on the part of the Forges masters. Haphazard cutting did not even allow for evaluation of the interval required for forest regeneration. The only concern was for the rising cost of transportation as timber had to be sought farther and farther afield.

Trees cut in due season and well conserved will grow back in fifteen years and even if they reach only two, three or four inches in diameter, they will make good charcoal. If care had been taken from the outset to cut in due season and conserve the regrowth, we would be in a position to judge at what interval to cut. We are convinced that this would be an interval of fifteen years at most. But even supposing that the wood did not grow back at all and that the seigneuries granted to La Compagnie des Forges were consequently completely exhausted, abundant supplies could be found on the seigneury of M. de Tonnancour at Pointe du Lac and on the seigneuries adjoining St Maurice on the Trois Rivières River [...].[109]

Such a pronouncement made in France at that time would have created a scandal, since the forest there was already at risk from the iron industry.[110] An interval of 15 years to regenerate a coppice forest would have required exemplary conduct on the part of the Forges masters, as the author of this memorial implied. It would seem rather that a cutting interval of about 20 years was generally observed. This is what Mathew Bell reported in 1829,[111] and manager Jeffrey Brock in 1852. It was Brock, moreover, who provided us with the only known estimate of the acreage required:

A lot of 100 acres, with a wood of 20 years growth should produce from 2600 to 3000 cords; seven Lots would be required annually to keep the Forges supplied with wood, and as the trees would require about 20 years to become large enough for the axe, these works should have attached to them from 140 to 150 Lots of 100 acres each, in order to preserve and cultivate the wood.[112]

Table 2.10 — ANNUAL WOOD CONSUMPTION FOR CHARCOAL

Date	Cords of wood	Source
1742	11,282	Estèbe, *Estat général*
1764	12,000	Courval's memorial
1804	10,000	Selkirk, p. 231 (over 2,000 cords of firewood)
1819	10,000	JHALC, 1834, app. X
1852	12,000	Stuart and Porter, 23 June 1852
1852	18,000–20,000	Timothy Lamb, 31 August 1852
1852	18,200–21,000	Jeffrey Brock, 1 September 1857
1857	About 11,000	Statement of account for December 1857 (11,034 dressed cords)
1870–80	12,000–13,000	Dollard Dubé, 1933

Based on these figures, we obtain an average yield of 2,800 cords per 100 acres, or 28 cords per acre; cutting seven 100-acre lots annually on a 20-year cycle would therefore require a reserve of 140 lots or 14,000 acres (5,665 ha). Brock's estimate was based on an annual requirement of 18,200–21,000 cords of wood (seven times 2,600 or 3,000 cords). Brock and his clerk, Timothy Lamb, stated that this was roughly the requirement at that time. Lamb, who had worked for Mathew Bell, based his estimate on "long experience" and added that "this amount would now have to be greatly increased"; he was no doubt implying that the new blast furnace, double the size of the old, proposed by the engineer William Hunter, would increase charcoal consumption. Their figures contradict the estimates of their employers who, two months earlier, set annual requirements at 12,000 cords. A statement of account from 1857[113] indicates that 11,034 cords of wood were dressed at the charcoal pits that year. There is no mention, however, of the charcoal made in kilns at the Forges themselves. The last workers interviewed by Dollard Dubé stated that, from 1870 to 1880, from 12,000 to 13,000 cords were normally coaled annually, and sometimes up to 20,000 cords (Table 2.10). Brock and Lamb's estimates, produced in the context of land claims, could be realistic, but they reflect an exceptional consumption.

Harrington reported in 1874 that the traditional method of coaling in pits was used concurrently with the kiln method. The pit method yielded 34.32% charcoal, while the kiln method yielded 60.1%. The variation in the amount of wood required was thus a direct function of the coaling method used. While the yield from brick kilns was greater, kiln charcoal was poorer in quality and entailed the considerable cost of carting the wood to the Forges, while pit charcoal was made where the trees were actually felled.

To estimate the timber acreage required for fuel at the Forges, we took an average annual cut of 12,000 cords of wood. On the basis of an average yield of 28 cords per acre, reported in 1852, and using a 20-year cutting cycle, the acreage can be estimated at nearly 3,470 ha, or an area slightly larger than that of the fief of St Maurice (3,254 ha, according to the 1806 survey).[114] This acreage corresponds, within a few dozen hectares, to the average area (see Appendix 2) of land acquired under the McDougalls; the McDougalls were, in fact, the only proprietors compelled to buy all the land required for the supply of raw materials.[115] When the Forges were abandoned in 1883, George McDougall owned 3,823 ha of land.

A constant consumption of about 12,000 cords of wood throughout the entire history of the Forges is surprising. The kiln charring process, introduced by John Porter & Company shortly after 1850, nearly doubled the charcoal yield and should have reduced wood requirements. Some technical factors may, in part, explain why the volumes of wood remained the same. First, pit charring was still being practised even as the McDougalls were stepping up kiln charring.[116] Harrington testified to this, without giving any indication of how much wood was used by either method.[117] The fact that he mentioned that pit charcoal was considered of better quality than kiln charcoal would suggest that it was still being made in considerable quantities. On the other hand, after 1854, the capacity of the blast furnace was increased to 4 tons of pig iron per day, which must have helped to keep up the volume of wood consumed, certainly cancelling out a good part of the saved wood resulting from kiln charring. Non-carbonized pieces of wood were also used in the smelting process of the new blast furnace built in 1854. A statement of account for 1857 mentions the payment of workers for "small wood for use in the blast furnace."[118]

But the main reason for a constant cut was certainly the extension of the annual campaign, which seems to have been eight months long until the 1850s. Then the blast furnace was rebuilt with double the capacity and the campaign was extended. In 1874, Harrington reported that a campaign lasted 10 to 13 months.

Table 2.11

CHARCOAL CONVERSION COEFFICIENT AT THE ORE REDUCTION STAGE, ST MAURICE FORGES

Year[119]	Pig iron output	Amount of charcoal	Unit of measure for charcoal	kg of charcoal/ kg of iron
1740	800,000 Fr. pounds	15,000	pipe	3,300/1,000
1743	810,000 Fr. pounds	15,300	pipe	3,325/1,000
1756	800,000 Fr. pounds	12,500	pipe	2,751/1,000
1828	3024 lbs.	648	bushel (12 lbs.)	2,570/1,000
1868 hot blast	365,680 lbs.	26,272	bushel (12 lbs.)	862/1,000
1874 cold blast	8.960 lbs.	720	bushel (12 lbs.)	964/1,000

The annual volume of cordwood thus remained fairly constant, even though the charcoal yield increased. In addition, the charcoal yield compared to pig iron output also increased. After the rebuilding of the blast furnace in 1854, less charcoal was used to smelt the same amount of pig iron, but more pig iron was produced. This yield was calculated on the basis of the amount of charcoal used to make 1,000 kg of pig iron. The figures in Table 2.11 show that from 1740 to 1874 charcoal yield at the ore reduction stage improved by 70%.

We can see that the yield was clearly improved with the alteration of the blast furnace in 1854, from a square to a circular type. In addition, the height of the interior cavity was increased from 3.6 to 9 m, and two **tuyeres** were used to activate combustion.[120] These new features resulted in better thermal efficiency, so that the blast furnace used less charcoal and burned it more efficiently.[121] Lastly, a hot blast rather than the traditional bellows resulted in savings of as much as 25%.[122] A comparison of charges and yields for the two smelting methods used at the Forges, however, shows savings in the order of 10.5% for charcoal, 16.6% for ore and 33.3% for flux. The pig iron yield of ore was improved by 10% with the hot blast (Table 2.12).

The observations of the two Geological Survey of Canada geologists refer to a hot blast in 1868 and a cold blast in 1874: Sterry Hunt reported a daily yield of 5.35 tons,[123] while Harrington spoke of a yield of 4 tons. Lacking any additional information, we are unable to say whether the 1874 yield resulted from abandonment of the hot blast or from the practice of alternating between the two. André Bérubé supports the second hypothesis, specifying that at that time it was possible to alternate between a cold blast and a hot blast, "depending on the type of ore used and the type of iron desired."[124] Expert metallurgists of the time and Canadian manufacturers did not advise the use of a hot blast in the production of iron for the manufacture of railcar wheels. It was only after the Forges had closed that the hot blast system for this type of product was perfected, particularly at the Radnor Forges in 1892.[125]

It is rather interesting to compare the yield of charcoal at the St Maurice Forges with that of a European blast furnace, at the ore reduction stage. The comparison indicates that the Forges lagged far behind in terms of technical innovation to the blast furnace.

Table 2.13 shows that the charcoal yield at the two plants was similar in the 18th century.[127] By the 19th century, however, the St Maurice Forges was 30 years behind. Allevard changed to a circular type furnace in 1824, which enabled it, with the introduction of a hot blast in 1832, to bring its charcoal conversion coefficient well below the 1,000 threshold. Such a decrease did not take place at the St Maurice Forges until after 1854, when a circular type furnace was adopted.

Table 2.12

COMPARISON OF A FURNACE CHARGE FOR A HOT BLAST AND COLD BLAST

Smelting method	Ore (lbs.)	Charcoal (bushels)	Flux (lbs.)	Ore yield in tons of iron (%)	Charcoal per ton of iron (bushels)
Hot blast	500	16	25	33%	161
Cold blast	600	16	45	43%	180

From geologists Hunt (1868) and Harrington (1874). English measure

Table 2.13

COMPARISON OF THE CHARCOAL CONVERSION COEFFICIENT AT THE ORE REDUCTION STAGE, ST MAURICE FORGES AND ALLEVARD[126]

St Maurice Forges St Maurice Valley, Quebec kg of charcoal/kg of iron		Allevard Dauphiné, France kg of charcoal/kg of iron	
1740	3,330/1,000	1724–50	2,400–3,200/1,000
1756	2,751/1,000	1778	2,680/1,000
1828	2,570/1,000	1830	1,020–1.112/1,000
1868	862/1,000	1842	920/1,000

Charcoal Consumption

During the French regime, charcoal was measured by the *pipe*.[128] It is estimated that a cord of wood yielded 2.5 *pipes* of charcoal, with 12,000 cords of wood yielding 30,000 *pipes* of charcoal. Accounts from the era confirm figures in the neighbourhood of these amounts (Table 2.14).

In the 19th century, the unit of measure for charcoal was the bushel. The weight of a bushel could vary between 12 and 20 lbs. (5.4–9 kg), depending on whether the charcoal being weighed was made of softwood, hardwood, or a combination of the two. In 1874, Harrington gave the weight of a bushel of softwood charcoal of the inferior quality then being used at the Forges as 11–12 lbs (5–5.4 kg). Overall numbers of bushels for the era were not given, but based on the figures provided for several tons of pig iron, we can estimate that about 250,000 bushels were consumed annually.[129] On the basis of known figures, we have established annual consumption of charcoal in tonnes in order to clarify changes over the entire operating period (Table 2.8 above).

Table 2.14

ANNUAL CONSUMPTION OF *PIPES* OF CHARCOAL, ST MAURICE FORGES

| Year | Pipes of charcoal | | | Total weight* in tonnes |
	Blast furnace	Forges	Total	
1740	15,000	14,000	29,000	2,498
1743	15,300	12,750	28,050	2,416

The trend in the consumption of raw materials indicates that less and less charcoal was required to smelt more and more ore. This shows the cumulative effects of improved carbonizing and the 1854 alterations to the blast furnace. The gradual decline in wrought iron production, which was abandoned in favour of the exclusive production of pig iron in the 1860s, should also be noted. The volume of castings produced had already dropped by over 60% at the end of the French regime and was down to only 81 t by 1828. It is this decline in wrought iron production which explains the drop of 800 t in charcoal consumption between 1756 and 1828. Subsequent savings are the result of both the abandonment of wrought iron production and better yields at both the coaling and reduction stages. In 1874, the consumption of charcoal for a 12-month period was the same as it had been 50 years earlier (1828), even though two and a half times as much pig iron was being produced annually.

DRESSING THE RAW MATERIALS

Ore Prospecting and Mining

In a region where surface ore was abundant, prospecting was never a difficult task. Vézin wrote, concerning the area of the Forges, "All these iron mines are near the surface of the ground."[130] Ochre-coloured stream beds were enough to suggest the presence of iron ore, which was confirmed by tasting the water. Governor Frontenac experienced this for himself in the seigneury of Cap de la Madeleine in 1672 and wrote to Colbert that "he found it all strongly impregnated with rust and iron."[131] In 1732, the physician Michel Sarrazin, passing through as Francheville was setting up his establishment, mentioned to the seigneur the medicinal benefits of this ferruginous mineral water.[132] At a later date, Benjamin Sulte, who lived at the Forges, attributed the health of the workers to the water from the St Maurice Creek.[133]

More careful prospecting had to be carried out if the mines were to be worked. The prospector would survey the land for indications of ore beds. Reporting on the methods of Forges prospectors, Lieutenant Baddeley wrote that swampy areas covered in softwood (cedar, fir, spruce and poplar) were often for them a sure sign of the presence of ore; he specified that iron ore was not found in hardwood forests.[134] Once an ore field had been located, prospectors used a simple probe or sounding-rod to find the veins:

> by probing the ground with a short iron instrument when if ore be present it is ascertained by the gritty metallic sound returned.[135]

Prospectors thus relied on sound, but also on the feeling of resistance in the soil:

> If, on the other hand, [the miner] feels some resistance, as if he were trying to force a stick into a barrel of salt or a cup of coarse flour [...], then that is a sign that there is ore in the location being probed.[136]

The probe was a simple iron rod, about 1 cm in diameter by 1 m in length, with a handle. Such sounding-rods are mentioned a number of times in Forges inventories. Using this tool, the prospector delimited the extent of the vein and marked off its boundaries with stakes. A good prospector, who paid attention to the land configuration, could trace the source of several veins. Once the vein was marked off, exploratory "prospect pits" were sunk at various points to assess its depth. The area of the deposit was then measured and the amount of ore it contained evaluated in terms of how many blast furnace campaigns it was good for.[137]

Some local *habitants* were employed as prospectors, such as Joseph Jutras, who was Francheville's prospector and who reported the Cap de la Madeleine deposits to the new administration.[138] During the early years of operation, Vézin himself, clerks and sometimes important figures in the colony were involved in prospecting. This was the case with Grand Voyer Lanouiller de Boisclerc, the colony's chief road commissioner, who accompanied Vézin on his 1735 prospecting trip and made a survey in 1740. We saw above that the deposits prospected by Vézin in 1735 were even mapped (Plates 2.3 and 2.4). However, prospectors were usually the miners themselves, such as "master miner" Langevin, who accompanied Lanouiller de Boisclerc in 1740, or the mine labourers mentioned as permanent residents of the Forges in 19th-century census returns.

Plate 2.6
Mining bog ore in the 18th century.
DIDEROT AND D'ALEMBERT, *ENCYCLOPÉDIE [...], RECUEIL DE PLANCHES [...]*
(PARIS: BRIASSON, DAVID, LE BRETON, 1765), "FORGES OU ART DU FER," SECT. 1, PLATE II.

Since the ore was located in shallow surface deposits, it could easily be mined with shovels. The Pointe du Lac mine, prospected in 1735, and probably the ore bed that was mined the longest, was apparently so easily worked that it was said that "the carters themselves could dig it up with shovels and load it for carting to the Forges. It did not have to be washed because it was pure and free from soil."[139] The ore was in fact mined by miners, apparently working in teams, but in tandem with carters to keep track of the number of cartloads of ore collected at each mine. The miners used nothing more sophisticated than picks, mattocks and **miner's bars** to separate the aggregate from the ore. According to Dollard Dubé, they also used a sort of hoe to remove the surface layer of peat.[140] Mining did not, therefore, require any particular skill. After the Forges land was settled, the farmers themselves collected the ore found on their land.

Plate 2.7
Washing bog ore in the 18th century.
DIDEROT AND D'ALEMBERT, *ENCYCLOPÉDIE [...], RECUEIL DE PLANCHES [...]*
(PARIS: BRIASSON, DAVID, LE BRETON, 1765), "FORGES OU ART DU FER," SECT. 1, PLATE V.

Ore Washing and Crushing

Miners usually had to wash the clotted organic matter from the ore on site. The price paid to the *habitants* who collected ore themselves varied according to whether it was clean or dirty. In the 1850s, dirty ore earned 30% less to compensate for the additional weight of the gangue clinging to the unwashed ore.[141] Bog ore was washed right at the mine itself (Plate 2.7). Dubé reported that ore washers dug a trench in which water would accumulate naturally. Kneeling on a wooden deck, they held a riddle full of ore at arm's length, shaking it and then emptying it onto a bed of peat. When the Forges closed in 1883, 16 ore riddles were inventoried. **Buddles** operated by two men were also used, with four buddleloads making a *barrique* of ore.[142] "Meadow ore" mined from drained bogs was washed at the Forges. With the exception of the so-called pure ore, it would appear that all the ore was washed again in an ore washery or launder *(lavoir)* up on the St Maurice Creek. The creek was even known, at a later date, as the *ruisseau du lavoir*.

The documentary record reports on the installation of a washery at two different times: at the very start of operations, and at the very end. These two different machines were not set up in the same location. In 1735, Vézin's plans included the installation of a **stamp mill** or ore crusher and washery. In establishments of the period, the ore was crushed and washed in the same apparatus, which combined a stamp mill and a washery (Plate 2.9). The large chunks of ore had to be crushed to facilitate reduction in the blast furnace. According to a report from 1737, a stamp mill seems to have been set up on the post, which may very well have served

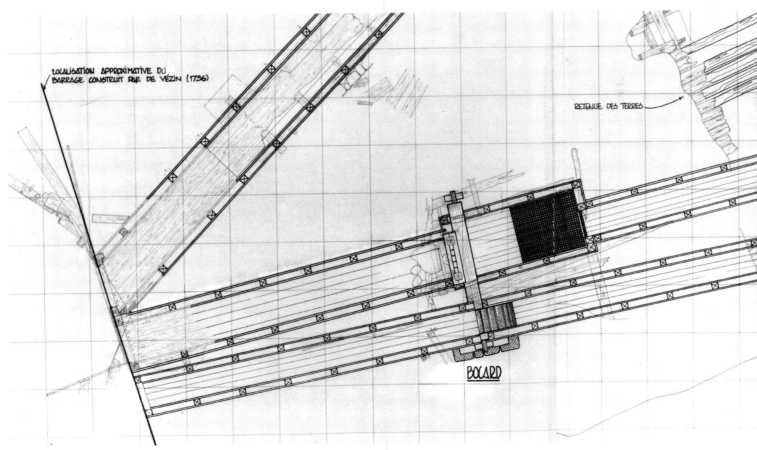

LOCALISATION APPROXIMATIVE DU BARRAGE CONSTRUIT PAR DE VÉZIN (1736)

RETENUE DES TERRES

BOCARD

Plate 2.8
Remains of a stamp mill at the St Maurice Forges,
with plan view reconstruction over the creek on which it was set.
ENGINEERING AND ARCHITECTURE, PARKS CANADA, QUEBEC.

Plate 2.9
Eighteenth-century stamp mill.
DIDEROT AND D'ALEMBERT, *ENCYCLOPÉDIE [...]*, *RECUEIL DE PLANCHES [...]*
(PARIS: BRIASSON, DAVID, LE BRETON, 1765), "FORGES OU ART DU FER,"
SECT. 1, PLATE IX.

to carry out contracts for crushing and washing ore in 1740 and 1741.[143]

Remains uncovered in the creek bed during archaeological digs show a pattern of pieces in a frame that resembles the base of a stamp mill (Plate 2.8). Such a machine is illustrated in Diderot's *Encyclopédie* (Plate 2.9). This is a device fitted with heavy iron pestles or stamps activated in turn by a camshaft driven by a waterwheel. It was usually installed at or slightly downstream from a **milldam**, where two **sluice gates** or **shuttles** abutted two parallel races. One race channelled the water towards the wheel and the other conveyed the ore shovelled into it. The gate to the wheel was raised to activate the stamps, and then the other gate was raised to propel the ore under the stamps. After crushing, the ore went into a race with a grating, which was the actual washery. There, washers sieved the ore again before removing it.

The second washery, described by Dollard Dubé, was installed up at the weir of the washery pond, that is, the pen pond located at the highest level, which was actually the reservoir supplying the four **millponds**. The washery was in operation during the Bell era. There was no stamp mill, but the washing was carried out in the same way as in the 18th century. It was a simple race or channel with a grating, 5 or 6 m long, with one end abutted to a sluice. The channel was filled with ore, and when the gate was lifted, the ore was scoured by the current of water while the washers raked it. Ore was washed in this way two or three times, if necessary.[144]

Table 2.15	WOOD CARBONIZATION PROCESS *	
Temperature	**Chemical transformation**	**Residue**
Up to 100°C	wood dries	water and oil (resins)
From 150°C	wood darkens decomposition begins	water and acetic acid
From 280°C	wood decomposes	2 liquid layers — black tar — pyroligneous acid# (acetic acid and methanol) gas: carbon dioxide, carbon monoxide, methane, other hydrocarbons
About 300°C	exothermic phase (risk of explosion)	
Between 350°C and 450°C	complete distillation	small amounts of gas and vapour
After cooling		black charcoal

* Based on Joseph Risi (1942), cited in Bérubé 1978, pp. 56–57

During the 1850s, there is mention of a "pyrolygneous acid machine" at the Forges.[148]

Limestone Breaking

Limestone for flux came from the quarries to the Forges site in large chunks that had to be broken down into smaller bits (about 10 cm in diameter). Like ore crushing, limestone breaking was intended to make it easier to smelt the mix of materials in the blast furnace. **Limestone breakers**, hired for each campaign, laboured in the great charcoal house at the blast furnace, breaking up the stone with an iron sledge hammer. This strenuous task occupied several workers. In the summer of 1742, six men were so employed, for a little over three months at 20–30 *sols* a day.[145] An iron **ram** was also used to break the limestone. In a photograph from around 1870 (Plate 2.10), a tripod-type structure can be seen near the blast furnace. This was a gin, a hoist fitted with a winch for raising and dropping a heavy drop ball on large pieces of rock. Dubé specified that the "great ram" consisted of a 225-kg block of cast iron winched up 5 m by a chain mounted on a pulley; the stone to be broken was placed underneath on a large iron slab.[146]

Charcoal Making

Carbonization may be briefly defined as follows: "conversion of a carbon-containing substance to carbon or a carbon residue as the destructive distillation of coal by heat in the absence of air, yielding a solid residue with a higher percentage of carbon than the original coal" (Table 2.15).[147]

Two wood carbonization techniques were used at St Maurice—pit coaling in the forest and kiln coaling at the Forges. Coaling in kilns, large brick or metal ovens, was introduced in the years following the sale of the Forges in 1846 and subsequently was heavily used, mainly during the last 20 years of the McDougall regime. Charcoal continued to be made in the pits, a technique which produced a smaller yield than the kiln process, but was said to produce better quality charcoal.

Pit Charcoal

Pit charcoal was made in clearings in the forest known in French as *ventes*[149] (Table 2.16) where the wood was cut. Charcoal was made in four steps: preparing the hearth, setting (or "dressing") the pit, leafing (or "feuilling") and charring, which were the responsibility of the pit setter (known as the "wood dresser" at the Forges) the leafer ("feuiller"), and the master collier, working generally as a team supervised by the master collier. Although the men worked in teams as part of the same crew, the three jobs carried different rates of pay.

Table 2.16

CHARCOAL PITS, 1740[150]

Location of pits	Number of pits	Number of cords in pits	Cords per pit (in round figures)
Behind the stables (on the post itself)	71	1,314	18
On the Forges road (from Trois-Rivières to the Forges)	172	3,502	20
On the road to the mines (from Pointe du Lac)	58	1,374	23
Total	**301**	**6,190**	**20**

The first colliers soon abandoned the **lump-sum**, turnkey method of payment (see Chapter 8). The master colliers, like the ironworkers, were recruited directly from France. The Aubrys were the best known, but others included Chabenac, *dit* Berry, Chaillot, Girardeau and others from Burgundy and Champagne, as well as Berry and Auvergne (Table 2.17). The requirements of charcoal making for the iron industry made these workers as indispensable as ironworkers.[151] The personality and productivity of each was taken into account, just as it was for the ironworkers. A memorial of 1743 described one of the Aubry brothers as follows:

> Claude Aubry, the most able of the colliers and the one who has produced the most charcoal of the highest quality, could be given a contract with his cousin, Jean Aubry.

The same report stressed the skill of the colliers:

> It is of great importance to have skilled charcoal burners who know how not to reduce the wood to breeze and to get all the charcoal it contains, can set the pits themselves and oversee other setters so that their pits are well set, leafed and banked. High quality charcoal depends both on the type of pit and the way it is charred. It is essential to have high quality charcoal, since less of it is used and this keeps expenses down [...].[152]

Table 2.17

COLLIERS AT THE FORGES, 1742–43 [153]

Collier	Charcoal made (pipes)	Price per pipe (sols)	Total cost
Berry	1,255.5	10	627 livres 15 sols
Mattenay	566.6	10	283 livres 5 sols
François Thomas	778.75	10	389 livres 7 sols 6 deniers
Chaillot	1,127.5	10	563 livres 15 sols
Aubry the elder	900	10	450 livres
Aubry the elder	2,958	9	1,331 livres 2 sols
Aubry brothers	4,729	10	2,364 livres 10 sols
Aubry brothers	7,224	9	3,250 livres 16 sols
Total	**19,539.25**		**9,260 livres 10 sols 6 deniers**

Plate 2.10

Ram used to break up limestone at the St Maurice Forges.
RECONSTRUCTED BY ILLUSTRATOR BERNARD DUCHESNE,
AFTER JOHN HENDERSON'S PHOTOGRAPH OF THE ST MAURICE FORGES (DETAIL),
CIRCA 1870, NATIONAL ARCHIVES OF CANADA, PA-135-001.

Charcoal pit locations were sometimes identified by collier (Table 2.18). In the same timber tract, colliers had a tendency to reuse the same locations several times to capitalize on these sites and the residual **breeze**, a mix of charcoal dust and earth that could be used for leafing. Pit sites were, in fact, selected on the basis of the following criteria: flat, well-drained terrain, near a source of water and sheltered from the wind. "Failing these conditions, the collier will make an artificial floor of branches and tree trunks covered with earth."[154] Once the site had been chosen, the pits were set.[155]

Setting and Leafing

During the French regime, pits were built to contain about 15 cords of wood. Later, 40–50-cord pits were built.[156] The wood used, 2–4 inches in diameter (French measure) came from trees about 20 years old. The wood delivered by woodcutters was cut into 1-m long lengths and assembled into cords (2.4 m × 1.2 m), so that the amount of wood charred in each clearing could be measured. Ideally, wood was cut on the diagonal to make it easier to stack the logs one on top of the other.

The first step was to prepare the hearth on which the pit would be built (Plate 2.11). The setter determined the circumference of the hearth, cleared and flattened the surface,

taking care to slant it down from the centre to let the carbonization liquids run off. In the centre of the circle, he placed the mast, which served as the lighting chimney. If the pit was to be set alight through the top, the setter erected a triangular cribwork that he filled with wood chips. If it was to be set alight from the bottom, a piece of wood exceeding the diameter of the base was placed on the ground against the chimney. Once the pit was set, this piece was removed and the space filled with dry wood for kindling. The worker thus built up vertically from the centre, using successive, closely stacked layers of billets slanted towards the chimney. Care was taken to use smaller billets for each layer, giving the pit its final, conical form. Then it was time for leafing.

Plate 2.11
Charcoal making in the 18th century.
DIDEROT AND D'ALEMBERT, ENCYCLOPÉDIE [...], RECUEIL DE PLANCHES [...]
(PARIS: BRIASSON, DAVID, LE BRETON, 1762),
"AGRICULTURE ET ÉCONOMIE RUSTIQUE, CHARBON DE BOIS," PLATE I.

Leafing involved covering the pit completely, to allow for braising. The feuiller, as the leafer was known at the Forges, preferred to use dead leaves to form an impervious layer, 10–13 cm thick. He might also use grass, straw, moss or peat. The leafing was topped off with a thick layer of charcoal dust (breeze). Once this operation was completed, only one opening remained, at either the top or the base, for lighting.

Charring

The collier lit the pit by introducing brands through the top or the base. He closed the opening with a clod of earth in 12 to 24 hours, when the smoke changed from white to brown, signalling the start of carbonization. From that point on, close supervision was required for 16 to 18 days. During this time, the collier lived in a hut in the clearing,[157] probably with other members of his family (Plate 2.12). A number of pits would be fired at the same time, giving the impression that some parts of the forest were on fire.

The collier's art was to control the combustion process by providing the fire smouldering inside the pit with sufficient oxygen to keep it going. This was accomplished by creating several vents in the pit, which he opened and closed as need be. Blue smoke escaping from the vents was a sign of good carbonization. It was important that the fire not be given too much oxygen, which would cause it to blaze up. The collier therefore had to ensure that the leaf covering remained impervious, and repair it as necessary. Since carbonization took place from the top, the collier made sure that the pit did not collapse too much. After about two weeks of charring, the collier plugged all the vents and allowed the collapsed pit to cool. To accelerate this process, he uncovered it from time to time and drew out small amounts of the cooled charcoal. He usually did this at night so that he could see and extinguish any sparks in the charcoal. The cooled and extinguished charcoal was then loaded into coal wagons. According to Bérubé, some charcoal was still alight, accidents were common and it was not unusual to see charcoal and wagon consumed in just a few moments.[158]

When a skilled charcoal burner was required to ensure that "the wood was not turned to breeze and all the charcoal it could produce was extracted," he was expected to see to it that all the wood used was carbonized, without it being reduced to breeze. The desired result was charcoal that was *clairsonnant*,[159] "brilliant and sonorous,"[160] in large pieces that could support the layers of ore with which it would be mixed in the blast furnace.

Plate 2.12
Collier's hut at the Hopewell Furnace, Pennsylvania.
PHOTOGRAPH FROM BOOKLET BY JACKSON KEMPER,
"AMERICAN CHARCOAL MAKING"
(HOPEWELL, PA.: HOPEWELL VILLAGE NATIONAL HISTORIC SITE).

Kiln Charcoal

The kiln charring process was introduced around the 1850s, at a time when the Forges land reserve was being restricted. Arson in the forest and the distance from the timber tracts led Timothy Lamb to recommend to John Porter & Company that wood be coaled on the post itself. Also, the high cost of carting the wood partially explains the fact that coaling in the forest continued. The first record testifying to the installation of kilns is a photograph showing the Forges shortly after they were acquired by the McDougalls in 1863 (Plate 2.13), with two kilns in the background. A few years later, in 1870, a journalist from *Le Constitutionnel* reported that the McDougalls had six kilns in operation at the Forges.[161] In 1874, Harrington reported the presence of these red-brick ovens and gave the dimensions of one of the largest: length, 15 m; width, 4.3 m; height at peak, 5.8 m; thickness of walls, 40 cm.[162] Later, Dollard Dubé mentioned the presence of six kilns,[163] clustered in groups of three around a mound where the kiln hut was located, and with a seventh kiln behind. Plate 2.14 shows Dubé's description, rendered into a plan and elevation by the architect Ernest Denoncourt in 1933.[164] The largest kilns described by Dubé had a capacity of 100 cords of wood. These were large, rectangular, vaulted brick kilns, with wooden tops supported on posts. Two heavy brick doors at either end, one at ground level and the other higher up, opened onto the mound, which enabled the oven to be filled to capacity. At the top was a chimney through which the kiln was lit. Inside, below the chimney, the wood was corded in a lattice pattern to facilitate lighting and combustion. It took six men an entire day to stoke the kiln.

Plate 2.13
Kiln coaling at the St Maurice Forges, circa 1870.
Stacks of wood ready for coaling can be seen in front of the two kilns.
RECONSTRUCTED BY ILLUSTRATOR BERNARD DUCHESNE,
AFTER JOHN HENDERSON'S PHOTOGRAPH OF THE ST MAURICE FORGES (DETAIL),
CIRCA 1870, NATIONAL ARCHIVES OF CANADA, PA-135-001.

To light the kiln, the collier, known at this time as the *gardien des kiles,* climbed onto the top and threw lighted chips into the chimney. Then, to make the kiln more airtight, he washed all the exterior walls with lime. It took one week for it to "come to post," but six times as much wood was charred as in a pit. As with pit charring, the collier had to control carbonization with vents, which were actually brick-sized openings 30 cm apart at various heights around the perimeter of the kiln. Floating bricks were removed and replaced to control the oxygen supply. Once carbonization was complete, the kiln stokers took an entire day to draw the charcoal. Kilns were fired, one every two days. The **stokers** emptied and loaded the kilns three times a week, which did not give them much respite, since they worked 10 hours a day.[165] The charcoal was then stored in large charcoal sheds or barns located at various places on the post. Construction and maintenance of the kilns called for a brickyard nearby, and a lime kiln was built north of the Grande Maison.[166]

Plate 2.14
Kilns at the St Maurice Forges, circa 1870.
RECONSTRUCTED BY ILLUSTRATOR BERNARD DUCHESNE.

Dubé also reported on the use of metal kilns, smaller in size and with a capacity of two or three cords of wood. These must have been portable kilns because, according to Dubé, "John McDougall had these kilns set up in the forest."[167] These very low capacity kilns were not used a great deal, and it is also known that, during the final years of operation, charcoal was still being made in pits in the forest.

TRANSPORTATION

Despite the current, the Trois Rivières River is easily navigable by canoe up to 100 or 150 toises below the Establishment.[168]

A fine road from Three Rivers crosses it [the Seigneury of St Maurice], leading mostly through the woods to the foundery.[169]

Communication and transportation were strategic factors in establishing the Forges in the heart of the St Maurice forest. An uninterrupted flow of traffic, both during the campaign and in winter, was kept up supplying the ironworks and delivering its wares. Raw materials and merchandise were transported by water and land, including over ice and snow, and also by rail in the later years. A number of forest roads were laid out, first to Trois-Rivières, and to the mines, timber tracts and charcoal pits. But the river was the favoured route from the outset.

The River

Francheville established his forge beside the only natural route available through his as yet uncleared land: the waterway that would later bear the same name as the seigneury, the St Maurice River, but was then called the Trois Rivières River. From just below the site, the current ran at about 3 **knots** to the river's mouth,[170] enabling travellers to reach Trois-Rivières in just a few hours. Embarking, however, was possible only 10 arpents (585 m) downstream because of rapids close to the site. A road was therefore built to the foot of the rapids. Upstream, the river was navigable for 9.7 km above the Forges, as far as the Grès falls, beyond the Gabelle quarry. In some places, the current reached a velocity of up to 5 knots.[171] This meant that the Forges could be supplied with limestone and building stone, brought downriver in **bateaux** or flat-bottomed boats, which landed "on the beach in front of the lower forge."[172] After 1760, charcoal from the seigneury of Cap de la Madeleine was unloaded at Pointe à la Hache,[173] slightly farther upstream, and hauled to the Forges on a cart road. The Forges wares were sent downstream to Trois-Rivières and from there to Quebec or Montreal on higher tonnage barques. The strong current of the river prevented boatmen from carrying heavy cargo to the Forges, although some accounts from the French regime indicate that food supplies were brought in from Trois-Rivières. The inventory of 1741 mentions "a large wooden canoe carrying 50 bushels of wheat."[174] Merchandise from Trois-Rivières was transported to the Forges mainly by land, which was why Francheville very soon set about building a cart road from the Forges to the town.

In his very first observations in 1735, Vézin was already making plans to transport limestone and building stone by "bateau" from the quarries he had prospected on the west bank of the St Maurice River.[175] It was later specified that early spring, starting in late March, was the best time to transport this heavy material, "as soon as high water admits of navigation from St Maurice to La Gabelle."[176] Following the 1741 bankruptcy, Vézin recommended:

> that the ironwares be carried on flat-bottomed double boats made of cedar, which will be navigable on the Trois Rivières River by three scowmen [...].[177]

He was thus recommending boats twice as large as those previously used, and strong enough to transport 5 to 7 t of iron, propelled by oars.[178] In Estèbe's accounts for 1741–42, a person called Champoux was paid 38 *livres* to supply two wooden scows "for use at the Forges," J.-B. Baron, 18 *livres* for 22 oars and six poles, and the Indian Polichiche, 12 *livres* for 24 oars.[179] We also learn that, in addition to transporting stone and iron, scowmen sometimes carried oats, hay, flour and victuals from Trois-Rivières (see Chapter 8). "Rowboats" were also mentioned in the inventories: one in 1746, three in 1748 and two in 1786.[180]

In the early 19th century, John Lambert[181] and Lieutenant Baddeley reported that sandstone, limestone and firestone were still being transported in "batteaux" from the Gabelle quarry and the surrounding area. These same boats were used to transport ironwares to Trois-Rivières. Baddeley observed that these 5-ton vessels made a return trip to Trois-Rivières in one day, returning empty, since supplies for the establishment were brought in on the cart road.[182] The employee roll of 1829 mentions four bateaumen employed during the shipping season. In 1860, Hamilton Rickaby stated that Forges wares were still being taken to Trois-Rivières by "bateaux."

A ferry service was also established between the two banks of the St Maurice. Early 19th-century census returns mention a "scowman" employed to operate the ferry. Francheville mined his ore on the other side of the river, in the seigneury of Cap de la Madeleine. When the ironworks was relaunched in 1735, there was talk of crossing the Trois Rivières River, which was frozen only from January to 15 March," it being "nearly impossible to take ore across in boats because of the strong current."[183] Once the Pointe du Lac mines were being worked, ore no longer had to be carried across the river. In 1760, the new British masters, planning to make charcoal in the seigneury of Cap de la Madeleine, estimated that there would be sufficient traffic between the two banks to set up a cable ferry. A ferry boat was built specifically to avoid the expense of bateaux and bateaumen. A cable 6 inches (15 cm) in diameter and 90 fathoms (165 m) in length was installed. The cable alone cost 606 *livres*, including its transportation from Quebec.[184] A road was then built from the Forges to the ferry, at Pointe à la Hache.[185] The road from the Forges to the landing stage where the bateaux were loaded at the foot of the rapids was also rebuilt. A statement of account in 1857 indicates that a scow was still in service between the two banks of the St Maurice River.[186] No doubt this was the same ferry that the McDougalls used at Pointe à la Hache to link the St Maurice Forges to the L'Islet Forges. They had a farm on the left bank of the St Maurice, where a couple ran the ferry service.[187]

In 1870, a journalist from *Le Constitutionnel* was one of 15 passengers on a cruise on a small steamboat up the St Maurice to the Forges. On the river, they met a boat coming from the Forges:

> Shortly afterwards, we met a sort of big scow, laden with iron. This iron was coming from the Forges and going to Montreal; it had been sold in advance to suppliers for the Grand Trunk. More freight for the Les Piles railway.
>
> At last we reached the foot of the Forges rapids; we felt the rocks which brushed roughly against the keel and felt it would be wise not to venture any further. Besides, we were only a dozen arpents from the Forges and it was not difficult to walk the rest of the way.
>
> [...]At half past twelve, we set sail again and were back at Trois-Rivières in less than an hour. (They had left Farmer's Wharf in Trois-Rivières at 8:30 a.m.)[188]

From Trois-Rivières, ironwares were sent to Montreal and Quebec on barques[189] and vessels such as *Le Manon*, Sieur Cugnet's schooner, during the French regime, and schooners and steamships at a later date.[190] The Forges masters chartered vessels, but also had their own craft. Monro & Bell purchased the schooner *L'Iroquois* in 1794, and the sloop *Abenakis* in 1795, and later bought shoreline lots and wharfs in Trois-Rivières and Quebec.[191] In 1868, the *St Maurice,* a sailing barge belonging to John McDougall & Sons, was registered in the port of Montreal. It was 2.3 m deep, 27.3 m long and 6.7 m wide, with a displacement of 114 tons.[192]

Roads

Roads. From Trois-Rivières to the Forges 2 ¹/₂ leagues
built by Sieur Francheville and widened
and causewayed in part by the last company,
from the Forges to the mine, to the rear
of Pointe du Lac 2 leagues, causewayed in part;
from the Forges to the foot of the rapids,
one quarter of a league, causewayed in part.[193]

The network of roads built and maintained for the Forges was three-pronged: first, connecting the Forges to Trois-Rivières; second, running parallel to the St Maurice River, of which it was an extension; and third, a multidirectional web through the forest, linking the Forges with the timber tracts, charcoal pits and mines (Table 2.18).

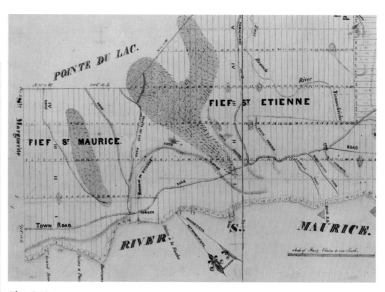

Plate 2.15
Figurative plan of the Fiefs St Maurice and St Étienne, by Joseph-Pierre Bureau, 1845. The St Maurice Creek is marked "Ruisseau St Étienne" [St Étienne Creek].
ARCHIVES NATIONALES DU QUÉBEC, E21, MINISTÈRE DES TERRES ET FORÊTS/ARPENTAGE/CANTONS NO. S.36D, 1845.

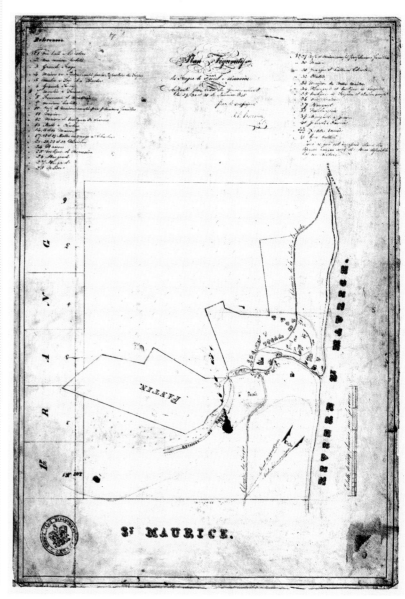

Plate 2.16
Figurative plan of the St Maurice Forges, by Joseph-Pierre Bureau, January 1845. The St Maurice Creek is marked "Ruisseau du Lavoir" [Washery Creek]. Every building shown is identified.
ARCHIVES NATIONALES DU QUÉBEC, E21, MINISTÈRE DES TERRES ET FORÊTS/ARPENTAGE/CANTONS NO. S.36B, JANUARY 1845.

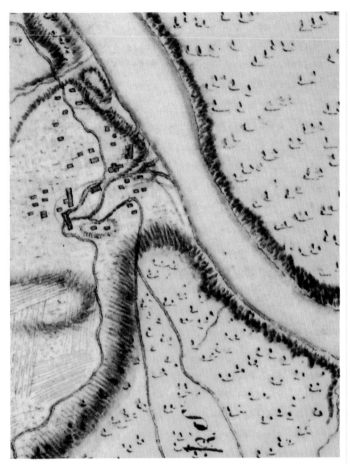

Plate 2.17
Detail of Murray's map of the St Maurice Forges, circa 1760.
NATIONAL ARCHIVES OF CANADA, MAP DIVISION, C-85809.

The Forges Road

The road to Trois-Rivières was opened by Francheville when he was setting up his establishment in 1732–33. It was a cart road 2 1/2 leagues (12 km) long, which led into the centre of Trois-Rivières (Plate 2.3 shows the route). In 1734, with the completion of the King's Highway between Quebec and Montreal,[194] the Forges were linked to the rest of the colony by a land route. Francheville's 10-foot wide road was cleared of stumps and widened to 12 feet when construction of the new Forges began in 1736. Five years later, in his inventory of 1741, Estèbe, quoted in the epigraph above, stated that it was causewayed in part with wooden corduroying wherever it ran through swamps, to keep the carts from getting bogged down. According to Vézin, maintaining a causeway or corduroy road was extraordinarily expensive, since it had to be redone almost every year.[195] The accounts for 1741–42 indicate that more than 450 man hours were paid out for opening and maintaining roads and bridges.[196]

At that time, it took from one to two hours to reach Trois-Rivières by cart. Visiting the Forges in 1752, the engineer Louis Franquet, who described the road as "good, broad and sandy," made the return trip in "five quarters of an hour."[197] Bouchette's maps of 1815 (Plate 2.1b) and 1831 show that the Forges Road was extended northwards inland, with new names *(côte rouge, côte du grand pont, côte croche, côte Turcotte, côte de 14 arpents, côte Jean)*. Bureau's map, drawn in 1845, shows that the road by that time went as far as Les Grès and beyond, all the way to Shawinigan (Plate 2.15).

The original line of the road between the Forges and Trois-Rivières was changed. The last workers interviewed by Dollard Dubé distinguished the Forges Road *(chemin des Forges)* from the *chemin des Français*, which was longer and more winding than the road used between 1860 and 1870, a small stretch of which wended along the St Maurice River. They reported that the new, straighter route was about 3 miles (4.8 km) shorter, by which Trois-Rivières could be reached in under an hour.[198]

Riverside Roads

In 1736, Vézin built a road leading to the foot of the rapids, 10 arpents below the establishment (Plate 2.4), where the Forges wares were loaded on to bateaux to be sent to Trois-Rivières and then transshipped for Quebec and Montreal. Estèbe's accounts for 1741–42 report on maintenance of this road. In 1760–61, what appears to be a complete overhaul of the *chemin du rapide* was carried out, while another road was built leading to the ferry landing stage above the Forges[200] (Plate 2.17). This road appears on Murray's map, drawn up between 1760 and 1762.[201] It originates on the Forges plateau and does not follow the bank of the St Maurice, but leads up to Pointe à la Hache. These two roads were used and maintained throughout the lifetime of the Forges.

On the east bank of the St Maurice, Bouchette's map of 1815 shows a colonization road following the line of the river up to the Gabelle Falls (Plate 2.1b). Part of this road past the Forges had been built when Mathew Bell was granted a strip of land along the river.[202] During the 1840s, a map drawn by the surveyor Legendre shows a road perpendicular to the St Maurice, linking the east bank, across from the Forges, to Montagne du Cap. This road led to the mines and charcoal pits used by Bell on that side of the river. At a later date, the McDougalls exploited the east bank extensively, particularly land in the parish of Mont Carmel, where they also ran the L'Islet Forges. In 1872, they began operating a tramway that ran on wooden rails, linking the L'Islet Forges with the St Maurice Forges. The wooden railway ended at the ferry landing stage, 3 miles (4.8 km) below the L'Islet Forges. The tramway that carried pig iron from L'Islet to the St Maurice Forges continued in operation until the L'Islet Forges closed in 1878.[203]

Table 2.18

ROADS AT THE FORGES[199]

In 1741	In 1870
Chemin de la ville	Chemin des Français
Chemin de la mine or de la minière	Chemin des Forges (Forges Road)
Chemin de la pinière	Chemin du roi (King's Highway towards St Étienne)
Chemin du rapide	Grande côte (towards the Forges)
Chemin de la vente des Aubry	Chemin du grand pont
De la vente à Berry	Cordon du curé
Chemin des ventes de la ville	Chemin de la petite savane
Chemin de la vente de l'étang	Chemin de la vente au diable or coteau de la mine
Pont du chemin de la mine	Chemins des bennes
Chemin de l'étang	Chemin des draveurs (bank of the St Maurice)
	Chemin de la Pointe à la Hache
	Roads on the Forges Post
	Chemin de l'empellement
	Chemin des kiles
	Chemin de la coulée
	Chemin du gros marteau
	Chemin de la grande maison
	Rangée du meunier
	Rangée de la cloche

Forest Roads

The oldest forest road was opened by Francheville in the seigneury of Cap de la Madeleine in the early 1730s, connecting the east bank of the St Maurice with the first ore mine worked. This jolting path was described as *fort rude et montagneux* in 1735.[204] The mines of Cap de la Madeleine continued to be worked along with those of Pointe du Lac during the French regime. The *chemin de la minière* and other roads leading to the Cap de la Madeleine charcoal pits were opened later, beginning in 1760, and especially during the Pélissier and Bell years, to make charcoal on a parcel of land leased from the Jesuits.[205] In the last 20 years, under the McDougalls, the many purchases of land for ore and wood in the parishes of St Maurice and Mont Carmel probably led to forest roads being built on that side, even though settlement of this area was already under way.

A number of forest roads were built on the west bank of the St Maurice, branching out from the Forges post or the road to Trois-Rivières. In 1736, a third road, 20 feet wide, was opened by Vézin between the Forges and the Pointe du Lac mines over a distance of 2 leagues (10 km). The work carried out by road builders in 1741–42 shows that this road was also a corduroy road. Table 2.18 shows that other roads, mainly leading to the charcoal pits, were also maintained. As the Forges territory expanded, mainly during Mathew Bell's time, new forest roads were built and later used for settlement. Bureau's map (Plate 2.15), drawn at the end of Bell's tenure, gives a very good idea of the system of forest roads that crossed the Forges land. The oldest roads (Pointe à la Hache, the Pointe du Lac mines) and those built later can be clearly distinguished. The list of roads identified by Dollard Dubé (Table 2.18) still shows the names of the old roads, and we can see that over the years the names had become more specific.

These same roads, especially those leading to the mines, were used during the winter. It was realized quite early on that it was easier to haul heavy loads of ore by sleigh over snowy or frozen roads. The ore was collected in swamps, where many corduroy sections were required to make them usable by carts. Francheville was probably the first to realize the benefits of the easy transportation afforded by Canadian winters. By 1734, Governor Beauharnois and Intendant Hocquart were aware of its advantages. Vézin even suggested that charcoal be carried by sleigh to avoid the high expense of maintaining corduroy roads during the summer. In order for the business to turn a profit, however, a strict schedule had to be adhered to for the transport of raw materials (Table 2.19). Vezin was the first to depart from the schedule, since he often had to transport ore in the summer. In the 1860s, Hamilton Rickaby made the point that ore had to be transported beginning in early December, when there was enough snow in the woods to make good roads.[206] Ore was always transported in the winter and, as we will see later on, sleighs were adapted and refined for this purpose.

Table 2.19

ANNUAL SUPPLY MAINTENANCE SCHEDULE, 1852–54[207]

Operation	Season
Woodcutting (hardwood)[208]	from Michaelmas (29 September) to the start of winter
Woodcutting (softwood)	from the start of winter to March
Charcoal making	summer
Charcoal hauling	fall (until late October)
Mining	summer (until the end of August)
Ore hauling	winter (from early December if there was enough snow)
Provisions (victuals and merchandise)	November (in the interval between charcoal and ore hauling)[209]
Ironware shipping	summer (by bateaux to Trois-Rivières)

Road Maintenance

Although it was done every year, road maintenance was rarely documented. A contract signed in 1806 by Monro & Bell with Jean Sauer, a worker at the Forges, does provide detailed information on the construction of forest roads. In particular, it refers to how corduroy roads were laid at that time. This was a two-month contract (16 August to 15 October) for £70 currency. The road, located in the St Maurice fief, was to start at the site of an old charcoal clearing northeast of the Grande Rivière Yamachiche, run to the river and continue on the other side for a distance of 2 arpents, as far as the hardwood forest of "Bois Franc." The line of the road had already been blazed. The detailed description of the work illustrates the traditional way of making roads:

> Which road is to be made as follows. Clear the land and remove the stones where the road is to pass through the aforesaid gully northeast of the said river, at least ten feet across, and causeway all along the said ravine with pieces of cedar and hemlock, the smallest to be six inches at the small end, squared on one face. Lay sleepers of the same wood thereunder, so that the water passing through the said gully can run its course without damaging the said road & causeway. Finish the causewaying with borders pegged every five feet to the sleepers, and clear the road of trees from the said gully to the river and causeway it with logs as necessary and sand it if need be; on the southwest side of the river, the road is to be cleared by the said Undertaker; all to be done for commodious passage with wagons and all the necessary causewaying to be made of logs, all to be ten feet wide. [210]

The total length of the road was not mentioned, but we can deduce that the work was carried out by a team of men directed by Sauer. At that time, £70 was enough to pay at least a dozen labourers each £2 to £3 a month for two months.

With the settlement of the Forges lands, after the sale of the establishment in 1846, the new owners faced some problems with the forest roads built for the needs of the business. Some new settlers claimed the right of way, and the company had to reach agreements with them. The roads laid out through the St Maurice forest were subsequently for the needs of settlers and timber merchants, who used the Forges roads before opening up their own roads.

Means of Road Transportation

Vehicles

Vehicles were adapted to suit the seasonal nature of transportation. According to Hamilton Rickaby, the transportation schedule was drawn up around the restrictions related to the successive employment of the same carters. In addition, we know that this schedule was not always followed to the letter, even in Rickaby's time (1850–60), when charcoal was often transported in winter, and ore in summer. The company kept a variety of vehicles suitable for the season: sleighs for winter, and an assortment of carts and wagons for summer and fall (Table 2.20).

Table 2.20

**VEHICLES INVENTORIED
AT VARIOUS TIMES, 1741–1883**[211]

Date	Winter transportation	No.	Summer transportation	No.
1741	sleighs on iron runners with harnesses	5	ore carts with iron wheels unmounted ore carts	3
	sleighs without iron runners	6	without iron wheels	3
	winter charcoal wagons	3	small carts	3
			charcoal wagons with iron wheels	3
			cutter	1
			calèche	1
1748	sleighs with harnesses winter and summer	20	ore carts with iron wheels	5
			carts with iron wheels	3
	charcoal wagons	14		
1760	high sleighs	4	wheeled carts	3
	low sleighs	4	big carts	2
1767	some sleighs			
1786	sleighs	2	mounted charcoal wagons with iron wheels	7
			ore carts	2
			carts*	4
1863			charcoal wagons with or without wheels	27
1883	double sleigh	4	double wagons	5
	sleigh with runners	2	charcoal wagon with wheels	1
	sleigh	3	charcoal wagon boxes	10
	sleighs	9		
	charcoal wagons on runners	2		

* Including a water cart, a pig iron cart and a hay wain

Since transportation requirements were considerable, an attempt was made right from the outset to attract and keep versatile carters by fully equipping them for summer and winter work. To encourage them to stay, it was even suggested, in 1739, that they, like other workers, be given a grant of one square arpent on which they were required to build a house within a year. Each carter wishing to settle at the Forges was to be set up with "two horses, two harnesses, a cart, an ore cart, a charcoal wagon and two sleighs."[212] In 1740, the idea of land grants was finally given up, but four or five carters were duly set up with the proposed material assistance. This was not a totally successful experiment, since some carters decamped with the horses and harnesses provided by the company. According to subsequent inventories, it would appear that the company later preferred to retain ownership of the vehicles. These inventories describe the various types of vehicle used, but this accounted for only some of the vehicles required for all transportation needs. Hauling the ore during the winter months required extra vehicles, belonging to "outsiders" who were provided with stabling and fodder during the period of their employment. These outsiders were local *habitants* whose services were engaged throughout all the years the Forges operated. Seasonal employment of these carters entailed strict scheduling of transportation. Since most of them were farmers, they had to be used during the off-season, outside ploughing and harvest times.

Sleighs were used to haul ore in winter. Of rudimentary construction, the sleigh was a frame mounted on two runners on which was set a box to hold the ore or charcoal.[213] The Forges sleigh would appear to have been refined to tolerate bumpy roads. Its design captured the interest of Montreal merchants in the late 18th century. In 1786, the Committee of Montreal Merchants contacted the manager of the Forges to obtain a model of this sleigh. A certain Mr Proust, Captain of the Trois-Rivières militia, provided them with details on its construction.[214]

It was quite a sturdy sleigh, consisting of a frame, set on runners 2–2.5 m long, which was not directly attached to the box, giving it some suspension. The description details the type of runner used and the way the shafts were attached to the runners. The runners were 23 cm deep and 10 cm wide in front, tapering to 8 cm at the back. They were attached in such a way as to be clear of the bottom of the box. The shafts were bolted to the runners with an iron bolt as used for trucks.[215]

Plate 2.18
Carts in the cartwright's yard at
the St Maurice Forges, 1845.
CAPTAIN PIGOTT, *THE FORGES
NEAR THREE RIVERS, 1845* (DETAIL),
ARCHIVES DU SÉMINAIRE
DES TROIS-RIVIÈRES,
DRAWER 258, NO. 48.

The fact that this type of sleigh was designed to handle bumps suggests that it may have been used to haul charcoal. Uneven roads were, in fact, bad for charcoal, which had to be kept in large pieces to support the burden of ore in the blast furnace.[216] The presence of "winter charcoal wagons" and "charcoal wagons with runners" in the first and last inventories of the Forges (Table 2.20) indicates that some charcoal was transported in winter. In 1832, Bouchette indicated that charcoal was still being carried by sleigh in the winter.[217] In the final years of operation, George McDougall had sleighs remade into double sleighs in order to compensate for recent increases in labour costs, and four examples of such sleighs were found when the Forges closed in 1883.[218]

It seems that, while it was preferable to carry ore in winter and charcoal in the fall, in actual fact both were transported in either season. Having to bring in one or the other out of season was regarded as a deviation from the schedule, but the fact that inventories show vehicles adapted to all seasons indicates that such deviations were expected. This was the case for ore carted in summer, preferably in the dry season,[219] using two-wheeled dump carts, which were emptied by tipping them on their axles towards the rear. The ore carts were also used to remove the slag and cinder produced in the blast furnace and the forges (Plate 2.18). In later years, as he had done for the sleighs, George McDougall made them over into double wagons. These were four-wheeled vehicles drawn by two horses.[220] Dollard Dubé, writing about the *chemin des brancards*, described the boxes used by local farmers to take ore to the Forges. These could be the removable boxes depicted in Bunnett's painting of 1886 (Plate 2.19), and which farmers set on carts to carry the ore.[221]

A wagonload or **binne** of charcoal held 8 $^1/_2$ *pipes*, weighing about 366 kg.[222] In winter, the box was mounted on a sleigh, and in summer on wheels. The summer charcoal cart, mounted on two wheels, resembled the ore cart and was drawn by two horses. It was emptied through a trap in the bottom. It also seems to have been mounted on four wheels, as we see in Captain Pigott's watercolour of 1845 (Plate 2.20) and as was also reported by the last workers interviewed by Dollard Dubé. Dubé spoke of "heavy wagons with four ordinary wheels supporting a box with sloping sides that held 75 to 125 bushels (2.8–4.6 m³)."[223]

Inventories also mention cutters and simple carts, large and small, used at the Forges. During the French regime, besides the vehicles used for raw materials, there were four horses, "one [...] at the furnace, one at the forges and two at the House."[224] The horses were harnessed to the different types of cart, as needed. A dump cart was used to carry slag and sand and probably also the limestone brought by bateau to the bank of the St Maurice. We also know that some carts were single-purpose, such as the hay wain, water cart and pig iron cart mentioned in the 1786 inventory (Table 2.20). Provisions, merchandise and building materials were also carried in carts, most of them belonging to independent carters employed in great numbers every year.

Horses

The vehicles used for ore and charcoal were pulled by teams of two horses. In the early years of operation, the provision was for "twenty horses for ten teams used to pull charcoal, stone and other carts, a horse at the furnace, one at the forges and two at the House, making twenty-four horses."[225] Table 2.21 shows the number of horses inventoried at various times.

During the French regime, the company kept 20 to 30 horses. In 1784, there were 22, and in 1804, Lord Selkirk mentioned 40 horses for 20 carters, specifying that there had previously been 28 (and thus 14 carters). Until the beginning of the 19th century, thus, about the same number of horses were kept as during the French regime. Selkirk reported a recent increase in the number of horses and the number of carters to drive them. This increase was directly linked to the employment of new permanent carters at the turn of the 19th century (see Chapter 8). According to the 1831 census return, the number of horses belonging to the company had risen to 55. From that census on, horses belonging to the workers themselves were also counted. The decline in numbers seen in 1851 and 1861 is linked to the temporary shutdown of the company at the time the census was taken. In 1871, numbers were back to pre-1800 levels, doubtless as a result of the larger share of transportation turned over to local inhabitants during the McDougall era, when raw materials were mainly collected from farther away, on the east bank of the St Maurice.

In 1742, fodder for each of the company's 24 horses consisted of "sixty bales of hay and ten bushels of oats per month." Combined with the fodder for the "outside" horses, annual consumption was estimated at 25,000 bales of hay (a bale was 15 lbs. or 7.3 kg) and 3,000 bushels of oats. During the 1850s, an average annual consumption of 40,000 bales of hay and 15,000 bushels of oats was reported. These figures imply that the number of company horses was around the same as in 1831, that is, 55.[227] Most of the fodder was bought in by the company, the Forges farm providing only a small proportion of the hay and oats, but all the straw.[228] The 1871 census figures indicate that the Forges farm produced only 1,000 bushels of oats and 500 bales of hay. The L'Islet Forges farm, also owned

Table 2.21

HORSES INVENTORIED AT VARIOUS TIMES, 1741–1871 [226]

	1741	1746	1760	1764	1780	1804	1831	1851	1861	1871
Company horses	24	30	6	6–7	22	40	55	14	1	33
Workers' horses	-	-	-	-	-	-	22	13	19	16

by the McDougalls at that time, produced 1,500 bushels of oats and 13,000 bales of hay.[229] The company was thus not self-sufficient from this standpoint, and a close eye was kept on purchases of fodder by the Forges proprietors, given that prices could vary considerably in the autumn.[230]

Plate 2.19
The Forges in ruins, painted by Henry Richard S. Bunnett, 1886.
In the foreground, the derelict blast furnace, and on the left,
in the background, a large tenement house.
HENRY RICHARD S. BUNNETT, *THE FORGES OF THE ST. MAURICE*, OIL ON CANVAS,
1886, MCCORD MUSEUM OF CANADIAN HISTORY, M-739.

Care of Horses and Vehicles

Haulage vehicles were made and maintained by one or more wheelwrights. The 1741 inventory describes a wheelwright's shop located in part of the building that housed the carpenter Bellisle. Accounts for the same year indicate that various workers were paid by the day or on piecework to repair cart wheels, make or repair cutters, or make sleighs.[231] In 1775, Laterrière reported employing four wheelwrights. Other documents and the 1851 census mention the presence of a master wheelwright. Wheelwrights appear to have turned their hands to various trades, however, since the same individuals are often censused as carpenters. Wheelwright's tools were also listed in a number of inventories, and a wheelwright's house was shown on the plan of the Forges drawn by the surveyor J.-P. Bureau in 1845 (Plate 2.16). On Captain Pigott's watercolour (Plate 2.18), painted that same year, two charcoal wagons and an ore cart can be seen close to this house. We also find a saddler, responsible for making and repairing harnesses. There was a farrier to shoe and look after the horses. Tools and utensils are listed in a number of inventories.[232] The inventory of 1741 lists "farrier's tools" in Marineau's shop, with Marineau being shown as an edge-tool maker. The duties of the farrier, like those of the wheelwright, appear not to have been exclusive, since, particularly during the French regime, workers paid for these duties were also designated as locksmiths, edge-tool makers or simply blacksmiths. This seemingly contradictory information is merely an indication of the versatility of these craftsmen in iron.[233] In the inventory of 1760, farrier's tools are still mentioned in the edge-tool shop.[234] In 1863, mention of a shop for shoeing horses would seem to indicate that this had become a specialized task.[235]

Plate 3.1

**MECHANISMS OF THE GREAT WHEEL
OF THE BLAST FURNACE, 1741**

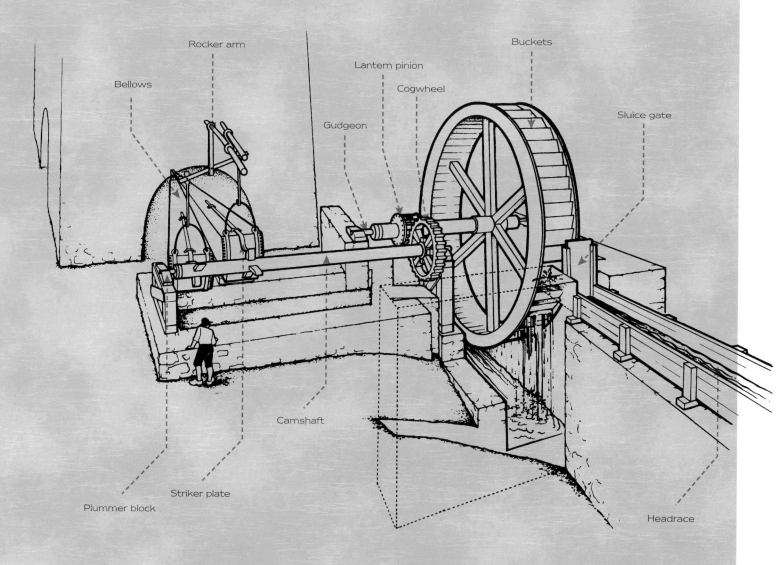

Rocker arm

Bellows

Buckets

Lantern pinion

Cogwheel

Gudgeon

Sluice gate

Camshaft

Striker plate

Plummer block

Headrace

RECONSTRUCTED BY ILLUSTRATOR YOLANDE LAROCHELLE.

III

Harnessing the Stream

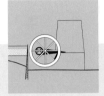

In order to describe the Forges, we need to begin with a discussion of how the component parts of the plant—blast furnace and forges—were set up along the St Maurice Creek. By looking at how it was harnessed, we will get a better idea of the site's distinctive characteristics. Each component of the plant was located to conserve and reutilize water power. The success of the ironworks' builders bears witness to their ability to use the topography of the little creek to maximize its power and drive several waterwheels. Studying the Forges gives us an opportunity to learn not only about the "art of **ironfounding**," the precursor of the iron and steel industry, but also the "art of water control," the precursor of the hydroelectric industry.

An analysis of the data available on the hydraulic and hydromechanical works will give us a better understanding of the technical knowledge that the ironmasters drew on to design the Forges.

THE ART OF
CONTROLLING WATER

*The foundery of St. Maurice is situated in this fief,
in a beautiful valley, at the confluence of a small
stream with the St. Maurice, about eight miles
above the town of Three Rivers; the high banks
of the river, embellished with every variety
of fine trees in groups on each side, the dark hue
of the large pineries and immense surrounding
forests, and the more distant and softened shades
of the lofty mountains that bound the view,
form together a bold and magnificent prospect,
when viewed from the place where the road ascends
the brow of the ridge that overlooks the valley.[3]*

The picture of a foundry, nestled so naturally along the little creek on which it is situated that it gives the impression of harmony and rural calm, contrasts sharply with the industrial reality of the 20th century. A look back today at how uninvasive industry once was is likely to arouse a comforting nostalgia in people and reinforce the conviction that industry once had a human side, in contrast to its modern complexity and environmental impacts. The difference in complexity between the two industrial eras is so great that we look on yesteryear's "engineering" with more benevolence than real curiosity. However, considering the means they had at their disposal, there is no denying that the ironmasters and their workers developed technologies whose daring and level of risk are worthy of our admiration. A close analysis of how the plant developed shows us how much the Forges' existence and long life were the direct result of how the water power problem was solved. The layout of the plant reflected the search for a lasting supply of water power, as an inadequate stream of water would have called into question Sieur Francheville's choice of site in his own seigneury, in 1729.

By looking in turn at the various facets of the water control works, we will see not only how the builders of the Forges capitalized on the particular features of the little creek's favourable location but also how they dealt with its constraints, all with a single aim in mind: to control a series of waterwheels driving the bellows and **forge hammers**.

We will first consider what was needed to operate the waterwheels, as their size and number affected the ironmaster's assessment of the flow and drop of the creek. We will then look at how the its potential was assessed and what characteristics governed the layout of the water control works. We will then turn to the works themselves and how they regulated the creek to dam and channel the water towards specific locations. That decided where the furnace and forges with their waterwheels would be located. We will look at how the water was channelled to the wheels in each case, according to the type of machine driven and how it was used. Along the way, we will come to see that the restrictions on the works were also a result of conflicting opinions by the various players involved—specialists, administrators and officials. Once we have described the water works, we will see how, by dint of repeated repairs and modifications, they were able to withstand the test of time and serve as a solid, lasting base for the company's growth.

HOW WATERWHEELS WORK

Types of Wheel

Waterwheels are hydraulic engines. Most mill and forge wheels were made of wood and mounted vertically. They were rotated either by the force of the water on blades or by the weight of the water in **buckets** around the wheel. Blade or paddle wheels were **undershot**, that is, driven by the action of a stream of water channelled towards its base by a **tailrace**. Bucket wheels were **overshot** or **breastshot**, in the first instance receiving the water at the top, in the second between the top and the base. Most of the wheels used at the St Maurice Forges were overshot or breastshot. It was the weight of the water in the buckets that turned the wheels. To turn an overshot wheel, either the top of the wheel must be located at the level of the stream—for example, at the crest of a waterfall—or a dam has to be built so as to raise the stream to the level of the top of the wheel. In the first instance, a natural waterfall is used. In the second, an artificial waterfall is created, as was done at the Forges to produce the required head of water.

The Head

In early hydraulics, the head was defined as the vertical distance between the upper level of the water, which is channelled towards the wheel, and the lower level (usually the stream bed) where the water is ejected by the wheel. In practice, it is the height of the head that determines the diameter of the wheel; the larger the desired wheel diameter, the higher the head must be. That is exactly what iron-master Vézin wanted for his wheels at the Forges. In the case of the forge's milldam, he wanted to obtain "chafery wheels as large as such a head would permit."[4]

Wheel Diameter and the Principle of Leverage

A wheel's power is a direct function of its diameter. That was the principle put forth—unsuccessfully—by Vézin to be allowed to build a 20-foot (6.5 m) high milldam at the single forge, to be sited at the bottom of the slope; with such a dam, he would have been able to install a large wheel. A wheel's power is explainable in terms of the principle of leverage: water-filled buckets form an arc of water around the rim of the wheel, and the weight of the arc creates an imbalance that causes the wheel to rotate. The farther the arc is from the wheel's axis, that is, the larger the wheel's radius, the greater the leverage created. Vézin, in one of his rare technical reports, explained to Intendant Hocquart that "the larger the diameter of the wheels of these kinds of forge movements, that is, overshot or undershot bucket wheels,[5] the less water it takes to make them turn, according to the rule of lever length."[6] Afraid that he might not have enough water for the forge, Vézin wanted to install large-diameter wheels to try and use less of it.

Gearing Mechanisms

Another factor to be considered is the speed of the waterwheel's rotation. In the 18th century, Borda established that a waterwheel produces maximum power when its velocity is half that of the watercourse.[7] In other words, beyond that velocity, the wheel does not gain any more power. Once the desired velocity is achieved, gearing makes it possible to control the speed of the machines to be driven. In the case of a blast furnace or forge blower, the wheel shaft drove a shaft fitted with **cams**, which acted directly on the bellows. Forge wheels could be **single geared** (direct drive) or **double geared** (spur geared). A single-geared wheel had one shaft, with cams at one end that acted directly on the machine—to raise the head of a **hammer**, for example. A double spur-geared wheel had two shafts: the wheel shaft with a **lantern pinion** at one end and a parallel camshaft. The camshaft was rotated by means of a **cogwheel** at the other end whose teeth meshed with the rundles of the lantern pinion. Vézin adopted this spur-geared system to drive the bellows of the furnace and the two forges (Plate 3.1). Depending on the diameter of each component, the lantern pinion-cogwheel arrangement made it possible to increase the wheel's power and control the rotation speed of the camshaft.

The two bellows, lowered alternately by a series of cams, were raised in cadence by a **rocker arm** (a mechanism based on the principle of the lever arm) suspended above the bellows. In 1737, the spur-geared rocker arm system was temporarily replaced by an *estrique* mechanism, which was probably a traction and compression system located below the bellows.

The potential of the St Maurice Creek was assessed with a view to installing such hydraulic mechanisms. Francheville left no trace of such an assessment; he probably left it up to a millwright to decide on the location of his bloomery, the size of its two wheels and the types of mechanism to be used. Vézin's analysis gives some clues to the characteristics of the first forge's water regulation system.

THE SITE OF FRANCHEVILLE'S BLOOMERY

Ironmaster Vézin was the first to quantify the flow of the creek, then driving the two wheels of Francheville's small forge built in 1733. He measured a flow of 240 **miner's inches** of water at the opening of the **flume** that conveyed the water to the bloomery's two wheels; he made no mention of a dam or millpond in connection with Francheville's system. Previously, it had been estimated that the creek could turn at least three mill wheels.[8]

Francheville built his forge close to where the creek emptied into the river. This location was certainly chosen deliberately, especially considering that Vézin endorsed it by building an even bigger forge there. Why then did he choose to build it at the bottom of a slope that had several other possibilities? The first reason that comes to mind is the proximity of the river, on which it was easy to transport ironwares to Trois-Rivières and bring in ore, which, at the time, came from the Cap de la Madeleine seigneury on the opposite bank. However, a site so close to the river had many major disadvantages as it would flood when the water levels rose in the spring.[9] Francheville maintained that he would have to spend a lot of money to build dams, most likely to contain the spring floods. Francheville, and Vézin after him, gave the same reasons for establishing the forge at the very bottom of the slope. Both of them toyed with the idea of building bigger forges that would use a larger part of the creek's slope.[10] For his part, Francheville held back on this for two or three years because he first wanted to check the quality and yield of the iron ore, having only limited capital at his disposal, while Vézin proposed to proceed without delay. In order to establish a complete ironworks, consisting of a blast furnace and the forge itself, a site that could accommodate the blast furnace had to be chosen. And, as the blast furnace produced the pig iron that would then be turned into wrought iron in the forge, it was preferable to site it higher than the forge on the slope, at an appropriate head. It would have been quite illogical to create a layout where iron pigs weighing almost a ton each would have had to be transported to the top of the slope!

By placing his forge at the bottom, Francheville left most of the slope free, thus giving him complete latitude to choose the site that offered the best head to drive the blast furnace's great wheel. The top-to-bottom traffic pattern simplified production operations, with the finished product ending up down at the riverbank, whence it was shipped to Trois-Rivières.

The creek's unusual drainage pattern meant that the ironworks would always be organized on a top-to-bottom pattern down the slope from the blast furnace to the upper and lower forges.

THE CREEK'S DRAINAGE PATTERN

The establishment [...] is [...] at the foot of a hillside next to which is a gully or ravine though which a stream from a spring flows [...].[11]

For an ironmaster or millwright, the most attractive feature of the St Maurice Creek was the steep drop over its last 600 m down to the river. Rising "along a very steep hillside to the south southeast side,"[12] the top of which was more than 20 m above the creek bed, the 4-km long creek flowed along the flat until it reached the Forges. From there, at 40 m above sea level, it plunged through 32 m down to the river.[13] This drop created a gully, which cut through a plateau overhanging the St Maurice River. The advantage of the gully lay in its width and graduated fall, which provided the conditions and space to build all the elements of a complete ironworks connected by a descending path. One hundred and thirty years after the Forges were founded, Inspector Symmes spoke of the remarkably accessible slope, which allowed the stream of water to be reused eight to 10 times to drive eight to 10 wheels successively:

[...] the water power, (altho. the volume of water is not great) is excellent. The whole length of the fall is remarkably accessible and the water easily controlled, it may be used 8 or 10 times over if required [...].[14]

A stream (even one with a small volume, according to Symmes) creates considerable kinetic energy when it suddenly rushes down a slope of more than 30 m. The art of controlling water consists in penning it up along its course in dams; this creates millponds fed by a constant stream of water that become veritable reservoirs of potential energy, large reserves of water ready to spill onto the wheels and make them turn. It was possible to dam the flow of water at different points on the slope, particularly where a fall created a head suitable for building a dam. The key in creating these reservoirs was to avoid being dependent only on the stream's natural flow, thereby ensuring that there were greater quantities of water at several places at a time. Once dams were erected across the creek bed, the stream of water could be used to fill the ponds so its natural energy could be channelled, controlled and harnessed.

The water level in these millponds was maintained thanks to a spillway that drained off the overflow. This occurred during the spring floods and when the water in the millponds was not being used to drive the wheels, for example, when the plant was idle.

We will see that the principle of controlling water through ponds was taken to its extreme in the system that channelled water to the wheels of the upper forge.

Francheville, who had plans for a complete ironworks, doubtless knew the potential of the St Maurice Creek. Such an inexhaustible stream, located right in the middle of timber and ore-rich country, could hardly have left the seigneur indifferent. Before Vézin, the expert, became involved in 1735, it had already been established that this "stream [was] able to turn at least three mill wheels at all times."[15] But after the failure of Francheville's bloomery in 1734—a failure unrelated to water—doubts were raised about all the advantages of its location. Before investing new capital, the government needed a new assessment.

THE CREEK'S ENERGY POTENTIAL

Fortunately there was a stream on the site that was sufficiently abundant, even during long droughts, to provide enough water for a foundry and finery.[16]

Vézin, the ironmaster from Champagne, was dispatched to the colony at great expense to make an accurate assessment. In addition to judging the location and efficiency of Francheville's bloomery, he was commissioned to compare the site with two others on the Batiscan River. Having spent five weeks surveying the territory, Vézin endorsed the choice of the St Maurice Creek, basing his observations on the detailed assessment of the creek that he provided. He established its source, measured its flow and specified the drop in its elevation.

The site survey was taken seriously and Vézin did not carry it out alone. He was accompanied by three mill experts, who would later become his leading detractors: Grand Voyer Lanouiller de Boisclerc, the colony's chief road commissioner, Leclerc, a Jesuit who was an expert on water mills, and Jean Costé, the millwright who had built the water works for Francheville's bloomery.[17] Such support for an expert sent at great expense seems excessive, but an expert evaluation was highly strategic in a colony pioneering its first industry. Furthermore, it was customary in France for an ironmaster to be accompanied by a millwright when the location of an ironworks was being chosen.[18]

Vézin measured a flow of 240 inches of water or miner's inches "at the mouth of a flume that conveys the water to the wheels of the said establishment," that is, at the bottom of the gully and upstream from Francheville's forge; where the creek emptied into the river, the flow reached 280 inches.[19] Later on, we will see how this measurement was arrived at, which was more akin to a sluice aperture than to a measurement of the stream's flow. Vézin then proceeded to distribute the 240 inches available to each of the six waterwheels needed to create a complete ironworks (Table 3.1). According to him, the six wheels would use only 190 of the 240 inches available. Thus, in his assessment, the ironmaster doubled the creek's capacity that had been estimated one year earlier.[20] His plan also allowed for a blast furnace (one additional wheel) and the future addition of a slitting mill (two wheels) and a grist mill (one wheel). His complete ironworks would have 10 waterwheels, at least six of which would be running at the same time![21]

Table 3.1

ENERGY NEEDS OF A COMPLETE IRONWORKS AS ESTIMATED BY VÉZIN, 1735

	Inches of water
For a forge hammer wheel	54
For two finery wheels and a chafery wheel	72
For the wheel for a plate mill hearth	24
For the tilt hammer wheel	40
Total	**190**
Surplus	50
Flow measured at the flume of Francheville's bloomery **Grand total**	**240**

The ironmaster's overly optimistic estimates came back to haunt him. The experts who accompanied Vézin disagreed with him, and their spokesman contended that there was only "enough water to turn two grist mills."[22] Later on, we will see that Vézin had to modify his initial plan, but he was not completely wrong about the energy potential of the creek.

Vézin also precisely measured the drop in the creek's slope (Plate 3.2). He noted the details of this measurement at the bottom of a map of ore mines in the vicinity of the Forges. The map shows a cross-section of six levels of the slope, beginning at a distance of 4 arpents, 9 perches, 13 feet above Francheville's forge, where the blast furnace would eventually be located.[23] In a note, he established that, from the lowest level represented on the cross-section, there was still a drop of "twenty-five feet, four inches beyond the forge" (levels not shown on the cross-section), making a total drop of 65 feet, 9 3/4 inches (French measure) down to the St Maurice River.[24] This measurement is a good indication of how the ironmaster was reading the creek's layout with respect to the plan he had in mind. He maintained that, on 5 arpents, "we can build a blast furnace, a slitting mill and a grist mill without the movements of the one interfering with the others."[25] Thus, above the site of his "complete ironworks" with its six waterwheels, he proposed adding a blast furnace, a slitting mill and a grist mill. He provided the cross-section map of the slope of the creek to demonstrate that there were enough falls of water over a sufficient distance to set up a complete ironworks along the creek. It was this idea that finally saved Vézin's plan.

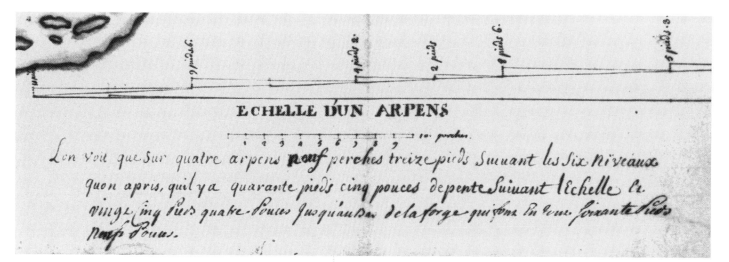

ECHELLE D'UN ARPENS

L'on voit que sur quatre arpens neuf perches treize pieds Suivant les Six Niveaux
quon apris, quil y a quarante pieds cinq pouces depente Suivant l'Echelle et
vingt cinq pieds quatre-pouces Jusquau Bas dela forge qui font En tout Soixante pieds
Neuf pouces.

Plate 3.2

Degrees of slope of the St Maurice Creek, with annotations by Vézin, 1735.
OLIVIER DE VÉZIN, PLAN OF THE MINES AT TROIS-RIVIÈRES (DETAIL), 1735, FRANCE,
BIBLIOTHÈQUE NATIONALE, PARIS, CARTES ET PLANS, PORTEFEUILLE 127, DIVISION 8, PIÈCE 50.

Having first checked that there was an adequate flow on a suitable slope, the next thing was to ensure that there was a constant flow, an equally important factor. Eighteenth-century ironmasters in France were well acquainted with the problem of low water levels, which left forges idle during summer droughts. In 1735, it was feared that the problem would occur with the St Maurice Creek, not because of any drop in water volume in summer, but rather because of timber felling right at the stream's source. The creek was thought to rise in a swamp fed by rainwater and spring runoff, and it was believed that the swamp would soon dry up once the trees growing in it had been felled to make charcoal.[26] Vézin maintained that felling in the swamp had so far had little impact on the volume of water and that the creek was fed by "a number of sources all along a very steep hillside to the south-southeast."[27] He was backed up by both the experts and the facts. Throughout the entire history of the Forges, the consensus was unanimous that the creek would never dry up, and indeed, the water-wheels could still turn today. That point was already made during the early years of operation, as seen from this 1742 memorial:

[...] this is not a risk for the St Maurice and St Étienne streams; for six years in a row they have not dropped significantly during low water, and they hardly rise at all during high water, which is almost a sure indication that they will never dry up [...].[28]

Recent assessments confirm that the creek is inexhaustible; it is a spillway from the water table of the St Maurice watershed. The creek's constant flow results directly from the permeability of the "very steep hill-side" at the foot of which Vézin "[saw] even more [streams] flow." The top of the hill is a 31-km^2 swampy plateau; rainwater seeps through its permeable layers to the imperme-able water table, which feeds the creek. According to engineer Achille Fontaine, the creek's flow "is a function of the permeability of the soil on the hillside; the head of water is constant, the seepage is always the same; it is as if the tap is always turned on at the same rate of flow."[29]

Thus, Vézin correctly assessed the source of the creek, but in the end his plan exceeded its actual capacity, as he wanted to turn too many wheels in the same place at the same time. He subsequently removed three wheels from his plan for the forge, which was originally supposed to include six, but he only succeeded in running two simultaneously during the first trials in 1738. Following this misadventure, the surveyor Champoux, who was sent to St Maurice, designed a daring channelling scheme (Plate 3.3) to swell the creek's waters. Fortunately, his very costly plan was not carried out.

Later on, in 1747, there were also complaints of "too little water in the stream to set up and support two other furnaces" that were to be used for cannon founding.[30] Beginning in 1740, some drainage work carried out on the

site may have raised the level of the creek slightly; a map from 1850 showing the line of a ditch joining the blast furnace millpond is evidence of this. Dollard Dubé later said that "an artificial gully collecting water from the land to the north increased the volume of water in the upper pond by a good third."[31]

In 1861, Inspector Symmes observed that, although the volume of water in the creek was not great, its water power was excellent and the water was easy to control. He noted that the existing works could be used with various types of mill and assessed the creek's capacity. These are the first documented figures on the flow of the creek since 1735[32]— it had a force of 20 horsepower and could turn two wheels for two pairs of millstones (*moulanges*): "—can work 2 prs: moulanges." It is striking to compare this observation with that of Vézin's detractors in 1735–36, who maintained that the creek could only turn two grist mill wheels. The creek's capacity was still the same 125 years later. But Symmes added that, owing to the remarkable accessibility of the

Plate 3.3
Surveyor Champoux's channelling scheme to increase
the flow of the St Maurice Creek, 1738.
"Mr De Léry's Plan of the Forges at Trois-Rivières," circa 1738,
National Archives of Canada, Map Division, C-8347.

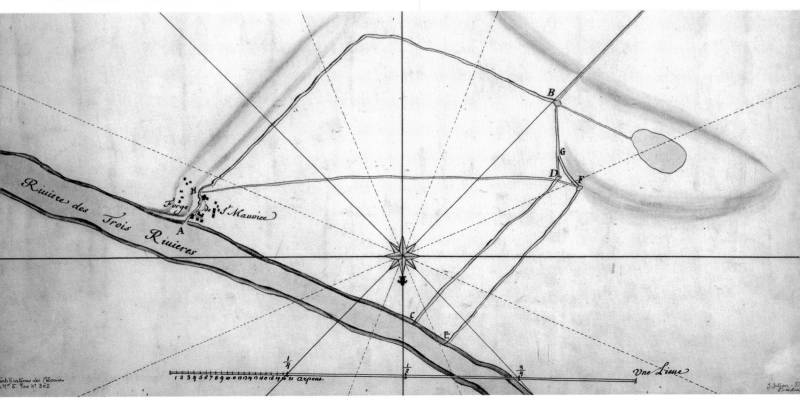

creek's slope, the water could be reused eight to 10 times. This was in fact the case, as at least eight wheels were used for upwards of 100 years, four times more than the mill experts estimated.

Vézin and his detractors were both right and wrong. Vézin rightly maintained that the creek had enough capacity to drive several wheels, but he was wrong in thinking that he could operate six at the same place, where Francheville's bloomery stood. He only succeeded in turning two wheels there at the same time, proving the mill experts correct. One hundred and twenty-five years later, when Inspector Symmes made an assessment similar to that of the experts in 1735, he no doubt used the same reference criteria. But he added that, by harnessing the creek at several points along the slope, its initial energy could be reused several times; by creating four dams, the creek's energy potential was quadrupled. Thus, Vézin's problem lay solely with the distribution of the wheels along the slope of the creek. His initial hydraulic system would also require significant adjustment.

But adjusting it was not easy. It monopolized some of the colony's best technical minds of the time (1735–40), including that of Intendant Hocquart, who proposed a model for the waterwheel. On the basis of his assessment of the creek's potential, ironmaster Vézin was to build an establishment that could produce 600 thousandweight (294 t) of iron annually, an output that would make the enterprise viable and allow the royal exchequer to be repaid. A technical solution had to be found to turn the six wheels needed for the plan, and it was the King's engineer, Chaussegros de Léry, who found it.

PERFECTING THE HYDRAULIC SYSTEM

Francheville's System

Vézin wrote that Francheville's original set-up included a flume that conveyed the water to the bloomery's two wheels: the bellows wheel for the chafery and the hammer wheel. The two wheels were suspended in a stone **wheelrace** flanking the forge itself. [33] Although we have no clear details about these wheels,[34] various clues suggest that they were overshot bucket wheels. First of all, there was a flume, which conveyed the water to the wheels. Had it been an undershot paddle wheel, the ironmaster would probably have called the channel that carried the stream of water under the wheel a tailrace and not a flume. Vézin also demonstrated in his analysis of the two sites at Batiscan that he was not in favour of undershot wheels, which in his opinion required frequent repairs, and he would have been critical had this been the type of wheel used at St Maurice. But the most convincing evidence remains the flume at Francheville's bloomery. When there was some debate over building a stone milldam above the new forge proposed by Vézin, Intendant Hocquart, anxious to reduce construction costs, asked the ironmaster whether he could not "do without a milldam by using the old bloomery's flume, of which there [were] still some remains."[35] Vézin was against this, maintaining that the flume did not provide enough water, thereby contradicting his optimistic assessment of it made six months earlier.[36] We know for certain that the wheels Vézin installed at his forge were overshot wheels. If Francheville's flume was able to drive the wheels of the new forge, that means it was designed to drive overshot wheels.

Although we know that Francheville planned to build dams, there are no documentary or archaeological traces of a milldam; moreover, the Intendant's comment as to whether Vézin could "do without a milldam" is a fairly clear indication that there was none. The flume must probably have received its water from a weir, perhaps a temporary one at no great elevation, which would at the very least have ensured that the water from the creek was channelled from a specific height.

Vézin's System

Vézin's system was based on the construction of two milldams to create two millponds: one for the blast furnace, located at the top of the slope (40 m above sea level) and one for the forge (hammer pond), built 23 m lower down and more than 300 m away. The two millponds were not directly connected to one another. The water in the upper furnace pond flowed back into the creek bed and collected 300 m farther down in the hammer pond.

When construction of the works began in the summer of 1736, Vézin had already modified his plans for the forge, proving his detractors right in part, without ever admitting it in writing.[37] He removed three wheels from the plan (those for a chafery, a plate mill and a **tilt hammer**), retaining only three to be used by two *renardière* chaferies and a hammer. Despite this scaling-back, he knew that he would have difficulty turning three wheels at once if he could not build a milldam at the right height. But he did not get the milldams he wanted, for either the blast furnace or the forge. The decision did not rest solely with him as to what kind of dams would be built and where. Intendant Hocquart himself was involved directly in these technical decisions, assisted by Grand Voyer Boisclerc and an officer, Sieur Demeloize. To get the milldams he wanted, Vézin had to convince none other than the Intendant of New France, the second

most important person in the colony, and two experts who had already contested his assessment of the creek's flow. His initial error was beginning to appear in the structures that he was obliged to build.

Let us first look at the milldam for the forge. Obviously aware of the problem of driving three wheels simultaneously, Vézin wanted to build a stone milldam that would produce a head of 16 or 20 French feet. This would allow him to use large-diameter wheels (for example, 12 feet for the hammer wheel) to compensate for the lack of water. Four arguments were advanced against this type of milldam. The Intendant told him that a masonry dam of that height would be too expensive; his two advisers added that, if it were made of wood, "it would have too much span" and could collapse; in addition, they added that a dam of that height, whether made of stone or wood, would never withstand "the effort of the water" because of the soil, which was too permeable at that location;[38] they also maintained that the same amount of water would be needed "for large-diameter wheels as for smaller ones." Vézin invoked his experience in vain—they were not convinced. His argument was simple and correct for anyone familiar with how waterwheels work. He wanted a dam that was high enough, and thus made of stone to be solid enough, to collect as much water as possible so as not to run the risk of emptying his millpond too quickly when the sluice gates of the three wheels were open simultaneously. At the same time, the high dam would have allowed him to use larger wheels, which would have needed less water to turn, contrary to what the Intendant's advisers believed. He was right, propounding that:

> [...] the larger the diameter of the wheels of these kinds of forge movements, that is, overshot or undershot bucket wheels, the less water it takes to make them turn, according to the rule of lever length [...].[39]

He would then go on to cite as an example the system at the second forge (the upper forge), which had been built three years later.

Vézin was right. It was in fact possible to compensate for a lack of water or insufficient head by installing wheels with a larger diameter whose lever arms were proportionately longer and more powerful. Hydraulic treatises confirm this. Very little water is needed to turn wheels that drive bellows. A comparable dispute in 1756, in this instance over the milldam for a blast furnace at an ironworks in France, gives us additional information on this point:

> We must not be persuaded that a furnace requires a prodigious amount of water. One only needs enough to turn the wheel that drives the bellows. One must even be careful that it does not flow too much or too quickly on the buckets around the wheel that receive it, otherwise they would turn it too quickly, which could cause the rockers and bellows to break.[40]

Vézin was proved right too late. He had to settle for a "small milldam" made of stone to turn 10-foot-diameter wheels.[41] And, to top off his bad luck and lack of foresight, he had to reduce their diameter to 8 feet because when the river overflowed the following spring, he was forced to raise the forge floor by 2 feet. When Vézin had to demonstrate to the Intendant and his advisers that he could actually make three wheels turn at once, he was hopelessly underequipped. When he opened the gates of the milldam, the water level in the millpond dropped visibly, as one of the sceptical advisers did not fail to see. Vézin was able to deceive the Intendant momentarily, but he knew very well that he could not turn three wheels at once. Accordingly, he abandoned one chafery, thereby cutting the anticipated output by half. As a result, he found himself with one too many crews of workers, who had been brought from France at great expense.

Vézin was unable to have his own way for the blast furnace either. When work began in the summer of 1736, the Intendant again was involved in choosing the location of the blast furnace. Vézin wanted to build a stone milldam "opposite the furnace" on the extension of a nearby spit of land. With a milldam right beside the blast furnace, he wanted to take "the water needed to operate the furnace directly from the millpond," no doubt to avoid having to build a long flume.[42] In comparison, in plans for French forges, some of which were in Vézin's own region,[43] there were several ironworks built next to milldams. Such was the case of the dam at Vézin's forge, which was only 9 m from the forge building itself. But the Intendant and his advisers again invoked the high cost of a masonry dam, preferring a wooden one designed by Sieur Demeloize that was built by Charlery the carpenter. Vézin felt that the dam was badly built and not very solid, and he informed them that he would not be held responsible for it. His concern proved to be well founded since it was later suggested that the dam be reinforced. Remains discovered upstream from the blast furnace attest to the reinforcement or replacement of the first milldam.[44]

Shortly after the Intendant's visit in the summer of 1736, Vézin changed the furnace's location on his own initiative. From "a spit of land where there was no risk" in building it, according to Cugnet, he moved it "to the edge of the stream's gully on sandy ground prone to weakening and overturning the furnace."[45] It is possible that Vézin altered the location because he had been unable to have the stone milldam built where he wanted it. The blast furnace and its millpond were thus so far from one another that a 76-m flume had to be erected to convey the water from the millpond to the furnace wheel.

During the summer of 1738, just when he was trying to hide his difficulties at the forge, Vézin was unable to keep the furnace in blast, once again because of a water problem—this time not because there was too little but because there was too much—in the foundations. That was the final straw. In the fall, the Intendant asked engineer Chaussegros de Léry to visit the Forges, and it was he who finally got the ironworks operating.

Chaussegros de Léry's Involvement

[…] I noticed […] that the stream could not supply water to turn both wheels […]. On my return, I gave an account of all of this to Mons. Hocquart, telling him that it was necessary to build a second forge on the same stream below the furnace, since the situation permitted it […].[46]

[…] The water that will be used to operate the [little] forge will drain down into the water of the lower forge and in this way, it will be quite unnecessary to convey any there.[47]

Chaussegros de Léry, who is best known for his work on fortifications,[48] had, in his own words, already seen "several forges in France."[49] His involvement is interesting both because he finally found a solution to turning Vézin's six wheels and because he left notes and calculations[50] (see Appendix 6) that point to the various tests and experiments that led him to propose and design a second forge and improve Vézin's forge.

These notes contain specific information about the technical details of Vézin's forge. They also give an account of the methods and formulas used at the time to calculate the velocity of a water current and its "expense" (outflow), as well as the "effort" (power) of a current with various heads. Finally, he provided valuable information on the role of **forebays** in a flume system, which were described in an inventory two years later.

Léry's notes, scribbled on the back of other documents, are true scientific records of the application of hydraulic engineering in Canada in the 18th century; Léry developed his plans for the forge through experiments that he translated into mathematical language. These notes are of such importance for the history of science in Canada that they are worth a closer look. By organizing the engineer's handwritten pages, we will try to gain an appreciation of the problems he was trying to solve. According to the dates of his notes, most of these observations were made between January and May 1739 (Table 3.2).

We know that Léry produced the plans for his forge in May. His notes also contain other calculations and drawings, including one of a wheelrace with three wheels (Plate 3.5) and a sketch of a waterwheel (Plate 3.4). Moreover, the first pages of the document include an itemized list of the oak timbers needed to build a forge gear mechanism (see Appendix 6).

Plate 3.4

DRAWING OF A FORGE HAMMER WHEEL, ATTRIBUTED TO ENGINEER CHAUSSEGROS DE LÉRY, 1739

Plan de la roue du marteau de la forge de St. Maurice

Celles des chauffries suivant la figure de celles du marteau de a b 14 pces 9 lignes de b c 8 pouces 8 lignes

Échelle de 2 toizes

UNSIGNED, 1739, NATIONAL ARCHIVES OF CANADA, MG 1, C¹¹A, VOL. 110, FOL. 199.

Table 3.2

CHAUSSEGROS DE LÉRY'S OBSERVATIONS ON THE HYDRAULIC MECHANISMS OF THE ST MAURICE FORGES, 1739[51]

Date	Handwritten page	Observations
14 January	185	- he notes the bar iron yield of a pig at Vézin's forge - he measures the amount of charcoal used in the finery process - he observes that the two finery wheels driving the bellows make 24 revolutions in 6 minutes, or 4 revolutions per minute
12 February	185–186	- he estimates that there is a 19-foot head available between the point where the water from the blast furnace is ejected and the horse pond lower down the slope where he plans to build the milldam for the second forge
12 February	185–186	- he adds that the drop below the proposed milldam will give him an additional 1 or 2 feet of head in the wheelrace - he measures the head of water on the three wheels of Vézin's forge, beginning from the bottom of the forebay that feeds them - he measures each wheel and the sluice apertures of Vézin's forge - he records that, when the sluice gates are equally open, the upper forge wheel uses more water than the lower one; he attributes the difference to the difference in their heads
22 March (before)	194	- he proposes to compare a 16-foot head with that of the existing forge's 82-inch head; he notes that the forge's forebay is 6 feet high
22 March	176–180	- by theoretically comparing various heads with those at Vézin's forge, he determines the differences in expense of water to measure the savings realized; he also plans to save water by using a lighter hammer
22 March	205 and 207	- he makes a theoretical comparison of different expenses of water at various speeds
25 March	191	- he writes that it is not difficult to build and install three 17-foot-high forebays, each containing 170 cubic feet of water
22–25 April	188–190	- at Quebec, he tests 3 wheels with the same head (6-foot-high forebay) at various sluice apertures
5 May	186	- at Quebec, he tests 2 wheels with the same head (6-foot-high forebay) at various sluice apertures
... May		- he submits the plans for the upper forge
10 October		- the newly built upper forge is in operation[52]

Léry's Scientific Approach

Léry took a logical approach. He first studied the existing forge and noted the yield of a finery powered by waterwheels (4 revolutions per minute) driving the bellows (14 January). He then selected a location for a milldam that would give him an adequate head (19 feet); he studied the existing wheels, how they were installed and how they were driven (12 February). On the basis of the available heads, he calculated the amount of water he could save with heads of 24, 28 and 30 feet (22 March). These experiments clearly suggest that Léry was trying to improve the yield of Vézin's existing forge and to increase the head available. He then suggested installing forebays (25 March) and finished up by studying the performance of three types of wheel.

Several undated pages in Léry's notes contain a number of calculations that shed light on the problems he was trying to solve. The purpose of all these calculations was to improve the performance of the existing forge wheels. As a result, we have detailed measurements of the forge's three wheels, the available head for each wheel, the sluice apertures of each forebay and the weight of the forge hammer. The engineer was trying to conserve water during the operation of the wheels.[53] To do so, he tried to increase the head of water to each wheel so as to reduce the aperture of the sluice gate that conveyed the water to each wheel.

Léry's Calculations

The engineer's method, as difficult as it is to decipher, can be described as follows: using the known components of the hydraulic engine of Vézin's forge (head and apertures), he calculated the velocity of the water based on the head and the effort supplied by the current at that height through a given aperture. The velocity of the current was obtained by computing the square root of the head,[54] and the effort was obtained by multiplying the head on the wheel by the aperture (Table 3.3).

With this information, Léry knew that an effort of 7,416 inches was needed to drive the existing hammer wheel. He hypothesized various heads with different apertures to obtain approximately the same effort[56] but with less expense of water. He did three series of calculations based on three possible heads of 24, 28 and 30 feet.

He formulated the problem as follows: with a 24-foot head, what expense of water is needed to obtain an effort of 7,416 inches on the hammer wheel? Let us analyse his method in detail.

The total height of head, that is, the vertical distance between the crest of the water in the millpond and the tail water ejected by the wheel, was 24 feet. Given that the wheel's diameter was 8 feet, this produces a 16-foot head on the wheel. This is used in calculating the effort on the wheel.

Léry computes the velocity of the water falling on the wheel:

head: 16 feet or 192 inches

velocity: $\sqrt{192}$ inches = 14

He then calculates a corresponding sluice aperture by dividing the effort required by the projected head:

effort desired: 7,416 inches

head: 192 inches

aperture: $\dfrac{7,416 \text{ inches}}{192 \text{ inches}} = 39$[57]

With these three givens, he seeks to measure the expense of water needed under these conditions, formulating his thesis as follows:

> The expense of water in the first instance will be to the expense of water in the second as the velocities and vice-versa.

He then compares the real figures with the hypothetical ones, which can in fact be compared as they produce the same effort (7,416 inches).

His ratio is expressed as follows:

> The velocity of 10 1/2 is to a velocity of 14 what the expense of 39 is to an expense of x, or:

10 1/2: 14 : : 39 : x,

> therefore

x = 54 inches

The expense of water desired is thus 54 inches. Léry concluded that the difference between the actual amount of water expended (72 inches) and the expense desired (54 inches) to drive the hammer wheel would allow him to save 18 inches of water (72 − 54 = 18) "by providing the suitable heads." It should be noted here that Léry equated "expense" with "aperture."

Table 3.3

CALCULATION OF THE EFFORT ON THE FORGE HAMMER WHEEL BY CHAUSSEGROS DE LÉRY, 1739[55]

Head on the wheel	Velocity	Aperture	Effort on the wheel
103 inches	$\sqrt{103}$ = 10 $\frac{1}{6}$	4 inches x 18 inches = 72 inches	103 inches x 72 inches = 7,416 inches

The heads envisaged by Léry (24, 28 and 30 feet) were hypothetical; he only used them to find an aperture that, with a "suitable" head, would provide an effort equivalent to the one he observed. He had to consider raising the milldam built by Vézin, which provided a head of 82 inches on the chafery wheel and 103 inches on the hammer wheel, the latter being situated farther along in the wheelrace.[58] However, it would have been unrealistic to raise the milldam to the hypothetical heads used, as it would have had to have been completely rebuilt, and would have entailed major alterations to the forge installations. Some of Léry's other notes bring us back to the reality of the existing forge. On page 207, he wrote: "head that can be given, 132." We can interpret this to mean that, taking into account the existing set-up and the lie of the land, he planned to raise the head on the two wheels by 50 inches, which would suppose an equivalent increase in the dam. Linking the calculations on this page with those on two others (pp. 201 and 204), we can see that Léry is estimating the amount of water saved at a head of 132 inches with the appropriate apertures to obtain approximately the same effort as before. He is redoing a calculation similar to the one reproduced above, for the chafery wheel and the hammer, using data from the existing works to determine the same parameters for a projected head of 132 inches. Table 3.4 compares the data for the existing works with those for the modifications that he was planning, using the chafery wheel as an example.

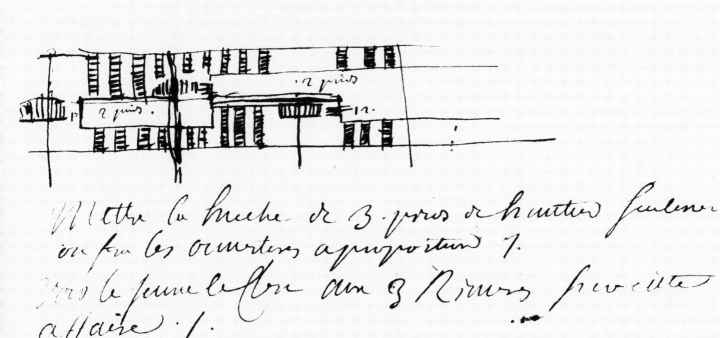

Plate 3.5
Plan of forge wheelrace, with series of three wheels,
attributed to Chaussegros de Léry, 1739.
UNSIGNED, 1739, NATIONAL ARCHIVES OF CANADA, MG 1, C¹¹A, VOL. 110, FOL. 177.

Table 3.4

**EFFORT OBTAINED FOR THE HAMMER
WHEEL FOR TWO HEADS ACCORDING
TO CHAUSSEGROS DE LÉRY**

	Head	Velocity	Aperture	Effort
Existing works	82 inches	$9\frac{1}{20}$	28	2,296 inches
Planned modifications	132 inches	$11\frac{1}{3}$	$17\frac{2}{3}$	2,331 inches

He obtains the expense of water for each wheel and finds the amount he can save by applying the formula "the expense of water in the first case […] etc.," concluding that:

> I will save 6 miner's inches on the first chafery
>
> I will save 3 miner's inches
> on the hammer wheel
>
> and by applying only three hundred pounds
> of weight on the hammer
> I will save another 17 miner's inches

The engineer's projected savings should be seen in conjunction with a plan and instructions on page 177 and a sketch on page 197 of his handwritten notes.

In the first case, there is a sketch of a wheelrace with three wheels preceded by forebays (Plate 3.5). Under the sketch are the following notes:

> Place the forebay only 3 feet high
>
> we will make the openings proportional.
>
> See young Le Clerc at 3 Rivières for this matter.

In the second case (Plate 3.6), we see the modification of the water's angle of arrival at the wheel. In our opinion, these two pieces of information indicate that Léry, after making his calculations ending in a "head that can be given" (132 inches or 11 feet), recommended raising the forebays 3 feet above the wheels in Vézin's forge, which would have modified (as his sketch indicates) the water's angle of arrival at the wheel.

This interpretation of Léry's notes seems to be confirmed by two subsequent pieces of information dated 1741. They are provided by Vézin and seem to fit this context. In a memorial in which they describe the repairs made in 1740, Vézin and Simonet refer to the "bottom of a [forebay] that was raised."[59] Elsewhere, Vézin gives the impression, although not a very clear one, that in October of the same year he had solved the problem of the lower forge "by raising its milldam and forebays, at little expense."[60] Moreover, in the fall of 1741, Estèbe, in his inventory, details the lower forge's milldam as "95 feet long," with no mention of either the height or the width.[61] This is 21 feet longer than the 74-foot milldam that had been measured in 1737. In our opinion, given the topography of the slope where the milldam was situated, the dam would have had to be extended in order to make it higher.

Comparing the 1737 description of the milldam (74 × 18 × 4 feet) with the height of the forebay and the wheelrace measured in 1741, we can see that the dam's height was raised by 6 feet (Table 3.5):

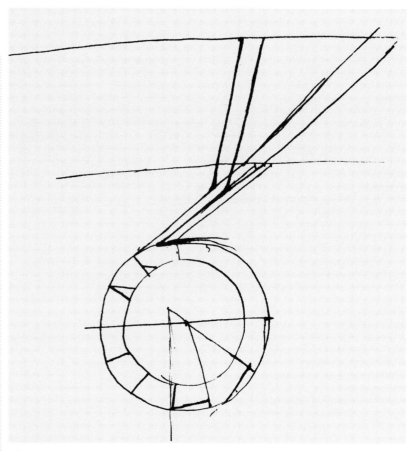

Plate 3.6
Sketch of waterwheel, attributed to Chaussegros de Léry, 1739.
The drawing shows two angles for water falling onto the wheel.
UNSIGNED, 1739, NATIONAL ARCHIVES OF CANADA, MG 1, C[11]A, VOL. 110, FOL. 197.

Table 3.5

RAISING THE LOWER FORGE MILLDAM BETWEEN 1737 AND 1741

Height of the milldam in 1737 (total head)	Height of the forebay retaining wall in 1741	Height of the forebay in 1741	Total head in 1741	Increase in height observed
18 feet	17 feet	7 feet	24 feet	24 – 18 = 6 feet

The height of the milldam (24 French feet) was again confirmed by a measurement taken in 1807, which gave it as 25 English feet.[62] Moreover, the remains found by archaeologists excavating the milldam of the lower forge also show evidence of the milldam having been raised, as two spillways were laid one on top of the other in the dam.[63] Léry's calculations may thus in fact have led to the dam being raised to increase the head.

These calculations thus led to the building of a second forge, the upper forge, being proposed and undertaken; we have found an elevation of the forge, doubtless drawn by Léry (Plate 3.7). Particulars of it are given in the 1741 inventory (see Chapter 4). A 20-foot-high wooden milldam erected at a horse pond at the foot of the blast furnace created a third millpond on the creek. From there, a **headrace** conveyed the water to three forebays in the forge's wheelrace above the three 10-foot-diameter wheels.

The engineer's calculations also allow us to clarify how Vézin arrived at the energy needs, in miner's inches, provided in 1735. The apertures, expressed in inches by Léry, who equates them with the "expense of water," are similar to Vézin's miner's inches deemed necessary in 1735 for the operation of each wheel; the engineer's "desired expense of water" for the hammer wheel corresponds exactly to the 54 miner's inches that Vézin estimated for a forge hammer wheel. Vézin thus did not calculate the latter figure using the contemporary measurement of outflow (miner's inch);[64] rather, he simply measured the size of the sluice apertures. The 240 miner's inches that he calculated at the mouth of the flume of Francheville's bloomery were only a measurement of the sluice gate's height and width, that is, the opening of the flume.

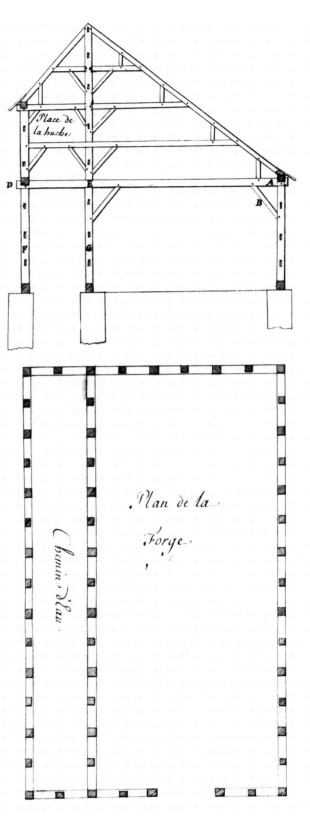

Plate 3.7
Plan and elevation of the planned upper forge by Engineer Chaussegros de Léry, 1739.
NATIONAL ARCHIVES OF CANADA, MG 1, C¹¹A, VOL. 110, FOL. 242.

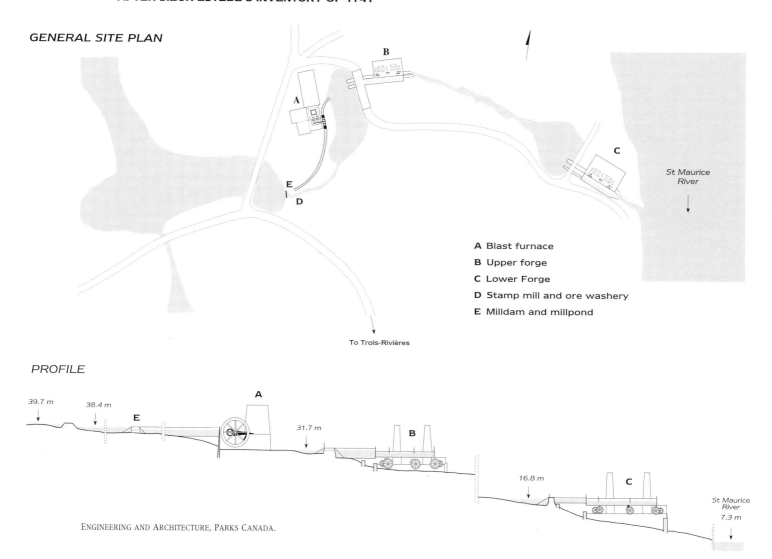

ENGINEERING AND ARCHITECTURE, PARKS CANADA.

LATER ADDITIONS

After the upper forge was built in 1739, the hydraulic system of the Forges consisted of three milldams, each with its own millpond upstream from the furnace and the two forges (Plate 3.8). The hydraulic mechanisms included seven waterwheels, five of which operated simultaneously (the blast furnace wheel and two other wheels in each of the forges); the two forges also had a back-up wheel (in case of a breakdown), which was also used by **sluicing** at high water. A fourth, bigger pond, known as the "washery pond" in the 19th century, was built at the top of the creek's slope, above the furnace millpond, and south of the King's Highway. We do not know when this pond was created; it must have been done fairly early, during the initial years of the Forges,[65] but we cannot prove it. Being at the highest point of the creek, the washery pond served as a reservoir or pen pond for the millponds. Throughout the entire history of the Forges, no significant changes were made to the hydraulic system designed and built between 1736 and 1739; the dams, furnace and forges remained in their original locations. Nevertheless, other installations were built on the creek.

As we saw in the previous chapter, an ore washery was built at the outlet of the washery pond, hence its name. We have also determined that a stamp mill was built downstream from the blast furnace milldam during the first few years. The creek was thus used to full advantage, but that was not all: three mills were added to the creek system, as we will see in greater detail in Chapter 9. The first sawmill was built in connection with the lower forge dam and underwent various improvements in the 19th century. Higher up in the gully, another brick building downstream from the upper forge was converted into a grist mill in the 1860s by Onésime Héroux;[66] the mill's vaulted foundation, which seems to have served as a dam, created a fifth millpond on the creek. Finally, at the end of the 18th century, another grist mill was built higher up the slope, on pilings, at the spillway for the blast furnace wheel. The mill's asymmetrical position in relation to the furnace wheelrace and the remains of an earlier tailrace running towards the base of the pilings suggests that water from the blast furnace wheel was conveyed to a paddle wheel underneath the mill.[67] These mills and the stamp mill were only used sporadically and were probably not operated at the same time as the furnace and forge machinery.

TURBINES

Archaeological excavations have brought to light four turbine caissons in the creek, of which there is no archival record. A caisson was found in the blast furnace wheelrace (Plate 3.9), one each in the wheelraces of the upper and lower forges, and a fourth one in a building erected after the Forges were closed and in which a preserved turbine remains to this day[68] (Plate 3.10).

Turbines may have been introduced along with new equipment such as air compressors and steam engines. It is entirely plausible that the machinery, such as the waterwheels and bellows, was not removed to make way for the new equipment. Before the early iron industry was completely overtaken by the modern one, ironworks operators tried to get the most from the old facilities with a minimum of investment. This phenomenon was observed not only in Europe, but on this side of the Atlantic as well, in the very vicinity of the Forges. For example, the Radnor Forges were powered by a water wheel, turbine and steam engine all at the same time.[69] In the 1880s, the Drummondville Forges were equipped with an air compressor "driven by a turbine and a steam engine."[70] And we will see in Chapter 4 that the second furnace built on the site of the upper forge in 1881 was also equipped with a steam engine to drive a compressor, so the turbine in the wheelrace may well have supplemented the new equipment.

Plate 3.9
Remains of turbine caisson in the furnace wheelrace.
QUEBEC, MINISTÈRE DE LA CULTURE ET DES COMMUNICATIONS,
MICHEL GAUMOND COLLECTION.

The turbine in the lower forge wheelrace must have powered the forge hammer and tilt hammers, as well as the millstone of the axe factory built in 1872 although, once again, this has not been documented.[71] Finally, the blast furnace turbine may have been installed at the same time as a hot-blast furnace and air compressor, some time in the 1850s. To use such a furnace, the wooden bellows had to be replaced with an air compressor as they would not have been able to withstand the heat. Removing the bellows may automatically have led to the great wheel being replaced by a turbine.[72]

When, however, the Forges were sold to John McDougall in 1863, there was still mention of the great wheel, which could only have been the blast furnace wheel:

> Said purchaser will be entitled to raise the milldam on said watercourse to the height needed to operate the great wheel; [...].[73]

We believe that this is the last explicitly documented reference to the blast furnace waterwheel. In 1874, the geologist Dr B. J. Harrington reported that water power was still being used to produce a cold blast, which suggests that either a wheel or turbine was in use, but definitely not a steam engine. We also know that the blast furnace at the St Francis Forges, which belonged to the McDougalls, had a 24-foot (7.3 m) waterwheel[74] at that time. The wheel at the St Maurice Forges may thus still have been in use, as Dollard Dubé mentioned a 31-foot (9.5 m) wheel in his notes for 1875–80.[75] The wheel may therefore only have been replaced by the turbine after the fire in 1881, which was caused by an explosion in the blast furnace.[76]

Installing the turbines in the old wheel-races designed in the 18th century illustrates how people were trying to make new technologies fit the old infrastructures rather than making wholesale changes. These improvements did increase productivity and keep production costs competitive, thus allowing the Forges to survive a few more years in an industrial environment where they would very soon become obsolete.

Plate 3.10
Hydraulic turbine found on the site during archaeological excavations.
It dates from after the Forges closed.
PHOTOGRAPH BY PARKS CANADA, 130/21.11/PR-6/S-16-3.

Upper forge, after Chaussegros de Léry's plan of 1739 and Estèbe's inventory of 1741.
RECONSTRUCTED BY ILLUSTRATOR BERNARD DUCHESNE.

IV

The Forges Plant

I consider the success of this operation so useful to the Colony and for Trade that I desire that you leave nothing undone that might secure it.[1]

Maurepas, 1735

ESTABLISHMENT OF THE FORGES

Sieur Olivier de Vézin who, I am assured, is very capable of undertaking this work, has decided to go to Canada for a consideration [...].[2]

Vézin's Arrival

In September 1735, when ironmaster Pierre-François Olivier de Vézin came on the scene, Francheville's forge had been closed for over a year and a half. In 1734, after operating for only two months, with disappointingly little in the way of output to show for it, the forge was found to be completely inoperable. The French experts summoned by Maurepas[3] had been severely critical of the set-up, as was Intendant Hocquart, who had visited the post in May 1734.[4] By the time Vézin arrived on his tour of inspection, the time was ripe for a radical change, and the colonial authorities were ready to consider putting the forge on a whole new footing.

Vézin, a native of Champagne, owned a forge at Sionne.[5] He was experienced in running an ironworks employing the indirect reduction method, and it would have been surprising had he favoured retaining Francheville's forge, which used the direct reduction process. In his "Observations," drawn up following his visit in October 1735, he, too, roundly condemned the set-up and quickly began to promote the indirect reduction process, which produces a larger quantity of higher quality iron.[6]

On the strength of his expertise and the prestige that his exclusive status as an ironmaster gave him in the colony, he drew up elaborate plans for the creation of a complete ironworks, which he offered to build himself. He suggested that the same location be kept, but that the small forge be replaced with a larger facility where the indirect reduction process could be used.

Setting Up the New Forges

According to the proposal, the new Forges, which would be equipped with a blast furnace and a proper hammer forge, would be able to produce a million pounds of pig iron (490 t), 100,000 pounds (49 t) of which would be earmarked for making castings, and 900,000 pounds (440 t) for conversion into 600,000 pounds (294 t) of bar iron.[7] The proposed facility would have at least 10 waterwheels distributed down the little St Maurice Creek; it would employ about 20 ironworkers, in addition to 100 other hands—miners, colliers, carters and bateaumen. Vézin's plans were so well received by the colonial authorities that they did not even wait for royal assent before beginning preliminary work on the project in the spring of 1736.[8]

Vézin's Plans

Vézin's "Observations" reflected a business-like professionalism, in keeping with what was expected of an ironmaster. He provided the first measurement of the flow of the creek; he plotted its fall (Plate 2.3); he drew up a map of nearby ore mines. This attention to detail was in contrast with the timid experiment hastily undertaken by Jean-Baptiste Labrèche upon his return from New England in 1732 and 1733! The colonial authorities were visibly impressed with Vézin and they were keen to launch an ironworks. His detailed plans, complete with figures (see Appendix 7), added to his credibility.

He presented a cost breakdown for building the Forges, accompanied by a statement of annual operating expenditures and returns. The new Forges would cost 36,000 *livres;* annual operating costs would be 61,250 *livres,* which would be offset by an estimated production worth 116,000 *livres,*[9] yielding a profit of some 54,750 *livres,* sufficient to break even and pay back any advances from the Crown.

Only a few days after Vézin submitted his plans, the partners in Francheville et Compagnie ceded their operating rights and assets to the King. That same day, they formed a new company—in which Vézin himself was a shareholder—and proposed to continue the operation.[10] The colonial authorities, in the expectation that the King would underwrite the venture, decided to keep Vézin in the country and began preparations with a view to starting construction the following summer.[11] They had every reason to expect a favourable response from the King, and it finally arrived in June 1736, but their haste would bedevil the conduct of work between 1736 and 1739.

Criticism and Modification of the Plans

Vézin's apparently precise calculations did not, however, particularly impress the local experts who accompanied him when he reconnoitred the site in September and October 1735.[12] These three watermill experts, including the millwright Costé, contested from the outset Vézin's claim that he could simultaneously operate the six waterwheels of his planned forge at the spot where Francheville's forge stood. And throughout that winter they continued to criticize Vézin's estimate of the flow, claiming that, at the proposed forge location, the creek could turn only two wheels at once.[13] As we saw earlier, Vézin was to change his plans for the forge:[14] he eliminated three wheels, which forced him to abandon the idea of a chafery, plate mill and tilt hammer (Table 4.1). The experts nevertheless continued to believe that the flow of the creek was insufficient to turn a third waterwheel at that location, but it was not until two years later, in 1738, that they were proved right.[15] Be that as it may, the result was that Vézin was compelled, much against his will, to accept a watchdog committee on the post from then on.

Table 4.1   **CHANGES TO VÉZIN'S PLANS**

Planned forge (1735)	Forge built (1736–37)
Finery 1	Finery 1
Finery 2	Finery 2
Forge hammer	Forge hammer
Chafery	
Plate mill	
Tilt hammer	

Carrying Out the Project

[...] with too great haste on the part of his partners, who were concerned only for their own interests and the lure of gain [...].[16]

[...] Sieur Olivier alone is responsible for the immense spending on this enterprise [...].[17]

It took three years before the Forges were finally established, since it was not until the fall of 1739 that output reached Vézin's initial projections.[18] According to the schedule, construction was to be completed in 1738, with production to begin in the spring of 1739,[19] so delivery was not too far behind schedule. The cost, however, was considerably more than estimated, and the haste and tension surrounding the project—which was implemented in a climate of suspicion, opposition and discord among the partners—resulted in errors, lack of foresight and shoddy workmanship. In 1741, on the inevitable bankruptcy of the fledgling firm, the establishment was valued at over 100,000 *livres*,[20] although, based on Vézin's estimates, it should have cost only 36,000 *livres* to build. Let us look at the saga of how it came to fruition.

Construction

1736

Vézin did not receive the go-ahead from Maurepas to begin work until June 1736.[21] By then, the oak timbers needed to build the gear mechanisms for the waterwheels and machinery were already on site. Hocquart, who anticipated that the King would underwrite the project, had had them cut in the Chambly area during the winter.[22] The post had been deserted since the spring of 1734; only the four buildings of Francheville's former establishment stood clustered at the foot of the creek: the forge, a blacksmith's shop, a stable which was about to be demolished and the forgemen's quarters, which were to be kept.[23] Otherwise, the site, encompassing the 5 arpents of the gully and the 60 ha[24] or so of the plateau on either side, was still a wilderness. Clearing was required:

> As soon as the orders were received, his associates hastily dispatched from Quebec a number of carpenters, masons, labourers, quarrymen, carters and others needed to build the establishment, before any clearing or opening up had been done to employ these workmen in the pursuit of their crafts, yet they were all earning big wages in addition to being fed at company expense.[25]

Even though his orders were slow arriving in the colony, which delayed the start of work, the Minister nevertheless required that the work be carried out with dispatch.[26] Consequently, instead of carefully planning and scheduling the hiring of workers,[27] there was a tendency to resort to improvisation, with the result that, right from the outset, the partners were little inclined to act in concert.

On site, Vézin had to make haste. He, too, did not want to displease the Minister. But we can well imagine all the headaches he had to contend with as, throughout that rainy summer, he not only had to ask skilled craftsmen to act as woodcutters and labourers, but he also had to see to their board and lodging in the middle of the woods, miles away from civilization. In addition, the river, swollen by the heavy rain, was not navigable for part of the summer. This delayed the quarrying and ferrying of the stone to build the blast furnace, which had to be brought from the Gabelle quarry, a little farther up the river. To deal with this setback, Vézin had the stone brought at great expense from near Quebec to Trois-Rivières. In the meantime, the river had become navigable again, so that stone could once more be quarried at La Gabelle, and the stone brought in was not used! This costly example is eloquent testimony to the highly charged atmosphere on the post during the summer of 1736,[28] to which Intendant Hocquart himself contributed by visiting the post in person.

Luckily for Vézin, a second ironmaster, Jacques Simonet, originally from La Bergement in Burgundy,[29] was dispatched from France with four assistants.[30] With their help, the work was "considerably" advanced during the fall, in the words of the colonial authorities themselves, who said that the ironmasters had assured them that the Forges would be "in working condition by next year before the ships sailed" and that the first iron made in the colony could be sent to France at that time.[31]

The merchant Sr Huguet has informed me that he has just received the implements and other utensils for an ironworks to be established in Canada from Sr le Blanc, ironmaster at Clavière, to have them sent with dispatch to Bordeaux, even overland, since this must be effected with haste. It would be absolutely impossible to use this route, since the utensils weigh 12 thousandweight [...].

In the fall of 1736, Vézin drew up a status report, accompanied by plans for the project (since lost), that reported considerable progress. Roads linking the post with the river, the mines and Trois-Rivières had been built, the foundations of the forge and its shed had been completed, the blast furnace was built and the frame of its milldam had been set in place. Simonet returned to France to report to Maurepas and bring back the necessary ironworkers, while Vézin remained in Canada to prepare for the following season.[32] No doubt it was Simonet who ordered the 12,000 pounds of tools, purchased from the ironworks at Clavières, which were readied for shipping to Canada in the spring of 1737.[33]

1737

During the summer of 1737, Vézin supervised the work alone. Simonet did not return from France with the skilled workers he had recruited until August.[34] Haste was still the order of the day, resulting in poor workmanship and technical problems. The discord between the partners—Vézin and François-Étienne Cugnet, the two powers in the company—was expressed through the hiring of separate carpenters. Vézin engaged Le Clerc, a Trois-Rivières carpenter who built the forgemen's quarters near the forge and some of the blast furnace secondary buildings; Cugnet used Charlery, a Quebec carpenter whom he hired in 1736, and who built the forge, stables, charcoal house and blast furnace milldam in 1737. Inexplicably, Vézin did not himself take the trouble to hire the workman who would build such critical structures as the forge and the blast furnace milldam. Similarly, he had the large bellows for the furnace built by a workman of his own choosing, yet he left the bellows gear mechanism to be built by a man hired by Cugnet. The employment conditions agreed to by Cugnet for Charlery and other workmen, involving large advances,

made things difficult for Vézin, who "had trouble keeping all these workmen at their tasks." Consequently, "Messrs Olivier and Simonet never took delivery of the buildings of the said Charlery," whose work they were not prepared to approve.[35]

By the fall, the complex was well advanced, but the blast furnace was not yet operational. The hydraulic system, although not in its final form, did function. It consisted of two basins or millponds created by two milldams, one made of wood, above the blast furnace, and the other of stone, above the forge; the great wheel for the furnace was mounted in its wheelrace, as were the wheels for the forge. The forge and its ancillary buildings, as well as the furnace secondary buildings, were completed. Housing for the workers and stables for the horses had also been built, and construction had started on the big house, the Master's House or Grande Maison.

By late fall, Vézin and Simonet, driven by "an ardent desire to demonstrate the success of their establishment" wanted to send samples of iron produced at the Forges to France. But, on the very eve that the King's ship was to set sail, the furnace was still not in a fit state to produce the pig iron needed to make the cast iron plates for lining the forge hearths. They quickly had stone **hearth plates** made and in this way managed to produce "four little bars of iron [...] from pure mine"—thus resorting to a direct reduction process—"which they had the honour of sending that same year to Monseigneur le Comte de Maurepas."

In November, Vézin and Simonet tried to commission the blast furnace "completed to the point of operating." Major problems were then discovered: when blown in, the furnace did not remain in blast. This was serious, since large amounts of charcoal were required to **season** the furnace every time it was blown in, a process that took five or six days. The iron-masters blamed the workman hired by Cugnet for not having set up the bellows mechanism properly. Sure of their diagnosis, Vézin and Simonet replaced the spur gearing with an *estrique* mechanism. Lardier, the founder, blew in the furnace again and was able to smelt an initial pig. But the fire went out again. This time, Lardier, who had been brought from France by Simonet, was accused of negligence. In the space of six weeks, the furnace was blown in three times, with no better results. Lardier nevertheless succeeded in smelting 150 thousandweight of pig iron, about 100 thou-sandweight less than the expected output for such a smelting period.[36] New trials were put off until the following spring, and the winter spent trying to "get to the bottom of the causes for this lack of success."

1738

Once spring came, the ironmasters made changes to the innards of the furnace and asked Paillé, a carpenter from Rivière-du-Loup, to replace the *estrique* with another spur-geared mechanism to meet the requirements of the founder, Lardier. Despite this, between late May and 20 August, six attempts to blow in the furnace failed. The ironmasters decided at that point to dismiss Lardier and replace him with Delorme, a forgeman with some experience as a furnace keeper, "who would soon bring them complete success." On 7 October, the successful blowing in of the furnace by Delorme on 20 August was notarized. This date marked the official start of production.[37]

In the words of Vézin and Simonet, in their subsequent reports on these problems, the failures to blow in the furnace were simply the result of the "lack of skill" and "negligence" of Lardier, the "wretched founder." But it seems rather that these accusations were merely an attempt to cover up mistakes. Lardier had not been taken seriously when he said that "the site of the furnace" made it difficult to blow in, pointing out that the bottom of the crucible sat on damp soil, the constant moisture from which repeatedly doused the fire. The iron-masters rejected this hypothesis, pointing out that the blast furnace was "scarped on two sides down to over fifteen feet below the bottom of the crucible," and taking great care to specify that they had not overlooked the installation under the hearth of "the necessary conduits, made of brick as usual, to draw off the damps."[38] According to Cugnet, the little arches under the hearth to dry it out had been omitted, and the new founder was not able to blow in the furnace until "this problem had been remedied."[39] This seems too gross an error to have been made by two ironmasters who should have known very well that the mill-pond above the furnace was likely to make the surrounding soil damp. The conduits, or "soughs," which they said they had installed, were not equal to the task. We have shown elsewhere that certain historical information, coupled with the remains of the furnace excavated by archaeologists, would seem to indicate that the bottom of the hearth had been raised by about 1.2 m, probably to insu-late it completely from the soil.[40]

Failure to blow in the blast furnace was a serious problem, which is probably why archival documents report on it in detail, albeit incompletely. This episode, analysis of which has proven challenging for archaeologists, historians and engineers, is also interesting for the fact that the various steps taken by the ironmasters left many traces, which can be seen in the remains excavated so far. These remains enable us to read the many ups and downs that finally led to the start-up of the St Maurice Forges.

A good part of 1738 was thus spent in remedying the construction errors of the previous year. In the spring, the charcoal house at the blast furnace collapsed under the weight of the snow. It had been built in 1737 by Charlery, and was rebuilt by Paillé, the same man who had replaced the bellows mechanism. Paillé, a millwright by trade,[41] would go on to complete the wheelrace for the great wheel, while Le Clerc would build the wooden headrace, 76 m in length, that conveyed the water from the millpond to the waterwheel.

At the forge, the floor of the wheelrace was raised. The joists had initially been simply laid on the ground, which had heaved with the frost. A lean-to was built to shelter the race and, at the same time, strengthen the weak frame of the forge. Lastly, the three waterwheels, initially built of pine, were replaced by three new oak wheels.

It was not only the blast furnace that could not be started during the summer of 1738. The forge also remained inoperable, owing to a water problem. The flow of the creek was insufficient to turn the three waterwheels simultaneously. Vézin therefore could only count on half the production capacity he had forecast for the forge.[42] We have already seen (Chapter 3) that Vézin had tried to save face with Hocquart during a demonstration and that it was Chaussegros de Léry who finally got him out of this embarrassing situation by suggesting a second forge. Despite all these problems, production at the blast furnace began in the late summer of 1738 and, shortly after, in the fall, the forge began operating with only a single usable chafery.

1739

Owing to the simple fact that the second chafery in the forge could not be brought into service, the production capacity for wrought iron was halved, and a whole crew of six forgemen who should have been working in three-man relays, remained idle. Brought back from France by Simonet during the summer of 1737, these workers nonetheless continued to be paid and lodged at company expense. In February, the partners decided, in the presence of Hocquart, to follow up on Chaussegros de Léry's recommendations and build another forge, and the following summer, the ironmasters asked Le Clerc to build the frame and Bellisle to make the gear mechanism. They also gave Le Clerc the job of building, at the location chosen by Chaussegros de Léry, a 6-m high wooden milldam. The forge began operating that fall, about 10 October,[43] but a water leak at the base of the dam on 8 November, not long after Vézin and Simonet had left for France,[44] caused a month-long interruption.[45] In their absence, Simonet's son, Jean-Baptiste, was in charge of the Forges.[46]

Between 1 November 1738 and 4 September 1739, the lower forge, the only one operating at that point, produced 227 thousandweight of wrought iron.[47] Consequently, it was expected that double this amount, if not more, could be produced annually once the second forge was in operation.[48] In early November, when the furnace was nearly 300 thousandweight of cast iron ahead of the forges, it had to be **blown out** because of a lack of charcoal,[49] despite the precautions that Vézin claimed to have taken before leaving for France.[50] It was not blown in again until 23 May 1740.[51]

1740

Vézin and Simonet were away for a whole year, and it was during this period that the Forges were at last repaired and completed. In March, Cugnet and Simonet *fils* set out the terms and conditions under which the young acting ironmaster, assisted by the clerk, Cressé, was to run the Forges. The memorial set out in detail everything that had to be done to finish construction and make the complex operational. The ironmaster was to make a written report each week.[52]

A large part of 1740, which normally should have been devoted to regular production activities, was taken up in repairing and completing the facilities built during previous years. The lower forge was, to all intents and purposes, rebuilt, and repairs were made not only to the building's structure (roof and charcoal shed), but also to the plant and machinery—the two chafery chimneys, the **hurst frame**, the bellows, the three waterwheels and the forebays, all of which were altered or replaced. At the upper forge, both the building (iron store and charcoal shed) and the machinery (a second chafery with a wheel and forebay, shafts and bellows) were completed.

All these repairs prevented the company from reaching its production capacity of 600 thousandweight of wrought iron (294 t). Only 355 thousandweight (174 t) were produced between October 1739 and October 1740, while the blast furnace yielded 674 thousandweight of cast iron (330 t) between 23 May and 30 September, despite the serious cracks that had been found in the wall facing the creek.[53] In late October, Hocquart was able to announce to Maurepas that everything was finally "moving along."[54] But things had not improved among the partners, who had split into two factions. The ironworks was also saddled with three ironmasters (Vézin and the Simonets, father and son), in addition to Vézin's brother, the Sieur d'Armeville, who had come from France to help him.[55] In his report, the Intendant made no more rosy promises for the future and even admitted to fearing that things "might fare badly" for a company that was divided and nearly 300,000 *livres* in debt. He planned to demand an accounting in January 1741 and would leave it to Maurepas to decide on the future of the enterprise.

1741

Even though the Forges were finally completed, the fledgling company that had struggled to establish them was on the brink of bankruptcy. Cugnet, the company treasurer, had, with difficulty, obtained credit to pay the workers and assemble the raw materials—ore, flux and charcoal—needed for the next campaign so that the Forges could operate in 1741. And he had undertaken to repay those loans from the 1741 operating revenues. The King's advances were exhausted and the company now had only its own resources—Cugnet's personal fortune—to fall back on; but, as we have seen, Cugnet, encouraged by Hocquart, did not hesitate to use other government monies for which he was responsible.[56] The 1741 campaign absolutely had to succeed.

Ever since they had become operable, the two forges had run all year, even though the winter months were very expensive in terms of charcoal. So much charcoal was used in the severe cold of the winter of 1741 that it had to be made even during the harshest weather and three times more wood was required than in summer to produce each cartload of charcoal. The year's campaign was therefore seriously compromised since, for lack of charcoal, the blast furnace could be blown in only in July, to be blown out again in August. The ironmasters estimated that 200 thousandweight of wrought iron was lost because of a lack of supplies. This mediocre performance during the 1741 campaign meant that the company had to grapple with mountains of debt and was short of funds to prepare for the next campaign. The inevitable bankruptcy came in the fall, after the resignation of all the company's partners.

Pending the King's orders, Hocquart placed Guillaume Estèbe, the keeper of the King's stores at Quebec, in charge of running the Forges, with Simonet *fils* as ironmaster. Estèbe arrived at the Forges in November and prepared a detailed inventory, which remains to this day the most complete document ever produced describing the Forges.[57]

The Forges in Operation

[...] this Establishment is worth infinitely more than what it has cost to date [...] the iron it produces is of higher quality than anything in Europe [...].[58]

Between Vézin's arrival, in the fall of 1735, and the bankruptcy of the Compagnie des Forges de Saint-Maurice in the fall of 1741, six full years had passed, four of them devoted to making the Forges operational. This was a short life for a business. In his autopsy on the company's bankruptcy, Intendant Hocquart wondered "why this establishment, which has now reached perfection, has not turned a profit" and concluded that "a lack of money is at the root of the whole problem."[59] By this he meant that the "truck" system, by which the workers were paid in merchandise, had become a source of discontent, leading to insubordination and ill will (see Chapter 8). There is no record of the workers' point of view and, despite this condemnation of their conduct, they were in the end the only ones unaffected by the bankruptcy, since they remained at their posts at the Forges. Throughout this whole episode, riven with conflict, miscalculation and shoddy workmanship, the workers nevertheless succeeded in making iron. And the criticisms made of their early iron no doubt prompted the necessary adjustments, since no suggestion of abandoning the venture was ever made.[60]

COMPONENTS OF THE FORGES PLANT

The Gully

Although the St Maurice Forges were equipped just like forges in France, the fact that the plant was clustered together in a complex made it a "considerable establishment" in contemporary opinion, even compared with ironworks in the mother country. In Europe, many plants had only a blast furnace or a forge; ironmaking and ironworking were split between separate locations and managed independently, and the workers lived nearby in villages. At St Maurice, the vertical integration of the Forges, encompassing the whole process, entailed the physical integration of the various elements of the complex on the site. The ironmasters had to make sure that the two main stages of the indirect reduction process were properly spaced out. They therefore sought to lay out the various components in the order that the work was carried out, taking into account the fact that the end products were always extremely heavy. It is this spatial distribution of the industrial process which makes the Forges site so fascinating. Such logical organization of space as existed at the Forges was without doubt found in the gully, rather than in the layout of the housing, which a witness correctly described as "higgledy piggeldy, with no symmetry."[61]

The parts of the complex were distributed down the gully through which the creek flowed, from top to bottom. At the top was the blast furnace; just below was the upper forge, so called to distinguish it from the second forge, built at the foot of the gully and known as the lower forge. Above each facility was its millpond, all interconnected and continually fed by the creek flowing down to the river. These millponds were the reservoirs needed to operate the waterwheels at each facility.

The crude iron ore was first charged into the blast furnace to be reduced to cast iron in the form of pigs. The heavy pigs were then taken down to the two forges to be converted into bar iron. The bars were then carted down to the banks of the St Maurice where they were loaded onto boats for Trois-Rivières. The complex was thus designed for the production of iron bars made in two distinct operations at two separate places. From the start, however, there were plans to manufacture other products, requiring the construction of other shops in the gully. Vézin intended, in fact, to build a hammer mill with a tilt hammer to make "small iron," as well as a slitting mill. And it was precisely for this reason that he chose to place the blast furnace at the top and the forge at the bottom of the gully and to earmark other locations down the slope where the fall would create a head of water sufficient to turn new waterwheels.[62] He did not expect that he would have to use this space so soon to build a second forge to correct his plans.

Plate 4.1
Remains of buttress broken away from blast furnace, 1921.
BIBLIOTHÈQUE NATIONALE DU QUÉBEC, MONTREAL,
MASSICOTTE COLLECTION, TROIS-RIVIÈRES.

In choosing the location for the upper forge, Chaussegros de Léry had also noticed, below the selected site, an 8-m fall suitable for a tilt hammer mill "to make small iron."[63] This location, where the gully narrowed, would in fact later be used for a sawmill and grist mill. The ironmasters were trying to exploit the basic hydraulic system to the full by making use of it, likely during down times at the furnace and forges, to drive other machinery required to run the operation—a flour mill, sawmill, charcoal mill, stamp mill and ore washery.

Let us now consider the particular features of each part of the complex so skilfully distributed down the gully.

The Blast Furnace

The blast furnace was the real heart of the Forges. It was a huge oven in which the crude iron ore underwent its initial conversion into a usable product: pig iron. We will discuss the operating principle of a blast furnace later, in the description of the manufacturing processes themselves, but here we will look at its installation and design.

Location

The blast furnace was located on the edge of the gully just where the falls began to be of respectable size. More precisely, at ground level, the east side[64] and part of the south side of the furnace stood nearly 7 m above the bed of the stream that ran down through the gully.[65] The east wall had to be buttressed shortly after construction to keep it from sliding into this ravine.[66] The earliest archaeological explorations of the site indicate that a large part of the buttress had tipped into the gully (Plate 4.1).[67]

Vézin had not originally planned to build the blast furnace here, where the land was so unstable. In July 1736, he had agreed with Hocquart, up from Quebec, on another location on a point of land overlooking the gully—likely on the most easterly point, where housing for the workers was later built. But Vézin changed his mind, probably because he was unable to get leave from Hocquart to build a stone milldam on the extension of a neck of land close to the planned site. Vézin hoped, by installing the furnace close to the milldam, to be able to "take the water required to power this furnace straight from the millpond,"[68] a practice that was common in France. His plans for the stone dam were rejected, no doubt because of the high cost of such a structure, so he opted for a location on the edge of the gully in order to make use of a sufficient head of water. Because of the layout and instability of the site, he was obliged to build the dam well upstream from the blast furnace. This was a wooden milldam only 4 m high by 16 m long, but it nevertheless created a large millpond, although a 76-m-long headrace had to be built to convey the water from the millpond to the wheel.

By placing the furnace on the edge of the gully, Vézin was able to mount a waterwheel nearly 10 m (30 French feet) in diameter in a wheelrace that flanked the furnace on the southeast corner. This required the construction of a whole series of walls, pillars and buttresses, the remains of which can still be seen today.

The Blast Furnace Complex

The blast furnace was, in fact, nothing more than a huge oven surrounded by machinery and ancillary buildings in proportion to its size. These ancillary sheds, in which the workers carried out their tasks, ran around three sides of the furnace—north, south and west (Plate 4.9). They were used to house the raw materials and the equipment for smelting ore and moulding cast iron. Their positioning was dictated by the location of the three openings in the furnace which were used for charging (**throat**), introducing the blast (tuyere) and emptying (**taphole**).

On the north side of the furnace, a large charcoal house was built to stock the raw materials (iron ore, charcoal and limestone) with which the furnace was continuously charged. The fillers climbed up a staircase inside the charcoal house to the **charging platform**, at the top of the furnace. Up there, they hoisted up the **charges** and tipped them into the throat. Prevailing wind direction (northwest)[69] had to be considered when choosing the site for the blast furnace, to make sure that the charcoal house was located upwind, away from the sparks that flew from the furnace chimney.

On the south side was the bellows shed, itself flanked by the wheelrace in which the great, spur-geared waterwheel was mounted. On this side, an arched opening in the furnace provided access to the tuyere, through which the bellows supplied the blast required for smelting.

On the west side was the casting house, where the founder and moulders worked. On this side was a second arched opening, the **tymp** arch or working arch, through which the founder gained access to the taphole, blew in the furnace at the start of the campaign, and tapped the furnace twice a day, at casting time. The floor of this building was made of sand, in which were formed the **pig beds** into which the liquid iron gushed from the taphole.

The three main sheds were connected by passages for ease of movement between the work areas or bays where the various stages of blast furnace operation took place (Plate 4.2). These passages were, in fact, part of other buildings which, joined to the sheds themselves, were used as moulding shops, stores and living quarters. In the southwest corner of the furnace, in the angle formed by the bellows shed and the casting house, was a building used as a storehouse and the founder's quarters. Not wanting to be far from the furnace, which operated 24 hours a day, the founder often walked from the casting house to the bellows shed during his shift to look through "the founder's eye" in the tuyere hole and check how smelting was proceeding inside the crucible. The common pillar between the two vaulted openings was known as the **pillar of the furnace**, probably because both sides allowed access to the crucible, the heart of the furnace, through the taphole or the tuyere. Between the bellows room and the founder's quarters was the master moulder's room.

In the other angle formed by the charcoal house and the casting house[70] was a **moulding shop**, where munitions were made during the French regime.

The blast furnace complex, with its three flanking sheds, to which were attached a moulding shop, storehouse and sometimes living quarters, was characteristic of early ironworks. When a furnace operated as a **merchant furnace** to cast directly on tapping, the moulding floor would be larger. At the Forges, the small moulding floor attached to the north side of the casting house was initially used to manufacture cannonballs. Around 1750, the founder's quarters were demolished to make way for a cannon **pattern** room, and a cannon moulding pit was dug in the casting house. Plans were being made at that time to manufacture arms at the Forges, and the construction of a second blast furnace was even being considered. Nothing ever came of this, however.

Table 4.2	MOULDING SHOPS AT THE BLAST FURNACE COMPLEX						

Period	Name	Location	Size (m)		Use	Equipment
			depth	width		
1738–48	1st moulding floor	North of casting house*	6	9	Munitions	Small stack furnace(s)
1748–60	2nd moulding floor (1st one extended)	North of casting house*	10	5.8	Munitions	Small stack furnace(s)
1750-52	Pattern shed	South of casting house#	19.5	8.4	Cannon founding	Moulding benches
1760-90	3rd moulding floor	North of casting house#			Unknown	Small stove
1790–1881	4th moulding floor (Monro & Bell)	South of casting house#	16	8.4	Sand, box and loam moulding	Stove, crane and pit
1830-54	Moulding shop adjoining 4th moulding floor	South of casting house; shop adjoining west wall of previous building**	10.7	7.6	Sand, box and loam moulding	Stack furnace (with cupola)
1854–81	5th moulding floor, similar to 4th, minus moulding shop	South of casting house#	16	12.5 *(approximate measurements)*	Sand, box and loam moulding	
1863–81	Pattern shed adjoining casting house	South of casting house# and adjoining it, west of the 5th moulding floor	Dimensions unknown		Box moulding	

Based on Barriault 1979. Dimensions correspond to remains unearthed or projections made from remains and archival documents.

* In the angle of the casting house and the charcoal house

Adjoining the casting house, into which it opens

** Possibly burned down in 1854

It was the casting of household items—stoves, kettles, and so forth—which entailed considerable alteration to the blast furnace premises around the end of the 18th century. Moulding space in the blast furnace complex was more than tripled, and a part of the upper forge was converted into a moulding shop. The casting house, which seemingly had not changed since the French era, was altered five times between 1790 and 1881; a sixth casting house was built following the fire of 1881.[71] During this same period, a large moulding floor was built, this time south of the casting house, to which it was joined; it was three times as large as the munitions moulding floor of the French era. A series of alterations was made to this moulding floor and its annexes (Table 4.2).

In Chapter 5, we will be looking at the casting processes for primary and secondary smelting (with **cupolas**) in use there.

Description

Surrounded by its three sheds and ancillary buildings, the blast furnace was largely concealed from view on three sides. At least that is the impression given by most illustrations of the period, where the east side, facing the gully, the only one not shielded, is never shown (see Chapter 7). The smoke escaping from the chimney of the furnace and the long headrace to the great wheel are the main signs of the presence of the blast furnace. It is only in illustrations produced after the Forges closed in 1883 that the massive furnace itself appears, shorn of its recently demolished sheds (Plate 4.6). The last blast furnace was taller than the first one[72] but was similar in shape, so that we can, to some extent, picture the blast furnace that was built in 1736 and inventoried in 1741.

Dimensions

The blast furnace was a square, almost cube-shaped, stone **stack**. Shaped like a truncated pyramid, it was wider at the bottom than at the top (Plate 4.3). The 28-foot-high sandstone furnace built in 1736 by ironmaster Simonet was 26 feet square at the base and 21 feet square at the top (French measure).[73] In 1735, Vézin had estimated that it would take almost 125 cubic *toises* (about 2,500 t in weight) of stone to build the blast furnace (not including the stone required to make the **inwalls** and the **hearth** itself).

Structure

It is easier to visualize the blast furnace if we keep in mind that it consisted of four sections, one on top of the other: (1) a solid masonry foundation 7 feet below ground; (2) from the ground up, the furnace stack itself, composed of thick sandstone walls (with a rectangular inner cavity forming the tunnel or belly of the furnace, lined with inwalls of stone; (3) the top of the tunnel walls projected several feet above the main stack, forming the **furnace top** around the central opening of the throat, through which the furnace was charged; (4) the charging platform itself, topped by a **penthouse** of masonry walls nearly 2 feet thick and $9\frac{1}{2}$ feet high for sheltering the fillers. Although there are few details on this subject from 1741, the furnace top was often capped with a chimney and the penthouse was roofed, as shown in 19th-century illustrations of the Forges (Plate 4.6).

Plate 4.3
View of blast furnace during French regime.
RECONSTRUCTED BY ILLUSTRATOR BERNARD DUCHESNE.

Openings

At ground level, two sides of the furnace had arched embrasures: the one on the west side provided access to the taphole, and the one on the south side, to the tuyere, which received the **nozzles** of the bellows. These funnel-shaped arches became progressively narrower as they pierced their way into the heart of the furnace through the massive walls; they thus had to be extremely rigid. The inner part of the arch was therefore not round, like the smooth inside of a funnel, but stepped in five horizontal levels, each supported by a thick iron pig known as a lintel or **morris-bar** (Plate 4.3).

The thick masonry of the blast furnace was able to withstand the pressure created by the heavy charges of ore being smelted at very high heat inside its belly. Nevertheless, to avoid any risk of explosion and ensure that the moisture contained in the masonry could escape, a system of brick flues or **vents** ran through the walls and supporting walls. It was believed that if evaporation was provided for, all would be well.

In the end, there are no established proportions for the bottom, top, middle and the position of the tuyere. This is where the mysterious industry of the founders comes into play.

There is perhaps some question as to why the interior cavity has this double funnel shape, and it would be difficult to get the workers to give a reason. However, it would appear that prior to the adoption of this shape, a number of others were tried and found wanting, and it would appear that this shape works very well.

Similarly, the foundation usually had arched conduits about 30 cm below the bottom of the crucible, "to draw off the damps."[74] As we saw above, all attempts to blow in the blast furnace in 1737 and 1738 were futile because the builders had forgotten to provide for these little arches. Not to mention the fact that the lack of vents in the stack was behind the fatal explosion of the new furnace, built in 1854.

The Belly

The stack was thus designed and built on a massive scale for the work for which it was intended: the uninterrupted production of a million pounds (490 t) of cast iron in seven months. From the technical standpoint, the real interest of the blast furnace lies in the shape of its belly or, rather, in the way this shape encouraged the consumption of raw materials. "A blast furnace," writes the *Encyclopédie*'s expert on iron, "is really a stomach which demands feeding steadily, regularly, and endlessly. It is subject to changes in behaviour through lack of nourishment, to indigestion and embarassing eruptions through too rich or voluminous a diet, and in such cases prompt remedies are to be applied."[75] The tunnel shape of this belly is the key to the design of the blast furnace, which was invented in the Middle Ages, and shows how well the master founders came to grasp the art of ironfounding.

The effectiveness of a blast furnace is directly related to the shape of its belly. In the 18th century, the design of its inwalls was still part of the "mysteries" of the master founder,[76] the sole repository of the knowledge handed down from father to son. After a very long process of experimentation measured in generations, the founders had perfected a tunnel shape that exactly reflected the thermochemical changes taking place in the smelting process. This is not how a contemporary founder would have put it, of course; instead, he would have said that the inside of the furnace had changed as a result of accident and chance. Let us consider, then, the ingeniousness of this shape more closely.

A cross-section of the belly shows that it has the form of two funnels, one on top of the other: a long funnel upside-down on a smaller one[77] (Plate 4.3). It is in the "stomach," created by superimposing these two funnels, that the secret of the blast furnace's effectiveness lies. We can see that, from the top of the crucible, the walls flare out to create the stomach itself. These walls, known as boshes, trap the heavy charge of raw materials, penning them up as in a funnel so that they are held there in the heat of the fire, fanned by the draft from the tuyere below, until they melt and drip down into the bottom of the crucible in a pasty mass.

Above the boshes, the long upside-down funnel of the upper tunnel also has a very specific function. It is narrower at the top than at the bottom because the materials fed in at the top are gradually heated by the escaping gases and expand as they drop down. This expansion creates the desired bottleneck above the boshes, which are designed to trap the charge.[78]

The interior cavity of the first blast furnace at the Forges, 15 feet 4 inches in height, was shaped like a rectangle 8 feet 1 inch by 7 feet.[79] These dimensions, provided by Estèbe in 1741, obviously refer to the chamber in which the founder created the inwalls of brick or firestone[80] in the form of two funnels. These inwalls were rebuilt regularly, as they were constantly deteriorating under the pressure of the charge and the heat of the fire.[81] The rectangular belly described by Estèbe suggests a quadrangular shape for the inwalls of the first blast furnace, rather than the round form found in the ruins of the last furnace. This was also the shape used at the time by founders in Champagne and Burgundy, where the original ironmasters and workers at the Forges came from.[82] The batter or slope of the boshes varied in accordance with the type of ore to be smelted and was determined by the founder.[83]

At the base of the interior cavity was the crucible, to which the boshes (the small funnel) narrowed to fit. These two parts, very clearly illustrated in the *Encyclopédie*, formed the *ouvrage*,[84] the whole being known as the hearth, which was remade by the master founder at the end of every campaign. "Laying the hearth" was one of his key responsibilities, for which he oversaw the work and selected the stone.

The Crucible

The crucible (also known as the well or receiver) was a rectangular receptacle about 4 feet (1.2 m) by 1 foot (30 cm) and 1 foot (30 cm) in depth.[85] Here the liquid iron and the slag from the smelting process accumulated. The bottom stone was a grey stone slab,[86] 8 to 10 inches (20–25 cm) thick, on which rested the sidewalls, made up of dressed limestone slabs, which formed the crucible sides, each with its own particular name. Opposite the fore hearth, at the far end of the rectangle, was the back wall. To the right, on the same side as the bellows arch, was the tuyere wall; to the left, opposite the tuyere, was the wind wall or stone rest. Opposite the back wall, supported on the tuyere and wind walls was the tymp stone. This did not close off the crucible, but served rather to support the thick, slanted iron tymp plate that rested on the last lintel of the arch and closed off the furnace. At ground level, the **dam** stone, protected by its angled iron dam plate[87] placed just behind the tymp, closed off most of the crucible. The gap between the top of the dam and the tymp created a small opening through which the slag floating on top of the liquid iron could be removed. The taphole, another small opening at ground level, created by the off-centre placement of the dam stone, allowed the iron to be removed from the well. During the smelting process, these two openings were plugged by a clay seal, which would be broken by the founder when it was time to tap the furnace.

The cone-shaped tuyere, designed to receive the nozzles of the bellows, consisted of a "piece of sheet iron" resting on the tuyere wall and projecting into the crucible about a foot (30 cm) from the bottom. The space created by the conical form of the tuyere was packed with clay of the same type used to seal the taphole. The founder sometimes opened this space to get a better view of the inside of the crucible during combustion.

The interior of furnaces in most of Champagne and Burgundy is an elongated square, although they differ according to the founders, who do not want to build something just like their neighbour's, and who, in similar mines, argue the quality of the mine.

The Bellows

Built to produce more than 2 tons of iron a day, the furnace was activated by a powerful blower, which ensured the uninterrupted combustion of the charge in its belly. Two large bellows, about 20 feet (6.5 m) long, 5 feet (1.6 m) wide and 3 feet (1 m) deep, were positioned side by side in the bellows shed (Plate 3.1), and fitted to the tuyere arch. The bellows were made of wood, usually by a bellows-maker.[88] The lower part of the bellows, the air **box**, was fixed, and sat on a support located on a level with the bottom of the crucible. The upper part, the **bag**, was movable and was continuously raised and lowered by a camshaft. The bottom of the air box was pierced by valves, "air inlets through which the outside air entered the bellows during inspiration." During expiration, leather strips sealed the bellows by closing over the valves.[89] The bag and the box were linked by a centre pin located at the **head**, just before it entered the nozzle. A flat piece of metal, the **striker plate** or shoe, was bolted flat onto the bag, projecting backwards into the trajectory of the cams, which struck it, lowering the bag. Each bag was lowered alternately by twin cams and then raised by another mechanism, the rocker arm, which was like a balance-beam suspended above the bellows. The bellows and the rocker arm constituted the blowing apparatus, which was activated by the spur-geared waterwheel.

The Art of Ironfounding

The design of an 18th-century blast furnace was based on the accumulated knowledge of the era on the art of ironfounding, an art that had been mastered, though not yet translated into scientific language. Many explanations existed, and analyses of the smelting process in the blast furnace already had a cautious scientific tone, although the science itself was lacking. They were not, however, without insight:

[...] the bellows were as good as if they had been made by a French bellowsmaker. They had only one defect, just like the bellows of the said forge, which stemmed from the quality of the wood, which should have cured for another year before being used.

One might say that it is not the ore particles that were melted, but the bodies which contain them, or with which they are mixed [...].[90]

As can be seen in the discussion on the chemistry of the blast furnace, this statement shows an intuitive understanding of the smelting process. Forgemen and ironmasters in the 18th century did not yet have scientific language to explain their art. Instead, industrial archaeology must look to their practices, recorded in the ruins of ancient blast furnaces, to decode the conditions and principles of such work. The sole repositories of this knowledge, ironworkers were able to master a difficult art without ever losing sight of the sense of fine workmanship that we appreciate in their creations today. The historian John Nef remarks that, prior to the greater production capability that came with the mechanization of the 19th century, there was an age of qualititative progress: a "quality economy" preceded, and may even have heralded, a "quantity economy." Here is how he describes the work of the 18th century:

The rhythm of work, as represented in the *Encyclopédie ou Dictionnaire raisonné des sciences, des arts et des métiers* (1751–66), was dictated by its nature, since the inventive talent of ingenuity sought continual improvement; this apparently ran counter to the manufacture of cheaper items, the main objective of the great mechanical inventions of the 20th century.

We sometimes forget that for Racine, as a hundred years later for Diderot (1713–84), co-editor with d'Alembert of the *Encyclopédie*, the word "invention" connoted craftsmanship. The qualitative economy of the Ancien Régime rested on such inventions.[91]

THE CHEMISTRY OF THE BLAST FURNACE

Knowing what happens inside a blast furnace makes it easier to understand its design, which is embodied in a particular shape, structure and charging method. A clear understanding of the smelting process overseen by the workers gives us a better appreciation of their know-how. Let us take a closer look at the art of smelting iron ore in scientific language.

All attempts to produce usable iron stem from the fact that iron is never found in nature in a metallic state (except for the rare meteoric iron). Iron, the chemical symbol for which is Fe, is always combined with other elements, chiefly oxygen, O, with which it has a great affinity. Rust, the cancer of iron, exemplifies this, since it results from the combining of iron, oxygen and hydrogen. Most commercially exploited iron ores are oxides of iron, and the iron obtained from them is usable to the extent that the oxygen is removed from the ore. In addition to being a chemical compound, iron ore is mixed with other, organic, matter. This earthy, sandy or clayey matter, or gangue, must also be removed in the smelting process.

Iron oxide thus undergoes a two-part transformation in the blast furnace: chemically it is separated from its oxygen, and organically it is separated from its gangue. This two-fold separation creates new bonds for the oxygen and gangue. To separate oxygen from iron, it must be brought into contact with an element with which it has a greater affinity: carbon, C. This element is found in concentrated form in charcoal. However, the oxygen in the ore cannot combine with the carbon in charcoal at temperatures below 800°C. The charcoal must therefore be burned, and combustion activated by a blast of air. Once the burning charcoal has reached incandescence, the carbon it contains combines with the oxygen in the ore and in the air to form carbon monoxide, CO, and carbon dioxide, CO_2, gases which escape up the chimney of the blast furnace. This chemical reaction is known as reduction, since the iron in the ore is reduced to metal. On reduction, the metal still has not been freed from the gangue, a process that takes place at a higher temperature (1,200°C–1,400°C), with the vitrification of the earthy matter of which it is composed. In order to get rid of the gangue, a fluxing agent (limestone) is added to the ore charge. This liquifies the vitrified matter at the required temperature. This mixture forms a residue known as slag, or scoria, which floats on the surface of the molten metal as a separate liquid mass.

But that is not the end of the chemical reaction. The carbon that combines with the oxygen in the ore does more than free it from the ore; it replaces it by combining with the iron. This chemical combination makes the iron oxide more fusible: fusion takes place at around 1,535°C. Lastly, a blast of air (laden with oxygen) decarburizes part of the liquified iron oxide in a final bonding of the oxygen in the air with the carbon in the iron. This is the combustion phase. The chemistry of the blast furnace is now complete.

When the resultant liquid iron has solidified in the form of a pig, the work of decarburizing this pig iron continues in a forge hearth, where it is resmelted to produce a more usable iron, becoming malleable and ductile as its carbon content is reduced.

In short, to smelt iron, the furnace must be charged, in the required order and proportions, with four types of raw material: the iron ore; a fuel containing carbon, such as charcoal; a flux, such as limestone; and air. Three products result from the blast furnace smelting process: liquid cast iron made up of an alloy of iron and carbon; slag, which is a combination of flux and gangue in liquid form; and gases that are given off into the atmosphere.

Changes to the Blast Furnace

Your memorialists find [...] further that no modern improvements in iron Manufactures have been introduced at the Forges of St. Maurice. [...] are daily expecting, from Scotland, an Engineer, fully competent to make the necessary improvements and conduct such a Manufacture.[92]

The design of the 18th-century French blast furnace did not change until the middle of the 19th century. In 1852, Étienne Parent, sent by the government to report on the state of the establishment, wrote in his report:

> Except one large brick building, intended for a sawmill, put up by Mr. Henry Stuart, and unfinished, and a hot-air Blow-Engine, also erected by that gentleman, everything bore the impress of age.[93]

That same year, John Porter & Company, the new owners of the Forges, announced major changes to the blast furnace. Henry Stuart, who had bought the Forges from the government six years earlier, had already installed an air compressor (a "Blow-Engine") and a "single" hot air furnace to replace the old bellows, which used ambient, that is, cold, air.[94] But his financial difficulties and poor maintenance of the facilities, under the direction of James Ferrier, blocked any further innovation.[95] These private businessmen already knew that, to keep the company viable, the productivity of the blast furnace had to be increased to meet the growing market demand for castings.[96] Preheating the blast would not only increase product volume, but also save fuel.[97]

In 1852, John Porter & Company resolved to innovate by hiring the engineer William Hunter, dispatched from Scotland, who would have to try and make the most of the existing facilities.[98] Hunter made specific recommendations with regard to the blast furnace.

> I would recommend a new Blast-Furnace of nearly double the capacity of the present one, with a double Hot-air furnace instead of a single one as at present.[99]

In 1854, therefore, the blast furnace was rebuilt to double its capacity, boosting the daily output from 2$\frac{1}{4}$ to 5 tons of cast iron.[100] These major changes were made, however, at the cost of two explosions at the new furnace (17 April and 16 October 1854), which killed two workers and seriously injured two others.[101]

The new furnace, rebuilt a second time[102] following the October explosion, was considerably different from the 18th-century furnace: the inner lining was now circular and twice as high as that of the original furnace. Dr B. J. Harrington provided the internal dimensions 20 years later:[103]

> The internal dimensions of this furnace are,

Height		30 feet	(9.1 m)
Diameter at	hearth	2 $\frac{1}{2}$ feet	(76 cm)
	boshes	7 feet	(2.1 m)
	throat	3 $\frac{1}{2}$ feet	(1 m)

Dollard Dubé questioned some workers who were at the Forges when Harrington visited, and his reconstruction, produced in 1933, is a second source of information on the blast furnace at that time[104] (Table 4.3).

Doubling the capacity of a blast furnace does not necessarily entail doubling its size. In the case of the blast furnace at the Forges, the shape of its belly and its charging height were changed without enlarging the perimeter of the base. Table 4.3 clearly shows this, as does a plan view superimposing the lining of the second furnace on the interior cavity of the first one in a further demonstration (Plate 4.4). The belly's shape and charging height make all the difference. The overall height of the furnace is not a good indicator, since it can be measured with or without the foundations, roof and chimney. Dubé, who drew a plan of the furnace "as it must have been during the French period—as it was in the time of Mr Bell (1845)," showed the charging height of the first and last furnaces as the same, placing the height of the throat at 36 feet (11 m) (Plate 4.5). At the time of his investigation, he did not know that the charging height had been doubled in 1854.[105] On the plan of the first furnace, he also superimposed the track for the mechanically operated charging wagon, which he had heard about from the last workers, but which was not installed until after 1854. This enabled raw materials to be winched up to the charging platform on a rail-mounted cart.

Plate 4.4

PLAN VIEW OF THE CIRCULAR BELLY OF THE BLAST FURNACE OF 1873 PROJECTED IN THE PERIMETER OF THE STRUCTURE OF THE BLAST FURNACE OF 1741

Belly of the 1873 blast furnace
(plan view)

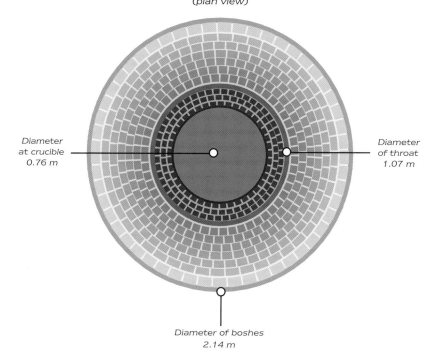

Diameter at crucible 0.76 m

Diameter of throat 1.07 m

Diameter of boshes 2.14 m

Belly of the 1873 blast furnace
projected in the footprint of the original furnace (1741)
(plan view)

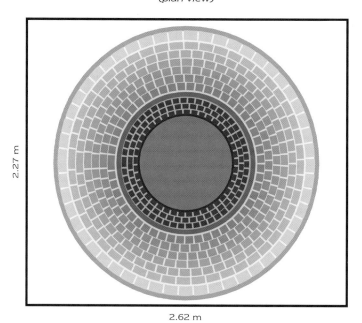

2.27 m

2.62 m

DRAWING BY ANDRÉE HÉROUX.

Table 4.3

**DIMENSIONS OF THE FIRST
AND LAST BLAST FURNACES** (English measure`)

	Perimeter at base	Interior cavity			Total height
		Height	Shape	Perimeter	
1741 Estèbe	30' x 30'	16'	rectangular	8.6'x 7.5'	34'#
1873 Harrington		30'	circular	7' diameter	
1870–80 Dubé	30' x 30'	36' (charging height)	circular		55'`

` The 1741 figures have been rounded off. One foot equals 30.5 cm.

Including the foundations (7.5 feet) and penthouse walls (10 feet).

` Including the stack, nearly 20 feet high, which capped the furnace top.

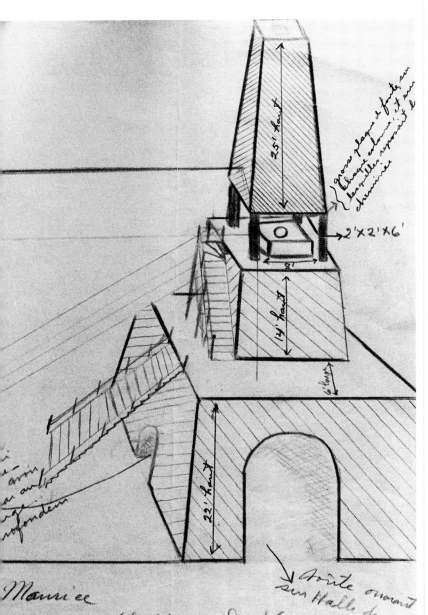

Plate 4.5
"Blast Furnace As It Must Have Been at the Time of the French," Dollard Dubé, 1933.
ARCHIVES DU SÉMINAIRE DES TROIS-RIVIÈRES, DUBÉ COLLECTION, N3 F11.

The last blast furnace, shown in a painting by Bunnet (Plate 4.6), includes buttresses on the three sides (east, north and south) adjacent to the gully. Formerly pierced by two arches, on the south (tuyere arch) and the west (tymp arch), the furnace now had a third arch, on the north side, for access to a second tuyere, opposite the first. The two tuyeres were connected to the new blowing engine.

This blowing engine was still activated by the great waterwheel and was located in the shed on the south side, but it now consisted of a compressor that blew air into the two tuyeres, placed opposite each other in the crucible (Plate 4.7). Harrington specified that the blast was cold, which means that—contrary to the statements in Hunter's report, Hunt's 1868 report, and Parent's 1852 report—the air was not first heated in an oven. These apparently contradictory statements suggest that it was possible to alternate between a cold blast and a hot blast,[106] depending upon the type of ore to be smelted. Based on the accounts of the last workers, Dubé produced a sketch of such an oven.[107] According to Bérubé, the air in this oven was heated by the hot gases from the furnace itself, which were recovered through a piping system linking the furnace top and the oven (Plate 4.8).[108] The tuyeres receiving this very hot air (which could reach 500°C) were cooled with water to prevent them from melting during blast. One of these tuyeres is still in place and was x-rayed to enable us to better understand how its cooling system operated. In addition, a plan by the architect Ernest Denoncourt, based on Dubé's notes which mention "a tun of water," shows a water reservoir, containing water for cooling the tuyeres, perched on a beam in the bellows shed.[109]

The new furnace also had two flues in the chimney: the regular one up the chimney capping the throat; and a second one, linked to the interior of the furnace through one side of the furnace top, for releasing the gases to which the workers were exposed during charging[110] (Plates 4.7 and 4.9).

The last furnace, as found by archaeologists, shows signs of modifications not documented by written and oral sources. The circular crucible unearthed is made of refractory brick (Plate 4.10). However, in 1878, five years before the Forges closed, sandstone slabs were still being used to build the crucible.[111] These modifications were probably made in 1881, following the furnace explosion that set fire to the sheds.[112] In addition, we have seen that the great wheel was replaced at the same time by a hydraulic turbine.

The final two years of operations at the Forges were especially productive, in part because of the improvements to the blast furnace and also because of the construction of a second furnace, known as the "new furnace" at the upper forge. Once this furnace began operating, the Forges had a production capacity of 7 tons of pig iron a day—an annual production of some 2,500 tons.[113]

The New Furnace

[...] After construction of the new furnace, we very certainly were able to produce twice as much at the Forges [...] as before. The new furnace is made of iron and is greatly superior to the old stone furnace. It also has the advantage of being steam operated, while the old one ran on water power.[114]

In the late 1740s, there were already plans to build a second blast furnace at the Forges, for cannon founding, but nothing came of this. It was not until 150 years later that a second

Plate 4.6
Derelict blast furnace, three years after the closing of the Forges. Buttresses can be seen against the north and south sides of the furnace.
HENRY RICHARD S. BUNNETT, *THE FORGES OF THE ST. MAURICE* (DETAIL), OIL ON CANVAS, 1886, McCORD MUSEUM OF CANADIAN HISTORY, M-739.

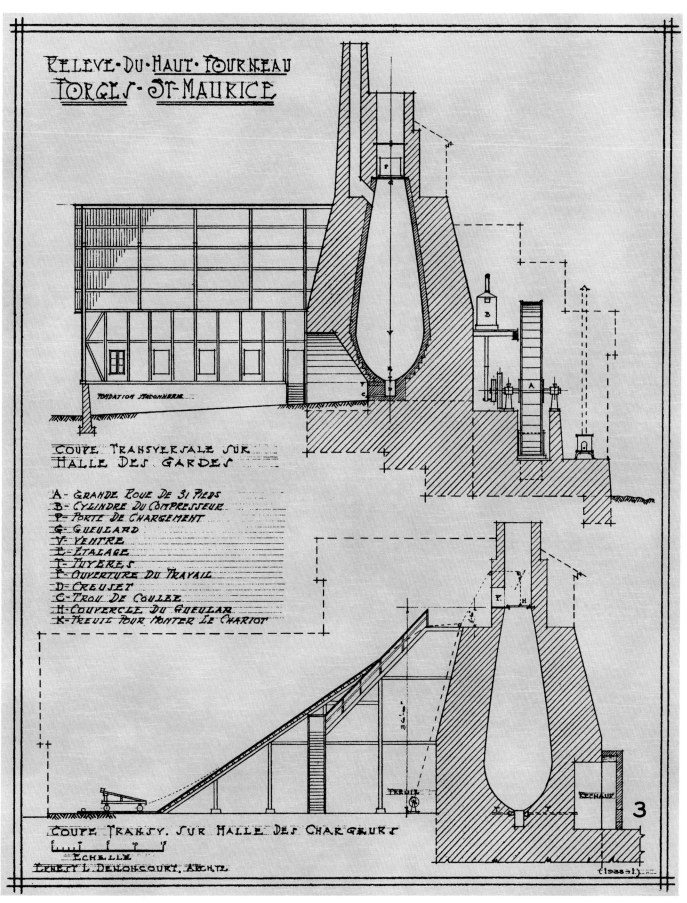

RELEVE·DU·HAUT·FOURNEAU
FORGES·ST-MAURICE

COUPE TRANSVERSALE SUR
HALLE DES GARDES

A- GRANDE ROUE DE 31 PIEDS
B- CYLINDRE DU COMPRESSEUR
P- PORTE DE CHARGEMENT
G- GUEULARD
V- VENTRE
E- ÉTALAGE
T- TUYÈRES
F- OUVERTURE DU TRAVAIL
D- CREUSET
C- TROU DE COULÉE
H- COUVERCLE DU GUEULARD
K- TREUIL POUR MONTER LE CHARIOT

COUPE TRANSV. SUR HALLE DES CHARGEURS
ÉCHELLE
ERNEST L. DENONCOURT, ARCH.TE

3

(1933)

Plate 4.7
Architect's reconstruction of the last blast furnace, based on accounts gathered
by Dollard Dubé in 1933.
DENONCOURT AND DENONCOURT, "RELEVÉ DU HAUT FOURNEAU, FORGES ST-MAURICE,"
RECONSTRUCTION DES FORGES, 1962 (1933), P. 3, PARKS CANADA,
CULTURAL RESOURCE MANAGEMENT.

furnace was finally built, this time to meet the needs of the railroad industry. Although the St Maurice Creek had been in use for 150 years, not all its possibilities had been exhausted. The upper forge, altered several times beginning in the 1850s, was located on a site suitable for charging a blast furnace. The forge had been built just at the foot of the north escarpment of the gully, about 30 feet (9 m) below the plateau. The new furnace was built just below, within the footprint of the old forge, and its charging platform was linked to the adjacent plateau by a covered bridge mounted on large trestles. The charges were conveyed along it to the furnace top and not, as was the case at the other blast furnace, up a slope (Plate 4.11).

Built of iron, with inwalls of refractory brick—as indicated by excavated remains— the belly of the new furnace was also circular in form (Plate 4.12). Its blower, probably a compressor using preheated air, was activated alternately by a hydraulic turbine and a steam engine.[115] Four firms were involved in installing the new furnace: Luckerhoff & Bros, Alexander McKelvie & Sons, Viger & Sayer, and Chantecloup.[116]

At this time, the technology used at the Forges was evolving fast, as journalists of the period were quick to point out. The *Journal des Trois-Rivières* of 14 March 1881 reported:

> Mr George McDougall, proprietor of the Old St Maurice Forges, has begun construction of a new furnace to smelt ore, based on the latest designs. It will be most interesting to see a new furnace, built to the most modern design, side by side with a furnace that has been in existence for two centuries.[117]

This change, however, did not result in a sustainable transition. The Forges closed their doors two years later, passing into history without mention in the press. The new furnace was demolished in 1887.[118]

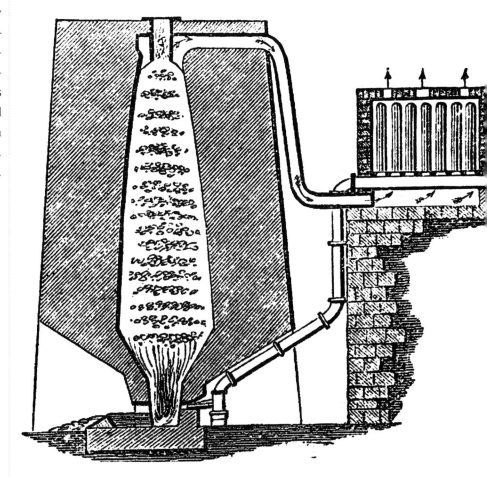

Plate 4.8
Gas recovery system from a hot-blast furnace.
BYRNE AND SPON, EDS., *SPON'S DICTIONARY OF ENGINEERING [...]* (LONDON AND NEW YORK: E. & F. N. SPON, 1874), VOL. 1, P. 347.

Plate 4.9
Blast furnace complex, circa 1880. In the foreground, on the right,
the casting house; on the left, the charcoal house with charging wagon.
RECONSTRUCTED BY ILLUSTRATOR BERNARD DUCHESNE.

Plate 4.10
Remains of circular firebrick hearth of the last blast furnace
built on the site of the original furnace.
PHOTOGRAPH BY PARKS CANADA, 25G-884X, JULY 1974.

81

55

39

Plate 4.11
The upper forge as it changed over time, with bridge
to the charging platform of the "new furnace," built in 1881.
RECONSTRUCTED BY ILLUSTRATOR BERNARD DUCHESNE.

Plate 4.12
Remains of hearth of "new furnace"
built at the site of the upper forge.
PHOTOGRAPH BY PARKS CANADA, 25G-813X, 1973.

The Forges

The place where the iron pigs are converted into
wrought iron is called the forge, and consists
of hearths and a hammer assembly, all contained
in a spacious building close to the charcoal house,
the workers' quarters and the worksite [...]

Encyclopédie, *1757*

The two forges, built at the top and bottom of the gully, were not simple village smithies; they were great forges (*grosses forges*), manned by **hammermen** and **finers**, assisted by **helpers** called *goujats*. These forgemen were headed by a working hammerman, the master hammerman, who was as important as the master founder and, like him, was in charge of the construction, maintenance and operation of his shop. We will look at the work of these men in more detail in Chapter 5. For the moment, it should be kept in mind that they carried out the second and last stage in converting the iron ore into a usable product: wrought iron, known as bar or **merchant iron**.

The great forge had one or two chaferies or fineries in which the heavy iron pigs made in the blast furnace were remelted. The iron, refined to a pasty state in the *renardière* chafery, was drawn into bars under a forge hammer weighing as much as 200 kg. These bars were the base material used by manufacturers of iron parts and objects—shipyards and nailsmiths, as well as village blacksmiths and farriers.

A typical forge was rectangular in shape with four work areas or bays: the forge proper was flanked on one long side by its wheelrace and on the other long side by a charcoal shed and iron store. Inside the forge, ranged along the wheelrace side, were the two chaferies on either side of the forge hammer, while the other side led to the sheds where charcoal and products manufactured in the forge were stored. From outside the forge, below its millpond, the position of the two chaferies was visible from the high, massive chimneys rising above the roof (Plate 4.13).

Plate 4.13
Upper forge, after Chaussegros de Léry's plan
of 1739 and Estèbe's inventory of 1741.
RECONSTRUCTED BY ILLUSTRATOR BERNARD DUCHESNE.

Each forge was an independent operation, with its own hydraulic system (dam, millpond, wheelrace, waterwheels and gearing), its own supply of fuel and its own storage areas, as well as its own crew of workers. And we shall see that the forgemen tended to develop a strong sense of belonging to their own particular forge.

The remains of the two forges tell two very different stories. The lower forge shows more clearly than do the remains of the upper forge what the original building was like. Our reading of the remains of the lower forge is certainly influenced to a large extent by the fact that one of the two chaferies has survived, but other elements also remain to flesh out the picture. And in attempting to describe this forge, we will see how a great iron forge was laid out and equipped. The remains of the upper forge, on the other hand, provide a better indication of the more substantial changes it underwent; it also lost its original vocation sooner than did the lower forge. Our discussion of the upper forge, built in 1739, will deal particularly with the technical changes of which the remains provide evidence.

The Lower Forge

[...] I noticed [...] that the stream could only provide enough water to turn two wheels and that they had built a building to contain six [...].[119]

Location

In his 1735 site survey, Vézin noted that "on a plateau [...] running along the said river," where Francheville's forge stood, "there would be enough land to build a great forge." This plateau along the river was only about 7 m above the water level, and measured about 90 by 14 *toises* (175 m by 27 m). It was, in fact, the last terrace crossed by the creek before it emptied into the river. It was from there that Vézin carefully established that its fall down through the gully was 65 feet, $9^3/_4$ inches (21.3 m) over a distance of 5 arpents (58 m). In fact, Vézin's whole plan centred on the location of Francheville's bloomery forge, which he intended to replace with a great iron forge. In his "Observations," however, he made no mention of the blast furnace that should have been part of his plans. Yet, the location of the furnace was critical for the subsequent distribution of the creek's water power, since the flow that Vézin measured in 1735 above Franchville's forge would presumably not be the same once part of the water was dammed for the furnace pond higher up the slope.

To level out the location for the forge and lay its foundations, "about 3,000 *toises* of clay and sand had to be removed," according to Vézin. Since he had been critical of the foundations of Franchville's forge, Vézin undoubtedly wanted to make sure that his forge was built "on the bedrock" which he said he had found at "10 feet (3.2 m) in depth" during his first visit, in the fall of 1735.[120] He therefore erected foundation walls that were 2 to 3 feet thick and from 6 to 14 feet high (1.8–4.3 m).[121]

The Building

When he realized, late in the day, that he had overestimated the energy potential of the creek at the foot of the slope, Vézin scaled back his plans for the forge. If Léry's comments, cited in the epigraph, are to be believed, however, he does not seem to have similarly reduced the size of the building itself. Before changing his mind, Vézin had had time to lay the foundations for two wheelraces, one on each side of the forge, but only the one on the south side was built. His initial plans for the forge proper were for a building 90 feet (29 m) long by 40 feet (13 m) wide by 10 feet (3 m) high to the eaves to accommodate "four fires" for two fineries, a chafery and a plate mill, a forge hammer and a tilt hammer.[122] The building measured by Estèbe had slightly different dimensions, but this was the building that Léry was referring to when he mentioned a forge that was too large for the equipment it contained. On the north side of the building, a lean-to running the entire length of the forge contained the charcoal shed (northeast) and the iron store (northwest). On the south side, also running the entire length of the forge, was the masonry wheelrace in which were set the forebays and the three waterwheels, 8 feet (2.6 m) in diameter. The topography and archaeological excavations indicate that access to the forge was from the east (the river side),[123] but Estèbe describes doors "in both gables," i.e. on the east (river) and west (milldam) sides. On the south side, the wheelrace under the forebays was bridged, so that the heavy iron pigs could be brought up to the openings in each chafery and fed in from the outside.

Layout

A chimney with its fire, furnished with its iron plates [...]. A chimney cracked in a number of places [...]. The hammer hurst frame in good condition, with a hammer in its hurst and an anvil in its timber block [...]. Two pairs of bellows in good condition, one pair mounted, with a tuyere, shafts and wheels for both pairs.

Estèbe, 1741

The forge comprised two hearths (both set up *en renardière*), two pairs of bellows, and a forge hammer, all activated mechanically by camshafts coupled to waterwheels. The two pairs of bellows fanned the fire in the hearths, and the hammer was used to work the pasty iron as it came out of the hearths.

The machinery was ranged along the entire length of the building, on the south side, parallel to the wheelrace in which the three waterwheels and their forebays were mounted. At each end was a *renardière* chafery with its bellows and, between the two, the hurst frame of the hammer. Originally, the chaferies were not placed symmetrically on either side of the hammer; both were set up the same way, with their bellows on the right, making the east chafery farther from the hammer than the west chafery. Several years later, when there were plans to rebuild the east chafery, mention was made of the fact that it was "located too far from the hammer [...], instead of which a hearth should be as close as possible to the hammer for the convenience of the forgemen and so that the **loop** has less time to cool down as it is taken from the chafery to the hammer."[124] The change was made, since the remains of the forge include, in the east corner, a second hearth base closer to the hammer.[125] It is difficult to understand why Vézin initially placed the hearths asymmetrically, which was not a very sensible arrangement. Perhaps he wanted to work around constraints imposed by the waterwheels or was simply following his original plans, which called for more machinery. Whatever the reason, the second chafery was never more than a back-up system, brought into service when the water was high or to replace the first one, since work stoppages, caused by all sorts of breakdowns, were common.

The Fining Process

The hearths are called chaferies, fineries or renardières, depending on the type of work [...].

Encyclopédie, *1757*

Vézin built the chaferies *en renardière*, that is, they were hearths in which the two operations required to convert the pig into wrought iron—fining and reheating—could be carried out alternately. In the other process, the Walloon process, also used in Europe at that time, the two operations were carried out in two separate hearths: the finery, which was used to refine the pig into a mass of iron known as a loop, and the chafery, which was used to reheat the **bloom** (hammered loop) during hammering. In the Walloon process, there was a strict division of labour between finers and chaferymen. In the *renardière* process, finers and hammermen carried out both fining and heating operations[126] alternately, in the same hearth. In this process, the loop was known picturesquely as the *renard* (the fox), formed during the fining process in its den or *renardière*, as the crucible was termed.

The *renardière* process or *méthode comtoise*, which originated in Germany, was very common in France because it used less charcoal. Franche-Comté forgemen familiar with it were much sought after by French ironmasters.[127] At the Forges, Vézin at first opted for the Walloon process,[128] but eventually settled on the Franche-Comté process, since it allowed him to reduce the number of hearths and thus, waterwheels. This decision explains why most St Maurice forgemen came from Franche-Comté.

From a technical standpoint, a *renardière* differed from a finery in the design of the crucible and the way in which the metal was worked. The bottom of the crucible of a finery was twice as deep (9–10 inches below the tuyere) as that of a *renardière*. In a finery, when the pig began to melt, a mass about 4 inches thick was allowed to form, made up of slag combined with breeze[129] (charcoal dust), and known as **dross**. It was on this bed of dross that the fining proper was carried out, with the molten metal being worked with ringers into a pasty mass, or loop. In this process, the slag given off in fining the cast iron ran off on either side of the dross to collect in the bottom of the crucible. In a *renardière* like the one at St Maurice, the fining was done directly in the bottom of the crucible, which was located 4 or 5 inches below the tuyere—at the same level as the top of the dross in the finery—except that in the *renardière*, the loop, or *renard*, was constantly bathed in the slag. In the 18th century, defenders of the *renardière* process saw only its advantages, claiming that the "iron [...] fattened and softened in the slag," while in ordinary fineries, the dross absorbed too much of the iron.[130] Iron from the Forges had the reputation of being soft and malleable; the ore no doubt accounted for much of this, but the work of the Franche-Comté forgemen also played a role.[131]

They say that the iron is fattened and softened by the slag: that is true when it is lacking, but in all cases and with molten iron always in the bottom of a renardière, the iron is more likely to absorb it than on the dross of a finery: has experience not shown us that, with the same quality of cast iron, renardière iron is the best?

Remains of the Chaferies

As much care had to be taken in building a forge hearth as in constructing a blast furnace. The hearth was an imposing structure 45 feet (14.6 m) high, and visitors to the site can see it for themselves, since the west chimney of the lower forge is still standing (Plate 4.14). Details of its construction would suggest that some parts of it are original. The chimney currently rests on four different types of pillar, including a recent one made of concrete and another one, probably original, made of thick slabs of cast iron. There is nothing left of the

renardière hearth itself, but it is easy to pinpoint its exact location, just below a group of stones bleached and reddened by fire, close to the cast iron pillar. The hearth of the extant chimney is open on all four sides, so that four pillars are visible. When it was in operation, however, it had only two pillars and was open only on two sides (north and east); the other two sides were closed, with only a small opening for the tuyere (west) and one for the pig to be fed into the hearth (south).

Plate 4.14
Parks Canada's reinforcement of the lower forge chimney.
One of the original pillars, the only one made of cast iron, can be seen.
PHOTOGRAPH BY PARKS CANADA, 25G-78, R13M, JULY 1978.

The St Maurice chaferies were built of the same type of sandstone as the blast furnace and, like it, rested on a solid base of masonry. The two chaferies of the lower forge, inspected by Estèbe in 1741, sat on a square foundation measuring 11 feet by 11 feet (3.6 m by 3.6 m) (Table 4.4). Apart from the west chimney, still standing, these masonry bases are almost the only remaining elements that bear witness to the two forge chaferies. Archaeological digs have revealed some variation in their sizes and, in particular, differences in how they were built.[132] The base of the chimney still standing (west chimney) is not of solid stone, but is instead made up of four half walls, slightly over 1.5 m high, set right on the ground around a hollow core, and the south wall is part of the foundation of the forge itself. The base of the other chimney consists of solid stone nearly 2 m thick, resting on a sill of wooden beams, clearly designed to insulate it from the damp soil. Unlike the base of the chimney still standing, it is not part of the foundation wall, but instead is supported by it. The third base, which indicates that the chimney hearth was moved closer to the hammer in the middle, consists of solid stone on top of a layer of fill. It also is supported by the foundation wall, on the south, and by the other base, on the east. We have seen that the latter base has a rectangular opening in the front (north side), which could well be the **cinder notch** for removing slag from the crucible, traces of which remain in the bottom.

Table 4.4

DIMENSIONS OF THE CHAFERIES OF THE TWO FORGES, 1741[133] (French measure)

Lower forge	Base	Chimney	Pillar	Hearth plates	Reinforcement
Chafery 1	solid 11'x11' 10' deep	width at base side 1: 9'6" side 2: 9'6" side 3: 8'6" side 4: 10' width at top side 1: 5'6" side 2: 5'6" side 3: 5'2" side 4: 5'2" height: 38'8"	(no mention)	"its hearth plates"	(no mention)
Chafery 2[134]	solid 11'6"x 11'6" 10' deep	width at base side 1: 10' side 2: 10' side 3: 9'6" side 4: 9'6" width at top side 1: 4' side 2: 4' side 3: 3'5" side 4: 3'5" height: 40'6"	(no mention)	(no mention)	(no mention)

Upper forge	Base	Chimney	Pillar	Hearth plates	Reinforcement
Chafery 1	solid 12'x12' 11' deep	width at base side 1: 9' side 2: 9' side 3: 9' side 4: 10' width at top side 1: 4'8" side 2: 4'8" side 3: 5'3" side 4: 5'3" height: 40'	7 squares of cast iron in one pillar	2 iron plates 9'6"x14"x2" and 1 plate 10'6"x14"x2" to support the lintel courses of the chimney; 1 iron plate 5'10"x14"x2" used as a lintel	5 iron bars with pins bracing the chimney
Chafery 2	solid 12'x12' 11'7" deep	width at base side 1: 9'10" side 2: 9'10" side 3: 9'4" side 4: 8'10" width at top side 1: 6' side 2: 6' side 3: 5'3" side 4: 5'3" height: 40'6"	(no mention)	"its hearth plates"	(no mention)

The Remaining Chimney

Unlike the remains of the blast furnace, which are difficult to read, the chimney still standing provides a clearer understanding of the design of an 18th-century *renardière* chafery (Plate 4.15). It just looks like an ordinary chimney, and that is what it was usually called. It had two parts—the base, or hearth, within the pillars and, above it, the pyramid-shaped chimney (which was once completely covered in rough cast), resting on the pillars. The chimney sits not on the pillars but on four cast iron lintels that span between them. There are two arches above the lintels on the north and east sides. These relieving arches, designed to spread the weight of the chimney to the pillars, show where the two openings in the hearth were—one in front, on the north side, and the other on the east, on the **fore spirit** side. There are no arches on the other two sides of the hearth, these being solid walls. Above the lintels, at the base of the chimney, are two tie-rods and two anchors for bracing the masonry, just like at the blast furnace.

Plate 4.15
Renardière chafery of lower forge, based on remains found.
RECONSTRUCTED BY ILLUSTRATOR BERNARD DUCHESNE.

The front of the original chimney was partly closed by a low wall, which was supported by an inclined lintel that ran between the two pillars. The wall served as a fireguard, screening the finer from the flames and sparks that flew out of the crucible as he made his loop[135] (Plate 4.15). The other opening, on the east side, enabled the helper to "tend the fire."[136] At the back of the hearth, on the wheel-race side, through a small opening in the hearth wall, the iron pigs were fed in from the outside to rest on the **hare plate**. The helper, working at the hearth, gradually slid the pig over the fire in the crucible, using his long **ringer** as a lever. In the west wall, another opening received the tuyere from the bellows.

The pillar of thick iron plates long puzzled researchers, but it is, in fact, the clearest indication of the location of the crucible. The texts to the illustrations in Diderot's *Encyclopédie* recommend that the two pillars be built of cast iron plates for greater strength.[137] In fact, the intense heat from the crucible would have weakened a masonry pillar, endangering the stability of the heavy chimney. An examination of the reddened stones of the existing hearth, near the iron pillar, indicates that the crucible was confined to the northwest corner of the hearth.

The Renardière *Chafery*

Inside the hearth, a set of iron plates formed the crucible (or *renardière*) on the tuyere side, and a cistern or wooden trough ran the length of the fore spirit side for "cooling tools and slaking the fire." The crucible was a rectangular receptacle (30 by 15 by 10 inches) made up of five iron plates, about 3 inches thick.[138] The bottom plate (30 by 15 inches) sat on two bed plates of iron to insulate it from the dampness of the soil and create a vent.[139] On the tuyere side (south side of the existing chimney) was the **tuyere plate** (6 inches high) on which rested the copper **muzzle** of the tuyere, which protruded 3 inches into the crucible.[140] Opposite the tuyere was the fore spirit plate (11 inches high), topped by a plate lying on the fore spirit that held in the charcoal covering the pig. The back of the crucible was formed by the hare plate (11 inches high), on which the pig rested as it gradually melted. The front of the crucible was closed off by the 11-inch high **fore plate**, which had two small openings, or slag holes, for runoff of the slag created during smelting. In front of this plate was the cinder notch, a small, sloping channel, edged by two supports, which drained off the slag. The supports held up the large front plate, in which was installed the "fork" used to clean the finer's ringer. The plates in the crucible could be ajusted "as required by the type of pig iron, on the principle that the pig is above the blast and the work is below it." Thus, in a *renardière* chafery, the work involves "sliding the pig into the crucible against the fore spirit plate, covering it with coals and working the bellows."[141]

The Bellows

In the initial forge layout, the two pairs of bellows were located one to the right of each chafery, on the same side as the tuyere. Shortly afterwards, the bellows of the east chafery, the lower chafery, were relocated close to the east wall after the chafery had been moved nearer to the hammer in the middle of the forge.[142] The forge bellows were the same as those at the blast furnace, but only half as large at 7 to 9 feet (2.2–3 m long); the location of the bellows, between the hearth foundations and the gable walls, made it impossible to install larger bellows, especially since room had to be left for the camshaft and rocker arm. The bellows were made of wood and, like those of the blast furnace, were lowered alternately by a spur-geared mechanism, activated by a water-wheel. The bellowsman had to ensure that the box of the bellows was level with the bottom of the crucible and that "the blast from the two pairs of bellows met in the middle of the hearth."[143]

Unlike the blast furnace, where the bellows never stopped, the forge bellows operated intermittently; they were activated depending on the type of work (fining or heating) and the stage of work. Accordingly, each waterwheel had its own forebay, a water reservoir located immediately above the wheel. The shuttle or gate of each forebay was raised by a lever from inside the forge, close to the chafery, by which the waterwheel could be activated at will. The speed at which the wheel turned depended on how high the shuttle was raised. Unlike the wheel of a forge hammer, the wheel that activated the bellows mechanism turned slowly, to produce an even blast. In 1739, Chaussegros de Léry reported that it made only four revolutions a minute:

The great wheel of the upper chafery turns 24 times every six minutes, at full water and with the millpond full. The wheel of the lower chafery makes the same number of revolutions in the same length of time, but using less water, with the shuttle not completely raised. But I noticed that the lower chafery bellows lose much more wind than those of the upper chafery.[144]

The Forge Hammer

The hammer rises and falls four times
with each revolution of the camshaft; and with
a good stream of water, the camshaft can turn
25 times a minute.

Encyclopédie, *1757*

The forge hammer was undoubtedly the most impressive piece of machinery in a forge. Sometimes the forge itself was simply called "the hammer mill," and we know that the man in charge, the "hammerman" took his name from this machine. In a great forge, the term "hammer" referred not only to the heavy hammerhead of iron in its haft, or helve, but also to the hurst frame, a massive structure that housed the hammer and straddled the middle of the forge itself (Plate 4.16). The first hammers installed by Vézin weighed 200 to 350 kg, and it is not difficult to imagine that the hurst frame had to be very strong indeed. In addition, it was set on its own foundation separate from that of the forge, since a single blow of the hammer on the **anvil** could shake the whole building, a fact that was mentioned at the time.[145] The encyclopedist Bouchu emphasized this point by stating that "to ensure great strength, all the parts have to support each other on a solid foundation."[146]

Plate 4.16
Hurst frame of lower forge hammer, based on remains found.
RECONSTRUCTED BY ILLUSTRATOR BERNARD DUCHESNE.

The hurst frame installed by Vézin was set on "a masonry foundation 14 feet long, 8 feet wide and 6 feet deep," against the south wall between the two chaferies. Since the wooden structure spanned the entire width of the forge, another masonry foundation, also 14 feet in length, but 3 feet wide and 3 feet deep, was set opposite the first. Finally, under the timber **block** on which the anvil was placed, lay another masonry foundation, 8 feet by 2 ½ feet thick. Only part of the foundations of the hurst frame can be seen today.

The hurst frame was an assemblage of nearly 30 interlocking wooden and iron parts (Plate 4.16).[147] A 1739 document details the various oak timbers required to build a hurst frame[148] (see Appendix 6). Generally speaking, the hurst frame resembled a huge sawhorse about 3.6 m high that spanned the entire width of the forge.[149] The massive horizontal main **drome-beam** was fitted at each end into two vertical **hammer posts**, themselves buttressed by great legs at the back and on either side. The great hammer post had a complex arrangement of vertical and horizontal parts surrounding the hammer, while on the other side, the lesser hammer post stood by itself. All the stout vertical timbers were slotted with mortise and tenon joints into a series of sills, supports and cross beams or cross bars, sunk into the foundations of the hurst frame. At the foot of the structure, parallel to the helve of the hammer, and supported on its **plummer block**, was the camshaft that activated the hammer. The forge hammer was thus a belly helve, since the four cams raised the haft from the side by striking an iron band, and not from the tail of the helve, as was the case with the smaller nose or frontal helve hammer, known as the tilt hammer. The hammer helve was fastened with wedges into the **hurst**, a wrought iron or cast iron collar with pivots fitted into sockets on the legs of the hurst frame to facilitate movement. Two and a half feet in from the great hammer post was another vertical post housing the **rabbet**, a heavy curved counter beam, above the helve and parallel to it. The rabbet acted as a spring, lowering the head of the hammer each time it was raised by the cams. Under the head of the hammer was the anvil, a block of iron weighing a ton, set into a solid frame, the timber block, the remains of which can be seen at the forge.[150]

Table 4.5

DIMENSIONS OF THE THREE WHEELS OF THE LOWER FORGE, AS REPORTED IN 1739[152] (French measure)

	Diameter	Width	Buckets*	Head†	Shuttle#
Upper (west) chafery	8'	2'	14"	10"	2"x14"
Lower (east) chafery	8'	2'	14'	15"	2"x14"
Hammer	8'	2'9"	18"	31"	4"x18"

* Depth of buckets in inches

† Vertical distance between the bottom of the forebay and the top of the wheel

Shuttle aperture

Like most parts of the forge, Vézin's hurst frame was entirely rebuilt in 1738, and only the main drome-beam could be reused. The axle-tree (3 feet or about 1 m in diameter) was not able to withstand the 350-kg hammer for even a year except for "the iron hoops with which it is girdled from one end to the other."[151]

Like the bellows wheel, the hammer wheel was operated from inside, by a clutch-like **rocker** and **connecting rod** that opened the forebay sluice. The camshaft was not activated by a gear mechanism, but was directly driven by the wheel, of which it formed the axle. A gear mechanism would not have been able to withstand the repeated hammer blows. The hammer wheel, although of the same diameter (8 feet) as the bellows wheels, was broader and its buckets deeper (Table 4.5 and Plate 3.4). Similarly, the forebay shuttle aperture was $2\frac{1}{2}$ times larger and the head of water to the wheel was greater. These special features, to give more momentum to the wheel so that it could raise the heavy hammer, entailed a greater expense of water. The hammer wheel could turn quickly—some 25 revolutions a minute, as reported by Bouchu in the epigraph. Trials with miniature wheels conducted by Chaussegros de Léry indicated that this was roughly the speed he was looking for. Experimenting with various wheel models, he achieved speeds ranging from 16 to 43 revolutions a minute, depending on the different sluice apertures.[153]

The Blacksmith's Shop

In his 1735 plans, Vézin intended to have a blacksmith's shop adjoining the forge, as well as another one for the slitting mill, which was never built.[154] The last part of the 1741 inventory of the lower forge contains a section entitled "Blacksmith's tools," which mentions "Marineau's shop," with anvils, timber block, and cowhide bellows, although no direct reference is made to a forge hearth. The shop planned by Vézin certainly included a "forge and chimney." Its cost (1,100 *livres*) was included in the total cost of the forge, and the wages of the blacksmith were also included in forge operating costs. The shop was therefore actually built, but it is difficult to describe it in the absence of any remains or specific record of the building and its layout. Other shops of the same type, as well as woodworkers' shops, were inventoried at the Forges at various times. There must, in fact, have been a permanent staff of craftsmen such as carpenters, wheelwrights and joiners in an establishment where most of the buildings, machinery and vehicles were made of wood and often had to be repaired. The shops of these craftsmen, most of whom could turn their hand to several trades, usually adjoined their living quarters (see Chapter 9).

Alterations to the Forge

We have seen that, even during its construction, the forge underwent a number of repairs and alterations, most of them as a result of poor workmanship in the beginning. Although by 1741 the establishment had, in the words of Hocquart himself, "come to perfection,"[155] an inventory taken that same year shows that some defects still remained, and work had not yet been completed. At the lower forge, one of the chimneys had several cracks and a pair of bellows (probably those belonging to this same chimney) were not in place. We know also that, shortly after this, the lower chafery was demolished and rebuilt farther west, not only because it would be more convenient, but also because, near the wall, the sparks from the hearth "were continually setting fire to the rafters." This roof was "too high" and was rebuilt in 1740 because it was "too close to the chimney pipe" and was truly a fire hazard. Cugnet said that during the winter of that year, he had "seen fire break out three times a week."[156] The workers used a copper syringe to put out small fires inside the forge.[157]

On 3 September 1746, a Saturday night, after the workers had left, the forge was completely razed by fire. The conflagration was caused by sparks from the hearth, which "set fire to the breeze scattered all over the building," despite the fact that everything had been "swept [...] as was usual on Saturday night."[158] The mention of breeze (charcoal dust) all over the forge speaks volumes about the black and dirty shop in which the workers laboured day and night, six days a week. During the summer of 1747, the forge was rebuilt and, at the same time, a tilt hammer, which had been part of the original plans, was installed.[159] On 19 October of that same year, Jean-Nicolas Robichon was identified in a baptismal record as the master tilter.[160]

The Tilt Hammer

Much smaller than the forge hammer, the tilt hammer was a quick-stroke, nose helve hammer for making round and small bars. The archival record would suggest that the tilt hammer was installed in the forge itself or nearby. During his stay at the Forges in 1738–39, which resulted in the plans for the upper forge, Chaussegros de Léry noted: "I saw by the level that, between the two forges on the same stream, we could build a tilt hammer to make small bars. There is a 24-foot head."[161] Could a tilt hammer, requiring at least one waterwheel, have been included in the forge, given the problems powering the machinery already in place? Be that as it may, the reconstruction of the forge provided an opportunity to reorganize the layout.[162] In the 1748 inventory, the "tilt hammer and tools" were listed without being attached in any obvious way to the lower forge; it was stated at that time that the building contained the hurst frame and its accessories, the masonry foundation under the hurst frame, the axle-tree, the bellows shaft with its gear mechanism and iron fittings, the double bellows, the sandstone chimney, lintels, a pig and parts for the hearth.[163] This description would seem to suggest a separate building and that two wheels were required for the tilt mill. In 1749, Pehr Kalm, visiting the Forges, noted that there were "two great forges, besides two lesser ones to each of the great ones, and under the same roof with them."[164] An inventory of 1760 provides more details on the location of the tilt hammer, since it mentions "a tilt hammer attached to the said forge [the lower forge],"[165] and, in an estimate for repairs in 1785, there is further mention of a tilt hammer and chimney, this time associated with the "lower forge."[166] The tilt mill must obviously have been part of the lower forge, where it may have taken the place of the second chafery.[167] In an inventory from 1767, signed by Pélissier, the first person to lease the Forges after the Conquest, we find:

> At the said Forge [the lower forge] a tilt hammer with iron fittings, wheels, bellows, two hammer heads and two hursts.[168]

The Lower Forge in the 19th Century

The lower forge was repaired several times, but it was not until the end of the 18th century and during the 19th century that substantial changes were made to it. An inventory from 1807 indicates that the forge was rebuilt by Munro & Bell, the masters at that time, and that the iron store and charcoal shed were buildings separate from the forge to the north (shed) and northeast (charcoal shed). The new layout of the buildings of the old forge complex erected 70 years before is confirmed by the remains. However, fragmentary remains on the north side of the new forge suggest a moulding operation, of which no record exists.[169] At the other end, in the forge wheel-race, a hydraulic turbine caisson was discovered, which may have been connected with the axe factory built on the forge site in the early 1870s. According to post-1860 transactions, the lower forge had already been abandoned by then.

The Axe Factory

This shop, the remains of which have been excavated, was in operation for only two years, 1872 and 1873[170] (Plate 4.17a). Some information does exist on its equipment and workers—**hardeners**, **sharpeners**, smiths and **strikers**—from the accounts of some of the last workers at the Forges, interviewed in 1933 by Dollard Dubé. The shop contains two treadle-operated hammers and a grindstone nearly 2 m in diameter, driven by a waterwheel or a turbine.[171] In 1874, Harrington, reporting on what he had seen during the summer of 1873, established that ten dozen axes a day were made from iron worked in "an old-fashioned hearth-finery." This finery was undoubtedly the last *renardière* in operation, located at that time in the upper forge.[172]

The Upper Forge

*Plans to build a small forge to produce
at least three hundred thousandweight of iron
annually above the lower forge and just below
the furnace.*[173]

Location

We have seen how Chaussegros de Léry carefully chose the location of the forge in the gully, below the blast furnace, based on a head of 19 feet (6.2 m), measured between "the level of the end of the furnace wheelrace and that of the horse pond below the furnace."[174] That was where the gully really began to deepen, and it was also from there that Vézin measured the drop of the gully to the river in 1735.[175] The remains show us the outline of the forge, and trace a series of changes to it, mainly in the 19th century. The remains of the original forge are less substantial than those of the lower forge, although the remains of the forge hammer are much more explicit, as we shall see below.

Plate 4.17a
Axe factory and millrace, painted by Bunnett, 1887.
In the foreground, chimney of old lower forge with cast-iron pillar.
Henry Richard S. Bunnett, *Three Rivers, Old Chimney*,
oil on canvas, 1887, McCord Museum of Canadian History.

The Building

The upper forge, with its milldam, was built in the summer of 1739, and by the fall, the first chafery was in operation, with a forge hammer. It was completed in 1740 (with a second chafery, forebay, "pig bridge" and sheds). This is the only building at the Forges for which we have the plans (plan of the forge and elevation drawing of the frame), accompanied by a short list of specifications drawn up by Chaussegros de Léry.[176] Thanks to this plan (Plate 3.7) we can imagine how the two forges looked at that time, since both were built to the same design.

In his specifications, Chaussegros de Léry provides precise instructions on how the frame was to be assembled, no doubt to avoid repeating the construction errors made in building the first forge, and to ensure that the frame supporting the heavy forebays (the posi-

*Five **goujats** have to be sent out from France, two for each forge and one for the third chafery run by sluicing.*

Plate 4.17b
Demolition at the lower forge, circa 1870.
An axe factory was built there in 1872.
COLLECTION OF LAWRENCE MCDOUGALL.

tion of which he shows in the wheelrace) was strong enough. The dimensions of the building inventoried in 1741 are quite close to those on the plan. Designed like the lower forge, with its wheelrace on the south side and sheds on the north,[177] the upper forge was smaller, even though it had two chaferies and a hammer of comparable size (see Appendix 9). It also had three 10-foot waterwheels, 2 feet larger in diameter and thus more powerful than those of the lower forge. This extra power, as Vézin himself realized,[178] would enable the second chafery to be used at the same time as the other "by sluicing,"[179] because filling of the forebays was separately controlled; the level of the millpond would be regulated by the number of chaferies (one or two) in use.[180] This is confirmed in a contemporary memorial describing both chaferies as being "designed to operate continually";[181] mention is also made several times of "the third chafery," and the need to assign it a permanent crew of workers.[182]

The charcoal shed and the iron store built one behind the other on the north side of the forge ("*du costé de la coste*") were separated by a stone partition. Large doors in the two gable ends "to bring in the charcoal" (west gable) and "take out the iron" (east gable) provided access to the two sheds, which also opened into the forge itself. The forge had "large double doors," as well as three other doors. Traffic around the building was facilitated by a "bridge to move the pigs" (south side) and "another bridge at the lower gable of the forge" (east side) "to enable vehicles to go around the forge." On the north side, access was by the hill on the stable side and, on the west side, access was provided by the milldam with its 25-foot embankment."[183]

Layout

As in the lower forge, the chaferies and hammer were ranged along the south wall of the building. A plan of the wheelrace (Plate 3.5), roughly drawn by Chaussegros de Léry, shows the wheels evenly spaced, indicative of a symmetrical arrangement of the chaferies, even though the water intake on the lower chafery wheel would seem to suggest a type of asymmetrical arrangement, similar to the original layout of the lower forge.[184] It is hard to believe that Vézin repeated the error he made in designing the lower forge by placing the east chimney against the gable end. His detractors would undoubtedly have reported this, as they did in the case of the lower forge and with regard to other details of the upper forge.

The remains bear the traces of a number of alterations that had to be made to the layout to accommodate changes in technology in the 19th century. We have already seen that the remains of the base of the new furnace, built in 1881 within the footprint of the old forge, are the most imposing evidence. The excavated remains of part of the original forge partially fill out what we have learned from the remains of the lower forge regarding the layout of a great iron forge.

Remains of the Original Forge

Despite signs of having been much altered, the foundations unearthed still sit on the original masonry and correspond more or less with the perimeter measured by Estèbe in 1741. Inside, only the remains of the base of the west chimney and the foundations of the bellows and the hurst frame have been excavated (Plate 4.18). Changes to the eastern part of the forge have wiped out any trace of the second chafery.[185]

Plate 4.18
Plan of archaeological excavations of upper forge showing remains of original forge.
DRAWING BY FRANÇOIS PELLERIN, PARKS CANADA, QUEBEC, 79-25G6-9.

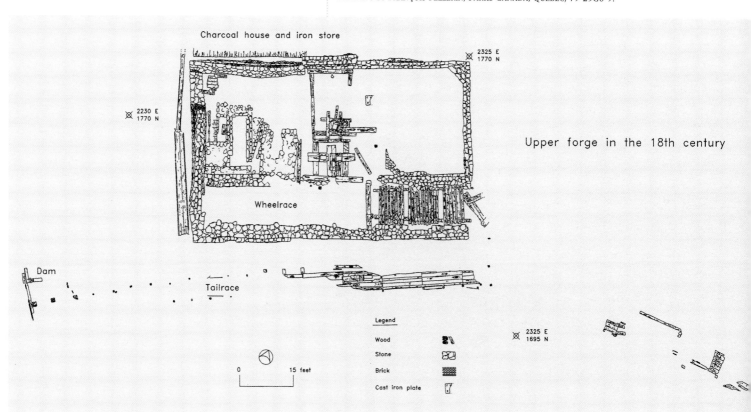

Charcoal house and iron store

2325 E
1770 N

2230 E
1770 N

Upper forge in the 18th century

Wheelrace

Dam

Tailrace

2325 E
1695 N

Legend

Wood

Stone

Brick

Cast iron plate

0   15 feet

The excavated base rests on the foundations of the south wall of the forge. It has a rectangular opening[186] on the north side (in front of the chafery) of the type found in the second, east chafery of the lower forge. This is obviously the cinder notch that formed the sloped extension of the crucible bottom. On the tuyere side, a shallow stone channel with a wooden drain, leading to the wheelrace, may have conveyed the water under the crucible to cool it.[187] On the same side, between the base and the west wall, are half walls that likely served as supports for the camshaft and the bellows.[188]

Remains of the Hurst Frame

The hammer assembly is made of parts that are concealed from view and parts that are in full view. The parts concealed from view are the frameworks that act as foundations.

Encyclopédie, *1757*

In the middle of the forge, the remains of the hurst frame show how its parts were arranged. Exposing the various members and joints of the hurst frame usually hidden from view lets us very accurately reconstruct the hurst frame[189] (Plate 4.16). Similarly, from the way in which the timber block supporting the anvil is aligned with the base of the hurst frame, we can guess the location of the camshaft, which was just to the east of the hammer. Slightly back from the timber block is an iron plate, probably the **loop plate** for working the loop just out of the crucible.

While not very impressive, the remains of the original forge allow us to visualize the machinery it housed very clearly and to imagine the space in which the forgemen carried out their work. If we put together the separate parts of the two forges, we can reconstruct fairly accurately the type of great iron forge on the site between 1736 and 1740 (Plates 4.13, 4.15 and 4.16).

Alterations to the Forge

As in the case of the lower forge, it would appear that the second chafery at the upper forge did not operate for long. When the plans to build the forge were submitted, the inclusion of a second chafery was justified by emphasizing that it would enable work to continue uninterrupted if the first chafery broke down.[190] We have seen that there was some thought of using the reserve chafery as a third chafery. A detailed report on work assignments from 1742 indicates that the forgeman Robichon was sent to work "at the second fire of the upper forge," but only for six days in June, out of an operating period of 10 months.[191] It would have been rather difficult to have two *renardières* running in the same forge at the same time, since in each team of forgemen there was always one man busy at the hammer while another fined the pig. Two teams working side by side would probably have created a bottleneck at the hammer, disrupting the rhythm of work and creating tensions among the workers, who were paid on piecework.[192]

It would seem, however, that at times when the second chafery could have been used as a back-up, this did not happen. A memorial by Intendant Bigot in 1748 mentions that work was interrupted "for nearly two months" to "rebuild the chimney of this forge"[193] so the second chafery was not used during that time. Later, in 1760, the lower forge crew came to the aid of the crew at the upper forge while their chafery was under repair, so that they would not have to share their facility with the upper forge hands.[194] In the inventories of 1764 and 1767, there was mention only of one outfitted chafery.[195] Finally, we learn that a chimney at the forge was demolished shortly afterwards. In the meantime, the second chafery bay was converted into a foundry and moulding shop.[196] Prior to 1850, alterations to the forge affected only the eastern end of the shop; the rest of the forge, including the upper (west) chafery and the forge hammer, remained untouched. After 1850, only the hammer had not yet been taken out of service. Everything else had been altered. In 1881, the hammer was removed and the building became a shed for the second blast furnace, known as "the new furnace."

It is not easy, at first glance, to trace the successive alterations to the forge, and one can become confused. Reconstruction of the various layouts is possible only by comparing several layers of remains with archival documents, which are not always a faithful record of events shown by the remains. Archaeological data are often confirmed by archival data. However, in this case, the clues found in the archaeological data attest to events that are poorly, or not at all, documented in the archives. These events testify to a large-scale reorganization of production during the last 30 years of operation, and the remains of the upper forge are the most revealing in this regard. We will attempt to shed light on the sit-uation by describing in stages the various changes to the forge.

Casting Cannonballs

[...] during the siege of Quebec in 1775–76, the assailants dug their trenches with shovels from the Forges [...] and the cannonballs they fired at the walls of Quebec came from the ironworks' furnaces.

Albert Tessier, 1952[197]

The collaboration of Forges director Christophe Pélissier in the American invasion of 1775–76 left its mark in the remains of the upper forge. Cannonballs and the moulds in which they were made, as well as the ruins of a small furnace that may have been used in their manufacture, were found at the eastern end of the forge, converted into a foundry and moulding shop.[198]

The remains show the foundations of a separate shop, carved out of the forge (Plate 4.19). The original forge lost a third of its length (to the east) when a wall with a solid foundation was erected along the entire width of one end of the moulding shop. The east wall of the old forge was demolished to expand the area of the moulding shop, which was bounded by a new wall about 6 m farther east. Also found was a corresponding extension to the foundations of the north side wall of the original forge, to match the size of the attached moulding shop. On the basis of these remains, the moulding shop must have measured 12 by 9 m. In one corner of this new shop (the northwest corner) are the ruins of a small cupola furnace (Plates 4.20a and 4.20b)

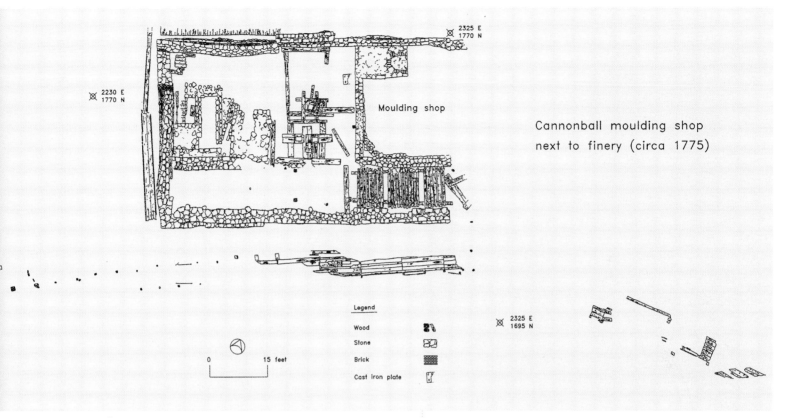

Moulding shop

Cannonball moulding shop
next to finery (circa 1775)

Legend

Wood

Stone

Brick

Cast iron plate

0   15 feet

Plate 4.19
Plan of archaeological excavations of upper forge showing remains
of partial conversion into a moulding shop in the late 18th century.
DRAWING BY FRANÇOIS PELLERIN, PARKS CANADA, QUEBEC, 79-25G6.

We do not know whether this new shop existed before Pélissier used it for military production, but the remains would seem to indicate that it was in operation for only a short period. The eastern end of the moulding shop was brought back to where it had originally been and the east wall of the forge was rebuilt in its original position.[199] The new shop, still separated from the forge by a wall, was then used to store charcoal. These archaeological data and their dating seem to be confirmed by an estimate for repairs in 1785. According to Mousseau, the high estimates made at that time for carpentry and masonry confirm the additional changes evident in the remains of the eastern end of the forge. In addition, an inventory of 1807 confirms the role and size of this charcoal shed.[200]

Before describing later changes to the forge, let us take stock of what happened prior to 1800: (1) demolition of one of the two chaferies (the lower, or more easterly one); (2) division of the building into two separate shops: the forge itself, minus the space of the demolished chafery, and the moulding shop, in the space thus freed, extending eastward beyond the original footprint; (3) conversion of this moulding shop into a charcoal shed.

These alterations do not necessarily mean a shift in the work of the forge, since fining and hammering continued unabated. Following the creation of an ancillary moulding shop, however, a new cupola furnace was introduced for remelting iron into, among other things, the cannonballs that Pélissier sold to the Americans in 1775. Other major changes would be made at the forge, but not until the mid-19th century. In the meantime (1800–50), the forge continued to operate as a finery as usual. The remaining chafery may have been replaced at this time by a smaller chafery, made of brick and located across from the first, along the north wall. Remains of a brick hearth

Plate 4.20a
Small metal cupola, Blist Hill Open Air Museum, England.
PHOTOGRAPH BY PARKS CANADA/ANDRÉ BÉRUBÉ,
SEPTEMBER 1974.

that was used as a finery have been discovered,[201] and a watercolour by Chaplin, dated 1842, shows the upper forge with only one chimney, corresponding to where the finery was located (Plate 9.1). The chimney would therefore have been built prior to that date, although we cannot say exactly when.[202] No further alteration was made to the forge, with its hammer, until the installation of the "new furnace" in 1881. Significant changes, however, were made to the charcoal shed that had been the moulding shop.

The Railcar Wheel Foundry

A brick mass containing five holes, eight feet high by three feet in diameter to chill the wheels.

E. Normand, 1857[203]

Starting in 1852, Stuart and Porter did more than change the blast furnace; in 1853, with no prior notice, they set up a foundry and moulding shop for the manufacture of railcar wheels, a move that makes it easier to understand their need to step up the production of pig iron. Only a few documents mention the manufacture of wheels.[204] From the remains, however, we can pinpoint the exact location of the moulding shop.

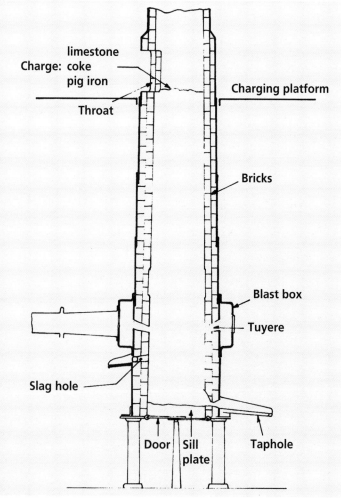

Plate 4.20b
Parts of a cupola.
DRAWING BY FRANÇOIS PELLERIN, PARKS CANADA, QUEBEC, 79-25G6-D 3.

The foundry was set up in the shed that had been used to mould cannonballs, and the moulding shop was built to the south of this building, on the other side of the creek. No trace of the foundry can be seen in the remains, all signs of it having been erased by the subsequent installation of the new furnace in the same location. According to accounts reported by Dollard Dubé, the foundry had two cupolas before it was replaced by the new furnace.[205] They were perhaps used during the short period when railcar wheels were manufactured[206] (1853–57). The foundry was connected to the moulding shop by a little bridge across the creek, while a system of cranes was used to transport the liquid cast iron to the moulding shop.

The remains of the moulding shop are highly instructive (Plate 4.21). The footprint of the building cannot be accurately determined, but we have been able to ascertain what was inside. In the beds of sand forming the moulding floor were discovered traces of moulding activity; in the southwest corner are the foundations of a furnace (the "complete furnace");[207] "five pits [...] for chilling the wheels" were also found; these are annealing pits; in fact, we found two other circular cavities, which could well be the remains of "annealing ovens."

This highly complex shop did not survive the administration that built it (John Porter & Company), and we have no indication of the number of wheels manufactured. We know, however, that the railcar factory operated by George McDougall in Trois-Rivières produced about 30 wheels a day, using pig iron from the Forges.[208]

Plate 4.21
Remains of car wheel annealing pits associated with a moulding shop attached to the original upper forge.
PHOTOGRAPH BY PARKS CANADA, 25G78, R43X-2, SEPTEMBER 1978.

Plate 5.1

Eighteenth-century blast furnace. A filler, at the window of the penthouse wall surrounding the charging platform, communicates with the workmen in the casting house below.

MR TRÉSAGUET, FORMER ENGINEER OF PUBLIC WORKS, *MÉMOIRE SUR LA FABRIQUE DES ANCRES* (PARIS: IMPRIMERIE ROYALE, 1737), MEMORIAL 2, PLATE 3, P. 46.

V

The Art of Ironworking

The founders are usually very mysterious about their work; this is the way they deal with questions they cannot answer: they only know this or that dimension by rote; they fear creating too many of their kind.

Ironmaster Bouchu, Encyclopédie, 1757

IRONFOUNDING

The quality of the product from the blast furnace depends on the evenness of the blast, the timing of the charges, the uniformity of the mine and charcoal and the master founder's smithcraft.

Encyclopédie, 1757[1]

The blast furnace was the master founder's domain, where he presided over a team composed of a keeper, four fillers, a **charger** and a helper. The founder was responsible for everything to do with the blast furnace, from its construction to the smallest details of its operation. He "laid the hearth," determining the design of the furnace's inwalls according to the type of ore used and selecting the size and materials for the boshes and the crucible. He installed and regulated the bellows. He determined the composition and timing of the charges, establishing the proper proportions of ore, charcoal, flux and clay according to the type of cast iron to be produced. He kept a constant vigil over the crucible to monitor the sparks, **scum** and colour of the flame, which were indications of whether the materials were being "digested" properly, and supervised tapping operations. He was always near the furnace and slept close by, either in an adjoining building[2] or nearby house. The importance of the master founder was well understood in New France, as Intendant Hocquart duly acknowledged:

> The master founder [...] is the most essential worker and without him, all work would come to a standstill.[3]

In short, the master founder was indispensable, and ironmaster Vézin recognized the importance of the founder's expertise in metallurgy:

> [...] the distinctive feature of the mine is to provide a cast iron to which the founder gives the quality it is to have by observing the workings of his furnace and acting accordingly.[4]

Vézin, however, dismissed the first founder, Lardier, unjustly blaming him for the initial problems with blowing in the furnace. Although the European ironmasters recognized the founder's importance, they often took a patronizing tone towards him, characterizing his skills as "mysterious," in a poorly concealed attempt to cover up their jealousy. Bolstered by their belief in Reason and Science, they scorned what they called the founder's "mechanical" rote knowledge. Indeed, ironmaster Bouchu recommends in the *Encyclopédie*[5] that a school be established to give founders an understanding of the "reasons behind their work." Bouchu was thus against the practice of family apprenticeships, or the craft being passed down from father to son, remarking that "founders feared creating too many of their kind."[6] The time had not yet come for metallurgical engineers to supplant these master craftsmen. This did not come about at the Forges until the mid-19th century and was marked by tragedy: the first explosion at the blast furnace, attributed to the incompetence of Hunter, the engineer. If there was any incompetence at the Forges, it should be ascribed to the ironmasters rather than the founders. The problem was particularly serious under the French Crown; in the 1750s, a competent ironmaster still had not been found.[7]

The work of the founder and his assistants was governed largely by the workings of the blast furnace. The work schedule and the timing of tasks were determined by the rate at which raw materials were "digested" by the furnace. The furnace had to be in continual operation due to its design, and the furnacemen were no doubt among the first industrial workers to work in **turns** or shifts[8] over a 24-hour work day.

Work Schedule

> *It is understood that, when the furnace is in blast, it has to be tended day and night without respite.*
>
> Encyclopédie, 1757
>
> *[...] to work [...] both day and night including Sundays and holidays according to the rules at the Forges post.*
>
> *Indenture of a keeper, 1805*[9]

Once blown in, the blast furnace ran for six to eight months straight; in the final years of the ironworks, the modified furnace operated for 10- to 13-month stretches at a time.[10] The period the furnace was in blast was called a campaign. Enough pig iron could be produced during a campaign to supply the two forges for an entire year, and even an uninterrupted campaign of eight months produced more pig iron than the two forges could use.[11] This relentless pace of production, however, soon caused the lining and crucible to deteriorate, and annual repairs to the belly were required. Consequently, the furnace was **taken out of blast** in late autumn[12] and was not blown in again until April or May, or even later, depending on the severity and length of the winter.[13] This operating cycle affected not only the preparations for each campaign (since enough raw materials had to be laid in to feed the furnace for six to eight months) but also the lives of the workers.

The company employed the master founder all year round, paying him on a piece-work basis (per thousandweight of pig iron) when the furnace was in blast, and a retainer or salary when it was not. His assistants (keepers, fillers and helpers) were employed only for the campaign.[14]

Information is scant on the pattern of the work day, although we do know that shiftwork was the norm. A visitor to the Forges in 1808 reported that the fillers, like the forgemen, worked six-hour shifts, while the moulders and finishers worked from dawn to dusk.[15] Evidently, then, it was the interval at which the furnace was tapped, and hence the production cycle, that governed shift length. Hardach notes that in 18th-century France:

> In the preindustrial age and the dawn of the industrial age, the work day was generally 12 hours long, although there were some differences from industry to industry. In many traditional trades, shift length was not defined in hours but rather on the basis of production: shift changes at blast furnaces took place after each tapping, which could be every 11 to 13 hours, depending on the quality of the ore, the smelting temperature, and other things.[16]

Describing blast furnace work in the *Encyclopédie,* Bouchu defined the work cycle in terms of charges, shifts or **tappings**, without specifying the actual number of hours per shift; however, the fact that the furnace was charged 20 times in 24 hours suggests that shifts were around 12 hours[17] (Table 5.1). Tapping was the critical event; the moment of tapping was determined by how long it took for the materials to be digested in the belly of the furnace, which in turn determined when charges were fed into the throat. Although the fillers could theoretically relieve each other any time between tappings, the founder and keeper had to be present at all times to control the smelting process. The time required for smelting varied, depending on the type of cast iron to be produced (white or grey). Under these circumstances, it was only natural that the founder was paid, not by the hour, but by the amount of iron produced, since the functioning of the furnace was variable, and the founder himself controlled the smelting process, which did not lend itself to a fixed schedule of hours.

Dr Harrington, who visited the Forges in 1874, reported that the furnace was charged every half hour (45 charges in 24 hours), and tapped every 12 to 18 hours. The furnace of Harrington's day, however, was twice as high as the earlier model and had two tuyeres, which allowed for a longer time between tapping, resulting in the more complete reduction of the ore and thus twice the output of pig iron.[18]

Table 5.1

COMPOSITION OF BLAST FURNACE CHARGE (in French pounds)

No. of baskets	Ingredient	Weight per basket	Weight per charge	Total weight per tapping	%
5	Charcoal	46	230	2,070	28.8
10	Ore	50	500	4,500	62.5
1	Flux	50	50	450	6.2
0.5	Clay	20	20	180	2.5
Total weight			800	7,200	100

From Bouchu, *Encyclopédie, Recueil de planches,* 1765, sect.3, plate VII.

The number of workers assigned to each station at the furnace also provides an indication of how working hours were distributed. The master founder and keeper were stationed on the ground floor of the furnace, where they took turns monitoring its operation, and the four fillers worked in pairs feeding the furnace. This arrangement reflects a rate of two tappings a day, allowing two shifts of 12 hours each or four shifts of 6 hours each. When interviewed in 1933 by Dollard Dubé, former workers at the Forges said that the fillers worked in 8-hour shifts[19] and some lists of workers in the early 19th century show two keepers rather than just one, a detail suggesting that they, too, worked shifts of 8 hours.[20] Dubé also reported that a bell was rung at 7:00 am, 12:00 noon, 1:00 pm and 6:00 pm, which would mean a work day of 10 hours. However, he stated that the blast furnace had both day and night shifts.[21]

Since the work cycle and the workload depended on the characteristics of the blast furnace itself, it was not always charged as often as Harrington observed in 1874. In Harrington's day, fillers handled close to 20 t a day, and this is no doubt why a loading ramp had been built so that the charges could be hoisted mechanically. The pre-1854 smaller furnace digested half as much raw material a day. Charging, which was done manually, was therefore required half as often, although the manual labour was not necessarily less. According to Bouchu, furnaces at the time were charged almost every hour,[22] meaning that the fillers had to handle close to 10 t a day, or the equivalent of 2.5 t per worker.[23] Given this heavy workload, it is quite conceivable that fillers worked 6-hour rather than 12-hour shifts, unlike the other furnacemen.[24]

Division of Labour

To keep a furnace going, at least three workers are required, a founder or furnace keeper and two fillers.

Encyclopédie, *1757*

The Furnace Crew

At least three workers at all times were needed to keep the blast furnace running: a founder or keeper and two fillers; since the furnace ran day and night, two shifts of three men each were required. The founder was a skilled craftsman in charge of ensuring that the furnace ran smoothly. Since his work involved a specific task, he was more of a team leader than a foreman or production supervisor. The other workers adjusted their work to the master founder's methods and style, which in turn depended on the type of ore used, the pig produced and the time between tappings.

After his shift was over, the founder was replaced by the keeper, with whom he shared some of his trade secrets.[25] There is some evidence that the keeper worked the night shift and, if required, could fill in for the founder to the extent that his knowledge permitted.[26] The keeper was apparently not considered an apprentice founder, and indeed, none of the keepers at the St Maurice Forges was promoted to founder. Belu (the Forges' first keeper) was said to be "capable enough to serve under a good master founder, but not capable enough to take the founder's place should he fall ill."[27]

The furnace crew also included four fillers working in pairs, the two teams alternating throughout the 24 hours. Two other employees were assigned permanently to the furnace: a carter to take away the slag and to bring the ore and sand to the furnace;[28] and a helper, to assist the master founder and keeper.[29] At the beginning and during the campaign, a charger, ore breakers and limestone breakers were employed to reduce the raw materials to the appropriate size for the charge.

Work Areas

The work space around the furnace was divided up around the furnace's three openings, through which it was charged, furnished with blast and tapped. As we have seen, the furnace was surrounded on three sides by separate but connected sheds that housed the raw materials, the bellows and the actual work areas. The master founder and his keeper worked on the ground floor, in the casting house adjoining the taphole, shuttling constantly back and forth between the casting house and bellows shed, where they monitored the blast through the founder's eye at the tuyere opening and made adjustments to the bellows. The casting house floor was made of sand laid out as a pig bed to receive the liquid iron tapped from the furnace. Some of the molten iron was also used to cast stove plates and other objects, which added to the bustle in the casting house when the furnace was tapped. There was also a moulding shed connected to the casting house, which was manned at different times by anywhere from 2 to 10 moulders. The moulders' work will be discussed in greater detail later in this chapter.

The fillers worked at the upper level of the furnace on the charging platform, or directly below it in the charcoal house; a filler down below, assisted by a charger, would fill baskets with ore, charcoal, limestone and clay, while the filler up above would hoist the baskets up to the platform and empty them at regular intervals into the throat of the furnace.[30] The charger and limestone breaker also worked in the charcoal house on the north side of the furnace.

A rare depiction (Plate 4.2) of the inside of the casting house shows the doors on either side, one to the bellows shed and the other to the charcoal house. To ensure that the work of the fillers and founder was synchronized, there had to be a way to communicate directly between the charging platform and the casting house. The plates in the *Encyclopédie* show a window in the penthouse walls that looked down into the casting house (see also Plate 5.1 taken from a memorial of 1737). There is mention also of an iron plate hung near the wall, which was struck to signal to the master founder which charge was being loaded into the furnace.[31] Sulte, who grew up at the Forges in the 1850s, describes a similar system of communication:

> At the blast furnace, commands were relayed by striking on a piece of sheet metal with an iron bar. No shouting or other verbal orders were given; rather three rings, or sometimes five were made with specific intervals between them, and the message was understood by everyone.[32]

Operations

Blowing In

When blowing in the furnace, the method of charging it, the quality, quantity and order of the charges, is different from that observed when the furnace is in blast.

Encyclopédie, *1757*

In 1738, Lardier, the Forges' first founder, was dismissed after he failed to blow in the furnace after several attempts. Thus, he was probably the first employee to be the scapegoat for ironmaster Vézin's errors, but probably not the last. He was replaced by Delorme, who had originally been recruited in France as a finer.[33] The saga of the blowing in of the blast furnace (recounted in the previous chapter) shows how difficult this operation was and why it was usually done only once a year in spring.[34] Blowing in was a delicate operation done in several stages. One only has to reconstitute the different stages from Bouchu's description in the *Encyclopédie* to understand, in the atmosphere of haste that reigned in 1737–38, how likely the operation was to fail.

To blow in the furnace, the walls of the belly and crucible had to be completely dry and the hearth had to be heated, or seasoned, so that it was ready to receive the molten iron. Depending on the level of moisture in the stone walls, this seasoning could take as long as 48 hours; only then was the furnace ready to receive the first charge of ore. At the Forges, the general rule was that seasoning took five to six days,[35] and usually 15 days were required before the furnace was properly blown in.

The process went as follows. First, the fillers fed the furnace with charcoal through the throat. Down below, the master founder plugged the tuyere with clay and inserted a shovelful of lit coals into the hearth through the taphole and waited until the charcoal had caught fire up to the top of the furnace. Once the fire had worked its way up to the furnace top, the next step was to allow the furnace to "swallow the charge": in other words the level of the charge had to drop about 36 inches (measured by the filler using a **gage** [gauge] that was inserted into the throat). This was when the first error was likely to occur, according to Bouchu: some ironmasters then introduced the first full charge of ore in order to save on charcoal. This was a mistake according to Bouchu, citing another ironmaster, Grignon, who maintained that "mine should not be introduced until the furnace was able to digest it properly; if the mine was introduced too early, the bottom of the furnace, the hearth, would not be hot enough to withstand the molten ore."[36] This scenario is a very likely explanation for the failed attempts at blowing in the furnace in 1737. Grignon recommended waiting 36 hours before inserting the first charge of ore, and in the meantime, making a **grate** to heat up the hearth:

> The grate consists of inserting ringers into the receiver, through the top of the dam, fairly close together, in order to prevent the charcoal from falling; any charcoal in the receiver is removed through the taphole and the heat is allowed to spread over the bottom. Keep the grate in place until one can see that the bottom is well lit, so that everything is in flames and sparks are being thrown.[37]

When the hearth was white-hot, the first load of ore (not a complete charge) was added to the charcoal in the belly. The number of baskets of ore was increased gradually as each charge was digested by the furnace. After 12 to 15 hours, "globules of imperfect molten iron" in the form of bright sparks appeared, heralding the arrival of the first few drops of fluid metal in the hearth. The master founder then made a "grate" one last time to clean the crucible and cover the base of the hearth with a layer of burning breeze[38] so the molten iron would not adhere to the bottom. The founder then removed the ringers and cleared the opening of the tuyere, leaving the covering of clay to protect it from the fire. He then installed the clay plug to close off the hearth and opened the sluice gate to start the great waterwheel that operated the bellows. The furnace was now in blast. Bouchu recommended that the proportion of ore be increased gradually until the furnace's maximum capacity (a complete charge of ore) was reached. The founder could tell if the furnace was ready by the colour of the flame, the consistency of the slag and the characteristics of the molten iron.

In the autumn of 1737, ironmasters Vézin and Simonet were in a hurry to get the blast furnace into commission, so they could send a few samples to France on the last ship leaving the colony that year. In the atmosphere of haste, there is some question whether the furnace was blown in according to the procedures described by their eminent colleagues (and critics) in the *Encyclopédie*, who came from the same region in France. Although they may indeed, as they claimed,[39] have insulated the bottom of the hearth from the damp ground, they certainly did not follow to the letter the procedures for blowing in the furnace.

Charging the Furnace

I was a filler at the blast furnace and it was one of the main jobs at the establishment and required a lot of care.

Antoine Mailloux père, *1860*[40]

The filler's task was an arduous one. Constant physical effort was required to prepare the raw materials, transport them to the charging platform and feed them steadily into the throat of the furnace. In the pre-1854 set-up, the platform was around 4.5 m high. The fillers had to handle nearly 10 t of raw materials in 24 hours, or over 500 baskets of ore (weighing 34 kg each), limestone (20 kg each) and charcoal (7 kg each). In 1854, the platform was doubled in height to 10 m, and a hoist was installed to winch the materials up to the platform; however, the quantity of materials to be handled also doubled to 20 t per 24 hours (over 1,100 baskets). This translated into 25 baskets, or over 380 kg, every half hour. According to the workers who were employed at the furnace in its last days, the hoist was raised every five minutes, which gave the fillers barely enough time to fill the baskets from the cart and empty them into the throat.[41] One of the first innovations to accompany the introduction of the taller coke furnaces in the late 1800s, which had to be fed more often, was fully mechanized charging, which also eliminated the filler's job. The last furnace at the Forges, which still ran on charcoal, represented a transition between the old manually charged furnaces and the new fully mechanized furnaces, and it still required a great deal of physical effort from the workers.

The fillers were also responsible for monitoring the rate at which the charges were being consumed by the furnace, no simple operation, as alluded to by Antoine Mailloux *père* in the epigraph to this section. As mentioned, during blowing in, the filler used a gauge to test the level of the charge to determine when the furnace needed charging, and the same operation was repeated once the furnace was in full blast. According to the fillers interviewed by Dollard Dubé, the order of charging was "a row of charcoal, a row of mine, a row of limestone" although this was no doubt a simplification.[42] In the *Encyclopédie*, Bouchu describes a more complicated two-step process that required routine observations by the fillers when it was time for another charge.[43] Chevalier Le Mercier notes that the filler could tell by the colour of the smoke coming from the throat whether there was too much ore (black smoke) or not enough (white smoke). Greyish smoke indicated that the burden was correct. Only one filler was responsible for monitoring the furnace, although both pitched in when it was time to load the next charge.[44]

Needless to say, there were significant health risks involved in the job. The fillers were exposed to noxious gases when they worked around the throat. Often, flames would leap up to the top when a charge was emptied into the throat, causing fires. The risks were even greater when the workers had to go down into the throat itself! In his account of his visit to the Forges in 1828, Lieutenant Baddeley told how, at the beginning of the campaign, just after the first charge of charcoal had been put in and the fire kindled, a worker climbed down 3 m into the furnace to clean the slag off the walls. This gave him hallucinations ("pleasurable sensations" in Baddeley's words) and he fainted. Three men were sent down in turn to rescue him; they, too, were overcome by giddiness and had difficulty getting back up again. For two of the rescuers, however, instead of enjoying "delightful sensations," they were left gasping and retching. They had a painful recovery, while the first man who fell (the youngest and healthiest according to the witness), although exposed the longest to the carbonic acid gas, was the first to recover.[45] The incident could have easily ended in tragedy according to Baddeley.

The only known accidental deaths to occur at the Forges were in 1854, when two fillers were killed by an explosion at the blast furnace, which had just been rebuilt to twice its size. The campaign was interrupted for three months. Six months later, there was a second explosion that severely burned two other workers.[46] In 1881, the fire caused by a third explosion left the keeper with very serious burns.[47] Dubé reports another incident resulting from the negligence of the workers that could have also had disastrous consequences.[48]

Smelting and Tapping

On entering the smelting forge, I was received
with a customary ceremony. The workmen moulded
a pig of iron, about fifteen feet long, for my special
benefit. The process is very simple: it is done
by plunging a large ladle into the liquid,
boiling ore, and emptying the material into
a gutter made in the sand.

Louis Franquet, on his visit to the Forges in 1752[49]

Franquet did not find the process of tapping the furnace and casting the pig especially impressive. Had he seen it at night, however, with the molten iron emitting an eery glow, it would have undoubtedly made a greater impression on him. Benjamin Sulte, who lived at the Forges in the early 1850s, describes the tapping process with a little more colour:

[The worker] set to the mouth of the furnace with an enormous *gentilhomme*, unplugged it in a quick motion, and the molten iron, white with an orange glow, gushed out into the avenues made for it.[50]

Tapping was the founder's final task, and it had to be done quickly lest the furnace cool down too much and waste valuable fuel. The bellows were halted during the operation.

The main responsibility of the founder and keeper, however, was to monitor and control the smelting process during the 12 to 15 hours before tapping took place. To do this, they observed the furnace carefully through the tymp and tuyere openings, and sometimes intervened with a few rudimentary tools, which consisted basically of long iron pokers, either straight or hooked on the end, with names such as ringer and rabble, dam hook and cinder hook. There was also the ship, a triangular hoe with a wooden end that was used to make the V-shaped furrow in the sand for the pig, after which the walls of the mould were firmed up with a shovel (Plate 5.2). Sulte relates some of the unusual names the Forges workers used to refer to their tools:

> Gigantic tools hung from the centre of the archway on slender iron chains. They rejoiced in the names of *demoiselle*, *gentilhomme* and *prince*, and the expertise with which they were wielded was something to behold![51]

As we saw earlier, the founder used his ringer to make the "grate" when the furnace was in the process of being blown in. He also used it every hour or so when the furnace was in blast to **rabble** the mixture. To do this, the founder inserted the tool into the opening between the dam and the tymp either to loosen the slag adhering to the sides of the hearth and mix it in with the iron, or to remove it.[52] The ringer was also used to facilitate the descent of the charge in the crucible.[53] The master founder also monitored the tuyere and used another tool, the **placket**, to plug it with clay and keep it from melting.

The consistency and colour of the slag, which the founder removed regularly from the hearth, provided a wealth of information on how the furnace was running and smelting progressing: slag that was too liquid meant too much limestone, while slag that was too sticky meant too much clay. The founder also used the colour of the smoke, flames and cooling metal to tell what needed to be done. The *Encyclopédie* and all the early treatises provide plentiful advice on how to conduct the smelting process, including what to look for, and admonitions and formulas for the burden; this profusion of advice reflects the empiricism of the time, before science had gained sway.[54]

Once the hearth was filled with molten iron, and the slag floating on the surface had begun to flow over the dam, it was time to tap the furnace. First, the founder had the mould for the pig prepared, a triangular furrow 4–5 m long in the dampened sand. A channel from the hearth to the mould (called a runner) was also made, which ran at a slightly downward pitch, reflecting the slope of the floor. Before tapping the furnace, the founder halted the bellows and plugged the opening of the tuyere with clay, "to prevent the flames from being urged by the blast and burning the workers."[55] Evidently, it was understood that allowing the bellows to operate during tapping would send a spray of molten iron out of the hearth. The founder then drew out the slag through the slag notch in the dam, and his helper removed it with a hooked ringer. At this time, as on any occasion when slag was being removed from the hearth, breeze was thrown on the slag as the keeper (or founder) was removing it, as a precaution against the excessive heat.[56] Then, with a poker, the founder knocked out the clay plug from the taphole, so that the molten iron ran out from the hearth into the mould, forming a pig weighing slightly over a tonne.[57] Immediately afterwards, the pig was covered with dry breeze so that it would not blister when exposed to the air, and care was taken to ensure no remaining bits of slag went into the mould. Then, the tuyere opening was unplugged, the taphole was plugged up again, and the bellows were started.

Throughout the operation, the workers had to cope with intense heat, donning appropriate protective clothing (gauntlets, anklets and aprons) to cover the most exposed parts of the arms and legs. Sulte reports that a heavy leather apron was worn to protect the front of the body, and an asbestos mask over the face.[58] The furnace was particularly dramatic at night, and the painter Léonard Defrance from Liège, who painted a number of pictures of furnace-men in action (see cover page), evokes the flavour of the scene in his notes, in contrast to Franquet's colourless description:

> What a difference there is between the glow from a blacksmith's forge and the glow from a smelting furnace [...]. The first is yellowish with hints of red, while the latter is milk white, turning the faces of the workers a livid white, like a sick man with no strength left.[59]

White, Grey, Black and Mottled Cast Iron

[...] the founder must know how to produce
grey iron to be used in the forges and to be able
to make stoves, pots, kettles and other items
for domestic use.

Vézin, 1740[60]

When drawing up his plans for the Forges in 1735, Vézin expressed a preference for **grey cast iron**. He noted that, when refined at the forge, it produced a "soft, strong, yielding, malleable, fine-grained iron," adding that these qualities were entirely the product of the master founder's skill and work, requiring just the right mixture of ore and flux.[61] Indeed, producing the type of cast iron suitable for the intended use was an important aspect of the master founder's craft. Vézin also described white cast iron as "**cold-short**, brittle and flawy, with a high yield in the finery but that tears in the chafery" and **black cast iron** as "fairly good, when the finery hearth is of sufficient size to work it, such iron costing

much to produce and never having the quality of grey cast iron." The main difference between these types of iron is the content in carbon (higher in grey cast iron than in white), silica (absent in white and high in grey), and manganese (absent in grey and high in white).[62] Although master founders, with their empirical mind set, were unable to express these differences in terms of their chemistry, they knew by experience that the difference between the types of iron lay mainly in the burden, or the ratio of charcoal to ore in the charge. As Pierre Léon explains, this ratio is responsible for whether the furnace is **cold working** or **hot working**:

> [...] with a hot working furnace, in other words, by increasing the proportion of fuel in relation to that of ore, grey cast iron is obtained and [...] conversely, with a cold working furnace, or by increasing the proportion of ore, white cast iron is obtained.[63]

In the first case, to produce iron for casting (grey iron), the founder increases the tempo of the bellows and the duration of smelting, thus increasing the carbon content of the iron. In the second case, to make iron for fining (white iron), the heat is reduced by slowing the tempo of the bellows, resulting in a product that has less carbon and is easier to melt, and thus easier to work in the forge.[64] A report on Delorme referred to this practice, which the master founder could sometimes turn to his own advantage:

> Messrs. Simonet and Cressé will see to it that De Lorme produces good cast iron, and that he does not rush smelting in his own interest, being paid by the thousandweight, or does not delay tapping out of negligence.[65]

By rushing smelting to produce white cast iron, which is heavier, the founder could increase his income, since he was paid by the thousandweight of cast iron produced.

The founder's art was necessarily limited by the characteristics of the ore used. Bouchu acknowledges this in his article in the *Encyclopédie*:

> Having acquired some knowledge of the best mixture for the smelting of the mine, I must admit, however, that I have not been able to determine what, with equal working, distinguishes the different irons. One must content oneself with saying in general that mines are of different sorts, and consequently, their products must differ.[66]

As we saw in Chapter 2, the bog ore used at the Forges was analysed by geologists to determine its chemical composition. In 1874, the geologist Dr B. J. Harrington reported that the cast iron produced at the Forges was 10% white and 90% mottled[67] —with mottled cast iron, which was used at the time to make pig iron, being an intermediate between white cast iron and grey cast iron. Harrington also claimed that, in general, most cast iron produced from bog ore was indeed white or mottled, like that produced at the Forges, because of the manganese and phosphorous content of the ore. Because of the presence of phosphorus, bar iron from bog ore had the reputation of being cold-short. However, the bar iron produced at the Forges was, on the contrary, famed for its softness and malleability; this, according to Harrington, was due to the fact that the ore used at the Forges had only traces of phosphorus.[68] It was probably also due to the Franche-Comté method (*renardière* process) used at St Maurice, which was based on the use of grey iron[69] to produce the "soft, strong, yielding, malleable, fine-grained iron" described by Vézin (see also below in the section on "Casting"). At the close of the 18th century, the Forges masters were still using this as their trademark, as this announcement in the *Quebec Gazette* of 26 August 1784 testifies:

> We deem it necessary to inform the public that the St Maurice Forges is now producing bar iron that is in no way inferior to the best iron imported from Europe. It is soft and malleable, manufactured from grey iron from the local ore [...].

Indeed, experts have been astonished at the quality of the wrought iron and cast iron artifacts found at the site, according to a metallurgical analysis carried out in 1979:

> [...] whether for cast iron or steel, the main elements vary in a fork that compares very favourably to those made with modern materials, which is quite surprising.[70]

When Harrington visited the Forges in 1874, wrought iron production had been practically abandoned in favour of the large-scale production of mottled pig, which now represented 90% of the Forges' annual output. According to Harrington, the remaining 10% was white iron; it was probably used to manufacture axes, a line of business begun in 1872.[71]

Therefore, St Maurice ironworkers had the advantage of acknowledged high-quality ore; indeed, from the beginning, the first bar iron made from the ore compared favourably with the best French irons, those from Berry.[72] Harrington also remarked that the high proportion of volatile matter (water and organic matter) in the bog ore (17–24% at St Maurice) made the ore easier to reduce. Laboratory reduction of the ore used at the Forges gave yields of 45–55% but, in the blast furnace, the yield was 30–40% because of the high silica content (found in the form of sand) in the ore.[73]

Casting

*After this operation, I was shown the mould
for a stove that had been stamped in the sand,
ready to be filled. One of the workers went to get
a ladle of molten iron and emptied it carefully
into the bottom of the mould, and then filled up
the mould to the top, so that the indentations
in the bottom of the mould would form the raised
parts of the object.*

Louis Franquet, 1752[74]

According to the epigraph, which is a continuation of the earlier one by Franquet describing the tapping of the furnace, the same cast iron was used for casting objects and for fining in the forge. Indeed, the engineer is describing, as part of a single sequence, the casting of the pig and, immediately afterwards, the use of open sand moulding to make a stove plate in the sand on the casting house floor (the box moulding technique for making three other objects is subsequently described).[75] Therefore, the same grey cast iron was apparently used for fining and casting, as ironmaster Vézin had prescribed.

In the original plan, small castings were to account for only a small part of production (10%). Indeed, out of the million pounds (490 t) of cast iron to be produced annually, a mere 100 thousandweight (49 t) was reserved "for the Colony's consumption."[76] The actual figure in the first few years was around 60 thousandweight (30 t),[77] representing objects such as stoves, stove plates, pots, kettles and other articles. Therefore, in the beginning (1740–45), only a master moulder, two assistants and a helper were required for the moulding work.[78] In 1745, two other moulders joined the team to cast domestic articles and cannons.[79] The production of castings gained steam in the last quarter of the 18th century, to the extent that it made up two-thirds of total output at the beginning of the 19th century.[80] At this point, nine moulders were employed; in 1851, when part of the upper forge had been converted into a moulding shop, 26 moulders were employed. After 1865, when the focus was on producing pig iron, castings represented only around 8% of production.[81]

Over the years, four different moulding techniques were used at the Forges: open sand moulding, box moulding (Plate 5.3), loam moulding and chill casting. These techniques required different types of moulds (wood, clay or iron), which were made by the moulders to shape the cavities into which the molten iron was poured. Some of these moulds were as intricate as sculptures.[82] Two methods were used to fill the mould with molten iron: either a **ladle** was used to scoop the iron directly from the hearth and carry it to the mould; or a channel, or runner, was made in the sand either directly from the furnace or from a pig to the mould or casting pit.

Open Sand Moulding

Open sand moulding is the technique described by Franquet in the epigraph. The moulder (also called a sand moulder)[83] carefully levelled the sand bed, and covered it with a layer of finer sand,[84] into which the wooden pattern for the object was stamped. Then he removed the pattern, sprinkled the resulting depression with breeze (charcoal dust) to prevent the metal from cooling too rapidly, and tamped down the breeze with the pattern. After making small holes (or whistlers) in the surrounding sand to vent any escaping gases, he poured the iron into the mould with a ladle[85] (Plate 5.4) or led it from the sow channel to the mould directly. Once poured, the top of the mould was covered with breeze to slow cooling.

Plate 5.3
On the left, box (flask) moulding; centre, open sand moulding; on the right, removal of mould from casting.
DIDEROT AND D'ALEMBERT, *ENCYCLOPÉDIE [...], RECUEIL DE PLANCHES [...]* (PARIS: BRIASSON, DAVID, LE BRETON, 1765), "FORGES OU ART DU FER," SECT. 3, PLATE IX.

Moulding scraps found during the excavation of the floor of the casting house reflect both methods of pouring the moulds.[86] Droplets of cast iron found in several places and ladles associated with the period from 1740 to 1780 indicate that the process described by Franquet in 1752 (involving ladles) was being used. Scraps produced by filling the mould from a runner or channel are associated with the following period (1780–1850), during which moulding blossomed. Moulding, although known to the French, was more typical of the British moulders employed at the Forges beginning in the 1780s.[87]

Box Moulding

Some of the stove plate fragments point to the use of box (or flask) moulding techniques. The fragments show traces of the scars left when the risers, runners and gates, through which the molten iron is poured into the mould, are cut off or fettled after the casting has cooled, a phenomenon characteristic of box moulding. In this procedure, the pattern is stamped in compacted sand and the mould is enclosed in a box; the molten metal is then poured through the gates and runners with a spout called the sprue. Unlike open sand moulding, in which the uncovered side is always rough, box moulding produces smooth-sided objects. However, the iron must be more fluid, so that it flows into the mould evenly and fills every tiny crevice exactly. This fluid iron was obtained by smelting at higher temperatures, achieved by modifying the bellows mechanism and height of the boshes and installing a double tuyere. These modifications were indeed made in 1854 to the blast furnace.[88]

Box moulding was also used to produce pots and other containers of various sizes. The 1741 inventory of the moulding floor included wooden moulds for "pots, stove, kettles, bake kettles, plates, saucepans, porringers, mortars and tackle-blocks" as well as "116 moulding flasks and associated iron fittings." A combination of box moulding and loam moulding may have been used to produce this hollow ware; loam moulding would have been used to make the core to form the empty space inside the pot or kettle.[89]

Plate 5.4
Open sand moulding.
LÉONARD DEFRANCE, *INSIDE THE FOUNDRY* (DETAIL), OIL ON WOOD,
MUSÉES ROYAUX DES BEAUX-ARTS DE BELGIQUE, BRUSSELS.

Plate 5.5
Cannon foundry in Douai, circa 1770.
PAINTING BY HEINSIUS, COLLECTION OF MR DE CASTEX,
SAINT-OUEN-DE-TOUBERVILLE, PHOTOGRAPH BY ELLEBE, LA DOCUMENTATION FRANÇAISE.

Loam Moulding

Loam moulding was used to produce large objects like cannon and potash kettles. These objects were moulded in specially created pits in the floor of the casting house; indeed, the remains of casting pits used to mould cannons and kettles have been found (Plate 6.3).

The skilled moulders that arrived at the Forges in 1745 were hired to found large cannon and, in fact, a second blast furnace was to be built for this purpose. In 1750, an officer, Chevalier François Le Mercier, was dispatched to France to study moulding techniques (Plate 5.5) and to report thereon. Although the second blast furnace was never built, the archival record shows that, between 1747 and 1752, many different kinds of armaments were produced at the Forges: gun carriages, bombs, cannonballs, mortars, shot, perriers, grenades, mushroom bullets and small-calibre cannon.[90]

Casting a cannon or potash kettle using loam moulding techniques was a fairly complex operation. Creating the clay mould was a three-step process involving the construction of a **core**, inner mould or **shell**, and outer mould or **mantle**. A technique similar to one used in pottery was used to make the mould (Plate 5.6). The moulding bench was open in the middle to accommodate the horizontal spindle carrying the **strickle** or sweep, to which were attached the straw plaits that applied the successive layers of loam. The

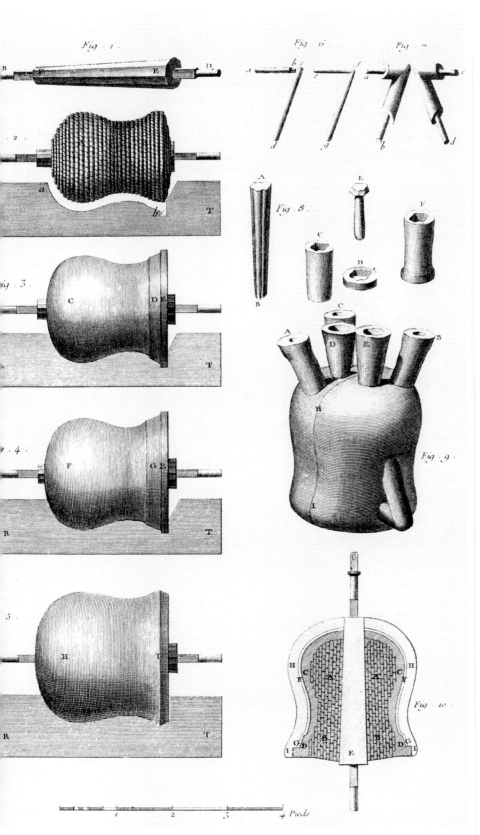

strickle was swung 360 degrees manually, sweeping the soft loam to a smooth surface of the size and shape required. The piece's calibre was inserted on the edge of the opening. The moulder formed the loam while his assistant turned the strickle board, ensuring that the mould was symmetrical (a little like a potter on his wheel). At various points in the process, the moulds were taken to dry in an **oven** on the moulding floor.[91] A way had to be devised of transporting these very heavy moulds back and forth between the oven and the casting pit, which was in the casting house floor. Although no remains of an oven have been uncovered, there is other evidence of its existence, including a brick structure with rails and a wooden structure that may have been a base for a crane, which were both discovered near a casting pit for kettles. These remains suggest that the moulds were rolled on a cart on rails into the oven and then to the casting pit[92] (Plate 5.7).

Two obvious casting pits were found in the casting house floor: one for cannon, which still exists (Plate 6.3), and another for potash kettles. According to the layout of the pits, these objects were moulded vertically. The cannon pit, which is roughly 1.5 m deep, shows particularly detailed workmanship, with its cut-stone walls lined with vertical boards and a drain in the bottom to remove moisture.

Plate 5.6
Stages in loam moulding of a large pot-bellied kettle.
DIDEROT AND D'ALEMBERT, *ENCYCLOPÉDIE [...], RECUEIL DE PLANCHES [...]* (PARIS: BRIASSON, DAVID, LE BRETON, 1765), "FORGES OU ART DU FER," SECT. 3, PLATE IV.

The usual practice was to fill the mould, which was placed vertically in the pit, through a **gutter** connected to the runner from the hearth. The casting pit for kettles, which was found near the cart rails and is approximately 1.2 m deep, shows the shape of the last kettle that was cast. According to archaeologist Monique Barriault, the absence of clearly identifiable gutter fragments, and the distance between the casting pits and the hearth (approximately 4.6 m) suggests that the moulds may have been filled using ladles.[93] The casting of potash kettles was temporarily halted in the 1830s, or perhaps earlier. Lieutenant Baddeley, who visited the Forges in 1828, related that large sugar and potash kettles, as well as iron gear for steamboats, were being made at Mathew Bell's foundry in Trois-Rivières from St Maurice pigs and old iron. This foundry was equipped with two cupola furnaces. Barriault has also shown that, at this time, part of the casting pit-oven-crane area contained a block of sandstone used as a base for a cupola furnace in the moulding shed adjacent to the casting house.[94] Furthermore, the discovery of other casting pits in a workshop adjacent to the cupola furnace suggests that the layout was perhaps rearranged a few years later to allow large kettle casting with remelted pig iron.

Plate 5.7
Trolley on rails between casting pit and oven, at the blast furnace interpretation centre, Forges du Saint-Maurice National Historic Site.
PHOTOGRAPH BY PARKS CANADA/JEAN AUDET, 1985.

Chill Casting

The chill casting process was probably used to make cannonballs (Plate 5.8a). Iron **chills** have been found in the moulding shed next to the blast furnace[95] and at the upper forge, part of which had been converted into a moulding shop around 1775. Cannonballs of different shapes (round and two-headed) and calibres (1 to 24 pounds)[96] were made at the furnace moulding shed from sometime around 1744 until possibly 1760. As we saw in Chapter 4, cannonballs and bombs were manufactured by Pélissier, who was in collusion with the Americans who marched on Quebec in 1775. Munitions were also supplied during the War of 1812, as reported by Lieutenant Baddeley in 1828; and according to Baddeley, the establishment could still produce gun carriages and shot at that time.[97]

Plate 5.8b
Cannonballs and bombs made at the St Maurice Forges.
PHOTOGRAPH BY PARKS CANADA/JEAN JOLIN,
130/ACM/PR-7/SPO-00057, 1980.

Plate 5.8a
Cannonball chill-casting mould from the St Maurice Forges.
PHOTOGRAPH BY PARKS CANADA/ANDRÉ BÉRUBÉ.

Second Fusion

Shaft Furnace

As we have seen, the excavation of the upper forge has revealed the remains of a small stack furnace, built in the last quarter of the 18th century. Significant amounts of furnace scrap were found in the moulding shed next to the blast furnace casting house, suggesting that a furnace of this type was used during the French regime to manufacture cannonballs. According to Barriault, it may have been a kiln or shaft furnace, a structure that could be taken apart, as described by Réaumur in 1761. The furnace consisted of a crucible arranged in a shovel or ladle head enclosed in an old cauldron with a curved-back lip. Its chimney was made from pots with the bottoms taken out or sheet iron cylinders. Where the cauldron and chimney met, an opening had been fashioned for the tuyere, to which were attached manually operated bellows. The furnace was used strictly to melt old iron that was broken up and placed in the furnace interspersed with layers of charcoal. Once the material had been melted, the furnace was taken apart and the crucible full of molten iron was levered up and tipped with a long handle to empty the contents into the chills.[98] The chills were hollow moulds made of cast iron, one male and one female, clamped together; each mould contained the halves of one, two or three cannonballs.[99] When the iron in the mould cooled, the halves of the balls were removed by separating the two chills (Plate 5.8b).

Cannonballs were most likely manufactured like this from remelted cast iron, although there is no documented proof of this. The only piece of evidence is Intendant Hocquart's reference to "the construction of two workshops for manufacturing cannonballs."[100] The furnace described above does not seem sufficient for the large-scale production that occurred during the French regime, of around 4,000 to 5,000 balls annually, and over 10,000 in 1745, according to the archival record.[101] The workshops referred to by Hocquart must have housed larger furnaces or several small furnaces.

Cupola Furnace

Around the last decade of the 18th century, the production of castings increased significantly, transforming the Forges into a foundry in the true sense of the word, with the blast furnace running as a merchant furnace, making castings directly rather than pigs for later conversion into bar iron.

The same moulding techniques described earlier were used, including loam moulding, which was abandoned or moved elsewhere in the 1830s. Aside from the iron produced in the blast furnace, remelted cast iron was also used, mainly to manufacture domestic articles requiring this type of iron. Therefore, the production of second fusion iron was stepped up. There appears to have been a remelting furnace, probably of the cupola type,[102] in the moulding shed south of the casting house—according to the remains of a limestone foundation found there, which appears to correspond to the chimney shown in a Pigott watercolour painted in 1845 (Plate 9.2). According to Bérubé, the cupola furnace is a British invention dating from the 1790s (Plates 4.20a and 4.20b):

> It is a long steel cylinder set vertically on pillars, lined inside with refractory materials. Loading is done from the top and there are tuyeres and a taphole at the base. In the cupola, the iron is in direct contact with the fuel, generally coke.[103]

Hunter may also have been referring to this furnace in 1852 when he wrote that "the cupola Furnace requires to be rebuilt."[104] It was probably used to provide molten iron for open sand and box moulding as well as for objects moulded in casting pits, as the remains of the floor of the nearby moulding shop show.[105] This activity seems to have been discontinued after 1860, when the McDougalls took over the Forges.[106] No traces after this time remain.

There were 23 moulders at the St Maurice Forges in 1851 according to the census of that year, making them the largest group of skilled workers at the works and reflecting the greater focus on moulding around the blast furnace and cupola. The large-scale production of castings continued until 1858, when the Forges shut down for six years.[107] When the ironworks reopened in 1865, moulding was cut down drastically, with the main activity being the production of pig iron, for remelting at a Montreal factory. Moulders were no longer needed, and there were only four listed in the 1871 census.[108]

Pig Iron Casting

Casting pigs was much like casting the sow. According to the former Forges workers interviewed by Dollard Dubé, the **guttermen** who saw to this job were known locally as *faiseux de beds* (literally bed makers), referring to their task of preparing beds in the sand floor of the casting house for the iron pigs. The French word for an iron pig, *gueuset* (or slut)—*saumon* was also used — refers to the shape of the pieces once cast, while the English term is inspired more by the technique itself.

▼

This technique consisted of digging in the sand the mould for the main bar (or sow), to which were connected perpendicularly a series of smaller moulds for the pigs. From the top, the arrangement looked like a sow suckling her young, hence the terminology. Casting scraps found in the floor of the casting house include fragments of pigs that bear witness to this technique. According to Barriault, an analysis of the fragments shows that two types of casting layout were used, the first corresponding to the one just described. In the second layout, however, medium-sized pigs were connected perpendicularly to the sow at 2-m intervals, with the medium-sized pigs, in turn, branching out into a network of smaller, more closely spaced pigs. Some documentary and archaeological evidence suggests that wooden forms were used to cast the pigs.[109] In the second type of arrangement, some of the branches from the sow could snake all over the casting house floor, the furnace being of large enough capacity to permit an extensive network of channels for moulding pigs and other items. Excavations of the floor of the last casting house, rebuilt after the 1881 fire, have revealed that the entire surface was covered with black sand.[110] Indeed, the term *faiseux de beds* evokes what a task preparing the moulds, or beds, in the casting house floor must have been. When the furnace was tapped, the entire floor must have been transformed into a fiery surface.[111]

Railcar Wheels

According to the scant information we have on the railcar wheel shop (see Chapter 4), the wheels may have been made from remelted cast iron produced in one or two charcoal-fired cupola furnaces. The layout of the shop suggests that a fairly complex manufacturing process was involved, consisting of many steps, from the creation and filling of the wheel mould to the various processes involved in cooling the wheel.

Excavated remains and a few archival documents provide only sketchy information about how the wheels were manufactured. The wheels were produced using box moulding.[112] Iron strips, known as **chills**, were used to cool the outer rim and the hub of the wheel. When the iron was poured into the mould and came into contact with the chills, the rim and hub cooled and hardened: these parts had to withstand direct contact with the axle and rail while the body of the wheel had to remain somewhat flexible.[113] After casting, the wheels were allowed to cool gradually in annealing pits much like the ones unearthed during the digs (Plate 4.21), which were used to re-establish the heat balance between the outer rim (already cooled by the chill) and the centre.[114] As metallurgist Marshall Kirkman explains:

> When the wheel is cast the outer rim or chilled portion (on account of its greater density and hardness) shrinks relatively more than the center (or plates) and is more over cooler. When the wheel is put into the annealing pit the heat throughout becomes equalized, after which all parts are cooled at the same time.[115]

Although skilled workmen were no doubt required for railcar wheel production, the trade was not carried on long enough at the Forges for records of their names and training to remain.

FORGING

[...] the hammerman is in charge of his renardière *or chafery, of keeping the equipment in good repair, and takes his turn with a finer [...].*

Encyclopédie, *1757*

Hammermen are a class of workers that must be skilled, hard-working, loyal and gentle.

Encyclopédie, *1757*

Chaillé [...] a good hammerman, loyal and sweet-tempered, but become a drunkard these past two years.

Les ouvriers actuellement employez à St Maurice, *1743*

At first glance, it seems rather surprising that one would want to assign to loyal and gentle hammermen the task of supervising workers who were said to be "brutal, intractable, independent, fickle, dissolute, libertine, drunken and altogether a bad lot."[116] In the master founder's case, the emphasis was put on his skill, while for the hammerman, his gentleness, loyalty and ability to maintain discipline in his shop were stressed. In other words, the qualities required of the former were technical in nature, while those expected of the latter had to do more with labour relations—loyalty to the ironmaster on one hand, and firmness and gentleness with the workers on the other.

The qualities sought in these two key workers reflected the roles they occupied in their respective departments. The founder was in charge of preparing the basic raw material (cast iron) for processing by the other categories of workers (moulders and forgemen) into their products. The quantity of cast iron produced by the founder depended mainly on the nature of the ore and the capacity of the blast furnace; his productivity was limited by physical constraints, and his main task was to monitor the smelting process to produce cast iron of the best quality possible.

The situation for the forgemen was different. The pace and intensity with which they performed their manual labour had a direct effect on both the quantity and quality of the wrought iron.[117] Therefore, the hammerman, the master of the forge, had to set an example to the men that he managed.[118] He had to be not only a skilled craftsman, but also a leader of men. The founder's authority lay in his trade secrets, while the hammerman's authority lay in his superior skills and leadership.[119]

When the upper forge was built in 1739, there were two hammermen at St Maurice, one for each forge. Each supervised a crew of three finers and two helpers. Rivalry between the two forges developed very early on, which the management naturally encouraged, and the work environment at the forges would henceforth be animated by this friendly spirit of competition.[120]

Work Schedule

[...] it is impossible for this work to be continuous.
There is more advantage to stopping the forges
during the coldest season since the amount
of charcoal required to thaw out the movements
and heat the forge is more expensive
than the worth of any iron produced.

Mémoire concernant les Forges
de St. Maurice, *1743*[121]

Annual Production Cycle

In France, forges and blast furnaces did not run year-round and the workers were routinely laid off in summer, when the rate of flow in the streams or rivers fell too low to provide power to the waterwheels. At the Forges, the regular flow of the St Maurice Creek throughout the year[122] opened up the possibility of year-round production at the forges; indeed, the construction of a second chafery (*renardière*) at each forge to be used as a back-up bears eloquent testimony to the ironmasters' intent to produce wrought iron year-round. During the Forges' initial years of operation, the forgemen worked all year, but after a few years, there was no denying that in Canada, the winter cold, and not the drying up of the rivers in summer, made it impossible to run the forges year-round.

The harshness of the winter in January, February and March, the three coldest months, resulted in decreased volume and skyrocketing production costs at the establishment. Production levels fell in the winters of 1740 and 1741, particularly in January and February (Figure 5.3), but were still respectable. However, levels during the winters of 1742 and 1743 were dismally low. This early experience showed that, although the upper and lower forges could be operated in winter, production levels would drop to well below the expected capacity of 25,000 pounds (12 t) per month for each forge,[123] and the costs would be astronomical. The 95,635 pounds (47 t) of iron produced in the winter of 1741 (1 January to 2 April) were so expensive in terms of charcoal that both forges had to be shut down altogether in April for lack of fuel. According to Vézin and Simonet, this had been done:

> [...] against their recommendations and against the practices of the trade to operate the forges for three straight months during that winter which was so cold [...].[124]

The ironmasters were referring to the wastage of almost 800 cartloads of charcoal:

> [...] both by the exceptional consumption at the chaferies as well as the quantity that was burned in each forge to prevent the movements from freezing, which they did nonetheless despite the continual fires that were kept burning everywhere. [125]

The winter of 1742 was equally disastrous. Only 10,510 pounds (5 t) of bar iron were produced in February and March and production had to be shut down altogether from April to 10 May.[126] To make matters worse, to operate the forges in winter, the buildings, forebays and wheelraces had to be covered with wattle-and-daub (*bousillage*) to insulate and protect them from the cold.[127] In fact, the author of the 1743 memorial (cited in the epigraph) no doubt drew his conclusions from the costly experience of the previous winters. Subsequently, there is very little mention of any work in winter, and some sources would seem to suggest that it was given up altogether.[128] Henceforth, the forges were generally operated only nine months a year, from April to December.

Work Week

The forgemen generally worked six days a week, with time off on Sundays and holidays; on Saturday night, the premises had to be swept meticulously to remove the charcoal dust that covered the floors and walls, as a precaution against fire.[129] There were exceptions to the six-day work week, however. Frequent small fires in the shops were the main cause of downtime, but mechanical breakdowns (of the camshafts, and the gear mechanisms of the waterwheels and hammer), and accidents (the anvil or milldam breaking) due to wear and tear, freezing temperatures or worker ineptitude were also common. Furthermore, the men, who faced very difficult working conditions and hours, often fell ill, which also disrupted the regular cycle of shifts. Data on the monthly production at the two forges from 1739 to 1741 show that, in general, worker productivity was directly related to the number of days worked (Figure 5.1). The two forges produced on average 31,233.5 pounds (15 t) of bar iron; the weekly average, taking into account the six full weeks without production, was 7,349 pounds (3.6 t).

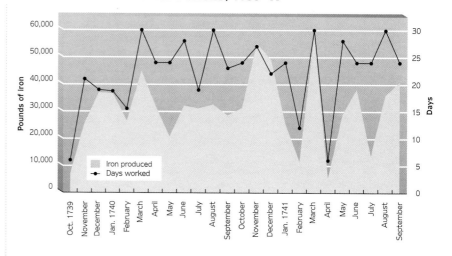

Figure 5.1

WROUGHT IRON PRODUCTION AND DAYS WORKED PER MONTH, ST MAURICE FORGES, 1739–41

Work Schedule and Division of Labour

[...] to man a renardière *operating without a break takes six workers—the hammerman, three finers, two helpers [...] the hammerman [...] takes his turn with a finer; two men generally produce six, sometimes eight, loops a shift; when their shift is over, they are relieved by two other finers and a helper, and so on.*

Encyclopédie, *1757*

All the evidence suggests that the forgemen worked in threes, relieving one another every six hours, as a visitor to the establishment in 1808 observed.[130] Therefore, there were four shifts every 24 hours, since the forges ran around the clock. The six-hour shift, which is quite short, is one of the distinctive features of the early iron industry, and probably corresponds to the rate at which the requisite number of loops were made. According to an early 19th-century French student of the iron industry, the workers made eight loops per shift, which took between five and six and a half hours in the case of 60-pound loops;[131] the weight of the loops varied, however. As in the case of the furnacemen, the number of workers attests to this schedule. Several sources indicate that there were 10 skilled hands (two hammermen and eight finers), assisted by four

TEAMS OF FORGEMEN, 1743 *

Forge	Hammermen	Finers	Remarks
Upper forge	*Marchand*		
Chafery 1		*Robichon*	
		Lalouette	
		Théraux	
Chafery 2		*Michelin*	Ill and infirm
		Ambleton	Extra hand
Lower forge	Chaillé		
Chafery 1		Dautel	
		Godard *père*	Unreliable
		Godard *fils*	
		Mergé	

* It is assumed that, because of their status and condition, Michelin and Ambleton were assigned to Chafery 2, which operated only by "sluicing." They were used as relief staff. Similarly, at the lower forge, one of the Godards was used as a relief worker. The workers kept on at the Forges after the Conquest are shown in italics.

helpers (two at the lower forge and two at the upper forge). Estèbe's records show that, during the 10-month period between October 1741 and August 1742, eight finers and two hammermen were employed at the Forges.[132] The 1743 memorial, which was indeed based on Estèbe's trusteeship of the previous years, lists 11 workers (two hammermen and nine finers), although one finer was described as an extra hand. Table 5.2 shows the teams at the two forges, based on the list of workers and recommendations contained in the 1743 memorial.[133]

The author of the memorial therefore wanted to dismiss Ambleton, the extra forgeman, and assign the two workers who were less reliable (Michelin and Godard *père*), along with a fifth helper, to Chafery 2 at the upper forge as needed. The use of relief staff is not surprising. As early as 1739, Vézin asked for two finers to replace workers who were ill.[134] When not on replacement duties, the relief workers could be assigned to Chafery 2 at the upper forge, which had been completed in 1740 and was operated by "sluicing."[135] This would result in four skilled workers (one hammerman and three finers) at each chafery, except at the back-up chafery at the upper forge, where only two finers were needed.[136]

Therefore, with the four skilled craftsmen and two helpers at each chafery, two teams of three men were formed, including the hammerman, who worked with one finer and one helper. Therefore, each team completed two shifts in 24 hours on a rotating basis, since the forges were also operated at night.

The production figures for the years 1739–41 point to night working. The contemporary compiler of those figures noted, on some occasions, an interruption of work during the night shift. The daily output of wrought iron at each forge also suggests that production continued day and night without a break. Based on the daily production capacity of each *renardière* (from 1,200 to 1,500 pounds or 0.6–0.7 t),[137] it can be inferred that, when things were running well, with no breakdowns, average production was near each forge's theoretical capacity.[138] Later accounts, by Lord Selkirk, in 1804, and John Lambert, in 1808, also indicate that both forges were running round the clock, just like the blast furnace:

The two forges each 4 men & 2 boys-half day-half night.

Selkirk, 1804.

The forges are going night and day, and the men are relieved every six hours. But at the foundry, only the men employed in supplying the furnace work in the same manner; those who cast and finish the stoves, &c. work from sun-rise to sun-set, which is the usual time among the French Canadians all the year round; a great advantage is therefore derived by carrying on any work in summer instead of winter.

Lambert, 1808[139]

In his comments on the benefits of summer rather than winter working, Lambert was referring to the European practice of operating forges in winter rather than summer, when many works were forced to close down and lay their workers off because of the low water levels in the streams and rivers.[140]

Operations

With a well-tended fire, four workers can make
12 to 15 hundredweight of iron in 24 hours.
A single hammer can serve two renardières.

Encyclopédie, *1757*

Fining and converting the pig into wrought iron involved two finers and an assistant working around the hearth, or *renardière*, during each shift. The fining process produced a pasty mass of iron called a loop. The loop was then hammered and reheated repeatedly through the bloom, **ancony** and **mocket head** stages into an iron bar. The two finers worked on different tasks simultaneously: while one fined the iron pig to make the loop, the other **shingled** (hammered) the loop with the hammer, taking turns reheating and drawing the iron out until it had assumed its final form. Because of the ceaseless toil required at the hearth and hammer, these shifts, though only six hours long, were exhausting.

Fining

The first operation was fining the pig, performed by the finer and a helper. One end of the heavy pig was fed from outside the forge through an opening into the hearth, so that it rested on the hare plate. The helper, who was stationed at the fore spirit plate side near the cooling trough, shoved the pig bit by bit into the charcoal fire, using his ringer as a lever (Plates 4.15 and 5.9). From time to time, the helper also covered the pig with charcoal, sprinkled it with water "to concentrate the heat," and adjusted, with the lever arm, the opening of the shuttle that released the water from the forebay, to drive the bellows. As the pig melted, the finer—who was positioned at the fore plate protected by the sloping mantle—used his ringer to gather the fused metal at the bottom of the hearth into a semisolid mass (the loop). He also used his ringer to work the metal, exposing it to the blast from the tuyere to decarburize it so that it would **come to nature**. The characteristics of the metal clinging to the ringer allowed the finer to determine how far advanced the operation was; he also made sure the metal was properly bathed in the slag and, when necessary, removed excessive slag from the hearth through the slag hole. When the loop was ready, the helper threw on a shovelful of moistened hammer scale to harden the surface. Using large ringers, the finer and his assistant then hoisted the loop onto the iron plate at the front of the hearth and slung it with hooks onto the iron loop plate set in the forge floor for preliminary hammering. The entire operation took about an hour, during which the workers had to manipulate in torrid heat, from only a few feet away, a formless mass of glowing metal weighing approximately 30 kg.[141]

Plate 5.9
Refining. On the right, the helper lifts the iron pig and gradually exposes
it to the flame; on the left, the finer pokes the heat-softened parts of the pig;
in the foreground, a hammerman pounds the loop to remove the slag.
DIDEROT AND D'ALEMBERT, *ENCYCLOPÉDIE [...], RECUEIL DE PLANCHES [...]*
(PARIS: BRIASSON, DAVID, LE BRETON, 1765), "FORGES OU ART DU FER,"
SECT. 4, PLATE IV.

Heating and Hammering

Once on the loop plate, the loop was
shingled with a sledge hammer to consolidate it
and shape it so that the "mordens," or **tongs**,
could grip it, and then it was back to the hearth
for another **heat**.

The loop was then taken to the anvil
and shingled into a bloom. The great forge
hammer, weighing almost 200 kg,[142] was set in
motion by tripping a rocker connected to the
sluice gate. The blows, first slow, accelerated as
the iron cooled. In the *Encyclopédie,* Bouchu
describes this delicate operation:

> [...] the first blows must be soft, since a strong
> blow would shatter the loop into a hundred pieces,
> greatly endangering the workers [...] the speed
> of the hammer is increased gradually as the
> different elements in the loop are consolidated
> and it acquires the shape and compactness of
> the bloom [...].[143]

Plate 5.10
Working the ancony.
DIDEROT AND D'ALEMBERT, *ENCYCLOPÉDIE [...], RECUEIL DE PLANCHES [...]* (PARIS: BRIASSON, DAVID, LE BRETON, 1765), "FORGES OU ART DU FER," SECT. 4, PLATE VI.

To complete the transformation from a bloom into an iron bar through the ancony and mocket-head stages took four or five heats, followed by a hammering each time[144] (Plate 5.10).[145] Depending on the size of the bloom, this required between 1,200 and 1,500 blows of the hammer:

> Depending on the weight and speed of the hammer, and the size of the loop and of the end product required, between 400 and 500 blows are required to forge the ancony; between 350 and 450 to forge the mocket, and between 450 and 550 to forge the bar.[146]

Once the loop had been formed, the two finers took turns at the hammer and hearth, working continuously. The only time the finer had a break was if he had to wait to reheat his bloom while his fellow worker was finishing fining his loop.[147]

Wrought Iron Production

The archival record tells us relatively little about the productivity of the ironworkers. The early years are the best documented, and this is particularly true in the case of the forgemen. A weekly report on wrought iron production from 15 October 1739 to 1 October 1741 (see Appendix 10) does, however, provide a very accurate picture of production at the two forges.[148] These data, already drawn on for the purposes of Figure 5.1, allow us to analyse other aspects of wrought iron production.

Figure 5.2

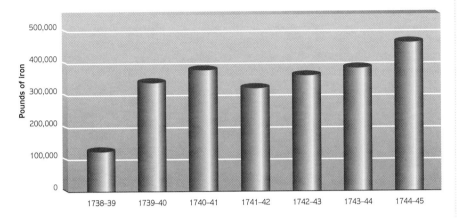

ANNUAL WROUGHT IRON PRODUCTION, ST MAURICE FORGES, 1738–45

Figure 5.3

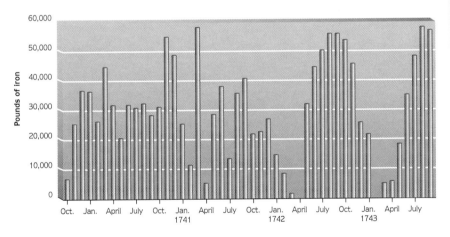

MONTHLY PRODUCTION OF THE LOWER AND UPPER FORGES, ST MAURICE FORGES, 1739–43

Figure 5.4

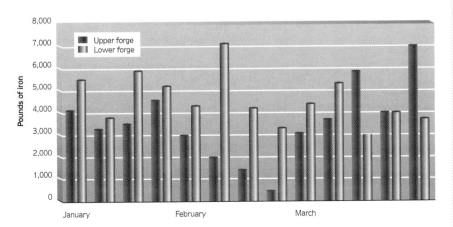

WEEKLY PRODUCTION AT THE LOWER VERSUS THE UPPER FORGE, ST MAURICE FORGES, JANUARY TO MARCH 1740

The production figures for the first seven years of operation (Figure 5.2) show that actual wrought iron production never reached the theoretical capacity of 600,000 pounds (294 t) for both forges together. The average was around 350,000 pounds (171 t), with a peak of 480,000 pounds (235 t) in the 1744–45 season. The poor results in 1738–39 are due to the fact that, during most of this period, only the lower forge was operating, since the upper forge did not begin production until 15 October 1739.

Although, as we saw, the winter months were not very productive and resulted in very high charcoal consumption, the most important factor negatively affecting productivity was the idling of the forges, often because of equipment breakdowns. Two- or three-day shutdowns (with the evocative name of *journées de débauche*) were common, occurring when the bellow shafts, anvils, hammer helves or any of the other parts subjected to stress or repeated blows had to be repaired or replaced. Sometimes, too, a worker would fall ill and this would disrupt work; the finers and hammermen, who worked in teams, were especially difficult to replace at the time.

Particularly in the early years, production sometimes had to be suspended because of a planning error by the ironmaster. In March 1741, because of exceptionally high production (58,407 pounds or 28.5 t), the charcoal supply was exhausted. (Figure 5.3 shows the extremely low output of the following month.) This shortage of fuel meant that the blast furnace could not be blown in until July instead of May. This poor planning led directly to the bankruptcy of the Compagnie des Forges de Saint-Maurice in the autumn of 1741.

Holidays were also a cause of work stoppages, and the ironmasters often complained that the festivities surrounding the Fête de St Éloi (St Éloi was the patron saint of smiths and forgemen) extended well beyond the actual day itself, 1 December.

Rivalry between the Upper and Lower Forges

We have already mentioned the friendly rivalry that the ironmasters tried to foster between the workers at the two forges. The weekly report of production at each forge prepared from 1739 to 1741 was no doubt for this purpose. Indeed, separate reports for each forge were written up from 15 October 1739 to 23 June 1740, but after that, only a single report was made, except for the period between December 1740 and March 1741. However, it would be risky to conclude that the differences in production levels at the two forges were due to rivalry, and a comparative analysis of the two teams' weekly production for three consecutive months in 1740 shows that the differences were caused instead by equipment failures or winter weather (Figure 5.4).

During the three months in question, the forgemen at the lower forge produced almost 14,000 pounds (7 t) more than those at the upper forge.[149] However, nearly 80% of the difference can be attributed to the seven and a half days of lost production in February at the upper forge because of repairs. Over the same number of working days, the men at the lower forge were slightly more productive, except during late March. The reason for the discrepancy is not known, but may well be due to the fact that, during the spring floods, the second chafery at the upper forge was operating.

The significant variations observed during the years from 1739 to 1743 (Figure 5.3) are clearly due to these frequent shutdowns in production that were characteristic of the early iron industry. Our totals for 1739–41 show that, out of the 618 working days in the period (thus excluding Sundays), the men at the upper forge worked 523 days and those at the lower forge, 543.5 days. The total duration of the shutdowns was 95 days at the upper forge and 74.5 days at the lower forge, which, taken together, average out to four months (85 working days) for the entire period, or two months per year.[150] Holidays only accounted for 16 days of the total (see Appendix 10).

Although it is difficult to determine which workers were the most productive, one cannot help wondering about the criteria used to select the five forgemen, all of them from the upper forge, ordered to remain at their posts by the British in 1760. As we will see in Chapter 8, it had more to do with drinking and discipline than individual productivity. The men at the lower forge—who, as far as we know, were just as productive, if not more so, than their fellow workers at the upper forge— were perceived as being headstrong and rebellious. In any case, the more amenable, apparently sober workers at the upper forge, who were the ones chosen to stay on, no doubt achieved acceptable productivity levels, since they and their descendants would form the basis of the labour force that the masters of the St Maurice Forges would rely on for generations, allowing the Forges to stay in business for another hundred years!

Single box stove manufactured by the St Maurice Forges.
The trademark FStM can be seen on the front.
PHOTOGRAPH BY PARKS CANADA/JEAN JOLIN, 25G-130/ACM/PR-6 /P-591.

VI

The Forges Wares

The ironwares produced at the St Maurice Forges reflected the social and economic context of the times. The Forges wares were designed to meet the specific material needs of a colony, itself in the process of development. This convergence of the company's production and the history of the country would be all the closer because the Forges were to remain the sole ironworks on Canadian soil for a long time to come.

As the history of the Forges' operations reveals, the iron-works was never to be solely an **iron mill** making bar iron: it would always operate simultaneously as a foundry. Indeed, the initial process of smelting could be immediately followed by founding or casting by running the molten iron into moulds. The blast furnace was then said to be working as a merchant furnace, meaning it was being used to produce castings:

> Iron destined for other uses than that of being converted into wrought iron, is known as *fontes marchandes*; instead of going to be worked in the forge, it is poured directly into moulds in its molten state.[1]

In the beginning, it was estimated that in a 7- or 8-month campaign, the blast furnace would produce more pig iron than the two forges could convert into bar iron in an entire year.[2] The blast furnace would thus operate in alternation between the production of pig iron earmarked for the finery (and subsequent second fusion) and the production of castings. The history of the Forges' production tells the story of the relative importance that was successively accorded products manufactured during the first stage (cast [pig] iron) and the second (wrought [bar] iron), respectively.

Only during the French regime, when nearly 90% of the pig iron was being made into bar iron (Figure 6.1), were the Forges strictly speaking an iron mill. Even then, part of the pig iron was earmarked to make castings. After the Conquest, production was refocused along new lines that were maintained for the next hundred years. Bar iron was made until nearly until the end of the Forges' history, but in such small quantities as to be of secondary importance. It was thus as a foundry that the ironworks operated for most of the company's existence. Casting was abandoned during the firm's final 20 years, with the year 1854 marking this pivotal change in direction, the last the company was ever to make. That year, for the first time since the start-up of operations at the St Maurice Forges, modifications were made to the blast furnace. Its capacity was doubled[3] (Figure 6.2), and the company focus was henceforth on mass production of pig iron, rather than on the manufacture of a variety of castings.

Within a few years, the Forges ceased nearly all production of castings, becoming a pig iron supplier to industry in Montreal.

The directions adopted by the company over the course of its history can be grouped into three main categories—iron for munitions and shipbuilding, iron for domestic use, and iron for industry—which we will discuss in turn. Each of these directions attests to the role that the Forges would be called upon to play in the country's defence, in its settlement, and in its industrialization.

Figure 6.1

PIG IRON MADE INTO BAR IRON AND CASTINGS AT DIFFERENT TIMES, 1742–1874

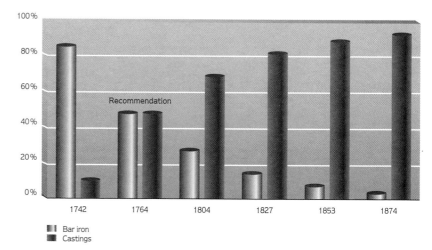

From Boissonnault 1981, fol. 59

Figure 6.2

DAILY CAPACITY OF THE BLAST FURNACE BEFORE AND AFTER 1854, ST MAURICE FORGES
(in tons of pig iron)

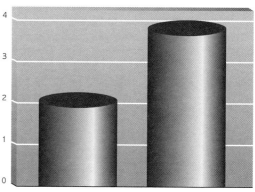

THE FORGES IN THE SERVICE OF THE CROWN

*Sieur Francheville expects to begin
his establishment in this fashion, and [...] hopes
within two or three years to be able to set up
a foundry like those in France. This will be both
to his benefit and that of the Colony.*

Beauharnois and Hocquart to Maurepas, 1732[4]

The first forge, constructed in 1733 by Francheville, was merely a pilot project that served to test the quality of the region's iron ore. The seigneur of St Maurice and the colonial authorities had greater ambitions in mind:

> [...] the ground on which the forge stands was as shifting and moving as a rope bridge [...] such that the entire works, which ran so smoothly at the beginning, today hardly serves for a trial run [...].[5]

Although the little forge had, with experienced workers, produced the expected output (400 pounds of iron per 24 hours), it could never have met the needs of shipbuilding, as the authorities had intended.[6] What is more, as Vézin stressed, along with the experts at the Rochefort Arsenal, the available ore was undoubtedly of high quality but the direct reduction process used at Francheville's forge "could not produce that wrought iron of high quality that the mine allows."[7] A more productive process was necessary for making the most of the ore, according to Maurepas:

> I have also caused tests and assays to be made at the Rochefort Arsenal of the 3 bars of flat iron sent by you and made from this mine, and you will see [...]that it has been deemed proper for certain works, and that it is believed that if the precautions noted herein are taken, [...] this iron may be rendered better in quality.[8]

The indirect reduction process proposed by Vézin made it possible to attain not only the level of quality desired but also the necessary output. With a production capacity of 1,000 pounds of iron a day, the Forges could supply both the specific needs of shipbuilding[9] and the colony's essential requirements (ploughshares, cast iron stoves and pots, and munitions). The Ministry of Marine's arsenals and storehouses thus became the first major customers of the St Maurice Forges.

Iron for Shipbuilding

*The iron which is here made, was to me
described as soft, pliable and tough, and is said
to have the quality of not being attacked
by rust so easily as other iron; and in this point
there appears a great difference between
the Spanish iron and this in ship-building.*

Pehr Kalm, 1749[10]

According to Bertrand Gille, few forges at that time in France's history specialized in the fabrication of iron for shipbuilding, an industry that required "special bars of precise dimensions, and thus difficult to execute."[11] Nor could France produce a twentieth of its navy's requirements in iron. It was forced to rely heavily on Spanish and Swedish imports to meet its needs. We have seen that Maurepas, the Minister of Marine, consulted the experts at the Council of Commerce before undertaking the venture, and that they in fact believed the Forges' output would make it possible to reduce imports of iron.

The Forges were thus established mainly to supply bar iron for building and fitting out ships both in France and at Quebec. The tests made by the Rochefort Arsenal on the initial bars and the high quality demanded of that iron speak eloquently of the use that the Ministry of Marine intended to make of them.[12] Without such a motivation, it is unlikely that the Ministry would have agreed to advance such large sums of money for construction of the Forges. The loan repayment terms were also directly linked to the specific needs of the Rochefort Arsenal. The colonial authorities had already agreed to this in 1735: "The iron to be delivered in repayment of His Majesty will be furnished to the Port of Rochefort in the proportions specified […]."[13]

Plate 6.1

An unfinished ship can be seen in the royal shipyard at the foot of Cap Diamant, 1760.

N. BENAZECH AND HERVEY SMYTH, *VIEW OF QUEBEC, CAPITAL OF CANADA* (DETAIL), ENGRAVING, 1760, ARCHIVES NATIONALES DU QUÉBEC, PICTURE LIBRARY, ORIGINAL COLLECTION, P600 S5, PGN40.

Table 6.1

BAR IRON ORDERED BY THE ROCHEFORT ARSENAL, 1738[14]

Category	Thickness (in *lignes*)	Quantity (in pounds)
Square iron	6	60,000
	7–8	100,000
	9–10	80,000
	11–40	60,000
Total		300,000
Flat iron		100,000
Grand total		**400,000**

As can be seen in Table 6.1, the "proportions specified" by Rochefort in 1738 for the year 1739 were 400,000 pounds of bar iron, a full two-thirds of the 600,000 pounds estimated by Vézin in his initial projections. Vézin made an initial promise of only 200,000 pounds, but several work stoppages limited production in 1738–39, so that only a little over 140,000 pounds of bar iron were actually produced.

The St Maurice Forges and the royal shipyard at Quebec were both started up in 1738. From their inception, the Forges supplied much of the iron to build the 500-ton royal storeship *Le Canada*, on which work was begun at Quebec between 1739 and 1742[15] (Plate 6.1). In October 1738, barely two months after the official blowing in of the blast furnace, Intendant Hocquart prepared a "return on the iron necessary for the construction of a *flûte* (storeship) to be built at Quebec for the King."[16] Seven months later, 1,813 bars of square, flat, and round iron weighing some 58,513 French pounds (29.25 t) had been delivered at Quebec,[17] representing 42% of the Forges' output of bar iron in 1738–39, the first year of operations.[18] Local shipbuilding thus seemed a promising market. The ironworks had demonstrated that it could meet the demand and standards of the shipbuilding industry.[19] The quality of St Maurice iron was not irreproachable in the eyes of French experts, however, who went so far as to accuse the Forges of negligence.[20]

Royal shipbuilding at Quebec did not develop at the pace predicted, despite shipbuilder René-Nicolas Levasseur's best efforts.[21] The Forges nonetheless probably supplied the iron for the 14 other royal vessels built at Quebec between 1742 and 1759.[22] According to Intendant Hocquart, construction of the *St Laurent* absorbed the bulk of bar iron output in 1747. Intendant Hocquart outlined the types of ship's iron manufactured:

> I trust that from now until the end of Xember [December] the two forges will render 350 thousandweight of bar iron. They might have given more, had it not been necessary to make a goodly quantity of iron in complex shapes in order to bring the construction of the *St Laurent* to a close. Rudder fittings, transom– or stern–knees, flat iron for masts, etc., are now in stock. I have lately received 1,000 pounds of new-made round and square iron, which I will proceed to have loaded onto the frigate *La Martre*.[23]

The small privately owned yards, which had been thriving at Quebec since the early 1700s, probably bought up a large part of the bar iron produced at the Forges.[24] According to Brisson, such yards built no fewer than 47 ships in the five years (1738–43) following the opening of the Forges. This included mainly small boats, schooners, and ships of 30 to 100 tons burden, as well as higher-tonnage vessels (three of which ranged from 150 to 200 tons burden, and five, from 250 to 300 tons). After 1743, these shipyards were for the most part shut down, due to poor business conditions, and, in particular, to the opening of the royal shipyard, which employed nearly all the shipwrights available at Quebec. Thereafter, until 1763, most shipbuilding would be concentrated at the royal shipyard, which launched, after *Le Canada* in 1742, a total of 14 vessels of 60 to 800 tons burden.

Figure 6.3 shows that during the first four years of production, 57% of the bar iron was shipped to France, mainly to the Rochefort Arsenal.

Combining the bar iron shipped to Rochefort with that delivered to the royal stores in the colony, we can see that 67.6% was put to the service of the Crown—or more precisely, the Ministry of Marine. The colonial market thus absorbed only 31.8% of bar iron output (Figure 6.4).

The large quantities of bar iron and munitions supplied to the French royal artillery and navy during the French regime encouraged the Forges to specialize. After the Conquest, in 1764, Claude-Joseph Courval[25] (whose father, Claude Courval-Cressé, had been director of the Forges for 13 years) confirmed the main focus of production. In an estimate of operating expenses and receipts given to the new British masters, he wrote:

> [...] the excellent quality of our iron, whose superiority to that of Europe has been proved many times, is such that only it was used for the building of the King's ships and for his artillery [...].[26]

Courval thus attests to the particular focus of the Forges as being that of supplier to the French King. During the Seven Years' War, the Forges were also called upon to produce large quantities of munitions.[27] Courval vaunts the superior quality of St Maurice iron, "oft proved" in comparative tests with European wrought iron.[28] Pehr Kalm himself (cited above in this section's epigraph), a native of Sweden, the world's largest iron producer at that time, had acknowledged this during a visit to the Forges in 1749.

The British also intended to make bar iron for their navy. They ascertained the quality of the Forges' stocks before giving orders that master founder Delorme, his keeper and five forgemen remain at their posts after the Conquest. During the four years that the Forges were under military occupation (1760–64), they were assigned to melting down old ordnance, producing nearly 500,000 pounds of bar iron, but only 280 cast iron stoves.[29]

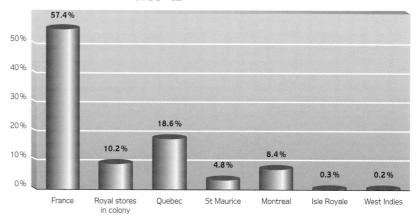

Figure 6.3

SALES OF BAR IRON FROM THE ST MAURICE FORGES, 1738–42

Source: NAC, MCI, C¹¹A, vol. 111, fol. 305 and vol. 112, fols. 82–83

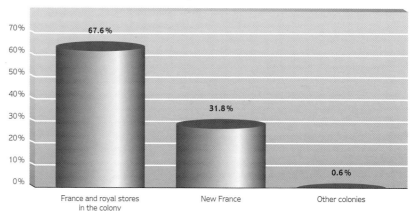

Figure 6.4

MAIN CUSTOMERS OF THE ST MAURICE FORGES, BY PERCENTAGE OF SALES, 1738–42

Source: NAC, MCI, C¹¹A, vol. 111, fol, 305 and vol. 112, fols. 82–83

Colonel Ralph Burton, the first British governor of Trois-Rivières, wrote in 1763 that:

> The iron made from this ore is so excellent in quality, that in a late trial made by order of his Excellency General Amherst, it was found superior to any made in America, and even exceeds that imported from Sweden.

He added that:

> All these, if thought proper, may certainly be greatly improved to the advantage of the crown by supplying his Majesty's navy with proper iron for shipbuilding.[30]

As with the French regime, little trace remains of what became of the bar iron sold to individuals and industry in the early days of the British period. However, taking into account the volume of certain sales by the British military,[31] the hypothesis cannot be excluded that St Maurice iron was used to build and refit naval vessels during the post-war period and to restore civilian and military infrastructures that had been damaged during the war.

Unlike the evidence available for the French period, there is no documented evidence that the Forges furnished bar iron to the British Royal Navy for shipbuilding. At any rate, the acknowledged quality of St Maurice iron for royal and civilian shipbuilding under the French leaves no doubt as to the use to which it would be put, at least in the colony's shipyards. In 1808, John Lambert observed that the five shipyards at Quebec and Montreal had used bar iron from the St Maurice Forges almost exclusively for a good twenty years.[32] Charles Robin, a major exporter of dried cod from the Gaspé who had his own shipyard, was also a customer of the St Maurice Forges (Table 6.2). However, he was critical of the product, which was known for its resistance to heat, but which, he claimed, was cold-short or made brittle by cold.

The construction of a ship of any size required large quantities of wrought iron. Woronoff reminds us that, at the end of the 18th century:

> [...] a 200-ton ship required between 8,000 and 10,000 kg of wrought iron, a good third for the hull, the rest for rigging, anchors, chains, and cables [...].[34]

In order to be better equipped to meet the French navy's other requirements, plans were also made to found cannon for ships. In 1744, engineer Chaussegros de Léry had proposed this.[35] A few workers were brought in and some trials were run, but the project, which would have required the construction of a second blast furnace and the hiring of skilled cannon founders, came to naught. Other weapons and large quantities of ammunition were produced, however.

Weapons of War and Ammunition

The iron industry is primarily an industry of the means of production, and only accidentally an industry of the means of destruction.

Denis Woronoff[36]

At certain points in its history, St Maurice furnished part of the bar iron, weaponry, and tools needed by the navy, and the military in general. A few documents available from the French regime indicate that bar iron was delivered to Fort Chambly, and that tools (sledgehammers) were delivered to Fort Carillon.[37] The abortive plan to establish a cannon foundry did produce some small pieces of artillery used in the defence of certain positions held by the French on the borders of New France.[38] In the end, mostly ammunition was produced

Table 6.2

BAR IRON ORDERED BY CHARLES ROBIN FOR HIS PASPÉBIAC SHIPYARD, 1796[33]

Quantity		Category		Dimensions			Thickness	
24	T	flat iron		3	In. wide		1	In. thick
20	T	D°	D°	3	D°	D°	³/₄	In. thick
60	T	Square	D°	2	Ins			
1	Bar	D°	D°				¹/₂	In
6	D°	D°	D°				¹/₄	D°
12	D°	D°	D°					In
6	D°	flat	D°	2 ¹/₂	Ins wide		¹/₂	Ins thick
3	Bars	best Germany Steel						
1	Bar	Square Iron		57	Ins long		3	Ins.
1	Twin saw							
2	Bars	flat iron		3	Ins wide			p. In. thick

D°: ditto

T: long tons (2,240 lbs.)

In: inch

(Tables 6.3 and 6.4), which was used both to defend Quebec (1759) and, ironically, to bombard it (1775–76). Archaeologists have uncovered many cannonballs, as well as the chills for casting them (see Chapter 5).

Military Production Under the French

From the outset, there were plans to found cannon and other munitions for the King's ships built at Quebec.[39] Cannon founding would no doubt have promoted the expansion of the company, since naval demand in wartime was particularly high. The weight of cannon on a man-of-war could amount to several hundred thousand pounds.[40] Even a storeship such as the *Le Canada*, launched in 1742, which was not, strictly speaking, a warship, could in wartime carry some forty cannons.[41] But to manufacture artillery pieces, skilled moulders had to be hired. Intendant Hocquart managed to have two moulders sent from France, who began the experimental manufacture of cannon and ammunition in 1747. The first artillery pieces sent to Brest, in France, failed to pass quality tests, but the Forges were nonetheless called upon to produce small cannon and ammunition in 1748, under the supervision of the Chevalier de Beauharnois, commander of the Royal Artillery in New France. These pieces were tested in Montreal, and then sent to Forts Frontenac, St Frédéric and Niagara.[42]

Table 6.3

AMMUNITION AND ARTILLERY PIECES MANUFACTURED IN 1748

Bombs				Cannonballs	Cannon		Mortars		Shot
Calibre	5"8	9"	12"	(all calibres)	2#	4#	6"	for grenades	
Quantity	161	110	144	2700	6	6	5	6	2

": inch
#: pound

Table 6.4.

CANNON AND GUNS* SENT TO THE FORTS OF NEW FRANCE, 1748

	Quantity	Type	Calibre
Fort Frontenac	2	cannons	4#
	2	cannons	1#
	10	guns	
Fort Niagara	2	cannons	1#
	10	guns	
Fort St Frédéric	10	guns	

* boucaniers or buccaneers' guns
#: pound

The colonial authorities, who were planning to produce heavy guns, decided to send an artillery officer, Chevalier François Le Mercier, to France for training in cannon founding. Le Mercier visited several foundries (Plates 5.5 and 6.2), and in 1750, he wrote a memorial detailing his observations at the Rancogne Forges.[43] In it, Le Mercier sounds very confident:

> [...] I make so bold as to assure you, Monseigneur, that I am able to have 8-pounders and smaller cannon cast at the St Maurice Forges, without need of workers other than those presently there.[44]

Upon his return, the possibility of building a second blast furnace was entertained, in order to produce pig iron in the requisite quantity and quality for a cannon foundry.[45] But the St Maurice Creek could not power a second blast furnace at the Forges and skilled workers were hard to recruit from France so the project was abandoned. Le Mercier nonetheless had a casting pit dug for 8-pounders[46] which was never used (it was excavated during the digs at the casting house; Plate 6.3). Le Mercier had

Plate 6.2
Drying a cannon mould.
DIDEROT AND D'ALEMBERT, *ENCYCLOPÉDIE [...], RECUEIL DE PLANCHES [...]* (PARIS: BRIASSON, DAVID, LE BRETON, 1767), "FONDERIE DES CANONS," PLATE XV.

certain pieces cast, including gun carriages, mortar bombs and cannonballs, which were sent to Isle Royale.[47] Henceforth, in its production for the military, the Forges would concentrate on these latter types of munitions. After the Conquest, John Marteilhe, a British merchant interested in re-opening the establishment, informed the authorities of the large quantities of munitions that had been produced there during the Seven Years' War:

> [...] the quantity of Cannon shots & other work delivered for the King's account at Quebec & Montreal was an Extraordinary produce of that year (1756) [...].[48]

Plate 6.3
Remains of a cannon pit in the blast furnace casting house at the St Maurice Forges.
PHOTOGRAPH BY PARKS CANADA, 25G-760X, 1973.

The American Invasion
and Its Consequences

The Americans, advancing on Quebec,
arrived at Trois-Rivières in the autumn of 1775.
Pélissier colluded with them, and was accused
of having supplied stoves, cannon, cannonballs,
and other articles to Montgomery's army,
which spent the winter on the Plains
of Abraham.[...] On the 1st of May [1776],
claims Badeaux, Pélissier had bombs cast
at the Forges measuring "13, 9 and 7 inches."

Benjamin Sulte[49]

The first civilian master of the Forges after the Conquest, Christophe Pélissier, was a liberal and a supporter of the radical ideas of British parliamentarian John Wilkes. It is thus not surprising that he sided with the Bostonians, the Anglo-American revolutionaries who tried to take the colony from the British in 1775.[50] His position as director of an ironworks made him a person of significance to the Americans, who badly needed war materiel during their march on Quebec. The invaders could not have hoped for a warmer welcome, since Pélissier had already been won over to their side. He not only supplied them with munitions, including cannonballs, but even went so far as to suggest a strategy for taking Quebec.[51]

The failure in 1776 of the American invasion forced Pélissier to flee across the border. A short time thereafter, the US Continental Congress officially thanked him for his loyal service, awarding him a commission with the rank and pay of Lieutenant-Colonel. Pélissier nonetheless returned to Canada in 1778, apparently without hindrance, in order to prepare for a final return to France.[52]

Pélissier's fleeting "collaboration" was apparently of no consequence for the workers and employees at the Forges, with the exception of Laterrière, Pélissier's inspector:

> [...] they taxed me, albeit falsely, with having betrayed the interests of the King in favour of that of the Bostonians. At the Château Saint-Louis, I was vilified as a traitor who had ordered the making of cannonballs and shot for the express purpose of breaking down, they said, the gates of Quebec, during the winter of the American blockade.[53]

He was detained for an entire month during the winter of 1776; the military authorities of New France apparently laid all the blame at the feet of the "gentlemen of the Forges."[54]

The Forges may thus have turned to manufacturing munitions because of the American invasion, although it is also possible that production was merely restarted at that time, sparked by the event. Notary J.-B. Badeaux reported, however, that munitions had not been cast since the time of the French, since the workers proclaimed themselves unable to produce good bombs, "not having the proper tools for their perfect manufacture."[55] It is surprising that the British authorities did not use the Forges to satisfy their own requirements for ammunition and arms. A short time later, in 1778, Lieutenant William Twiss of the Royal Engineers reported to the Governor of the colony that the Forges would be able to manufacture cannonballs if the necessary wooden patterns were forthcoming.[56] Pélissier had thus succeeded in drawing the attention of the new British masters of the colony to the fact that the Forges could participate in the war effort, either on their side or against it.

The Americans were also to have a taste of the cannonballs from the St Maurice Forges. According to a report written at the beginning of the 19th century, by that time the manufacture of munitions was a regular part of the company's output. Lieutenant Baddeley of the Royal Engineers reported that the British forces had used munitions manufactured at the Forges against the Americans during the War of 1812, in particular during the naval battles on Lake Champlain. Sent on a reconnaissance mission in 1828 by the Board of Ordnance, Baddeley informed his superiors that the Forges could supply gun carriages, shot and other supplies as needed.[57]

Digs at the upper forge have shown that the "American" cannonballs, as well as those manufactured later, had obviously been moulded in the eastern part of the forge, which had been converted into a moulding shop.[58] In 1785, it was reported that a chimney from the upper forge had been demolished, which is evidence that a chafery had been taken out of service and the forge converted into a moulding shop.

In sum, if only the instruments of war strictly speaking (ammunition, cannon, etc.) are taken into account, the military production of the company was significant only during wartime, when the Forges were called upon to participate in the war effort. But it must not be forgotten that the St Maurice Forges also supplied massive amounts of iron for shipbuilding for the French navy, thereby contributing to creating war materiel at other times.

THE FORGES AND THE COLONY

Supplies for Settlement

It was not so much the Forges' bar iron, destined or not for the French navy, that won the ironworks renown among the colony's *habitants* but its castings. Bar iron, though termed "merchant iron," was intended for use in shipbuilding and other industries that made parts for machinery, tools and other basic articles rather than goods for domestic use. Furthermore, St Maurice iron had to compete with imported iron from France (see Appendix 11), and the competition was so fierce that, to protect the Forges, Intendant Hocquart had to lower the price of local wrought iron repeatedly. The competition from European iron, the extent of which is difficult to quantify, did not abate when New France passed to the British. On the contrary, the Forges even had problems adapting quickly enough to British quality standards, and could not fill some orders. It was not until much later—under Conrad Gugy (Tables 6.5 and 6.6) and particularly under Monro and Bell—that the Forges were managed with the necessary drive to deal with their English and Scottish competitors, particularly in the case of bar iron and finished castings.[59]

Table 6.5	

ADVERTISEMENT IN THE *QUEBEC GAZETTE*, 1784

St Maurice, 20 August 1784

We deem it necessary to inform the public that the St Maurice Forges are now producing bar iron that is in no way inferior to the best iron imported from Europe. It is soft and malleable, made from grey iron from the local ore and does not contain scrap iron of any sort. Aside from bar, flat or square iron, it can be supplied in all forms such as sock plates, cranks for sawmills, gudgeons, millrinds or any other parts, by sending the dimensions.

Cast-iron anvils and andirons, fish kettles, pots, *culplats* and kettles of all sizes are also made: the latter appropriate for use as sugar kettles etc.

Box stoves are also sold at the following prices:

No. 1	23 inches	£2/6/8	or 2 guineas
No. 2	29 inches	£3/10	or 3 ditto
No. 3	32 inches	£4/5	or 17 dollars
No. 4	36 and a half	£5	or 20 dollars

N.B.: Both sides of stove tops are moulded in sand. St Maurice offers an advantage that cannot be found elsewhere, which is, if a stove plate breaks, you can return it to the Forges and exchange it for a new one. Bar iron and stoves are sold at Quebec by Mr. A. Proust, *fils*, and at Montreal by Mr. Uriah Judah, near the Market.[60]

Table 6.6

ADVERTISEMENT IN THE *QUEBEC GAZETTE*, 1794

The Lessees of the St Maurice Forges are desirous of informing the public that, due to the improvements and expansion executed last winter and the new patterns of iron stoves and implements obtained, they can claim that the articles currently being manufactured at the Forges are in no way inferior in quality or in substance to the same items imported from Great Britain. The list hereunder is a catalogue of the articles manufactured at the Forges and their prices for the current year.

	No.	Gal.	Price S	d		No.	Gal.	Price S	d
Bar iron and sock plates cwt			25		Cast-iron socks	1		8	
					Ditto	2		9	6
Kettles	1	6 $\frac{1}{4}$	10 ea.		Axle boxes	1		2	9
Ditto	2	9 $\frac{1}{4}$	13		Ditto	2		3	4
Ditto	3	10 $\frac{1}{2}$	14		Ditto	3		5	
Ditto	4	11 $\frac{1}{2}$	15	6	Ditto	4		7	3
Ditto	5	16	20		Ditto	5		7	6
Ditto	6	22 $\frac{1}{4}$	27		Anvils	1		28	4
Kettles with lids		1 $\frac{7}{8}$	3	4	Ditto	2		35	
Ditto	2	1 $\frac{3}{4}$	4	6	Double kettles with lids	1		13	
Ditto	3	2 $\frac{1}{4}$	6	6	Ditto	2		15	6
Ditto	4	2 $\frac{1}{2}$	7	6	Andirons			11	8
Ditto	5	3 $\frac{1}{4}$	8	4	Pestles	1		2 6 ea	
Ditto	6	5 $\frac{1}{4}$	13	6	Ditto	2		3	9
Pots		$\frac{4}{8}$	1	6	Stoves	1		55	
Ditto	2	$\frac{5}{8}$	2		Ditto	2		75	
Ditto	3	$\frac{6}{8}$	2	3	Ditto	3		85	
Ditto	4	1 $\frac{1}{4}$	3	4	Ditto	4		100	
Ditto	5	2	4		Scottish stoves	A		90	
Ditto	6	2 $\frac{1}{2}$	4	6	Ditto	V		80	
Ditto	7	3	5		Ditto	C		85	
Ditto	8	4	6	8	Ditto	D		90	
Ditto	9	4 $\frac{1}{4}$	7	6	Ditto	F		120	
Ditto	10	5 $\frac{1}{4}$	8	9	Ditto	FU		120	
Bake kettles	1	$\frac{3}{4}$	3	6	Double stoves	B		130	
Ditto	2	1 $\frac{1}{2}$	6	6	Ditto	N		180	
Basins	1	$\frac{1}{4}$	1	6	Stove plates for fireplaces			28 p cwt	
Ditto	2	$\frac{1}{2}$	2	3					
Ditto	3	$\frac{3}{4}$	3	4					

Handles	No.	Price		Handles	No.	Price
For kettles	1	12 d ea.		For *culplats*	1	5 d
	2	15 d			2	6 d
	3	18 d			3	6 d
	4	18 d			4	8 d
	5	20 d			5	9 d
	6	21 d			6	15 d
For pots	1	3 d				
	2	4 d				
	3	5 d				
	4	5 d				
	5	6 d				
	6	6 d				
	7	7 d				
	8	8 d				
	9	9 d				
	10	10 d				

The Three Rivers Store is still run by Mr. Jos. L. Leproust who takes orders and fills them as per the above.
In Montreal, Mr. James Laing has retail stocks for sale; and at the Quebec Store, items can be had for retail or wholesale by contacting Mr. Thomas Naismith, next door to the Undersigned. A discount will be granted to those who purchase for resale.

Quebec, 1 June 1794. MONRO & BELL

Abbreviations: No. = Item No.; gal. = gallon; ea = each; cwt = hundredweight.
Prices are in shillings (s) and pence (d).

The St Maurice Forges' main contribution to supplying the settlers was through the provision of **manufactures** as they were called at the time—mainly tools and other implements. Close to 125 products made at the Forges, mostly castings, have been identified.[61] Indeed, the company changed its production focus radically during the last quarter of the 18th century[62] from bar iron to manufactures. To appreciate the extent of the change, which occurred in just 10 years, one only has to consult the newspaper advertisements of the day (Tables 6.5 and 6.6). The 1784 advertisement emphasizes the quality of the bar iron, citing characteristics—"soft and malleable, made from grey iron"—that were of interest mainly to artisans and trades in which wrought iron was used as a raw material to make various objects; manufactures from wrought or cast iron were of minor importance. In contrast, the 1794 advertisement suggests that the firm made mainly manufactures for the *habitants*. It describes a wide range of implements and articles, in wrought iron or mainly cast iron, including both tools used by individual artisans and small shops as well as articles used in transportation, farming, cooking and heating. During the French regime, most of the pig iron produced was reserved for bar iron for the home country, thus preventing expansion of the castings side of the business. The blast furnace was operated part of the time as a merchant furnace to produce cast-iron stoves and pots, but only a limited amount of cast iron was reserved for this purpose (Figure 6.1). After the Conquest, wrought-iron production remained significant, but, since the company no longer had to supply France with low-price bar iron, it could concentrate more on the production of manufactures, mainly of cast iron, which were much cheaper to produce and therefore more profitable.[63] British control of the colony also gave the Forges access to the British market, both in Britain and her colonies, which stimulated the economy by providing new markets for local products. However, the ironworks profited only indirectly from this situation, by producing the machines for mills that supplied the new market.[64] It is difficult to assess how much of the Forges' output was exported to this market. Some information in the early 19th century suggests that 100 to 200 stoves were shipped from Quebec, mainly to other British colonies rather than to Britain itself.[65] According to Laterrière, pig iron was sent to London in 1771, and Bouchette also reports this export in 1815.

In Canada, the growing population in the colony and the concomitant expansion of settlement spurred the production of castings to such an extent that most of the pig iron output went to manufactures rather than to the production of wrought iron strictly speaking.[66] Now, the company's chief focus shifted from bar iron to a wide array of castings. The floor of the casting house was covered with moulding boxes, with the blast furnace taking on the appearance of a foundry.

During the French regime, the Forges employed only two moulders, who made cast-iron stoves,[67] pots and cannonballs. After the Conquest, the British kept on the French forge-men but let both of the moulders go. Apparently, they preferred English, Irish and Scottish moulders, who began to appear at the Forges after the ironworks reopened in 1767, evidently so that it would be able to compete successfully, in terms of manufacturing processes and quality standards, with products imported from England and Scotland (Table 6.7). Consequently, while forging continued in the French tradition, a new British tradition of moulding techniques took hold at the casting house towards the end of the 18th century. British moulding had an excellent reputation, and by the end of the 1700s, French industrialists were singing the praises of the moulders' ingenuity.[68]

Uses for Wrought Iron and Cast Iron

Almost every processing activity uses iron as a basic material. Be it cooking or manufacturing, farming, fishing, smithcraft, construction, transportation, hunting or war, there are few machines, implements, tools, instruments, buildings, bridges, vehicles or weapons that are not made at least partly from iron. Therefore, the chief interest in analysing the production of an ironworks like the St Maurice Forges—which could supply the iron for each of these spheres of activity—lies in the fact that its priorities closely mirrored the basic material needs that arose gradually from settlement. In other words, the Forges manufactured few objects that were not directly linked to specific needs for housing, cooking, work, industry and transportation.

Table 6.7

NAMES OF THE FIRST KNOWN BRITISH MOULDERS AND FOUNDERS[69] (from civil records)

Name	Craft	Years mentioned
Thomas Lewis	moulder	1771–85
John Slicer Sr.	moulder	1771–89
John Anderson	founder	1789–96
William Kenyon	founder-moulder	1799–1809
John Cooper	master moulder	1811
John Anderson Galbraith	iron moulder	1828
James McOwen	iron founder	1833

Heating

[...] we are dressed warmly enough, and our rooms are heated by stoves, and, to sum up, I was colder in France each winter than I ever was in Canada.

Jesuit missionary Luc-François Nau, 1735[70]

The product for which the St Maurice Forges were most famous in the colony was undoubtedly the heating stove (Plates 6.4a, 6.4b and 6.4c). In a country with long, cold winters, these stoves quickly became an absolute necessity. Replacing the traditional hearth, the stoves made life for the *habitants* more comfortable and, indeed, transformed domestic life in the colony.

During the ironworks' early years, 60 stove plates and 100 stoves were produced a year on average. By around 1800, this was up to 1,000 heating stoves a year, and this pace was kept up until the 1850s.[71] In the 1820s, the Forges also began to produce cooking stoves.[72] The plates for the cooking stoves, which bore the Forges moulders' most fanciful and ornate designs, were eagerly copied by local competitors and sometimes even by foundries farther afield. The Carron Iron Works in Scotland made "Canada stoves" beginning in 1775.[73] To arm themselves against foreign competition, the Forges lessees went so far as to publicly warn their customers, in the newspapers, about the inferior quality of these imitations.[74] The competition also forced the Forges under Mathew Bell to improve their products and introduce

Plate 6.4a
Single box stove manufactured
by the St Maurice Forges. The trademark
FStM can be seen on the front.
PHOTOGRAPH BY PARKS CANADA/JEAN JOLIN,
25G-130/ACM/PR-6 /P-591.

Plate 6.4b
Double stove manufactured by the
St Maurice Forges.
PHOTOGRAPH BY PARKS CANADA/JEAN JOLIN,
25G-130/COH/PR-7/SPO-0008.

Plate 6.4c
Single box stove made at the Forges under Monro and Bell
(1800–16), as indicated by the mark M & B on the front.
COLLECTION OF THE HÔPITAL GÉNÉRAL DE QUÉBEC.

new stove models, as shown by the advertisement published by Monro and Bell in 1794 (Table 6.6). In this advertisement, Monro and Bell take on their competitors in Scotland head on by offering "Scottish stoves."

By 1820, competition centred on who could make the finest and most lightweight articles. That year, Mathew Bell went to Britain to hire skilled moulders. He was apparently so successful that when he returned, he published the following advertisement in the *Quebec Gazette* of 17 July 1820:

> [...] the beauty of the work has been greatly increased, particularly the hollow ware, which, for its lightness and elegance, is the equal of similar articles manufactured in Great Britain. The St Maurice stoves are famed for their superior quality.

In the following years, the competition from Europe remained intense. Since imports of European stoves did not decrease, prices began to be affected. An 1833 circular from the agents Woolsey & Son at Quebec testifies to the veritable dumping that was occurring at the time:

> STOVES. - These indispensable articles in Canada were imported in such immense quantity the past season, that daily sacrifices took place at Auction, and at the close of the navigation, when higher prices are generally realized, fully 50% cent under the usual rates were obliged to be submitted to.[75]

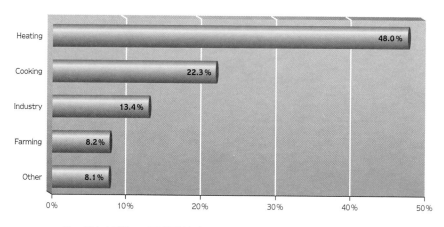

Figure 6.5

SALES OF FORGES WARES AT MONTREAL AND QUEBEC, BY USE, 1852–53

From Bédard 1986, pp. 114–15, Table 4.

During the French regime, the Forges only made two types of stoves, a large and a medium-sized model.[76] By 1784, four models were offered (see advertisement in Table 6.5) and, by 1794, 12 models,[77] including double stoves (Plate 6.4b). In 1823, cooking stoves were offered, and from then on, the ironworks produced a wide variety of heating stoves[78] for every room in the house. Financial accounts for 1852 and 1853 show that 23 stove models, including 7 cooking stoves, were produced.[79] By that time, stove production made up close to half (48%) of total production, both in value and in volume, representing 293 t out of the total 600 t of iron produced in those two years (Figure 6.5).[80]

Plate 6.5a
Cast-iron kettle manufactured by the St Maurice Forges.
PHOTOGRAPH BY PARKS CANADA/JEAN JOLIN, 25G-130/ACM/PR-6/P-415, 1981.

Stove accessories were also manufac-tured—ashpans, stovepipe collars and grates, for example—as well as doors, top and bottom plates, which were sold separately, and fire-place accessories such as andirons, firebacks and footmen.

To deal with the competition and main-tain profits, the ironworks decreased stove production.[81] In 1854, Forges superintendent William Henderson decided it would be just as lucrative to sell plain castings as **core castings** and stoves. That same year, blast furnace capac-ity was doubled, probably with a view to the new production focus Henderson was contem-plating.[82] Indeed, in 1870, only 200 stoves were being produced a year at most, and stove making would be abandoned altogether two years later.[83]

Plate 6.5b
Three-legged cast-iron kettle manufactured by the St Maurice Forges.
PHOTOGRAPH BY PARKS CANADA, 25G-130/ACM/PR-6/P-417-7, 1984.

Cooking

The greatest variety of castings manu-factured at the Forges was cookware. Kettles, pots and bake kettles were some of the first items to be made, and figure most often in the Forges' records (Plates 6.5a and 6.5b). Various models in different sizes were produced (Table 6.5). Other items of cookware included basins, teakettles, saucepans, porringers, *culplats* (flat-bottomed pots without legs), skillets, fish kettles and coolers, and the moulders also turned out utensils such as spoons, mortars and pestles. Here, too, as with stoves, the Forges' products had to compete with British products; in the 1770s, some buyers favoured British products over Canadian ones, finding them finer and lighter, albeit more expensive:

> [...] what makes me ask the price is that there are some very well made ones from the Forges that sell here at a lower price. They do not have lids, in truth I believe those from London are better since they are finer [...].[84]

The moulders who had been recruited in England, Ireland and Scotland were soon able to turn out products that compared favourably with the British ones. Later, in the early 19th century, the Batiscan Iron Works would also have to compete with British ironworks, a situation that entailed similar problems.[85]

Farming

[…] Nos plaines sont sans borne et le fer
de nos charrues ouvre des chemins aux moissons […]
[Our plains are limitless; the iron of our ploughs
opens the way to the harvest]

Pamphile Lemay[86]

Wrought-iron parts for ploughs (sock plates and plough points) were among the first items made at the Forges for Canadian farmers. The two-wheeled plough, which was brought over to New France at the colony's very beginnings, remained the main implement for cultivation until the late 19th century and even later.[87] This type of plough (Plate 6.6) was particularly suited to the heavy soils of the St Lawrence Valley. The key element in its design was the ploughshare, or sock plate, which dug the furrow, and only iron was resistant enough to withstand the stress and shock of ploughing. The sock plates were made from hard wrought iron in the forges[88] themselves, in the same way as bar iron, that is, they were heated in the chafery and pounded with the forge hammer:

> Plough irons are forged separately, passing through a series of heats and turns at the hammer, depending on their strength and size; models from 8 to 15 pounds are made.[89]

From the late 18th century onwards, cast-iron sock plates were also manufactured, as newspaper advertisements at the time attest.[90]

In the Forges' early years, close to 800 sock plates were made every year,[91] which represented 2.38% of the total volume of production in terms of value. Within a hundred years, this figure had more than tripled,[92] with sock plates making up 8% of the value of products manufactured at the Forges.[93] This large output reflected not only the needs of the expanding colony, but also the fact that broken or damaged sock plates had to be replaced. We do not

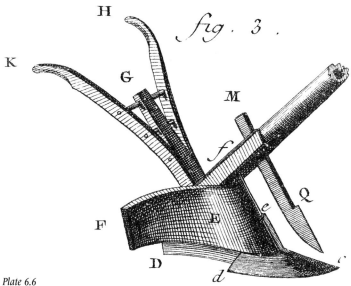

Plate 6.6
Plough with ploughshare.
DIDEROT AND D'ALEMBERT, *ENCYCLOPÉDIE […], RECUEIL DE PLANCHES […]*
(PARIS: BRIASSON, DAVID, LE BRETON, 1762), "AGRICULTURE
ET ÉCONOMIE RUSTIQUE. AGRICULTURE, LABOURAGE," PLATE II.

Plate 6.7
Ploughshare manufactured by the St Maurice Forges.
PHOTOGRAPH BY PARKS CANADA/JEAN JOLIN, 25G-130/ACM/PR-6/P-256, 1980.

yet have adequate data on wrought-iron use by farmers in the colony, but French sources suggest that, at the close of the 18th century, farmers cultivating heavy clay soils went through two or three sock plates a year.[94] Robert-Lionel Séguin compiled references on replacing damaged sock plates in notarial deeds in Quebec (in lease agreements for farms, for instance) and references to sock plate repairs in the accounts of a village blacksmith in the late 19th century.[95] Only one example of a complete plough with all its cast-iron parts made at the Forges has been unearthed, although such manufacture went on at the Forges for over a hundred years (Plate 6.7).

Tools and Equipment

Supplies for the Forges

Tools and equipment were first produced at the Forges for the ironworks' own use, since its operations required parts that could only be made in a blast furnace or iron forge. The first thing Francheville did when setting up

his forge was to order two chests of tools, along with anvils and hammers, from France.[96] Similarly, in 1737, 6 t of tools and implements to equip the new plant were ordered from the ironworks at Clavières, in the department of Loire-Atlantique, France.[97] Although the basic equipment to get the operation started had to be brought from France, the ironworks was subsequently self-sufficient.

In addition, essential structural elements for the chaferies—such as the cast-iron lintels and square iron pillar plates—could only be made at a blast furnace. These parts, along with the various plates making up the chafery hearths and anvils and hammers were undoubtedly among the first castings made at the Forges.[98]

The inventories also include a number of spare parts that were made on the premises; the heavy equipment must have broken down frequently.[99] The 1741 inventory makes mention of no less than 20 t of such materials in stock at the blast furnace iron store alone (Table 6.8).

All the implements for the operation of the blast furnace and forges were also made on the premises (Table 6.9).

Table 6.8

CAST IRON PARTS FOR THE FORGES IN THE BLAST FURNACE IRON HOUSE, 1741 INVENTORY[100]

Use	Item	Quantity	Weight (pounds)
Chafery	pillar plate	2	600
	bottom plate	1	300
	fore spirit plate	2	400
	tuyere plate	2	300
	large fore plate	1	350
	broken plates	50	3,200
Hammer	hammer	2	900
	hammer (broken)	23	10,350
	anvil	8	16,000
	anvil (broken)	3	6,000
Waterwheel	gudgeon	6	2,050
	plummer block	1	100

Table 6.9

WROUGHT-IRON TOOLS IN 1741 INVENTORY

Department	Implement	Quantity	Weight (pounds)
Blast furnace	ringer	17	
	quarrier	2	
	torchett	1	765
	cinder hook	3	(altogether)
	dam hook	2	
	sledge	3	
	iron fittings for rocker arm	1	
	shovel	3	
Moulding shop	casting ladle	3	
Upper and lower forges	cold chisel	4 + 4	
	dam hook	3	
	quasse	2	
	dipper	2 + 1	
	counterweights	1 + 1	2,000 + 1,000
	straw hatchet	1	
	hand hammer	4 + 4	
	sledge	4 + 3	
	cinder and slag shovel	5 + 2	
	ringer	9 + 10	350 + 275
	riole	1 + 1	
	great tongs	6 + 4	
	shingling tongs	4 + 6	
	forming tongs	11 + 15	
	hurst wedges	1	

The Forges also made and repaired tools for other artisans on the post—carpenters, joiners, cartwrights and edge-tool makers, for example, allowing for this in their indentures:

> [...] any tools of the aforesaid Baudry that need mending will be repaired by the forgeman of said establishment free of charge.[101]

Trade and Industrial Equipment

As we have seen, smiths such as blacksmiths, farriers, edge-tool makers and nailsmiths used some of the wrought iron produced at St Maurice in the pursuit of their trades, and were also supplied with anvils and hammers by the ironworks. Manufacturing firms were also customers for tools and implements. The Forges provided them with the means of production, and thus had to manufacture custom-made articles, as shown in newspaper advertisements from 1784 (Table 6.5) and 1817:

> All types of movements and engines for mills, etc., will be made from your models at the shortest notice, of a quality hitherto unseen at the Forges [...].[102]

Furthermore, with the development of the forest industry in the early 1800s, the ironworks began to supply equipment for sawmills. It would appear that the "engines" mentioned in the 1817 advertisement also included steam engines. Well before the St Maurice Forges had equipped their own plant with steam engines, they made parts for steam engines by special order, as certain documents show. Indeed, the Forges made the parts for the 6-hp *Accommodation*, the first 100% Canadian-made steamship, which was launched at Montreal on 19 August 1809 (Plate 6.8).[103] In 1820, the ironworks signed a contract with François Jérémie to supply parts for a ship's 40-hp steam engine and, in 1824, with Alexis Rivard and Olivier Larue to provide parts for a steam-powered sawmill.[104] Very early on, the St Maurice Forges played a significant role in

Plate 6.8
The steamboat *Accommodation*, launched by John Molson at Montreal in 1809.
MOLSON'S PRESENTS OLD MONTREAL
(MONTREAL: GAZETTE PRINTING CO. LTD., 1936), P. 15.

ushering in an industrial era that, ironically, would eventually make them obsolete.

From the early 19th century onwards, the ironworks also produced sugar kettles with a 6–32 gallon capacity (27–145 litres), for maple sugar producers. In 1829, cauldrons for making pitch were shipped to the Lachine Canal.[105] The firm also produced huge kettles for the potash industry, which exported large quantities of potash to Britain. This industry had grown up—along with the timber trade—as a result of the extensive clearing of forests brought about by settlement. Potash, which was made from the ashes of hardwoods, forms a lye that was used not only to make soap and glass but also to dress wool and cotton.[106] Potash kettles were produced in four different sizes and were undoubtedly the largest cast-iron containers made at that time, holding up to 190 gallons (864 litres) and weighing over 1,000 pounds (.45 t).[107] These large kettles were poured on the casting house floor using loam moulding, and indeed, one such casting pit has been excavated in the casting house floor. John Lambert witnessed the casting of these kettles on a hot August day in 1808:

I saw the process of modelling and casting, which is conducted with much skill. It was a remarkably hot day, and when they began to cast the heat was intolerable. The men dipped their ladles into the melted ore, and carried it from the furnace to the moulds, with which the floor of the foundry was covered. After they were all filled, they took off the frames while the stove-plates and potash kettles were red hot, and swept off the sand with a broom and water.[108]

Large sugar and potash kettles were also made in Mathew Bell's foundry in Trois-Rivières; according to Lieutenant Baddeley in 1828, the foundry had two cupola furnaces which were used to recast pig iron from the Forges to make potash and sugar kettles.[109] A list of workers drawn up the following year attests to the fact that five moulders and two forgemen belonging to the Forges were actually working at the Bell foundry in Trois-Rivières.[110]

Plate 6.9
Railcar wheels, manufactured by the Montreal Car Wheel Works owned by J. McDougall & Co., 1867.
FORGES DU SAINT-MAURICE NATIONAL HISTORIC SITE.
PHOTOGRAPH BY PARKS CANADA/ JACQUES BEARDSELL,
130/13/ PR-6/S-38-2, 1991.

In the 1870s, when the St Maurice Forges had more or less abandoned the production of manufactures, they kept a small factory making axe heads on the site of the former lower forge. About 10 dozen axe heads were made a day and, although they were very popular with lumberjacks,[111] this line of business was abandoned after only two years of production (1872–73).[112] Axe heads would be the last manufactures made at the Forges, which in its final years concentrated solely on making pig iron for the iron and steel industry in Montreal.

Public Works

[...] the iron castings from the St. Maurice foundry are very suitable for heavy work [...]

Colonel Durnford, 1830[113]

There is very little on record about the ironworks' participation in large-scale public works projects. A few letters from the Corps of Royal Engineers written between 1829 and 1830 refer to the supply of castings for the construction of the Rideau Canal in Ottawa. This matter is also mentioned in correspondence with Edward Grieves, the Forges' agent in Trois-Rivières, in reference to the renewal of a contract for this purpose.[114] Although the type of castings ordered is not specified, they were most likely parts of lock gate mechanisms (gears, cranks, lock fittings, etc.).[115] Colonel By, the engineer in charge of the canal project, had a very high opinion of St Maurice iron, preferring it to British iron. The way in which he justified his choice suggests that the officials in charge of procurement for the project were reluctant to renew the contract as Edward Grieves had proposed. Lieutenant Luxmann relays the reasoning behind By's choice:

Experience has shown, that the iron of this country is much superior to the English which makes me anxious that the iron required for the various services of the Canal should be procured in Canada; my preference arises from the metal in this country melted with charcoal, renders it tough and more malleable than English iron which is melted with Sea coal.

N.B. Lt. Col. By, in the paragraph of his letter from which the above is an extract is speaking of the ironworks on the St. Maurice near Three Rivers L. Canada, Thos. Luxmann, Lt. Rl. Eng.[116]

Another contract obtained in the autumn of 1854, just after the blast furnace was rebuilt, provided for the supply of 16-20 t of castings, probably in the form of gas pipes, to the Three Rivers Gas Company. These pipes were to be used to supply gas to streetlights on Notre Dame, Platon et Forges streets in Trois-Rivières.[117] The contract was almost lost to an upstart, the Radnor Forges, a new rival in the region that was established the same year.[118]

Transportation

The first bar iron produced by Vézin in 1737 was used to shoe a carriage wheel in Paris. As we have seen, in the fall of that year, the ironmaster hastily produced three small iron bars for testing in France. When one of the bars was found to be the equal of the famed Berry iron, Louis Fagon, the head of France's Council of Commerce, had one wheel of his carriage shod in St Maurice iron and the other in Berry iron to see which one was strongest.[119] We do not know the end of the story, but it is a reminder that iron from the Forges shod most of the wheels of all types of carriages, as well as the horses hitched to them. Furthermore, large numbers of axle boxes were also manufactured at the ironworks, pointing to the extensive use of iron in transportation at the time.[120]

The Forges also played a pivotal role in the development of other modes of transport. For example, with the development of the railroads, railcar wheels had to be manufactured (Plate 6.9). When the blast furnace was rebuilt in 1854,[121] a railcar wheel shop was established next to the upper forge and a cupola furnace was installed that August to remelt pigs to make the wheels.[122] The charcoal iron produced at the Forges, with its low sulfur and silica content, was ideally suited for this type of production.[123] Although railcar wheels were only made for three years (1854–57), this initiative would completely change the production focus, as we will see below.

Steelmaking

The St Maurice Forges attempted to produce steel on a very small scale, perhaps using the natural steel technique in one of its fineries. According to Bérubé, samples of steel were sent to France in 1747 for testing.[124] The tests were inconclusive, however, based on Pehr Kalm's account of his travels in Canada two years later:

> They have likewise tried to make steel here, but cannot bring it to any great perfection, because they are unacquainted with the best manner of preparing it.[125]

In 1771, a Mr Humfrey appears to have made some steel, but there is no conclusive evidence that it was made at the Forges. Nor do we know the use for which it was intended.[126] In 1828, Baddeley wrote that steel was not being made at the Forges, explaining that this type of production was problematic (as was plate iron and wire production) because of the nature of the bog ore. According to Baddeley, the metal obtained from the ore was probably too brittle to produce these products:

> [...] owing to the presence of the phosphoric acid which is always found in metal obtained from this ore.[127]

THE FORGES AND THE IRON AND STEEL INDUSTRY

Changing Times

> *The iron trade has had a term, we can no longer trust to making Box Stoves to be sold by auction never exceeding 12 sh and sometimes even under 7 sh p cw—that will not pay— great quantities of casting are now required for new work we must lay ourselves out to obtain a share of these and of plain (not core) castings to a large amount can be got even as low as 12 sh [...] it is infinitely better than making stoves [...].*

William Henderson, Forges superintendent, 1854[128]

The 1850s would be a crucial time for the St Maurice Forges. Henderson's recommendations, cited in the epigraph, mark a change in the firm's basic direction, a change that was made necessary by the increasingly fierce competition of the previous decade. Around 1840, the number of foundries in Lower Canada, and Upper Canada especially, began to increase rapidly, resulting in a more competitive market for finished castings—particularly stoves (Figure 6.6). Competition would also come from local sources with workers trained at St Maurice starting their own businesses. For example, Louis Dupuis, a former Forges employee, established his own foundry in Trois-Rivières in 1843, and went on to found the L'Islet ironworks in 1858:

> M. Dupuis learned his trade at the St Maurice Forges and only had the idea to start his own foundry three years ago. His establishment is now in a most prosperous state, and he has more work than he can take on. *His stoves are excellent and cheaper than those made at the St Maurice Forges* (our emphasis).

Gazette des Trois-Rivières, 1 October 1846

Although previously the Forges had been able to compete with foreign producers (mainly the United States and Great Britain), now the ironworks seemed to be unable to deal with domestic competition. In asserting that "the iron trade has had a term," Henderson was clearly establishing that the time had come for the Forges to abandon finished castings and to concentrate instead on supplying pig iron or semi-finished castings. Henderson was keenly aware of the highly competitive nature of the market, particularly in Montreal, and indeed his superiors would demand that he be more discreet when giving information on prices to retailers:

> If as you say I opened myself too freely to our agents in Montreal, it was under these circumstances. I knew that many competitors battle here and in the U States were in the field, that tact, zeal and discretion was needed to get anything [...].[129]

The rapid increase in the number of foundries was due in large part to the expanding market, which in turn was stimulated by construction of the railroad network in Canada. From that point on, railway construction would be a determining factor in the market, as was the upsurge in trade between Upper and Lower Canada, itself made possible by construction of the rail network. The intense competition among Canadian foundries, however, obscured another, even more damaging, level of competition—the competition with foreign blast furnaces (British and particularly American), which were the main suppliers of pig iron to Canadian foundries.[130] In the medium term, this foreign competition would lead to the demise of the iron industry in the St Maurice Valley.

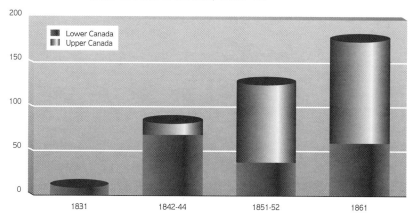

Figure 6.6

FOUNDRIES IN LOWER AND UPPER CANADA, 1831–61

Source: Census of Canada, 1665–1871, Statistics Canada, vol. IV (Ottawa: Taylor 1876)

Fortin and Gauthier have shown the extent to which market penetration by foreign pig iron was facilitated by the absence of adequate tariff protection for Canadian iron until 1879.[131] In 1854, under the Reciprocity Treaty with the United States, the meagre customs duties on pig iron and scrap iron were completely abandoned. This meant that Canadian foundries could henceforth buy American pig iron more cheaply while still enjoying protection from competition from imported castings. This contradictory tariff policy was a boon to Canadian foundries but it was detrimental to Canadian smelters, which logically should have been supplying these foundries with pig iron.

Despite the apparent upsurge of activity in the St Maurice iron industry in the 1850s, resulting from the Forges' participation—along with the Radnor Forges and the L'Islet Forges—in the railroad boom, the combined effects of American competition and the tariff policy would be devastating. In a short time, all three concerns, which were very fragile financially, would go bankrupt for the first time (St Maurice in 1859, L'Islet in 1862 and Radnor in 1866). The industry would suffer another serious blow in the 1870s, when the dire economic circumstances brought the price of Canadian pig iron plummeting down from $45.60 a ton in 1873 to $20 a ton in 1877.[132] The St Maurice Forges reopened in 1863, but were shut down again in 1877 for four years (1877–81), before closing their doors for good in 1883.[133] The new tariff policy, adopted in 1879, came too late to save the ironworks.

The End of Manufacturing

The groundwork for the change in direction recommended by Henderson in 1854 had already been laid by a series of technological modifications under Henry Stuart and John Porter & Company. Indeed, Henderson's proposal came in 1854, just when the new blast furnace was blown in. On the strength of its new production capacity, he sought contracts to supply pig iron, just as lucrative a trade as the old manufacturing business but much less expensive in terms of manufacturing costs.

Plate 6.10
Trois-Rivières car-wheel foundry, circa 1887.
It was founded in 1865 by the owners of the Radnor Forges.
NATIONAL ARCHIVES OF CANADA, PA-126950.

Around the same time, the Radnor Forges came into operation. Radnor was not just another foundry, dozens of which had been launched during the past two decades, but an iron smelter too, just like the Forges. Aside from the brief incursion of the Batiscan Iron Works Company into the market at the turn of the century, it was the strongest competition that St Maurice had faced in its history. The Three Rivers Gas Company contract episode would be the first indication, and the second would be the luring away of several Forges employees by the new firm.[134] Radnor had a slightly greater capacity and could compete effectively with St Maurice, particularly in manufacturing railcar wheels, a market that St Maurice intended to enter. Radnor also produced nails and stoves similar to those made at St Maurice.[135] The L'Islet Forges, which came into the picture in 1858, would also increase the local competition.[136]

The effects of this competition have yet to be well defined.[137] The launching of the Radnor Forges in particular no doubt weakened the venerable Old Forges, as they had become known, which were being restructured at the time. Barely three years after Radnor opened, the St Maurice Forges closed down for five years (1857–62).

In the three years before the shutdown, the railcar wheel trade—although it did not bring about new prosperity—did allow the business to break even, winning it a reprieve for a time, producing the pig iron sought by the railway industry.

Plate 6.11
Clendinneng's Foundry in Montreal, 1872.
CANADIAN ILLUSTRATED NEWS, MONTREAL, 4 MAY 1872, P. 277.

John McDougall acquired the Forges in 1863[138] but only operated them as a manufacturing concern for one year. It is not known whether he tried to start up the railcar wheel business again, but we do know that McDougall quickly understood that he could not, due to production costs and volume, compete with the Radnor Forges and the large railcar wheel manufacturers in Montreal, Toronto and other major centres (Plate 6.11). The ironworks could, however, produce the basic pig iron. McDougall decided to concentrate on supplying this raw material to other wheel manufacturers:[139]

The McDougalls have made it into a significant concern. It supplies pig iron to the main foundries of our country, particularly the McDougall foundry in Montreal. This iron is reputed to be the best in North America for railcar wheels and other goods that require first-class iron. American companies seek it out, despite the huge customs duties on iron at the US border and extremely high freight costs.

Under the five-year contract signed in 1865 with John McDougall & Company of Montreal, almost all of the pig iron produced at St Maurice and L'Islet would go to the Montreal manufacturer. In February 1871, the contract was renewed for another two years, setting the direction of production at the Forges until it closed in 1883.[140] Under the contract, St Maurice kept a mere 150 t of pig iron a year[141] to make a small number of stoves (around 200), axes and a few other articles; however, this line of business would be abandoned for good in 1873.[142] There is no direct evidence of the contract with John McDougall & Company being renewed after that date although there is indirect evidence (from the Quebec Bank) that John McDougall & Sons were probably still a major client of the Forges in 1875 and 1876.[143] Subsequently, as we have seen, poor markets and disagreements between George and Alexander Mills McDougall would lead to the Forges' shutdown in the fall of 1877 and to the closure of the L'Islet Forges in 1878. St Maurice was reopened one last time in 1880, by George McDougall, supplying the bulk of its output to the railcar wheel foundry in Trois-Rivières (Turcotte and Larue), which he himself leased (Plate 6.10). The construction of a second blast furnace at the Forges on the site of the old upper forge was perhaps not unrelated to the appearance of this new outlet for their wares. However, the venerable ironworks would not survive George McDougall's debt problems. During its last years, it would be dependent on the fortunes of the iron and steel industry in the wake of the development of the railroads, and the inevitable ups and downs of this industry would constantly threaten to bring about the Old Forges' downfall.

THE QUESTION OF PROFITABILITY

How could this Establishment generate profits, given all the wasteful expenditures occurring in its Operations?

First, there are a number of useless clerks earning considerable wages, who seek only to enrich themselves and who are present at these Forges only when the Intendant and the Company are assembled there.

Vézin and Simonet, 10 June 1741, a few months before the bankruptcy [144]

Were the St Maurice Forges profitable? Given the changing fortunes of the iron trade, were production costs low enough to provide a sufficient profit margin? Were certain products more profitable than others? To answer these questions with any kind of accuracy, one would need to look at the account books. Unfortunately, these have never been found, and we only have a few figures on operations, primarily from the French regime. This lack of archival material means that the profits generated by the ironworks are without a doubt one of the Forges' best-kept secrets. There are some figures only from when the ironworks was under Crown control. However, the periods when the Forges were administered by lessees and private owners provide few numbers and those available are vague and approximate, apart from the limited information in legal proceedings instituted by former partners. Therefore, our examination of the Forges' history must be done without the systematic measuring instruments and long-term data that would help make sense of the little information there is and provide a clear picture of the establishment's profitability. This makes it all the more difficult to carry out a financial analysis and assess the relative importance of factors such as price and cost fluctuations, effects of competition and market slumps, and frequent

▼

work stoppages. Nevertheless, we will try to make the best of the data available, most of which date from the French regime and the latter half of the 19th century.

In 1735, Vézin forecast a dazzling return of approximately 50% on revenues, but it was not until after Estèbe's trusteeship that the expected profit was reduced to the more realistic figure of approximately 10–15%. In a balance sheet he drew up on the Compagnie des Forges de Saint-Maurice immediately after its bankruptcy in 1741, Intendant Hocquart observed that the company "had made absolutely no profit." The following year, under the frugal stewardship of Estèbe, the merchant appointed as trustee to run the Forges, the establishment barely covered its costs.

Looking at the company's operating results after five years in operation, Hocquart estimated that, with a total production worth some 110,000 *livres*, the ironworks would never generate a net profit of more than 12,000 to 15,000 *livres*, assuming a yearly output of 600,000 pounds of iron.[145] A year later, trying to find arguments to convince potential entrepreneurs to take over the Forges, Hocquart provided more details. Using figures from Estèbe's one-year trusteeship, he projected a profit of 13,640 *livres* on proceeds of 87,800 *livres* (Table 6.10). However, Hocquart admitted that such a small profit, 15.5% of revenues, "would never convince a Company to take on such an Establishment, given the matter of paying off its expenses." This was a reference to the cost of setting up and operating the business since 1735, some 350,000 *livres*, more than half of which were royal advances. A new company would have to pay back this money and, as we have seen, the Crown finally recouped its costs by taking over the Forges (see Chapter 1).

Hocquart also identified other potential sources of profit. He estimated that an additional 15,000 *livres* in revenue could be generated if a new Forges operator were to be granted exclusive rights to the trade in metals in the colony, including those not made at St Maurice (steel, lead, tin and copper). This suggestion was not well received by Governor Beauharnois and was never approved. Another potential source of profit were the Forges workers themselves. Hocquart noted that the company store sold 45,000 *livres'* worth of merchandise each year which, with a return of 25%,[146] would yield an annual profit of 8,000 *livres*, once the wages of a clerk and other expenses had been deducted.[147] With these three sources of revenue, he intimated, the establishment could produce potential yearly profits of around 35,000 *livres*. In a later memorial, it was estimated that the establishment could yield a total annual return of 20,000 *livres* by combining earnings from the sale of its products, some 12,000 *livres*, with the 8,000 *livres* in profits from the company store.

Table 6.10

**COST OF PRODUCTION,
ST MAURICE FORGES**

Year	Source	Type of data	Production (pounds of cast iron)	Period	Value (livres)	Cost (livres)	Profit (livres)	Profit† (%)
1735	Vézin	estimates	1,000,000	8 months	116,000	61,250	54,750	47.2
1739	Chaussegros de Léry	estimates			105,000	45,000	60,000	57.1
1740	Anonymous	estimates	1,200,000	8 months	110,000	80,000	30,000	27.3
1742	Estèbe	accounts	655,660	4 months 25 days	65,186	65,208	-22	0.0
1742˙	Project based on figures from Estèbe et al.	estimates	810,000	6 months	87,800	74,160	13,640	15.5
1745	Hocquart#	accounts			133,180	133,180	12,608	9.5
1756	Marteilhe	accounts			141,432	127,170	14,262	10.1
1762				6 months	13,721	5,957	7,764	56.6
1764	Courval##	estimates	800,000	8 months	145,000	110,708	34,292	23.6

† In this column, the profit is calculated as a percentage of total revenues.

50,432 *livres* = profit for four years (1742–45); the profit is based on the value of the inventories.

These costs include 31,300 *livres* spent on maintenance.

˙ These estimates were obtained from Estèbe's accounts for 1741–42 and contemporary projections for six months of operation.

This profit was not considered excessive, given the inherent risks of running an ironworks, which was prone to frequent shutdowns due to such factors as work stoppages, accidents, forest fires and problems with supplies. In fact, this return would be clearly insufficient, considering the outstanding debt of 350,000 *livres* owed to the King and various other creditors. In the end, the memorial recommends that the King recover his advances by repossessing the Forges and the St Maurice seigneury and by having an "agent" operate the business for 25 years, with exclusive rights to the trade in metals. The author of the memorial nonetheless estimates that the earnings generated under such market conditions (35,000 *livres*) would still be "too modest" and suggests that the new operator also be granted the trading concessions at four posts for 25 years, which could bring in more than 30,000 *livres* in annual profits. The enterprise that took over the Forges would thus earn a net profit of approximately 65,000 *livres*.[148] History shows that this objective was never realized. The sale of foodstuffs and merchandise to the Forges workers and of the Forges wares themselves remained the only sources of profit.

The data in Table 6.10 clearly indicate that, prior to 1742, the estimated profit margins were greatly inflated due to overestimated production and underestimated costs. Indeed, 1742 was the first year that accounts were kept properly and realistic production figures given. The projected figures for 1742 cover six months of operation and take into account a number of improvements. The estimates are still very optimistic, but one only has to examine the accounts to see that production goals were rarely achieved. However, Estèbe's frugal stewardship proved that the Forges could be profitable. And, according to Intendant Hocquart, total profits of more than 50,000 *livres* were generated in the next four years.

The line of business the Forges were in at that time would never allow the ironworks to achieve significant profit margins. After all, during French ownership, nearly all the pig iron was converted into bar iron, making the Forges first and foremost an iron mill. The process of making bar iron was very expensive in terms of production costs. Since the pig iron was used primarily as raw material to make bar iron, most of the Forges' profit had to come from the sale of the finished product, which was sold to the Rochefort Arsenal and elsewhere in France at preferential rates. It is no wonder that profit margins were so slim. The enterprise would have been much more profitable had most of the pig iron gone into castings, rather than being converted into bar iron. The additional cost, in terms of charcoal and labour, to make bar iron added significantly to the cost of production.

Another point to remember is that the return on the Forges' obligatory trade with France was minimal while the domestic colonial market was much more lucrative, even for bar iron. The figures projected from the 1741–42 operating results are particularly eloquent in this regard. Sales within the colony generated a profit of 40% on bar iron, and 80% on cast iron, while the bar iron shipped to France was sold at a loss (Table 6.11). Although the Forges sold very little on the colonial market, it was these sales that enabled the enterprise to maintain a profit margin of 15.5% (26%, in fact, after adjustments).

The masters of the Forges were very aware of this fact. A production report from 1747 noted that:

> It is the castings that will generate the most profit because they are sold within the Country itself; on the other hand, we cannot produce castings alone because pig iron is needed to maintain the 2 forges [...].[151]

It was not until after the Conquest, when the Forges were no longer in thrall to the French market, that bar iron output was scaled back and production geared more to the more profitable colonial market. The period of British military administration quickly showed the wisdom of this shift in focus. Even concentrating heavily on bar iron, in 1762, for example, the Forges generated profits of 50%. This was admittedly an exceptional year because the pig iron was produced by melting down old ordnance, and so was not included as an expenditure.

Table 6.11

COST AND REVENUE PROJECTIONS, ST MAURICE FORGES, BASED ON ESTÈBE'S 1741–42 ACCOUNTS (in *livres*)

	Destination	Production (t)	Cost (per t)	Total Cost	Price (per t)	Revenue	Profit	Return (%)
PIG IRON	Both forges	375	82	30,750	82		0	0.0
	Colony	30	82	2,460	400	12,000	9,540	79.5
BAR IRON	Colony	60	287	17,220	500	30,000	12,780	42.6
	Rochefort Arsenal	100	287	28,700	260	26,000	-2,700	-10.4
	France	90	287	25,830	220	19,800	-6,030	-30.5
TOTAL						**87,800**	**13,640**	**15.5**

PROJECTIONS AFTER ADJUSTMENT[149]

	Destination	Production (t)	Cost (per t)	Total Cost	Price (per t)	Revenue	Profit	Return (%)
BAR IRON	Colony	60	287	17,220	400	24,000	6,780	28.3
	Rochefort Arsenal	100	287	28,700	320	32,000	3,300	10.3
	The Islands	90	287	25,830	360	32,400	6,570	20.3
TOTAL						**100,400**	**26,190**	**26.1**

From "Dépenses générales des Forges dont le fourneau doit supporter un tiers y aiant deux forges entretenues qui doivent en supporter chacune un tiers," 1742.[150]

Courval's 1764 estimates show a shift in focus to make the ironworks profitable. His experience as inspector of the Forges during the British military administration, combined with that of his father who had preceded him, shine through in his recommendations. Contrary to the estimates made under the French, Courval suggests, for the first time, that half the pig iron be sold locally in the colony in the form of castings and the other half, as bar iron. His projected profits were nearly 24%.

During the next few years, the proportions of castings and bar iron produced were completely reversed, in favour of castings. From then on, the bulk of the pig iron was turned into castings, while bar iron production was cut back to only 10% of total output. Castings cost less to produce and sold at a good price, with a doubtless concomitant increase in revenues and profit margins, although no precise figure can be put on these. According to Laterrière, the inspector and director of the Forges from 1775 to 1779, profits could have been as high as 33%.[152] Evidence suggests that, given the same market conditions, the profit margin was no lower than the 24% that Courval had estimated in 1764. Furthermore, since the Forges had increased their production of castings, Courval's projected value of production (145,000 *livres*) could even be considered a minimum; this figure would almost double by the turn of the century, before achieving new highs during the heyday of Mathew Bell. Although we will discuss some figures on value of production later in this chapter, it is impossible to determine the Forges' profit margins with any accuracy. The amount by which Monro and Bell's shares appreciated within a mere 30 years (see Chapter 1) suggests that profit margins were considerable, especially taking into account the extra profits generated by supplying goods to the population at the Forges, which had doubled.

In 1804, Lord Selkirk reported that the produce, or annual value of production, was between £10,000 and £12,000. If the conversion rate of the time is used (24 *livres* equals £1), this is nearly double (264,000 *livres*) the amount forecast by Courval in 1764. In 1833, Mathew Bell estimated that annual production was worth £30,000, triple the figure of 30 years earlier. This spectacular increase was no doubt due in part to the greater emphasis on castings, and it also mirrors the increase in the value of Bell's shares during the same period.

Later, both clerk Hamilton Rickaby and superintendent William Henderson reported that earnings from production were approximately £7,000 (£6,872). Rickaby set the average net proceeds of the Forges at £12 2s 10d per ton for the years 1852–54; with an average yearly production of some 566.8 tons, earnings totalled nearly £7,000 (£6,872). Henderson came up with a similar figure, £7,000, which he termed a "large profit" after examining the accounts for the year 1853.[153] However, there is no indication of the corresponding revenues. The 1853 balance sheet, while not explicit, suggests that production for the year was worth between £20,000 and £25,000.[154] If we take the higher figure of the two, the profit margin would be 28%.

In the early 19th century, French ironworks that, like St Maurice, had both a blast furnace and forges, had similar profit margins. Denis Woronoff mentions four establishments that generated profits of 18–25%. A survey conducted in 1805 showed that the iron industry in the French department of Mont-Blanc generated profits of 24%; a similar return was observed at ironworks in Burgundy and Franche-Comté. As was the case at St Maurice, only cast iron generated high profits. In the Haute-Saône region, it was said "the only thing that makes a profit is cast iron."[155] Woronoff goes on to discuss variations in annual profits, due not only to economic conditions and stocks of raw materials in hand, but also to the management policies of the companies in question.

At the St Maurice Forges, up to the 1860s, earnings were generated primarily from the sale of manufactures. Casting required skilled workers, who increased production costs, even though, as we saw earlier, it was more profitable to produce castings than bar iron. After 1860, the decision to stop making castings and to concentrate primarily on producing pig iron resulted in significantly lower production costs and a higher profit margin. The figures compiled below in Table 6.12 illustrate that, under the McDougalls, the Forges generated average net proceeds of 46% on the sale of pig iron.

When the Forges reopened in 1880 after being shut down for over two years because of a downturn in the economy, it still cost between $18 and $20 to produce a ton of pig iron. Now, however, it only fetched approximately $24–27, leaving a profit of only 25–26%.[156]

Breakdown of Production Costs

The operating costs for the year 1741–42 were used to estimate production costs, broken down by category (Table 6.13). Raw materials accounted for half of production costs, and wages, for nearly a quarter, while expenditures for administration and maintenance, including freight costs, accounted for the remaining quarter in more or less equal proportions.

Table 6.12

COST OF PRODUCTION AND RETURN ON PIG IRON, ST MAURICE FORGES, 1865–70

Year	Cost of Production ($ / ton)	Selling Price ($ / ton)	Profit (%)
1865–70	15.00	28.00	46.4
1871–72	15.00	28.00	46.4
1873	18.54	40.00	53.6
1875	18.54	30.00	38.2

From Bédard 1986, p. 178, Table 8.

Table 6.13

BREAKDOWN, BY CATEGORY, OF ESTIMATED PRODUCTION COSTS AT THE ST MAURICE FORGES, BASED ON 1741–42 OPERATING COSTS[157]

Category	Blast Furnace		Upper and Lower Forges		TOTAL	
	(livres)	(%)	(livres)	(%)	(livres)	(%)
Raw materials*	22,290.0	67.2	15,147	21.1	37,437.0	50.5
Workers' wages	5,182.5	15.6	11,100	15.5	16,282.5	22.0
Administration#	3,700.0	11.2	7,400	10.3	11,100.0	15.0
Maintenance	1,980.0	6.0	7,360	10.3	9,340.0	12.6
TOTAL	33,152.5	100.0	41,007	57.2	74,159.5	100.0
% of total cost	44.7		55.3		100.0	

* Includes collection, dressing and transportation.

Administration costs are apportioned equally among the three departments.

Table 6.14

COMPARISON OF PRODUCTION COSTS AT THE ORE REDUCTION STAGE, ST MAURICE FORGES AND FRENCH IRONWORKS (%)

	St Maurice Forges 1741–42	France 1813†
Ore*	15.9	19.7
Charcoal#	51.3	71.2
Labour	15.6	3.7
Maintenance and administration	17.1	5.4

* In the case of the St Maurice Forges, this figure includes limestone.

In the case of the St Maurice Forges, this figure includes labour and transportation.

† The Nivernais region; Woronoff 1984, p. 472.

If we compare these costs with those incurred at 18th-century French ironworks, it is clear that, generally speaking, workers and administrators at the St Maurice Forges were better paid than their French counterparts (Table 6.14). It is particularly revealing to compare the cost of producing pig iron with that of producing bar iron, although the numbers for the Loire region are quite similar (Table 6.15).

Table 6.15

COMPARISON OF PRODUCTION COSTS AT THE FINING STAGE, ST MAURICE FORGES AND FRENCH IRONWORKS (%)

	St Maurice Forges 1741–42*	France 1750#	France 1761–89†	France 1811
Pig iron	43	60	46.9	64
Charcoal	21	24	16.7	26
Labour	15	10	15.6	6
Maintenance and administration	21	6	20.8	4

* Rounded figures. The proportionally lower cost of the pig iron at the St Maurice Forges is probably due to the fact that it was produced in the blast furnace on site.

Forge de la Frette, Seine-et-Oise region; Gille 1947, p. 141.

† Forge Neuve, Loire region; *Les Forges du pays de Châteaubriant* (see note 97), p. 133.

· French Department of Forests; Woronoff 1984, p. 473.

Table 6.16

COST OF PRODUCTION OF PIG IRON, ST MAURICE FORGES, 1873·

Raw materials	Quantity (per ton of cast iron)	Cost ($/ton)	% of total cost
Iron ore	6,750 lbs. (3,092 kg)	7.53	40.6
Limestone	506.25 lbs (229.6 kg)	0.21	1.1
Charcoal	180 bushels (6.6 m³)	10.80	58.3
		18.54	100.0

· From Harrington. Pig iron sold for $40/ton in Montreal during the summer of 1873.[160]

At the St Maurice Forges, raw materials cost less, but expenses were higher. If these comparisons are valid,[158] the higher costs of labour and administration at the St Maurice Forges would support observations made during the French regime that Forges workers were overpaid and that there were too many clerks and administrators (see Chapter 8). Pehr Kalm and engineer Louis Franquet both commented on this fact. Furthermore, in 1808, when John Lambert was told about the old days at the Forges, he also heard that an inspector and 14 clerks had all managed to enrich themselves at the King's expense.[159] That this story was still being told after 50 years lends credence to the impression of a bloated French administration.

Existing documents do not provide any details on production costs for later years, with the exception of the very last years of the Forges. Using figures provided by Dr B. J. Harrington for the year 1873, we do know how much it cost to produce a ton of iron, but it is impossible to break down the costs of labour or administration (Table 6.16). The cost of raw materials includes mining or wood cutting, dressing and transportation, but as we saw earlier, charcoal still accounts for the lion's share of the costs.

THE IRON TRADE

Wares from the St Maurice Forges, near Trois- Rivières, will be sold by Mr. Alexandre Dumas at Quebec, Mr. Dumas St-Martin, Esq., at Montreal, and Mr. Christophe Pélissier at Trois-Rivières, both wholesale and retail, at very reasonable prices.

Quebec Gazette, *15 October 1767.*

Agents

In the early days, the trade in iron from the St Maurice Forges was based in the three major cities of Canada at the time: Montreal, Quebec and Trois-Rivières. The partnership agreements for the first two companies to run the Forges, Francheville et Compagnie and Cugnet et Compagnie, stipulated that sales would be handled by the partners living in each of these three cities, at no profit to themselves (see Appendix 12). During the French regime, Forges iron was stored in company warehouses in each of these cities, while the iron for export was kept in the King's stores. [161] After the Conquest, the partners in the Pélissier syndicate continued this system for several years, using their own warehouses. Starting in 1769, sales agents were appointed,[162] a practice that lasted until the 1860s, when the McDougalls put an end to practically all wholesale and retail sales by shipping the bulk of their pig iron directly to Montreal factories, although they did keep their sales outlet in Trois-Rivières.

Merchants

Very little business in the way of sales was conducted at the Forges themselves. However, the company did sell directly to rural merchants (outside the three-city network) and to customers who placed special orders—shipyards and the Royal Engineers for the construction of the Rideau Canal, for example. Although there is little documentary evidence of sales to merchants outside the network, this practice was probably quite common. During the 19th century, Soupras et Franchère, in St Mathias sur le Richelieu, was one such merchant.

In addition, the company also engaged in barter on occasion, swapping its products for other merchandise. It was a common commercial practice at the time and like many such practices, a way of boosting profits. Payment in kind was already common during the time of Cugnet et Compagnie. However, Vézin and Simonet were against this practice and criticized partners who sold Forges wares at Quebec and Montreal by this method. Vézin and Simonet actually claimed that these "gentlemen" would, in payment for ironwares, deliver to the Forges merchandise "at inflated prices" that they had purchased for much less, and pocket the difference, since "it suited them to carry on their own trade."[163] The ironmasters were also opposed to this practice—even though Vézin himself had been the first to institute it—because paying the workers in merchandise created problems for them (see Chapter 8). Merchants quickly realized that Forges wares could generate other types of profit.

It is also important to remember that, throughout most of the history of the Forges, the ironworks' masters were merchants, for whom the Forges lease was just one aspect of their business. Monro and Bell, for example, were very open to bartering, as shown in this proposal addressed to a merchant in Upper Canada in 1807:

> We will cheerfully receive produce in payment either at a fair market price or on consignment for sale in the disposal of which to advantage every exertion will be made and no commission charged on the sales to the amount of the iron delivered and then due.

By exchanging goods, the parties avoided having to pay a commission. There are few documented cases of this practice, although according to Bédard, the Batiscan and Marmora ironworks also engaged in barter. We also know that, in 1829, Mathew Bell accepted a shipment of wheat from Soupras, the merchant, in payment of a £306 debt.[164]

Commissions and Discounts

Agents for the Forges received a commission, which was generally a percentage of the total value of the goods sold. Once again, few sales agreements are documented, and those that are (Table 6.17) show that the terms varied depending on the year, the agent and even the city where the sale was made. In addition to the commission on the sale itself, these agents sometimes billed the Forges for transportation and advertising. Although such expenditures were no doubt included in the higher commissions charged by certain agents, generally speaking, these expenses were paid by the Forges, a practice that started during the French regime. Newspaper advertisements from the late 18th and early 19th centuries (see the epigraph) show that the Forges lessees saw to their own advertising. They also set the prices (Table 6.17).

The case of Leproust, the Trois-Rivières merchant who received a flat fee for his services in 1793, is exceptional. An agent since Conrad Gugy's time 10 years earlier, he had been paid on commission during the brief tenure of Alexander Davison & John Lees that had followed. Monro and Bell set these terms and conditions and required mortgage security because of accumulated debts. Leproust was, however, paid a 5% commission on any supplies and foodstuffs he bought to supply the Forges during the same period.[165]

Merchants outside the three-city network who dealt with the Forges, either directly or through an agent, were entitled to a discount. Unlike the agents, who merely held the iron on consignment, these merchants bought the products for resale (Table 6.18).

The discounts were either negotiated directly with the Forges lessees or simply advertised in the newspapers. Usually based on volume, discounts were also given to merchants who paid cash or agreed to pay their invoices promptly. Sometimes, the Forges decided to reduce prices simply to get rid of excess inventory due to slow sales or increased competition, as suggested by the following advertisement in the *Quebec Gazette* of 24 October 1799:

> The Lessees of the St Maurice Forges would like to inform their Customers that various circumstances have induced them this year to significantly reduce the prices of their cast iron wares, as is apparent from the following Schedule of Rates, which is equal to approximately 12 ½ per cent, over and above the discount normally granted to rural Merchants who purchase an assortment.

Table 6.17

TERMS OF SALE GRANTED TO AGENTS OF THE ST MAURICE FORGES

Agent and city	Year	Storage	Commission (% of sale)	Fees	Other expenses or conditions
Leproust, Trois-Rivières	1793	£20 per year		£55 per year	mortgage security
Gillespie & Moffatt, Montreal	1852	2s 6d per ton	6.5 (auction) 5.5 (retail)		5s per ton*
Weston Hunt & Company, Quebec	1852		10		
J. W. Leaycraft, Quebec	1856		10		

From Bédard 1982a.

* Wharf fees, haulage, advertising and insurance.

The Forges paid transportation, advertising and insurance.

Table 6.18

DISCOUNTS GRANTED TO RETAILERS

Year	Discount	Conditions
1769	2s 6d per 100 lbs.	with the purchase of 500 lbs. or more of bar iron
1794 1799	10%	with the purchase of 6 stoves or more
1807	5%	merchants from Upper Canada
1827	7.5–15%	wrought iron and pig iron (quantity not specified)
1832	7.5%	(quantity not specified)

From Bédard 1982a.

Several years later, Monro and Bell admitted that they had built up an "immense" inventory at that time, and cited several reasons to justify their decision to cut prices. First, they pointed out that they had acted this way because their lease at the Forges could be handed over to someone else.[166] Indeed, in March 1799, their lease had only been extended for another two years instead of the seven they had requested; furthermore, the partners had already begun negotiations for a new lease.[167] Second, Monro and Bell cited the competitiveness of a limited market, with the advent of the Batiscan Iron Works, which went into business that same year[168] and helped to push down prices. Lastly, they cited the years of poor harvests, which caused slumping sales and falling prices.

According to Bédard, who examined the firm's correspondence, the Batiscan Iron Works Company, which operated from 1798 to 1815, granted discounts varying from 7.5 to 12.5%; some merchants even received a discount of 15 to 17.5% from this competitor of the St Maurice Forges. The competition was not just between the Batiscan and St Maurice works. Around 1810, discounts of 14.5% were being given on European stoves. However, the clerk at Batiscan stated that he could not grant a discount of more than 12.5%.[169] These discounts show, as we saw earlier, that profit margins were very slim.

Payment

An announcement that appeared in the late 1760s specified that Forges wares were payable in cash. According to the scant archival material available from the St Maurice and Batiscan ironworks, this requirement was due to the economic conditions of the times, the company's debt situation and the scarcity of some products.[170] During the 19th century, the iron trade, like other industries, generally operated on credit. For example, in the 1820s, Mathew Bell offered wholesale merchants credit for a period of 12 months. Ten years later, agents also agreed to sell on credit at auctions, but offered much shorter repayment periods, usually only two to four months depending on the value of the goods purchased. Similar terms were advertised in the 1850s,[171] and the accounts of Weston Hunt & Company for the years 1853 and 1854 show that stoves were sold on credit for terms of three to six months.[172]

Wholesale Versus Retail

A distinction was made between wholesale and retail sales, as the advertisement in the epigraph of this section illustrates. The discount and credit conditions mentioned above indicate that the wholesale merchants were preferred clients, enjoying a wide range of advantages. Shipyards and processing industries—nailsmiths, farriers and blacksmiths—were also important wholesale customers, purchasing large quantities of iron. However, the Forges also wanted to reach the retail market, as the detailed catalogues of Forges wares in newspaper advertisements, which featured a wide variety of stoves, show. There is little information available (aside from auction documents) about retail sales figures compared with wholesale. Statements of account from Weston Hunt & Company for 1852 to 1854, agents at Quebec, and from Gillespie & Moffatt, in Montreal, show that these agents made a

Figure 6.7

PRICE OF BAR IRON, ST MAURICE FORGES, 1739–1853 (shillings/cwt)

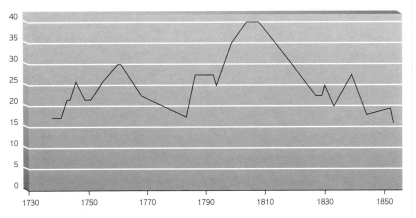

From Bédard 1982b, pp. 17–27.

The French *livre* has been converted into shillings and the French hundredweight (*quintal*) into the British imperial hundredweight (1 cwt = 112 lbs.).

distinction between auctions and private sales. For example, at Gillespie & Moffatt, 56% of the company's revenues in 1852 were from auctions, while 44% came from private sales. The following year, on the other hand, auctions accounted for 93% of revenues. Obviously, these proportions could vary considerably, but according to William Henderson, the Forges superintendent at the time, retail sales were very flat during the 1850s. Henderson also maintained that auction sales, held primarily during the months of September, October and December, were a well established tradition. According to Bédard, however, the earliest mention of this method of selling St Maurice iron goes back only as far as the 1830s—the 13 August 1833 issue of Montreal's *Daily Advertiser*.[173]

OVERVIEW OF PRICING

As was the case with production costs, the lack of systematic data on prices and price trends makes it impossible to conduct a serious analysis of price fluctuations. More importantly, the wide variety of products further complicates any attempt to obtain systematic data. However, we can look at a few typical products to illustrate price trends over the years.

Apart from production costs that determined the cost price of each product, at certain points in time we can identify other factors that affected prices. We will try to list these factors by taking a look at the specific products chosen for consideration.

Bar Iron

Initially, the price of bar iron made at the Forges was based on the price of the French bar iron sold in the colony. During the first three years of production (1739–42), bar iron generally sold for 20 *livres* a *quintal* (Figure 6.7). In 1744, however, Intendant Hocquart raised the price to 25 *livres*, claiming that the price of imported wrought iron sold in France justified the increase. Three years later, the price rose again, this time to 30 *livres*, because of the war in Europe.[174] In 1750, Intendant Bigot had to reduce the price to 25 *livres* once more, to reflect the price of imported French iron. Bigot wanted to "prevent the shipowners of France from sending bar iron, heating stoves and sock plates to this country"; he even said this lower price was essential to the Forges' profitability. He was subsequently proven wrong when the price was driven back up to 30 *livres* by the Seven Years' War, which had already broken out in the colony in 1754.

After the Conquest, under the British military administration, the price shot up to 36 *livres* (or 30 shillings), and then dipped once more before beginning a slow steady climb, starting in the 1780s, up to 40 shillings a cwt around 1810. This increase of 15 *livres* a *quintal* in 10 years (also observed by Ouellet during the same period)[175] was most likely due to the resumption of shipbuilding at Quebec, itself driven by the thriving timber trade.[176] Following the general market trend, the price of bar iron declined during the 1850s, and had dropped below 20 shillings by the time wrought iron production was being phased out at the Forges. The price of sock plates followed the same general trend (Figure 6.8).

Stoves

The prices of finished castings such as stoves apparently did not follow the same general trend observed in bar iron prices. Unlike bar iron, there was no significant price increase in castings at the turn of the 19th century. Subsequently, prices did begin to fluctuate more, but it is important to remember that competition stiffened due to the arrival of imported products on the Canadian market and an increase in the number of foundries in the second quarter of the 19th century.[177] However, generally speaking, the first half of the 19th century saw a general downward price trend (Figure 6.9).

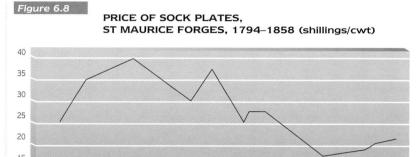

Figure 6.8

PRICE OF SOCK PLATES, ST MAURICE FORGES, 1794–1858 (shillings/cwt)

From Bédard 1982b, pp. 17–27.

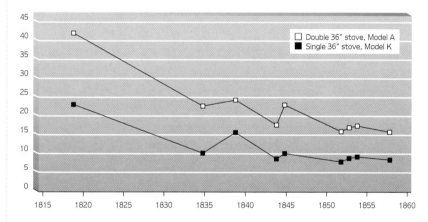

Figure 6.9

PRICE OF TWO STOVE MODELS, ST MAURICE FORGES, 1819–58 (in dollars)

☐ Double 36" stove, Model A
■ Single 36" stove, Model K

From Bédard 1982b, pp. 17–27

As for the last 20 years of the Forges' operations, as we have seen, the McDougalls stopped manufacturing wrought iron and castings, except briefly for axe heads, which, in the two years they were made, sold for $1 each.[178] The prices fetched by pig iron made it much more lucrative. We have already mentioned the profits generated by the huge sales of iron to the Montreal railcar industry. According to prices quoted on the Toronto market, pig iron sold for around $25 a ton at that time, except from 1872 to 1874 when it topped $35, rising to an average price of nearly $45, then quickly dropping to somewhere between $20 and $25. For a few months in 1872 and 1873, the prices in Toronto even exceeded $50 a ton. The price of St Maurice pig iron followed market prices, which at that time were strongly influenced by pig iron imports from the United States. As of 1855, imported pig iron was no longer subject to duty. Relief from imports was late in coming, and it was not until 1879 that the Canadian government imposed a duty of $2 per ton on imported pig and scrap iron, in an attempt to protect the Canadian iron industry. Although this protection no doubt encouraged George McDougall to reopen the Forges and build a second blast furnace, the new policy was also implemented when the economic recovery of the late 1870s was just around the corner.[179]

Plate 7.1
The Forges seen from the St Maurice River, by Joseph Bouchette, Jr., 1831.

VII

The Working
Population

*The works (proper) employ 24 or 25 hands,
besides the people employed for getting ore,
cutting wood, charring, washing ore—
all of which is done by Habitans from
the neighbourhood, also the horsemen.*

Lord Selkirk, 1804

DEFINING THE POPULATION OF THE FORGES

It should be easy to define an isolated population with a skilled work force at its core, but it is not as easy as it seems. A distinction must constantly be made between the "employed population" working for the company, as determined from the archives of vital records, and the "resident population" living in the industrial village, as listed in the censuses. For this reason, it is worth going into the methods used to define the population of the Forges.

Because the people who lived at the time are now dead, a historical demographic analysis such as this must examine a reconstruction of the population. Demographic historians use a variety of methods to check facts and reconstruct the "real" population of the time as accurately as possible on the basis of surviving archival documents, which reflect the state and behaviour of the population at certain very specific points in its history. But all sources have their limitations, and the reconstruction depends entirely on the information historians have managed to extract from the documents available.

Any study of the population of the Forges is subject to limitations of this type, magnified by various factors. For example, people who said that they were connected with the ironworks at the time a baptism or marriage was registered were not necessarily among those "domiciled," or residing, there. Sometimes they were only "employed." And even if they were actually domiciled at the Forges then, there is nothing to say that they stayed for long. In this chapter, we will therefore attempt to paint a picture of the population of the Forges and describe the differences between the results based on the registers of births, marriages and deaths and those based on the censuses.

DEMOGRAPHIC PORTRAIT OF THE FORGES DEPENDING ON SOURCES USED

Registers of births, marriages and deaths (parish registers) and population censuses are the essential sources upon which demographic historians rely in painting a picture of a population in the distant past and tracing its development. They can use the written records in the registers to track natural demographic changes (births, marriages and deaths) over time by calculating the annual fertility, birth, marriage and death rates of the population under study. The censuses tell them about the state of the population—its size and composition by age, sex, marital status, occupation, and so forth—at various well-defined moments.

Registers of Births, Marriages and Deaths

In the specific case of the Forges, the methods used to extract information from the parish registers have had a considerable impact on the results of the demographic studies based on that information. Using the parish registers posed a number of problems, since the "village" was never raised to the status of a parish and so never had its own exclusive registers. The post was served as a mission of the Parish of l'Immaculée Conception de Trois-Rivières from 1730 until 1857, when it was absorbed into the Parish of St Étienne des Grès.[1] The lack of parish status might have compromised demographic studies or discouraged people from conducting them, since the parish is generally the basic geographic unit for such studies, but the population of the ironworks left such a mark on the Trois-Rivières area that it can easily be traced through the various local parish registers. The Parish of l'Immaculée Conception de Trois-Rivières did in fact keep a register of Forges workers from 1740 to 1764, but after that period, this parish no longer kept a separate register, and neither did any of the others. The records of baptisms, marriages and deaths of families at the Forges are thus scattered throughout the registers of the Parish of l'Immaculée Conception de Trois-Rivières and of other neighbouring parishes.[2] A large number of records for the population of the Forges can be found in these registers, however. Depending on the whim of the officiating priest, and depending on the individual worker's sense of belonging, records, especially those of baptisms and marriages, sometimes mention people's trades and that they live at the Forges. There are enough such cases for them to be compiled systematically. Researchers have also managed, through censuses and lists based on other sources (marriage contracts, employee rolls, and so on), to associate a good many records with people who

Table 7.1

RECORDS OF BAPTISMS, MARRIAGES AND DEATHS FOR THE POPULATION OF THE ST MAURICE FORGES, 1730–1883

Baptisms	2,356
Marriages	556
Deaths	1,054
Total	**3,966**

belonged to the St Maurice Forges.[3] Through this selective extraction of information from the registers, close to 4,000 records over the Forges' entire lifespan have been compiled (Table 7.1).

But it must be acknowledged that not all the records in this corpus concern people who actually lived on the post, and the demographic historians who have described the population based on this set of records have felt bound to make this statement:

> All these observations suggest that the population described by the records selected according to the above criteria is not really the same as the one residing in the Forges village; on the contrary, we feel that our research has covered a huge population consisting of everyone who had any sort of link with the St Maurice Forges.[4]

To be more specific, the population concerned by the records is not "only" that of the Forges village. Clearly the records also include many people who lived at the Forges at some point, but for an unknown length of time.

As defined on the basis of these vital records, the population then is made up of people "employed" by the St Maurice Forges for any length of time. But the registers alone will certainly never be enough to enable us to trace with sufficient accuracy the changes in the "resident" population at the Forges. We could even hypothesize that the records compiled are only a sample of the records of births, marriages and deaths for the Trois-Rivières area between 1730 and 1883.[5] But except for earlier periods, when the size of the Forges population would have carried a greater statistical weight, it would be practically impossible to determine the specifics of the working population of the Forges within such a group. That said, it is still useful to analyse the records compiled this way, but caution must be exercised in interpreting the results.

In any case, the fact that the vital records of people connected with the ironworks are scattered throughout various registers highlights an inescapable reality. The "population" of the Forges was not made up exclusively of people living on the post itself; it also encompassed others from nearby or surrounding localities employed seasonally but regularly for a variety of jobs, as well as their families. The "population" of the Forges as reconstructed from the registers therefore includes *both* **inside** and **outside workers**, as they were known.

Employed Population Much Larger Than Population Censused

Censuses are the other main source for demographic studies. They provide a snapshot of the population at the time the census was taken. With most of the censuses we have available (9 out of 12), it is possible to identify residents of the post, either because the Forges are named specifically or because people in occupations carried on at the Forges are grouped together. The population defined by the parish register data is larger than that obtained from the census data; the parish registers tend to lead to an overestimate of the population, by comparison with the censuses. This can be shown by a variety of methods, including calculations of crude birth, marriage and death rates. The demographers acknowledge that, according to the records compiled,

> [...] only the decades 1820–29 and 1840–49 show plausible crude [birth] rates [around 50 per 1,000]. For all other 10-year periods, crude rates are far above the acceptable threshold. Crude marriage and death rates are also too high.

> It is therefore quite certain that the births, marriages and deaths are not all from the population of the St Maurice Forges alone.[6]

To provide another demonstration and to calculate the actual overestimate attributable to the method of extracting vital data from parish registers, the demographers have tried to determine the state of the population at the date of the census from the vital record data. To do this, they chose the 1851 census, the first complete nominal, or name-by-name, census, which made it possible to isolate the village at the Forges. They arrived at the conclusion that "parish registers yield twice as many families as the census, due to the extraction methods used," and that "half the population defined by vital record data resides outside the boundaries of the St Maurice Forges."[7] By this means, they "uncovered the exact population" to which the examined records refer.

The population of the Forges is thus quite different depending on whether it is reconstructed from parish registers or census data. Birth, marriage and death records concern a population twice the size of the village, comprising not just the workers living in the village itself but those from the surrounding area as well. They also show that many workers, even those whose families had been there for generations, did not always live in the village.

As disconcerting as these findings might seem at first, they do corroborate first-hand accounts from different time periods, which show that the ironworks recruited most of its labour from among the *habitants* living nearby. Data from parish registers suggest that many of them were employed regularly enough there that they thought of themselves as "belonging" to the Forges, just like the many skilled craftsmen who spent their entire lives there did. The demographers' choice of method has indeed borne fruit, then. It has not only given us a better idea of the population defined by the selective compilation of data from the parish registers, but has also confirmed that workers who said they belonged to the Forges did not necessarily live there year-round.

Now that we have managed to get a better grasp of the population defined by the birth, marriage and death records, we can analyse its demographic characteristics more closely. We will see that they sometimes provide evidence of the status of the workers that made up the population, and that some aspects of natural demographic trends mirror developments at the ironworks itself.

NATURAL POPULATION TRENDS

Annual and 10-Year Trends

The 3,966 records of baptisms, marriages and deaths examined with respect to the "population" of the Forges cover an observation period of 158 years, beginning with a death recorded in May 1733 and ending with a wedding performed in August 1891. Examination of the year-to-year changes in the population—which, as we have seen, does not mean solely workers living on the post—shows that the population's growth and decline closely mirrors that of the Forges.

The annual figures for baptisms, marriages and deaths fluctuate several times, but do follow certain long-term trends, which become more obvious when transferred to a 10-year scale. Figure 7.1 shows that, generally speaking, all these vital statistics kept increasing until 1850, then began to drop almost steadily until 1890.[8] To take a closer look at this general tendency, we will divide it up into five separate trends, which correspond directly to five major periods in the history of the ironworks. From 1730 to 1765, the number of vital events rose rapidly, then dropped sharply. This is the period when the Forges were founded—and thus when the first workers began to arrive (especially beginning in 1736) and some families began to settle—and when the Forges had to close temporarily (1765–67) after the Conquest, which caused a number of families to leave. The second period, from 1765 to 1790, marked a new beginning. The company drew on the few French families that remained after the Conquest to try to rebuild its work force, hiring Canadian workers as well as British immigrants. The new rise in vital events between 1790 and 1820 resulted from an increase in the number of workers at that specific time, with the addition of skilled moulders and unskilled labourers who, once seasonal, were now hired all year round. This

Figure 7.1

**BIRTHS, MARRIAGES AND DEATHS
AT THE ST MAURICE FORGES, BY DECADE,
1730–1890**

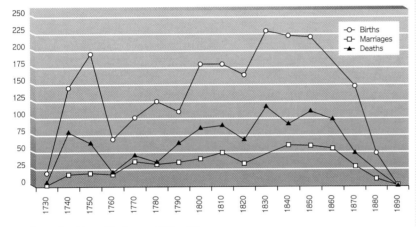

Source: Tremblay and Charbonneau 1982, p. 104.

influx of workers and their families doubled the population of the village at the Forges and led to a rise in the number of all vital events over the years 1820–50. The first impact was a significant increase in the number of marriages, which had a direct effect on births and deaths. Between 1850 and 1890, the final decline in the number of vital events was initially due to the departure of many families that had been hired by the Radnor Forges (1854) and the temporary closing of the St Maurice Forges from 1857 until 1863, when they were bought by John McDougall. Operations then resumed again, but a new production focus and another shutdown (1877–81) led to a further reduction in the labour force that was maintained until the Forges closed down for good in 1883.

The major natural population trends thus correspond to significant moments in the company's history,* in which temporary shutdowns, hiring practices and shifting production focus had a direct impact on both the growth of the employed population and its demographic behaviour.

Seasonal Changes in Births, Marriages and Deaths

Examination of seasonal changes in births, marriages and deaths reveals the way of life at the Forges. The time of year people got married or had children is a good indicator of prevailing traditions in previous historical periods and of the yearly cycle of their activities. We know, for example, that farmers had fewer children during the planting and harvesting seasons. But the seasonal pattern of births among the workers at the Forges is not very pronounced (Figure 7.2). It is similar to that of a town population with steady work tied, in this case, to the cycle of plant operations, which were spread out over the entire year.

These findings suggest that the lives of the *habitants* of the surrounding area—the outside workers who were hired at different times of the year to collect, prepare and transport raw materials or products and goods— were structured more by the work cycle at the Forges than by the seasonal cycle of the farmland on which they were settled. Although they did not "belong" to the ironworks in the same way as the inside workers living on

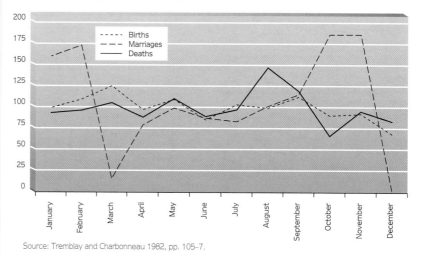

SEASONAL CHANGES IN BIRTHS, MARRIAGES AND DEATHS AT THE ST MAURICE FORGES, 1733–1890
(proportional figures)

Source: Tremblay and Charbonneau 1982, pp. 105–7.

Table 7.2

**MEAN AND MEDIAN AGE
AT FIRST MARRIAGE, BY SEX**[10]

Sex	Age		Number of instances
	Mean	Median	
Male	23.7	23.1	163
Female	20.1	19.2	161

Table 7.3

**MEAN AGE AT FIRST MARRIAGE,
BY PERIOD AND SEX, EXACT AGES ONLY**[11]

Sex	Periods		
	1740–99	1800–39	1840–90
Male	25.1	23.7	23.4
Number	21	59	83
Female	20.0	19.7	20.4
Number	28	66	67

Table 7.4

**COMPARISON OF MEAN AGE AT FIRST
MARRIAGE OF COUPLES ENUMERATED
AT THE ST MAURICE FORGES
AND IN ST MAURICE COUNTY, 1851**[12]

Sex	Forges	St Maurice County
Male	22.8	25.7
Female	20.2	24.6

the post did, their lives were structured by the same pattern of work, which was determined by plant operations. It is no wonder contemporary observers and census takers had trouble finding terms to distinguish between the two types of employees and that they sometimes differentiated "persons belonging" to the Forges from "employed workers." This simply reflects the fact that, in reality, Forges workers did not all live in an easily definable geographic area.

Seasonal patterns in marriages are no different from those of other populations of the time, whether in Quebec or France. They show that the teachings of the Catholic Church, which forbade marrying in Advent (December) and Lent (March), were followed. At the Forges as elsewhere, weddings were more frequent in winter, especially in November, January and February. Generally speaking, seasonal patterns in deaths followed those of births, due to high infant mortality. The demographers say that "the high mortality rates in late summer were usually caused by digestive disorders and epidemics of childhood diseases."[9]

Fertility, Marriage and Death Rates

In comparison with other subpopulations of the colony, the workers at the Forges seemed to tend to marry at a slightly younger age, especially in the 18th century. The average age of the men at their first marriage was 24, while that of the women was 20 (Table 7.2).

If we group cases by period, however, we see that the age difference between spouses dropped from five years in the 18th century to three in the second half of the 19th century (Table 7.3). This trend seems to be due to the fact that in the 19th century, men married earlier, because the age of women at marriage remained virtually unchanged over the same period.

Data from the 1851 census can help refine our analysis, showing that people born at the Forges married younger than did those of St Maurice County (Table 7.4).

The early marriages at the Forges can probably be explained by the fact that sons born there could get a job at a young age and would thus be in a position to start families of their own sooner. We will see, however, that these young men did not necessarily have jobs when they finished their apprenticeships.

It has been possible to determine the proportion of endogamous marriages at the ironworks, that is, marriages between people born there. The specialization and relative isolation of the working population are two reasons why the incidence of endogamous marriage was high. The data from the 1851 census show that in 33 out of 70 cases (47%) *both* spouses stated that they had been born at the Forges. Totals based on a sample of marriage records in the registers for the entire period of study, which put the relative number at 44%,[13] appear to confirm this percentage. These figures may be considered a minimum. Of the 33 couples in 1851 who said they were both born at the Forges, only 6 were actually mentioned as natives in the parish records. These mentions became increasingly infrequent as the 19th century advanced.

Although we have little data on the age of mothers at the Forges, which means that our analysis is based on only a small sample, we can see that the fertility rate of these women is similar to fertility rates observed elsewhere at the same time. They gave birth to an average of six children, although the women who were under 20 when they got married had one more, on average. The figures also show that these women could give birth to as many children as nature allowed.

Table 7.5

DEATH RATE OF CHILDREN 15 AND UNDER (PER 1,000), BOTH SEXES COMBINED, FOR VARIOUS POPULATIONS [16]

Age	Forges	Quebec, 1831	Canada, births before 1720	Quebec City, 1821–30
0	213	181	211	268
1	127	137	94	173
5	31	37	38	47
10	14	23	30	36
Life expectancy	39 *	39	36	29

* Based on the corrected mortality table for the Forges.

A study of death rates at the Forges is significant only for children 15 and under, because the children were less mobile than the adults. To calculate death rates, you have to "be sure that the individuals are continuously in the same area" and, according to the demographers, "the crux of the problem is that we cannot establish the adult population base exposed to the risk of death."[14]

The death rate of children 15 and under was strongly influenced by the high infant mortality rate: 64% of deaths occurred among children less than a year old. At the time, infant mortality, at 213 per 1,000, was extremely high, even when compared with rates in the poorest countries in the world today.[15] This rate translates into a life expectancy at birth of 39 years. These figures are consistent, however, with those calculated for other populations at the same time (Table 7.5).

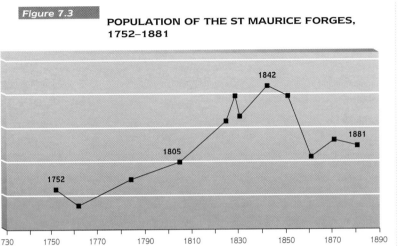

Figure 7.3

POPULATION OF THE ST MAURICE FORGES, 1752–1881

Source: See Appendix 13.

The slight seasonal variation in patterns of births in the Forges population is thus comparable to that seen in towns. The same is not true of death rates. The statistics show that "mortality at the Forges followed a much more rural than urban pattern," while the figures we have for Quebec City are typical of "the high mortality usual in old towns." The demographers also note that "endogenous mortality ravaged the Forges less than it did Quebec City between 1771 and 1870." This observation brings us back to the reality of our parish registers, which define a broader population than that actually domiciled at the Forges. The population in question does not consist solely of the population confined to the 160 ha of the industrial village, the density of which could have encouraged epidemics. Some working and living conditions on the post, such as housing families together in tenements, might have or may have in fact helped spread disease, but with no significant impact on mortality rates. Keep in mind, too, that regardless of the living and working conditions, the Forges were out in the country.

In short, given the limitations of the data gathered chiefly from parish registers, the demographic study "has nonetheless produced clear conclusions about the main aspects of the three major demographic phenomena" of marriage, fertility and mortality. According to the demographers, "the population at the St Maurice Forges tended to marry young and give birth to as many children as nature allowed, losing no more than did the population as a whole at the time."[17]

SIZE AND COMPOSITION OF THE POPULATION OF THE INDUSTRIAL VILLAGE AND HOW IT CHANGED

Population Size

By considering a number of enumerations, censuses and employee rolls, we can estimate the size of the "resident population" of the industrial village at various points in time. Figure 7.3 shows the population trend.

Four general observations can be made from this graph. First of all, there is no enumeration or official census for the 30 years of the French regime (1730–60), although there are some estimates, including one for 1752, and we will come back to those later. Second, we have no official figures for the 41 years between the 1784 enumeration and the 1825 census, during which time the population doubled, but we have taken into account Mathew Bell's estimate from 1805. Third, it should be pointed out that the population increased between 1762 and 1842, after which it dropped and then remained fairly stable until 1881, two years before the ironworks finally closed down. Last, the population was never more than 425.

Looking at the general curve of the graph, we might be tempted to jump to two hasty conclusions. First, we might think that the increase in the size of the population (from 1762 to 1842) was due to normal demographic growth, which would have meant that this small population increased sixfold in 80 years solely from natural causes. We might also tend to link this growth and the decline that followed to developments in the business, which initially enjoyed a long period of expansion, entailing an ever-expanding work force, then decreased, with a concomitant reduction in the work force. But such is not the case. To interpret the variations in population size, we have to take into account the needs of the company, particularly in terms of hiring. But first, we need to make a clarification about the industrial plant, as well as its production capacity and focus, and the associated labour requirements, if we want a better understanding of how the size of the village population changed.

The Population and the Ironworks

It should be recalled that the Forges were built from scratch between 1736 and 1739. The original plant, which remained essentially unchanged throughout the major part of the company's existence, comprised four main departments: the furnace (with its sheds), the lower forge, the upper forge and the moulding shop. The production capacity of the forges and moulding shop were directly dependent on the output of the blast furnace, which remained steady at about two and a quarter tons of pig iron per day until 1854. Since the technology did not change over that time, the same number of workers were always needed in the shops. As we have seen, only the moulding operations were stepped up beginning in the late 18th century, requiring extra moulders to be hired.[18] In 1854, the capacity of the blast furnace was doubled to 4 tons of pig iron per day, but more workers did not have to be hired. On the contrary: According to the 1861 census data, the number of workers had been halved. This was largely due to the switch from forging bar iron to making plain castings, for which skilled moulders were no longer required. Nor did the addition of a second blast furnace in 1881 affect the total number of workers employed by the Forges, which remained the same until 1871. As we can see, the increase in production capacity starting in 1854 did not bring about an increase in the population living at the Forges, which had peaked between 1830 and 1840, well before the furnace was modified. The explanation for the population growth lies elsewhere, as we shall see.

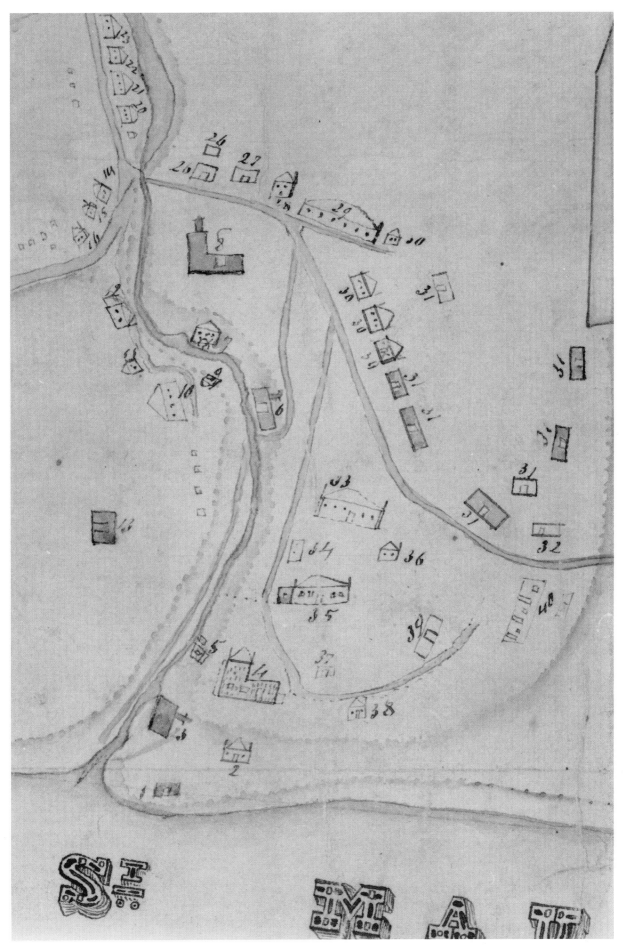

Plate 7.2
Detail of figurative plan by Surveyor Bureau, showing the main
workers' residential area on the plateau north of the gully.
ARCHIVES NATIONALES DU QUÉBEC, E21, MINISTÈRE DES TERRES
ET FORÊTS/ARPENTAGE/CANTONS No. S.36B, JANUARY 1845.

We have no enumeration or official census for the French regime (1730–60), but we can still estimate the size of the population residing at the Forges at the time, thanks to some specific information we have about the number of families and workers living there. We can establish a correlation between heads of families and jobs. Table 7.6 shows that, during the French regime, the resident population consisted solely of inside workers and management. The few first-hand accounts we have from later periods indicate that the situation remained the same until the early 19th century.

Except for the years 1760 to 1764—a period in which the ironworks was not in regular operation, although seven iron-workers were kept on by order of the British military government—the resident population of the Forges in the 18th century was about 150.[19] In 1805, according to Monro and Bell themselves,[20] there were about 200 people living there. The population reached 321 fairly quickly, over the next 20 years, and the ironworks began to take on the appearance of a real village.

A Village Takes Shape

This sudden rise in the population was not due to natural growth during the period,[21] nor to any sort of expansion of the company that might have called for a bigger work force. The arrival of a few Irish and Scots moulders, which began as early as 1770,[22] and continued thereafter, had little effect on the growth of the village population. Rather, the increase can be explained by the absorption of some of the outside workers and their families. Especially after 1800, the company offered regular jobs to workmen—mainly carters and colliers—who up until then had been employed only casually. Data from a number of sources seem to corroborate this observation.

Table 7.6

NUMBER OF WORKMEN OR HEADS OF FAMILIES DOMICILED AT THE FORGES, 1742–1804

Date	Workmen or heads of families	Remarks	Source
1742	23	352 workmen employed, comprising 209 woodcutters and 143 hands, 23 of them full time	NAC, MG 1, C¹¹A, vol. 111, "Estat général de la dépense [...]," signed Estèbe, Quebec, 2 October 1742.
1752		"Employing upwards of 120 men"	Louis Franquet, Voyages et mémoires sur le Canada (1752–1753) (Montreal: Éditions Élysée, 1974).
1754	20	"A manager — a storekeeper — 2 employees — 20 heads of families"	NAC, MG 1, C¹¹A, vol. 99, fol. 529 (Microfilm F-99).
1760	7	Workmen	NAC, MG 23, G¹⁴, vol. 2, pp. 5–6, Bruyère, on behalf of Colonel Burton.
1762	11	72 inhabitants, including 11 heads of families	RAPQ, 1936–37, pp. 1–121.
1764	7	Workmen "for operations at present"	NAC, MG 21, B²¹⁻², fols. 139–44, (Microfilm A-615), Haldimand Papers, memorial from Courval, 20 September 1764.
1764	30	30 workmen necessary, plus an ironmaster and foreman	NAC, MG 21, B²¹⁻², fols. 139–44, (microfilm A-615), Haldimand Papers, memorial from Courval, 20 September 1764.
1784	30	30 married men and a total of 149 persons	AUM, Baby Collection, CC, box 48.
1804	25	"The works (proper) employ 24 or 25 hands"	NAC, MG 19, E.1, 1 (Lord Selkirk 1804).

The employee roll for the Forges in 1829 put us onto this track. On the roll, Super-intendent Macaulay noted how long workers not born at the Forges had been living there. By his count, 37 out of 88, or 42% of the employees then on the post, had not been born there. Close to two-thirds of them had arrived since 1814. The incomers and their families totalled 169, according to the roll. And if we add them to the figure from the 1784 enumeration, we get a population of comparable size (318) to the one enumerated in 1825 (325). Close to half (83) of those 169 people were families of carters and colliers, while only 12 belonged to the families of two inside production workers. The most senior of these newcomers was a carter who had arrived at the Forges in 1785. Another carter had arrived four years later, in 1789, and a car-penter at the end of the century, in 1799.[23] The 26 other families arrived at more or less regular intervals between 1804 and 1824. Notarial deeds offer evidence of new regular employment conditions associated with indentures between the Forges and carters and labourers at the time. In 1805, a group indenture set out terms of the year-long engagement of nine carters. And, in 1810 and 1811, two other group contracts were signed for the year-long engagement of 11 and 13 carters, respectively.[24]

Taking on the carters had a direct impact on the number of horses kept on the post, which jumped from 22 in 1784 to 77 in 1831, 55 of them company property[25] (see Chapter 2); three new stables were built between 1787 and 1807, under Davison and Lees and then Monro and Bell.[26] The acquisition of such a large number of horses clearly shows a change in how transportation was handled,[27] in contrast with the situation during the 18th century, for which no mention of more than 30 horses has been discovered.[28]

The turn of the century also saw considerable new construction, probably in proportion to the number of new families that moved in. An inventory of 1807 (see Appendix 4) shows that, under Davison and Lees (1787–93) and particularly as of 1793, the year that Monro and Bell took over the works, 33 buildings were erected or renovated, including 9 dwellings. One of them, a tenement, housed several families.

The new families, who significantly swelled the size of the resident population, also had an effect on the number of vital events recorded in the community, which increased not just when they arrived, but afterwards, too. The rise in the number of births, marriages and deaths registered (Figure 7.1) shows a trend that accurately reflects what was going on in the village. By the first quarter of the 19th century, the Forges had "altogether the appearance of a tolerably large village" than a mere foundry, in the words of Surveyor General Joseph Bouchette.[29]

A comparative study of the families of 1829 and 1851 shows that only a few newcomers arrived in the intervening years. The new families that had arrived at the turn of the century had come to stay, so the company had probably planned for their permanent settlement. Their migration therefore cannot be seen as part of a continuous trend, which would have periodically replaced workers who were both geographically and occupationally mobile. After increasing, the population of the Forges became stable again, since the new workers settled in with their families. Moreover, in doubling in size, it became more self-sufficient, making it henceforth easier for people to find a spouse within the community. By 1851, almost 50% of spouses were native to the Forges, and the entire native-born population of the village had risen from 58.7% in 1829 to 80%. With the assimilation of these new workers, the company drew increasingly on the industrial community for its labour.

Plate 7.3
Anonymous drawing
showing the Forges
circa 1880.
LA PRESSE
[DAILY NEWSPAPER],
7 AUGUST 1920.

Between 1825 and 1851, the population went from 321 to close to 400; the roll drawn up by Superintendent Macauley in 1842 put the population at 425, the highest figure in the history of the Forges. The gap between the figures of the 1825 census and of one taken soon after, in 1831, was very small, but between 1831 and 1851, 60 people joined the population of the Forges. The rolls drawn up by the superintendent in 1829 and in 1842 supply figures comparable to those of the two official censuses, although somewhat higher. They seem to be reliable as far as the list of workers and their occupations goes (see later in this chapter), but they are not as complete as the censuses, because they count only workers. They do, however, mention the marital status of the workers, and how many children of each sex they had. The census of September 1825 and the August 1829 roll are the only enumerations done while the furnace was actually in blast. Censuses were usually taken early in the year, in winter, and the 1842 roll was drawn up in December, all times when the Forges were idle.

The significant difference between the figures of the 1829 roll and those of the official censuses conducted very close to the same time, in 1825 and 1831, is probably due to the fact that the 1829 roll was made up in the middle of the campaign. A comparison of the figures from this list with those of the two censuses shows that some categories of workers on the roll simply do not appear in the censuses, or comprise many fewer employees. This is the case mainly of the extra bateaumen, ferrymen and carters, who were likely hired only for the campaign. There are also seven blacksmiths on the 1829 roll who were not counted in either census. It is possible that the Forges hired the blacksmiths—who usually made finer pieces or did assembling and finishing—only for the campaign. Seasonal and annual variations in employment, even among craftsmen like blacksmiths and moulders, are also probably attributable to fluctuations in orders received.[30]

Although similar to those of 1829, the off-season figures for 1842 and 1851 seem to provide evidence of a real increase in the population living at the Forges year-round. The numbers are higher than those of the 1831 census, chiefly because more moulders and labourers were counted in 1851. This increase can be explained, in turn, by the focus on the manufacture of castings that marked Mathew Bell's long tenure. The new moulders were recruited from families at the Forges.

Starting in 1861, the population declined steadily, by 100 or 150 inhabitants, until the ironworks closed down. The decline was largely a result of the shift in production focus beginning in 1854. Paradoxically, at the same time, the capacity of the blast furnace was doubled and the company began to manufacture railcar wheels. A good many skilled workers were no longer needed for the new type of production. Many of them, mostly moulders, went to work at the Radnor Forges, which opened in 1854. At the time of the 1861 census, the Forges had been closed for three years. When the works were reopened by John McDougall in 1863, the manufacture of castings and wrought iron was all but abandoned. The number of workers assigned to this type of production, which was still 39 in 1851, dropped to just 12 in 1861 and to 9 in 1871. By 1881, the Forges employed only one blacksmith and 41 labourers.

Plate 7.4
Parishioners in front of the chapel beside the Forges, early 20th century. Dollard Dubé interviewed some of them in 1933 to gather their recollections of the last days of the ironworks.
COLLECTION OF RAOUL RATHIER.

The changes that occurred in the last 20 years had an impact on not just the size of the population, but its make-up. The old families of craftsmen, who had been there since the French regime, disappeared completely from the Forges. From an "iron mill" and "manufacture," the company became a mere primary iron producer, or iron smelter, employing only labourers, most of them making charcoal and casting pigs, work that did not require the skills of the ironworkers of generations past.

Population Composition

A breakdown of the population by sex, age and marital status highlights a number of features characteristic of an industrial community like the Forges. It is interesting to try and understand how a population that never numbered more than 200 prior to 1800 or 400 thereafter found an equilibrium. Dependent on the company's inside-labour requirements, could the working population survive without turning to the outside world? Was the population at the Forges really self-sufficient and closed, as its image as a specialized industrial village might suggest? As we have seen, the population of the Forges as defined by the records of births, marriages and deaths was twice the size of the population censused on the post itself. That means there were as many people outside the village as there were inside who saw themselves as belonging to the ironworks. Analysis of the documentary evidence shows that the Forges could draw on a larger population—or rather population pool—that undoubtedly helped maintain a balance within the village's resident population.[31]

Our examination of the population by sex, age and marital status, as well as the composition of households and families, will concentrate on the first half of the 19th century, but we will also draw on some data from 18th-century enumerations. This approach was determined by methodological considerations. The censuses of 1825, 1831 and 1851 are the only official counts that specifically isolate the population at the works, which is considered as a distinct social unit, and the employee rolls of 1829 and 1842 show clearly which workers lived at the Forges. There are also several other grounds for treating the second half of the 19th century and the first separately. From 1861 onwards, for instance, the Forges were included in a larger census district, so that its population can no longer be isolated, although the census data on men's occupations allow us to determine the size of the community fairly accurately. Also, starting in the decade 1850–60, the ironworks underwent major technological changes. Our primary goal, then, is to observe growth and change in the population within the very same technological context, to arrive at a better understanding and comparison of the distribution of men throughout the plant at different times. Furthermore, when the 1861 census was taken, the ironworks was closed, having been seized by the government. It was bought by a farmer in 1862, then resold to another entrepreneur.

SEX RATIO, 1762–1851

Age	Census years						
	1762 *	**1784**	**1825**	**1829**	**1831**	**1842**	**1851**
13 and under	(64.3)	100.0	109.0	96.7	108.1	110.2	97.7
14 and over	(116.7)	117.5	138.8	127.3	110.4	110.5	97.4
All ages	80.0	109.8	122.9	112.4	109.4	110.4	97.5

* Age was not recorded in the 1762 census. We therefore decided to put "boys" and "girls" in the 13-and-under age group, and adults and servants in the 14-and-over age group.

These changes caused a drop in the resident population, largely due to the departure of people descended from several families of long standing at the Forges, and had an impact on the composition of the population. It can be seen that in 1875, the village was made up mostly of new families who had gradually replaced the old ones over the preceding 25 years.

By Sex

The male-to-female ratio[32] can be a good indication of the numerical balance between the two sexes in a population. It might be expected that in a working village the ratio would be high, especially among people of working age, since male workers predominated. At the Forges, the changes in the sex ratio must be looked at in conjunction with the size of the population. Overall, and especially among inhabitants 14 years old and over, there are three main phases and they correspond to three different demographic periods (Table 7.7).

According to data from the 1784 enumeration, which was taken before the wave of immigration of the early 1800s, the male-to-female ratio for people 14 and over was 117.5. Immediately after the wave of immigration—which consisted chiefly of young workers who would more than double the population—the sex ratio increased significantly, rising to 138.8 in 1825 and 127.3 in 1829. This was a passing phase, however. Once these newcomers had become part of the community, the ratio declined and stabilized around 100, indicating an almost perfect equilibrium between the sexes.[33]

Generally speaking, there is no direct correlation between population growth and a rise in the sex ratio unless, as in the case of the Forges, the rise is due to a mass influx of men. But the way in which the ratio changed indicates that the population growth recorded in the early 19th century was indeed associated with a change in the ratio between the two sexes. From that time on, a greater number of workers lived on the post, so that for a while the increase caused a rise in the sex ratio. Subsequently, as a result of population growth, which tends to expand the marriageable population, and because most workers were married, the equilibrium between the two sexes for all age groups in the village stabilized (Figure 7.4).

By Age

In terms of age, the population at the Forges is comparable to other population groups of the time, with an average of 45% in the 13-and-under age group (Table 7.8 and Figure 7.5). Moreover, from 75 to 80% of the population was under 35 years of age.[34]

The 1825 census data reveal the effects of recent immigration on the age structure of the population, just as they reveal a change in its sex structure. Figure 7.5 shows that the relative number of children aged 13 and under rose by nearly 10% from 1784, with a corresponding decrease in the 14-and-over age group. This growth stemmed from the fact that immigration brought complete families to the Forges. Later, ratios returned to the proportions observed initially in 1784, indicating that the members of the new families—who were under the same constraints as the original families as regards the number of their children who chose to leave or stay—were quickly assimilated into the village community.

The children of these newcomers would regenerate the community, helping broaden the marriageable population as they reached marriageable age. The doubling of the population by family immigration doubled the chances of finding a spouse within the community. Children in the 13-and-under age group, who were greater in number in 1825, would gradually move into the 14-and-over age group. They would marry and have children in proportions similar to those of the initial families, so that the age structure of the population would rebalance and return to its initial state after 1830.

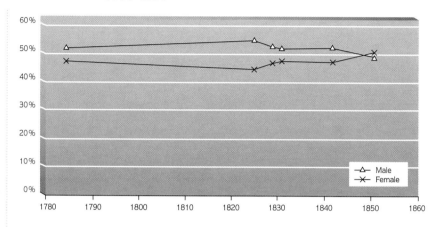

Figure 7.4

POPULATION OF THE ST MAURICE FORGES, BY SEX, 1784–1851

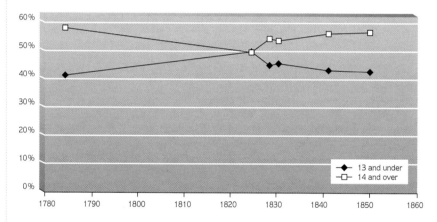

Figure 7.5

POPULATION OF THE ST MAURICE FORGES, BY AGE, 1784–1851

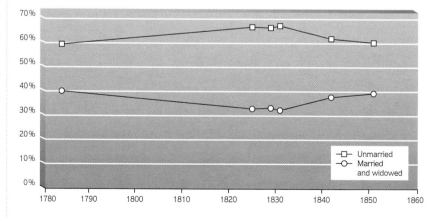

Figure 7.6

POPULATION OF THE ST MAURICE FORGES, BY MARITAL STATUS, 1784–1851

Figure 7.7

POPULATION OF THE ST MAURICE FORGES, BY SEX, AGE AND MARITAL STATUS, 1784–1851

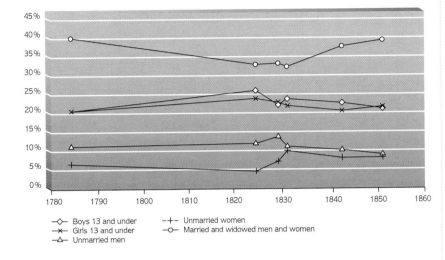

- ◇ Boys 13 and under
- ✕ Girls 13 and under
- △ Unmarried men
- ＋ Unmarried women
- ○ Married and widowed men and women

By Marital Status

Changes in the structure of the population by marital status before and after 1831 can likewise be explained by recent immigration. The population of the Forges was chiefly made up of unmarried people, actually unmarried children (Figures 7.6 and 7.7). Between 1784 and 1851, single people represented 64% of the population (45% 13 and under; 19% 14 and over), on average, whereas married or widowed people accounted for the remaining 36%.[35] The high proportion of unmarried children (70% of single people) indicates that the population of the Forges consisted principally of families, who, as we

Table 7.8

POPULATION OF THE ST MAURICE FORGES, BY SEX, AGE AND MARITAL STATUS, 1762–1851

Male		Single							Married and Widowers							Taken Together						
		1762	1784	1825	1829	1831	1842	1851	1762	1784	1825	1829	1831	1842	1851	1762	1784	1825	1829	1831	1842	1851
13 and under	Abs.	18	31	84	88	80	97	84								18	31	84	88	80	97	84
	%	25	20.8	26.2	22.3	23.9	22.8	21.3								25	20.8	26.2	22.3	23.9	22.9	21.3
14 and over	Abs.	3	17	39	55	38	44	36	11	30	54	66	57	82	75	14	47	93	121	95	126	111
	%	4.2	11.4	12.1	13.9	11.3	10.3	9.11	15	20.1	16.8	16.7	17	19.3	19	19.4	31.6	29	30.6	28.4	29.7	28.1
All ages	Abs.	21	48	123	143	118	141	120	11	30	54	66	57	82	75	32	78	177	209	175	223	195
	%	29.2	32.2	38.3	36.2	35.2	33.1	30.4	15	20.1	16.8	16.7	17	19.3	19	44.4	52.4	55.2	52.9	52.3	52.6	49.4

Female		1762	1784	1825	1829	1831	1842	1851	1762	1784	1825	1829	1831	1842	1851	1762	1784	1825	1829	1831	1842	1851
13 and under	Abs.	28	31	77	91	74	88	86								28	31	77	91	74	88	86
	%	38.9	20.8	24	23	22.1	20.7	21.8								38.9	20.8	24	23	22.1	20.7	21.8
14 and over	Abs.	1	10	15	29	34	35	33	11	30	52	66	52	79	81	12	40	67	95	86	114	114
	%	1.4	6.71	4.67	7.34	10.1	8.2	8.3	15.3	20.1	16.2	16.7	15.5	18.6	20.5	16.7	26.8	20.9	24	25.7	26.8	28.9
All ages	Abs.	29	41	92	120	108	123	119	11	30	52	66	52	79	81	40	71	144	186	160	202	200
	%	40.3	27.5	28.7	30.3	32.2	28.9	30.1	15.3	20.1	16.2	16.7	15.5	18.6	20.5	55.6	47.6	44.9	47	47.8	47.5	50.7

Both Sexes		1762	1784	1825	1829	1831	1842	1851	1762	1784	1825	1829	1831	1842	1851	1762	1784	1825	1829	1831	1842	1851
13 and under	Abs.	46	62	161	179	154	185	170								46	62	161	179	154	185	170
	%	63.9	41.6	50.2	45.3	46	43.5	43.1								63.9	41.6	50.1	45.3	45.9	43.5	43
14 and over	Abs.	4	27	54	84	72	79	69	22	60	106	132	109	161	156	26	87	160	216	181	240	225
	%	5.6	18.1	16.8	21.3	21.5	18.6	17.5	30.6	40.3	33	33.4	32.5	37.9	39.5	36.1	58.4	49.9	54.7	54	56.5	57
All ages	Abs.	50	89	215	263	226	264	239	22	60	106	132	109	161	156	72	149	321	395	335	425	395
	%	69.4	59.7	66.9	66.6	67.4	62.1	60.5	30.6	40.3	33.1	33.4	32.5	37.9	39.5	100	100	100	100	100	100	100

Abs.: absolute figures.

Note: Age was not recorded in the 1762 census. We therefore decided to put "boys" and "girls" in the 13-and-under age group, and adults and servants in the 14-and-over age group.

will see later, accounted for over 90% of the total population. Besides affecting the breakdown of the population by sex and age, the immigration of the early 1800s also had a temporary impact, from 1825 to 1831, on the ratio of single to married people; the initial equilibrium would be re-established 20 years later. The growth in the number of single people to the levels seen between 1825 and 1831 was due essentially to the number of children that arrived with the new families. The curves of Figure 7.6 show that from 1784 to 1825, the children of new arrivals would push up the number of single people by 7%, that between 1825 and 1831 the marital status of these children would not change—although some would have reached marriageable age in the meantime—and that after 1831 many would marry, thereby increasing the relative number of married people in 1842 and 1851. After a generation, the relative numbers of single (60%) and married people (40%) had returned to the ratio found in the initial population of 1784.

Changes in the number of single men and women 14 and over also play a specific role in this demographic transformation. The proportion of single men remained relatively stable during the period under study, whereas the proportion of single women rose significantly before stabilizing (Figure 7.7). The number of young women, who had become a minority as a result of the arrival of many young workmen prior to 1825, quickly caught up with the relative number of men of the same age, thus increasing the ranks of women of marriageable age in the community. As we will see, this equilibrium among young single people was no doubt the most striking consequence of the wave of immigration of the first quarter of the 19th century. The small number of people at the Forges in the 18th century and the gap between the numbers of single people of each sex observed in 1784 suggest that a balance was not achieved at that time. Since there were not enough potential spouses within the community, the young people had to look outside the village for someone to marry.

In the early 1800s the population of the Forges was therefore revitalized in a way that affected both its size and composition. But through a process of fairly rapid growth, it gradually returned to the demographic "model" of the ironworks, which counted on family vitality to ensure its long-term survival.

Emigration of Young Adults

The 1829 and 1851 censuses, which counted the same number of people (395), show that in both cases there were more women than men aged 20 to 29. The constancy of this imbalance would seem to indicate that young men sometimes had to emigrate, as their chances of employment were determined by the production capacity of the Forges, which could not hire all the young men who reached adulthood each year. When young men left, so

Table 7.9

WORKMEN, HEADS OF FAMILIES, FAMILIES AND HOUSES AT THE ST MAURICE FORGES, 1825–81[41]

	1825	1829	1831	1842	1851	1861	1871	1881
Men or workmen	71	88	95	126	110	53	65	70
Heads of families	52	77	60	84	82	52	52	48
Families	55		60		80	52	52	48
Houses	55		57		72	42	47	46

did some women of the same age, albeit in a lesser proportion.[36] On the basis of projections[37] for the 10-to-19 age group, we have estimated that half of the men and women emigrated when they moved into the 20-to-29 age bracket. Between 1829 and 1851, two or three men a year left the Forges during this decade of their lives, as opposed to only one or two women. Over 20 years, emigration remained at the same level for men—hardly surprising, considering the stability of employment at the Forges at that time—whereas it fell by half among women, who were then in a better position to find a spouse within the community.[38] The village had a higher number of endogamous couples according to the 1851 census, confirming the fact that the marriageable population had indeed expanded at the Forges; between 1829 and 1851, young people from the Forges were therefore more successful in finding a marriage partner within their own community.[39]

Households and Families

The population at the Forges was made up essentially of families. Some dwellings housed two families,[40] a fact that explains in part the difference between the number of houses and the number of families in Table 7.9. The "houses" that the census takers counted were not necessarily comparable to the single-family dwellings we have today. Instead, we know that they were often multi-family tenement houses (*corps de logis*) described in some inventories as "buildings providing lodgings for several families" (*bâtiments servant de plusieurs logements*) or else grouped into two or five "houses within a single building" (*maisons d'un seul corps*) (see Chapter 9).

Households that lived in these dwellings consisted chiefly of couples with children (Figure 7.8). Throughout the period under study, nuclear families accounted, on average, for 73% of households and for 85% of the total population. A few households were headed by a widow or widower, but single people were definitely the exception in this family-based society. The predominance of families is easy to explain in the general context of the labour market of the time, as well as in the specific context of the Forges. Because the labour market was relatively undeveloped (particularly in the iron industry), the company did not have the option of drawing freely from a labour pool consisting of young mobile men, and moreover it gave priority to families, who represented the best means of passing on the technical know-how essential to continuing its operations.[42] The characteristics of this family-oriented society at the Forges are worth examining in more detail.

Figure 7.8

COMPOSITION OF HOUSEHOLDS AT THE ST MAURICE FORGES, 1825–51

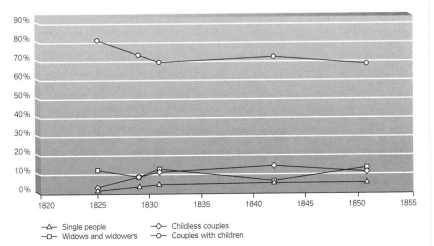

- —△— Single people
- —□— Widows and widowers
- —◇— Childless couples
- —○— Couples with children

Family Composition

In the first half of the 19th century, families represented 93% of the population, on average. The average family consisted of six people, four of them children (Tables 7.10 to 7.14), but this figure dropped in 1842 and 1851. By the latter date, there was one child fewer per family (3.51) than there had been in 1831 (4.42) (Figure 7.9). This drop is no doubt linked to the average younger age of workers as a result of the wave of immigration that occurred at the start of the century.[43] Examination of marriage statistics shows that the men's age at marriage fell from 25 to 23 starting in 1800. We have also seen that several marriages were entered into following the arrival of these new young workmen, and that the number of marriages between people from the Forges increased. Since these spouses were younger, their families were smaller.

The first column of Tables 7.10 to 7.14 shows that, in absolute terms, the number of families grew in the second quarter of the 19th century, rising from 50 in 1831 to 68 in 1842 and 1851. It can also be seen that it was principally the number of families with four children or fewer that virtually doubled, rising from 27 in 1831 to 50 in 1851, whereas the number of families with five children or more decreased. This illustrates once more how families at the Forges were getting younger. It is worth noting, however, that this rejuvenation was not simply and exclusively a direct consequence of growth in the population, which went from 335 in 1831 to 395 in 1851. In 1829, when the size of the population was comparable to that of 1851, the number of families had not increased significantly since 1825, and as many families had four children or fewer as had five or more. One generation, from 1829 to 1851, made all the difference. The young families of 1851 were not simply new additions to the total number of families;

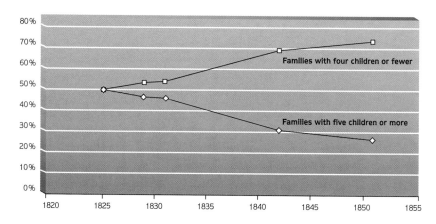

Figure 7.9

CHANGES IN FAMILY SIZE AT THE ST MAURICE FORGES, 1825–51

Families with four children or fewer

Families with five children or more

rather, they replaced many of the larger families, as a result of younger workers replacing the older ones.

In conclusion, studying the population of the Forges tells us a great deal about this special type of industrial community. Despite inherent documentary and methodological limitations, the demographic composition and trends can be determined and linked to the needs of the relatively isolated industrial establishment. It becomes clear that the term "industrial village" refers not just to its function and production facilities, but to the working families that adapted to it and brought it to life. The hundred and fifty years that the Forges were in operation take on another dimension when looked at as five generations of workers.

Table 7.10

SIZE OF FAMILIES AT THE ST MAURICE FORGES, 1825

Family category	Number of families			Number of children per family category		Number of people per family category			Mean number of children per family	Mean size of family
	Abs.	%	% Total	Abs.	%	Abs.	%	% Total		
One child	9 [1]	17.30	17.30	9	4.18	25	8.01	8.01		
Two children	3	5.76	23.06	6	2.79	12	3.84	11.85		
Three children	8 [2]	15.38	38.44	24	11.16	38	12.17	24.02		
Four children	6 [3]	11.53	49.97	24	11.16	35	11.21	35.23		
Five children	12 [4]	23.07	73.04	60	27.90	83	26.60	61.83		
Six children	8	15.38	88.42	48	22.32	64	20.51	82.34		
Seven children	4 [3]	7.69	96.11	28	13.02	35	11.21	93.55		
Eight children	2	3.84	99.85	16	7.44	20	6.41	99.96		
All families	**52**	**100**	**100**	**215**	**100**	**312** [5]	**100**	**100**	**4.13**	**6.00**

1. Including a widow and a widower.
2. Including a widow and a widower.
3. Including a widower.
4. Including a widow.
5. This total does not include the two couples without children (four people) and the five people counted at the Grande Maison.

Table 7.11

SIZE OF FAMILIES AT THE ST MAURICE FORGES, 1829*

Family category	Number of families			Number of children per family category		Number of people per family category			Mean number of children per family	Mean size of family
	Abs.	%	% Total	Abs.	%	Abs.	%	% Total		
One child	11	18.96	18.96	11	4.56	31	8.78	8.78		
Two children	4	6.89	25.85	8	3.31	16	4.53	13.31		
Three children	9	15.51	41.36	27	11.20	45	12.74	26.05		
Four children	7	12.06	53.42	28	11.61	41	11.61	37.66		
Five children	8	13.79	67.21	40	16.59	55	15.58	53.24		
Six children	9	15.51	82.72	54	22.40	72	20.39	73.63		
Seven children	7	12.06	95.78	49	20.33	63	17.84	91.47		
Eight children	3	5.17	99.95	24	9.95	30	8.49	99.96		
All families	**58**	**100**	**100**	**241**	**100**	**353**	**100**	**100**	**4.15**	**6.08**

*   Fifteen unmarried workmen are not counted among these families.
Including them would modify the family categories slightly.

Table 7.12

SIZE OF FAMILIES AT THE ST MAURICE FORGES, 1831

Family category	Number of families			Number of children per family category		Number of people per family category			Mean number of children per family	Mean size of family
	Abs.	%	% Total	Abs.	%	Abs.	%	% Total		
One child	5 [1]	10.00	10.00	5	2.26	14	4.47	4.47		
Two children	9 [2]	18.00	28.00	18	8.14	34	10.86	15.33		
Three children	5 [1]	10.00	38.00	15	6.78	24	7.66	22.99		
Four children	8 [3]	16.00	54.00	32	14.47	46	14.69	37.68		
Five children	5 [1]	10.00	64.00	25	11.31	34	10.86	48.54		
Six children	7 [1]	14.00	78.00	42	19.00	55	17.57	66.11		
Seven children	6	12.00	90.00	42	19.00	54	17.25	83.36		
Eight children	3	6.00	96.00	24	10.85	30	9.58	92.94		
Ten children	2	4.00	100.00	18	8.14	22	7.02	99.96		
All families	**50**	**100**	**100**	**221**	**100**	**313**	**100**	**100**	**4.42**	**6.26**

1. Including a widower with child.
2. Including a widower with children and a widow with children.
3. Including two widowers with children.

Table 7.13

SIZE OF FAMILIES AT THE ST MAURICE FORGES, 1842

Family category	Number of families			Number of children per family category		Number of people per family category			Mean number of children per family	Mean size of family
	Abs.	%	% Total	Abs.	%	Abs.	%	% Total		
One child	13 [1]	19.12	19.12	13	5.24	38	10.00	10.00		
Two children	7	10.30	29.42	14	5.64	28	7.37	17.37		
Three children	16 [2]	23.53	52.95	48	19.36	78	20.53	37.90		
Four children	11 [3]	16.18	69.13	44	17.74	65	17.11	55.01		
Five children	10	14.71	83.84	50	20.16	70	18.42	73.43		
Six children	2	2.94	86.78	12	4.84	16	4.21	77.64		
Seven children	6	8.82	95.60	42	16.93	54	14.21	91.85		
Eight children	2	2.94	98.54	16	6.46	20	5.26	97.11		
Nine children	1	1.47	100.01	9	3.63	11	2.89	100.00		
All families	**68**	**100**	**100**	**248**	**100**	**380**	**100**	**100**	**3.64**	**5.59**

1. Including a widower.
2. Including two widowers.
3. Including a widow.

Table 7.14 SIZE OF FAMILIES AT THE ST MAURICE FORGES, 1851

Family category	Number of families			Number of children per family category		Number of people per family category			Mean number of children per family	Mean size of family
	Abs.	%	% Total	Abs.	%	Abs.	%	% Total		
One child	16	23.53	23.53	16	6.69	48	13.15	13.15		
Two children	10	14.71	38.24	20	8.37	37	10.14	23.29		
Three children	13	19.12	57.36	39	16.32	65	17.81	41.10		
Four children	11	16.18	73.54	44	18.41	64	17.53	58.63		
Five children	5	7.35	80.89	25	10.46	32	8.77	67.40		
Six children	3	4.41	85.30	18	7.53	22	6.03	73.43		
Seven children	5	7.35	92.65	35	14.64	45	12.33	85.78		
Eight children	4	5.88	98.53	32	13.39	40	10.96	96.72		
Nine children	0	0	98.53	0	0	0	0	96.72		
Ten children	1	1.47	100.00	10	4.18	12	3.29	100.01		
All families	**68**	**100**	**100**	**239**	**100**	**365**	**100**	**100**	**3.51**	**5.37**

Exterior of a great forge near Châtillon-sur-Seine, by Étienne Bouhot (1780–1862).
Some of the original workers at the St Maurice Forges came from the bailiwick of Châtillon-sur-Seine, in France.
MUSÉE DES BEAUX-ARTS, MONTBARD, PHOTOGRAPH BY RMN—P. BERNARD.

VIII

The Industrial Community

But at night, when one sees
the flames continually rising several feet above
the furnace and spreading an eery glow
over the entire village; when one sees
the workers in this light, wandering like ghosts
around their old dwellings, their clothing
blackened by coal and smoke, and especially
when one thinks that a few years ago,
the village was completely surrounded
by several leagues of dense forest,
one's imagination runs riot,
and one involuntarily says to oneself:
"Strange things must have gone on here."

Napoléon Caron, 1889

COMPOSITION OF THE WORK FORCE

As with the overall population, a clarification is called for before we attempt to establish the composition of the work force at the Forges. The early iron industry was characterized by a labour hierarchy of various categories of workers, which brought together a variable number of individuals with neither the same skills nor the same status. In estimating the number of workmen employed at the Forges, outside witnesses of the period—visitors, commissioners of inquiry and census takers—did not always appreciate the distinction between the different categories of workmen (see Appendix 14). They saw the Forges as an exclusive world apart, and most often had only a superficial view of it, with little grasp of the subtleties of its internal structure. They were immediately struck by the large number of workers (approximately 350) the Forges employed annually—mostly on a seasonal basis. And these manpower requirements had a substantial economic and social impact on the whole Trois-Rivières region. As we have already seen in our discussion of the demographics of the working population (see Chapter 7), many workmen tended to define themselves, in official records, as workmen at, or employees of, the Forges, even if they did not live there permanently. In other words, the perception of the workers' environment went far beyond the bounds of the industrial village. Fortunately, however, we can gain a more accurate picture of the actual size of the Forges' work force and the different categories of workmen constituting it by consulting the reports drafted for the government by the establishment's directors, in the 18th and early 19th centuries in particular, in which they sometimes provide lists of the workmen assigned to the different positions.

But management was not alone in identifying its workmen in this way—the workmen were perfectly capable of attending to this themselves. In fact, they made a point of emphasizing their differences, apparently, since some of them would even become truly legendary characters. Historiographers have given prominence to the independent spirit, and even the arrogance, of the first French workers at the Forges, who were well aware that, in a colony pioneering its first industry, theirs was an exclusive specialty. Indeed, the first directors of the Forges would sometimes bemoan this attitude. The parish registers testify eloquently to the skilled workmen's desire to assert their difference. At christenings, weddings or funerals, they entered in the registers not only the fact that they belonged to the Forges, but also the craft they practised there. So the records of the French era in particular often have the following notations: "finer at the Forges," "carter at the Forges," "helper at the Forges" or "collier at the Forges"; and several workmen had "master" added to their designation of founder, hammerman, finer or collier. Furthermore, it goes without saying that skilled workmen and directors were not alone in wishing to differentiate themselves in this way, since workers of more modest station readily presented themselves as a "carter at the King's Forges," or better yet, a "carter employed in the service of His Majesty's Forges."[1] We shall see that the social status of workmen at the Forges would lose its distinction, later in the 19th century, to the extent that a growing number of workers would be identified only as mere "day labourers."

For an accurate picture of the working world at the Forges, at different times in their history, it is necessary to take into account the industrial processes used there and their concomitant manpower requirements. These processes, along with the industrial plant required to put them into practice, decided the number of workmen employed at the Forges, as well as the skills required of them. Between 1737 and 1850, there would be no substantial change in the processes, and the same skills would continue to be required of the workmen. What would alter the number of workers engaged in the different categories, rather, would be variations in the use of certain processes, such as casting. After 1850, the slowdown in and then abandonment of the manufacture of castings and wrought iron would have the effect of altering the size and composition of the work force at the Forges.

MANAGEMENT AND SERVICE PERSONNEL

Among employees at the Forges, we must first single out the staff assigned to administration and management. Their duties, pay and privileges, along with the mere fact that they were lodged at the Grande Maison, bestowed on them a special status. This group naturally includes the directors and their assistants, as well as the clerks and foremen (Table 8.1). Depending on the period, and on whether they were owners, public officials or lessees—or acting as their representatives—the administrators had the title of director, ironmaster, inspector, agent or manager. Serving under them and directing the workmen was a foreman, or "overseer," who had the authority of an ironmaster. Apparently, this position designated the person responsible for a campaign, who oversaw the operations of the inside departments in particular, but also supervised the outside work, in the forest. In fact, he performed the duties of the ironmaster, with the

To a manager/ironmaster (expert, also a skilled founder, hammerman and even a finer, a true workman, able to oversee all the workmen at the furnace and the forges, able himself to know when and how they are lacking, rectify their errors both to correct their work and show them the correct proportions and the degree of fire required to produce good pig iron and well-made wrought iron) at 2,000 livres per year, board not included [...].

Table 8.3

INSIDE AND OUTSIDE WORKMEN, ST MAURICE FORGES

	Department	Occupation
Inside workmen	Blast furnace	Founder
		Keeper
		Filler
		Helper
		Carter
		Charger
		Limestone breaker
	Moulding shop	Moulder
		Bomb maker
		Sand moulder
	Upper forge	Hammerman
		Finer
		Charger
		Helper
		Carter
	Lower forge	Hammerman
		Finer
		Charger
		Helper
		Carter
		Tilter
	Blacksmith's shop	Blacksmith
		Farrier
		Edge-tool maker
		Locksmith
	Finishing shop	Ware dresser
	Wheelwright's shop	Wheelwright
	Shop	Saddler
		Cobbler
	Sawmill	Sawyer
	Carpenter's shop	Carpenter
		Joiner
	Masonry works	Mason
	Grist mill	Miller
	Bakery	Baker/butcher
	Farm	Farmer
Outside workmen	Mines	Miner
		Ore washer
	Quarries	Quarryman
	Charcoal pits	Collier
		Pit setter
		Feuiller
		Garde-feu
	Forest	Woodcutter
	Roads	Road maker
		Carter
	River	Ferryman
		Bateauman, scowman

So, Lord Selkirk first indicates the 24 or 25 ironworkers and their assistants employed at the blast furnace and the two forges, and then mentions all the others who were recruited from among the *habitants* living in the vicinity of the Forges. Most of the outside workmen were woodcutters, employed during the winter. In 1741–42, the company employed more than 200 woodcutters, who would deliver no fewer than 11,000 cords of wood for coaling and close to 1,000 cords of firewood.[9] In the early 19th century, the directors of the Forges said they employed more than 150 woodcutters annually. In numerical order of importance, these were followed by the carters (40, according to Laterrière in 1775),[10] miners (some 20),[11] colliers and other day labourers, road makers and bateaumen.

During the French regime, the seasonal recruitment of a large number of outside workmen posed serious problems, since the population of the region was as yet small. Making use of the *habitants*—who could not always be available, owing to ploughing and harvesting, when the Forges needed them—was an expensive proposition, and it was decided to augment the Trois-Rivières garrison so that these outside workmen could be recruited from among the soldiers. After the Conquest, under the military administration, the British would enlist the *habitants* for woodcutting, using the captains of militia as recruiters.[12] And subsequently, the *habitants* from neighbouring seigneuries—Pointe du Lac and Yamachiche in particular—would gradually be called upon to form part of this outside work force. Certain categories of outside workmen, such as the colliers, carters and miners, enjoyed a degree of independence because they were paid by the job or load. And, since they were scattered over a vast area, it was somewhat difficult to exercise direct control over their work.[13] In the early 19th century, raw materials would be collected farther and farther away from the

▼

Forges,[14] and the cost of carting them would rise accordingly. Indeed, this was without doubt one of the main factors that led the directors of the Forges to further rationalize this type of work shortly after 1800, and thus to curb the seasonal hiring not only of a number of carters but also and more particularly of colliers and miners.

Among the inside workmen, the iron-workers—that is, the 24 or 25 hands of whom Lord Selkirk spoke, who toiled in the blast furnace, the two forges and the moulding shop—were set apart by their skill and specialization. Then there were the other smiths who worked on the assembly and finishing of certain products and provided domestic services (blacksmiths, farriers, edge-tool makers and locksmiths). The woodworkers (carpenters, joiners, wheelwrights and sawyers) attended primarily to the maintenance of the equipment and workshops, also carrying out construction work. There was also a mason charged with regular maintenance and repairing hearths and other masonry work, and a leather crafts-man (saddler) employed in making and maintaining all the harness gear. Finally, several carters were employed; these men, assigned to the forges and blast furnace, were more especially responsible for transporting slag and wares from the workshops.

THE WORK FORCE

Lists of Workmen

To establish the composition of the inside work force employed by the Forges throughout their lifespan, two types of lists are available to us: lists of positions or occupations, and lists of workmen's names with their occupations (see Appendices 13 and 14). The former were largely drawn up during the French regime, while the latter were put together by ironmasters and administrators, either to estimate the number of workmen necessary for operations or to take stock of the positions held at a given moment. The lists of positions in fact detail the man-power needs specific to the various industrial processes in use, depending on the period, but the names of the workmen holding the positions are not mentioned.[15] The other lists, which provide the workmen's names, with or without their occupations, essentially date from the 19th century;[16] they are taken from official censuses, parish censuses (tallies of parishioners or communicants), a citizens' directory, a militia muster roll, and employee rolls drawn up by the Forges superintendent.

A comparative analysis of these different lists is not without its difficulties, since they provide different levels of detail. The first nominal censuses in 1825 and 1831 clearly distinguish the population of the Forges, but list only the names of heads of families—and in the case of the 1825 census, with no mention of their occupations. The other official censuses (1851, 1861, 1871 and 1881), on the other hand, list all inhabitants by name but, with the exception of the 1851 census, do not treat the Forges as a distinct entity within the census district; the population of the Forges may, however, be identified by the occupations specified for the persons listed. The 1866 and 1875 parish censuses, for their part, distinguish the population of the Forges, but do not specify occupations; moreover, the former provides only the names of heads of household, whereas the latter gives the names of all the inhabitants. Lovell's Directory of 1871 gives the names and occupations of the workmen, but the list is apparently not exhaustive.[17] The 1835 militia muster roll for Captain H. Macaulay's company—named after the superintendent of the Forges—comprises only the names of the 79 militiamen between 18 and 60 years old and makes no mention of their occupations. The same Henry Macaulay drew up two other employee rolls, in 1829 and 1842, respectively. The latter mentions only the names of heads of household, whereas the 1829 list is without doubt the most detailed post-Conquest document we have for identifying the men and their occupations. Drawn up in August, in mid-campaign, it lists all the men employed that summer, clearly specifying their occupations, marital status, family situation and, for those not born at the Forges, how long they had been living there. It also contains the names of workmen—primarily colliers[18]—employed only during the summer months, as well as of workmen who, while born at and belonging to the Forges, were working at Trois-Rivières at that time.[19]

Underestimates

A number of pitfalls must be avoided in attempting to reconstruct the inside work force at the Forges. Thus, for instance, if we take into account only those men for whom specific occupations are mentioned in the censuses, we will generally underestimate by some 30% the total number of men from the village who were of working age (Table 8.5).

Generally speaking, the census takers tended to report occupations only in the case of heads of household (Table 8.4).

This was evidently the case with the 1825 and 1831 censuses, for which only the names of heads of household were given, but it is worth noting that the number of heads of household is quite close to that of workmen worthy of mention. We must therefore try to assess the underestimate of workmen resulting from the "selective" methods of census takers in other full nominal censuses.[20] In Table 8.5, we have totalled the number of men whose occupations were *specified* or *not specified* in order to obtain a more realistic assessment of the work force residing in the village. The men we counted for whom occupations were *not specified* were 14 or over, and were therefore considered to be of working age. In the 1851 and 1861 censuses, occupations are specified for boys as young as 14, 13 or even 12 years old.[21] Generally speaking, however, most men whose occupations were *not specified* were 16 and over.

Table 8.4

COMPARISON OF THE NUMBER OF HEADS OF HOUSEHOLD AND THE NUMBER OF WORKERS WHOSE OCCUPATIONS ARE SPECIFIED IN THE CENSUSES *

	1784	1825	1829	1831	1842	1851	1861	1871	1881
Heads of household (man or woman)	30	52	73	60	84	82	52	52	48
Workers	*30*	*43*	77	59	*98*	75	26	53	62

* When the figure is in italics, the occupation is not specified in the census. In the case of the 1825 list, the occupations of heads of household were attributed on the basis of the occupation reported in the 1829 list. In the case of the 1842 list, the 98 individuals for whom the occupation is not specified but who are considered to be workers comprise males of working age (i.e., 14 and over) and one widow.

Table 8.5

DISTRIBUTION OF MEN OF WORKING AGE, ST MAURICE FORGES, 1784–1881[22]

	1784	1825	1829	1831	1842	1851	1861	1871	1881
Men (and one woman) whose occupations are specified	30	43	77	59	98	75	26	53	62
Men whose occupations are not specified	12	28	11	36	28	35	27	12	8
OVERALL									
Men or workmen of working age	42	71	88	95	126	110	53	65	70
Percentage of men whose occupations are not specified	28.6	39.4	12.5	37.9	22.2	31.8	50.9	18.5	11.4

Note: In 1784 and in 1825, occupations were not specified; men considered here whose occupations are *specified* are either counted as heads of household or lumped in with them, whereas men whose occupations are *not specified* are other men aged 14 and over. The 1842 list of workmen makes no mention of occupation either; it includes 84 heads of household, 15 other single men and 28 other men 14 and over. A woman is listed as a schoolmistress in 1871 and 1881.

If this were any other village than the Forges, we might think these men for whom no occupations are specified were not company employees, so there would be no need to talk of underestimates. But in a village where the company, which owned or leased the site, provided room and board for the workmen and their families, it is hard to conceive that men without an occupation would be "kept."[23] These workmen whose occupations were not specified, who formed on average close to 30% of the manpower residing at the Forges, were probably assistants, apprentices and day labourers. In an industrial village where passing down the craft within the family was the rule, it is perhaps not surprising that there was a tendency to attribute an occupation worthy of mention primarily to heads of household. The 1829 employee roll remains one of the most reliable, with only 12.5% of men having no occupation specified. As this is a list of workmen, the 11 men for whom no occupations are specified were actually employed by the Forges. Furthermore, the order in which the workmen were listed may be indicative of the positions they held.[24] This list therefore allows us to confirm in another way the underestimate that occurred during the close-together censuses of 1825 and 1831. Only in the last two censuses of 1871 and 1881 did the underestimate decrease, doubtless because the census methods were applied more rigorously.

Child Labour

The mere fact that child labour was not taken into account no doubt also contributes, albeit to a lesser extent, to the underestimate of the work force in official censuses. We have seen that boys aged 12 or 13 were sometimes reported with an occupation in the censuses, but these cases are rare. However this may be, there is reason to believe that the Forges used child labour. During the French regime, it was indeed recommended that child labour be used for nail making.[25] We also know that in 18th-century France the employment of children was favoured for manufacturing cannon-balls, for instance, and it was recommended that children as young as 9 be used to make files.[26] The list of workmen at the Forges in 1829 specifically mentions employment of children: Among the workmen referred to were "Aug. Rivarre and boys, ware dressers." These children were assigned to the finishing (dressing and trimming) of wares. It was also mentioned that Rivard was born at the Forges, 35 years old and married with four sons and two daughters. Thanks to the parish registers, we have been able to trace, in August 1829, the family (six children) of Augustin Rivard, husband of Marie-Anne Reignière. The boys were named Antoine, Augustin and Joseph, and were 12, 10 and 9 years old respectively (the fourth boy was only 2 years old). Since the family tradition of the craft was so well entrenched in this milieu, we should not be at all surprised to find workmen's children taking part at an early age in some of the lighter work, such as charging coal or washing ore, or indeed in some of the helper's tasks.[27] Similarly, boys were apprenticed fairly young in the skilled ironworking crafts, since to be "a born forge-man" and to be able to succeed his father, a boy had to be immersed throughout his childhood in the rigours of the job.

Individuals' Occupations

The growth in the work force until the 1840s, followed by a reduction in, and stabilization of, the number of employees during the Forges' final 30 years, is in keeping with the observations we have already made concerning the population in general. We have seen that, from around 1800, the addition of 30 or so families to the industrial community helped double the work force residing in the village. The shift in production to finished castings increased the number of moulders, and a number of carters joined the inside work force. After 1860, finished castings were abandoned in favour of the production of pig iron, and this explains the reduction in the number of workmen, as well as their lower skill levels.[28]

Table 8.6 does not provide a clear picture of changes in the work force by occupation or by department. Here again, if the figures for certain categories vary, or if no workmen are shown in any of these categories at a given time, one should not conclude that the corresponding positions have decreased in number or completely disappeared. In fact, from 1851 onwards, the census takers would no longer note as precisely the occupations of the workmen listed. Thus, of the 21 occupations listed in 1831, only 13 would remain 20 years later. Only moulding, a skilled craft, would post an increase, from 9 to 26 moulders over the same period,[29] in parallel with the emphasis on the production of castings during the Mathew Bell era, which would continue into the late 1850s.[30] In the three subsequent censuses, most workmen would increasingly be described as day labourers, to the point that, in 1881, only a single workman would be designated as a blacksmith, whereas 45 day labourers were listed. Nor would this deskilling of workmen to day labourers spare the other ironworks in the region, as Table 8.7 shows.

The era of the *master* workmen, with pride in their craft and in belonging to what were formerly the King's Forges, was therefore well and truly past. Clearly, the shift from the large-scale production of finished castings and wrought iron from the early 1860s onwards had a direct impact on the disappearance of the crafts of forgeman, blacksmith and moulder. The Forges were no longer the famous "manufacture" of bar iron and castings. Reduced to an iron smelter producing pig iron for the Montreal railcar wheel industry, the Forges no longer needed skilled labour. With no further need for the crafts, the long-established families would also leave the works. The final 20 years of operation would see a complete change in the working families, since most of the old families—Aubry, Gilbert, Imbleau, Mailloux, Marchand, Michelin, Robichon, Tassé and Terreau—would move away from the Forges, and even the region, leaving behind a few isolated members among the new families—Beauchemin, Blais, Bouchard, Bourassa, Désaulniers, Gélinas, Garceau, Landry, Loranger et al. In Lovell's 1871 directory, there were still three individuals designated as blacksmiths and four as moulders; the workmen performing the duties of founder and keeper were simply called "furnace heater" and "furnace hand," and would never again be defined that way in censuses.[31] The working conditions and the lower level of skill demanded of the workers would alter the working world of the Forges and, from what the censuses show, would also alter the contemporary perception of workmen at the Forges. The workers' perception of themselves would also change. The last workmen interviewed by Dollard Dubé in the early 20th century would describe the colliers as *"chargeurs de kiles"* or *"gardiens de kiles"*—in reference to the kilns, brick ovens erected on the post itself for making charcoal. They would also qualify the moulders as mere *faiseux de beds* or "bed

Table 8.6

OCCUPATION OF RESIDENTS, ST MAURICE FORGES, 1825–81 *

Occupation	1784	1825a	1829b	1831	1842c	1851	1861	1871	1881
Superintendent (agent, merchant or manager)		1	1	1		1	1	2	1
Clerk						2			2
Foreman (overseer)		1		1		1			
Physician								1	1
Schoolmistress								1	1
Baker				1	1	1	Abs.g		1
Miller		1	1	1			1		1
Gardener			1						
Farmer		1	3	2		1	7	1	9
Collier		1		3					
Pit setter				3	1				
Feuiller					2				
Garde-feu							1		
Miner		2	2	1		1			
Ore washer		1	2	2					
Founder		1	1	1					
Keeper		2	2	1					
Filler		2		4					
Forgemand		7	10e	10		13	3	4	1
Moulder		9	9	9		14f	3	4	
Sand moulder				1	1				
Blacksmith		3	7						
Mason								1	
Carpenter (joiner)		4	3	3		2	Abs.g	4	
Sawyer				1					
Saddler				1	1	1			
Wheelwright						1			
Carter		5	14	9		4	Abs.g		
Bateauman			4						
Ferryman		1	5	1					
Cobbler							1	1	
Painter							1		
Day labourer		1	5	4		33	8	34	45
Workers whose occupations are specified	(30)	(43)	77	59	98	75	26	53	62
Men whose occupations are not specifiedh	12	28	11i	36	28	35	27	12	8
Men whose occupations are not specified (%)	28.6	39.4	12.5	37.9	22.2	31.8	50.9	18.5	11.4
TOTAL									
—Men or workmen	**42**	**71**	**88**	**95**	**126**	**110**	**53**	**65**	**70**
—Heads of household (men or women)	30	52	77	60	84	82	52	52	48
—Families j		55		60		80	52	52	48
—Houses		55		57		72	42	47	46

See note to Table 8.5.

* The St Maurice Forges do not have their own exclusive pages in the 1861, 1871 and 1881 censuses; the establishment is included in the St Étienne sub-district. Workmen were identified by the occupation reported. For each census, we took the workmen as follows:

1861:   pages 96–100
1871:   pages 57–70 (families 227 to 278)
1881:   pages 17–27 (families 88 to 136)

a   The 1825 census makes no reference to occupations; occupations were attributed here on the basis of the occupation reported in 1829 for men present at the Forges at the time of both censuses. The 9 unspecified were not present in 1829. Of the 71 men aged 18 and over who were counted, we have only the names of 52 heads of household. So there are at least 19 men of working age (including 17 single men) whose occupations are not specified. The sex of those aged 14–17 is not mentioned in the census.

b   The figures for 1829 exclude the following workmen: 5 moulders and 2 blacksmiths who were at Trois-Rivières at the time of the census, and 11 seasonal workers (8 colliers, 2 pit setters and 1 garde-feu).

c   The 1842 list contains 99 names, including one widow and Superintendent Macaulay. Among these 99 were 15 single men living alone. Subtracting these 15 single men leaves 84 names with which other individuals under 14 years of age or over 14 years of age are associated, who were therefore heads of household. The count shows 126 men aged 14 or over. So 28 (126 - 98) men aged 14 and over were not named.

d   Only the 1829 census makes a distinction between forgemen (finers/hammer-men) and blacksmiths.

e   Of the 10 forgemen, 2 were retired.

f   There are 14 specific references to moulders in the census; but 9 other people entered after a person whose occupation is specified may be taken as moulders, in view of the "do" (ditto) convention used to attribute to the next name in the list the occupation associated with the previous name.

g   Absent. Workmen reported absent at the time of the census.

h   Men of working age with no specific mention of occupation, or not identified as heads of household. Since the 1825 census makes no reference to occupation, we assigned the occupations reported in 1829; but 9 men from 1825 were not listed in 1829, so we had to add them to the 19 men aged 14 and over who were not on the list.

1825:   Men aged 18 and over
1831:   Men aged 14 and over
1842:   Men aged 14 and over
1851–81: Men aged 15 and over

The 1825, 1831 and 1842 censuses deal with the "over 14" age group. Including or excluding the 14-year-olds makes a difference of only a few people.

i   According to the order of listing, these 11 unspecified workmen could logically be 1 foreman, 3 colliers, 4 pit setters and 3 fillers.

j   Based on the number of heads of household in 1825 and 1831, the "number of families in the house" in 1851, and the "number of families living in the house" in 1861. In 1871 and 1881, each family had a number.

Table 8.7

INCREASE IN THE DAY LABOURER CLASSIFICATION IN ST MAURICE VALLEY IRONWORKS

Establishment	1861			1871			1881		
	Employees	Day labourers only		Employees	Day labourers only		Employees	Day labourers only	
		No.	%		No.	%		No.	%
St Maurice*	53 (53)	8 (36)	15.1 67.9	65 (65)	34 (46)	70.8	70 (70)	45 (53)	75.7
L'Islet	15	1	6.6	62	40	64.5	31	31	100.0
Radnor	165	11	6.6	30	17	56.6	91	58	63.7
St Pie				34	20	58.8	56	39	69.6
St Tite				18	7	38.8			
Grondin							20	17	85.0

* For the St Maurice Forges, the figures in parentheses include men for whom no occupation is specified. The figures for other establishments are taken from Hardy and Gauthier 1989; the latter stress that, compared with St Maurice, the number of men whose occupation is not specified is very low in the other establishments.

makers," thus reducing their role to that of the guttermen who prepared the pig bed in the sand of the casting house floor.[32]

The workmen who were there at the closing of the Forges were for the most part no longer descendants of those who had been in at its birth. Twenty years before they closed down, the Forges had already become the *"Vieilles Forges,"* or the Old Forges.

ORIGIN AND RECRUITMENT OF THE WORKMEN

Sieur Simonet shall therefore embark with him on the King's ship Le Profond *four workmen. He has engaged several others, who may be sent next year [...].*[33]

Estienne Gochereau [...] is established at the St Maurice Forges as master collier. We are pleased with him. I beseech you kindly to arrange passage for this woman and her three children on His Majesty's ship, next year. Her domicile is in the parish of St Barthelemy d'Etay, near Chatillon sur Seyne, diocese of Dijon, in Burgundy. I am writing to Mr Lefevre, Director of the Sampson Forges, who knows this family, so that he can advise this woman [...].[34]

The First French Workmen

The first ironworkers at the Forges were all recruited in France, except for one who was from Ireland. In 1730, Francheville first brought over founder François Trébuchet from Brittany and keeper Jean Godard from Normandy; these men seem, however, to have stayed in Trois-Rivières in 1731–32 only to assess the quantity and quality of the region's ore. The seigneur of St Maurice would subsequently engage three Montreal workmen (Labrèche, Jamson and Bellisle) to set up his modest forge and learn the craft, but the experiment was hardly a conclusive one. When the venture was relaunched in 1735, all the ironworkers, as well as the first colliers, would be recruited in France.

When Maurepas decided to send experts to New France to set up an ironworks again, he settled on ironmasters from French iron-making regions renowned for the skills of their workmen: Burgundy and Franche-Comté. As we have seen, this reputation was rooted in the Franche-Comté process of fining in a single hearth, which meant major charcoal savings. The origin of the first ironmasters, Vézin (Champagne) and his assistant Simonet (Burgundy), would therefore largely determine where in France they themselves would recruit their workmen (Table 8.8 and Plate 8.1).

The deeds of indenture of workmen recruited after 1735 have not been found, so we know the names and origin of the first workmen only from what can be gleaned from parish registers or marriage contracts. These traces, all signs of some degree of settlement, tell us which workmen remained in the employ of the Forges for some time; for lack of other sources of information, it is impossible for us to retrace the origin of the other workmen, who would stay only the space of one indenture. Nevertheless, the places of origin as reported, primarily upon marriage, by the more seden-tary workmen tell us the main areas where skilled workmen were recruited in France.

Table 8.8	ORIGINS OF THE FIRST WORKMEN RECRUITED IN FRANCE*	
Western France	Picardy	1
	Normandy	4
	Brittany	1
	Anjou	3
	Poitou	1
	Aunis	2
Eastern France	Champagne	6
	Burgundy	13
	Franche-Comté	9
Central France	Berry	2
	Bourbonnais	1
	Auvergne	1
Region undetermined		7
TOTAL		52

* Figures indicative of men of known origin only.

Based on cases of known origin, Table 8.8 shows that in the French era workmen were recruited from at least 12 different regions. It is worth noting that Western France, which provided most of the colonists for New France, also had major ironmaking regions—such as Normandy, which, along with Ariège and Franche-Comté, was one of the three main sources of recruitment for the French iron industry.[35] We were not, however, able to find occupations reported for most workmen recruited in this part of France, nor indeed for those rarer workmen from the regions of Central France. But we do see very clearly that the neighbouring regions of Eastern France—Champagne, Burgundy and Franche-Comté—were among the main sources of recruitment. The first ironmasters quite simply recruited workmen in their home regions (Plate 8.1). Vézin came from the ironworks at Sionne, in Champagne, and Jacques Simonet, who came to assist Vézin in 1736, was ironmaster at La Bergemant, near Dijon.[36] Simonet, in charge of recruitment for the Compagnie des Forges de Saint-Maurice, would bring "50 men, women and children" back from France in August 1737.[37] Most of those who became known as the "principal workmen," the furnace-men and forgemen, were thus recruited in Burgundy and Franche-Comté (Table 8.9 and Plates 8.2 and 8.3).

Table 8.9

ORIGINS AND CRAFTS OF THE FIRST WORKMEN RECRUITED IN FRANCE[38]

Region of origin*	Furnacemen (founder, keeper and moulder)	Forgemen (finers and hammermen)	Colliers	Other**
Anjou				3
Aunis				2
Auvergne			1	
Berry			2	
Bourbonnais				1
Brittany	1			
Burgundy	1	6	4	2
Champagne			6	
Franche-Comté	4	2		3
Normandy	1			3
Picardy				1
Poitou	1			
Region undetermined		3	1	
Other country		1		

* Figures indicate only the 52 cases where the origin and craft is known.

** Category mainly including workmen of unknown craft, not stated in the parish registers. Includes two helpers, one from Normandy, the other from Burgundy.

Plate 8.1

ORIGIN OF EARLY WORKERS AT THE ST MAURICE FORGES, 1731–60

Sources: Marie-France Fortier, "La structuration sociale du village industriel des forges du Saint-Maurice: étude quantitative et qualitative," Manuscript Report No. 259 (Ottawa: Parks Canada, 1977), pp. 178-82. Hubert Charbonneau and Normand Robert, "The French Origins of the Canadian Population, 1608–1759" in *Historical Atlas of Canada*, Vol. I *From the Beginning to 1800*, ed. R. Cole Harris (Toronto: University of Toronto Press), 1987, Plate 45.
MAP BY ANDRÉE HÉROUX.

The fact that ironmasters Vézin and Simonet maintained their links with their ironworks in France doubtless facilitated the recruitment of workmen for New France. During the summer of 1736, Simonet arrived with four workmen. And subsequently, on authorized trips to France to take care of personal business in 1737, 1739 and 1740, Simonet and Vézin completed the recruitment by bringing, not without some difficulty,[39] workmen back with them.[40]

Under Crown administration, the directors of the ironworks would have some trouble establishing their authority over the workmen, especially the forgemen from Franche-Comté. Thought would be given very early (1742–43) to diversifying the areas from which workmen were recruited, and it was proposed that workmen from the Ardennes be engaged, as they were "more amenable to control." But this was not followed up, quite probably because workmen from that region were not at that time using the *renardière* process.[41]

In a 1743 memorial concerning the Forges, the image of the workmen recruited by the first ironmasters was not a very positive one:

> [...] most workers brought from France by Sieurs Simonnet and Olivier are drunkards, mutinous and independent, and among them are very few who might be called competent.

The author of these lines[42] seemingly knew these workmen well enough to be able to describe their characters and personalities. He was somewhat indulgent towards keeper Belu (Bellisle), whom he considered "quite good"; finer Robichon, whom he also considered a "fair finer"; and Chaillé, hammerman at the lower forge, whom he described as a "good hammerman, loyal and sweet tempered, but become a drunkard these past two years." As to the others, he was scathing. In his eyes, founder Delorme, who had been

Plate 8.2
Interior of a forge near Châtillon-sur-Seine, by Étienne Bouhot (1780–1862).
Two men can be seen fining and hammering.
MUSÉE DES BEAUX-ARTS, MONTBARD, PHOTOGRAPH BY RMN—P. BERNARD.

recruited as a finer, was "absolutely not a competent founder" and "of uncertain temper"; Marchand, the hammerman at the upper forge, was a "poor hammerman," although "quite well behaved"; Lalouette was the finer "who [produced] the least and most inferior iron"; finer Théraux was a "bad character, mutinous and incapable"; the Irishman Ambleton was "quite a good finer, but unruly, restless, unable to get along with the workmen nor them with him"; Dautel, at the lower forge, was "a good finer but brutal drunkard, excessively flighty, and difficult to control," who had had "a very bad influence on Chaillé, his brother-in-law"; Godard *père* was "mutinous, always ready to complain and to incite others to complain"; Godard *fils* would have been "capable of being a good finer," but he was "a drunkard and most dissolute"; Mergé, finally, was also a "good finer, but a drunkard."[43]

These workmen, whom Simonet had nevertheless gone to great lengths to protect the previous year,[44] would for the most part remain employed by the Forges until the end of French rule. Once the first generation of workmen had settled, recruitment problems, due primarily to the backdrop of war that dominated the last two decades of the French regime,[45] would help entrench them. In 1748, in response to a request for manpower from Intendant Hocquart, Maurepas would write:

[...] you are well aware how much it costs to have workers from the ironworks in France where it is even difficult to find any who wish to go to Canada [...].

Plate 8.3
Exterior of a forge near Châtillon-sur-Seine, by Étienne Bouhot (1780–1862). Some workers at the St Maurice Forges, like finer Godard and colliers Aubry and Trotocheau, came from communities in the bailiwick of Châtillon-sur-Seine.
MUSÉE DES BEAUX-ARTS, MONTBARD, PHOTOGRAPH BY RMN—P. BERNARD.

Maurepas would instead suggest that workmen be trained on site. The same year, Hocquart's successor, Intendant Bigot, confirmed the lack of mobility of skilled workmen, who "have long been held back as if by force," as he would write to the Minister of Marine. A number of them wanted to return to France at the end of a summer during which they were hard hit by disease, which left them "unfit for shift work." And the Intendant would stress the "essential need" to bring over 15 new workmen from France to replace those who wished to leave and the "few bad characters who often cause disorder that tends only to interrupt the work."[46] Bigot would then suggest tapping a network, of which Marchand, one of the hammermen at the Forges, had spoken:[47]

> Pierre Deschelotte, hammerman at the Blanc ironworks, near Chatillon sur Seine, a good, respected man who can have workmen found, forgemen, moulders and founders, and all others suitable for the Forges.
>
> This Pierre DesChelotte is the brother-in-law of Marchand, one of our hammermen, who says that this man will come to this country.[48]

It is possible that such networks were used previously, particularly following the departure of Simonet and Vézin in 1741, but it is uncertain whether they always yielded results. In 1750, there were still problems with skilled manpower. Bigot would report that he "was obliged to allow three workmen to go to France for their affairs," adding that "they have promised to come back next year. These are bad characters whom I am forced to employ, for lack of any other."[49]

In 1752, the problem had not yet been resolved. Bigot would still talk of stoppages for lack of workmen. And when one of the best finers died,[50] he wrote:

> [...] they often fall sick, and some are even so old as to be no longer able to work, but we are obliged to employ them and pay them as if they were good.

He would once again ask for four good finers from Burgundy or Franche-Comté, a good keeper who could act as founder, a good sand moulder and a good moulder's assistant.[51] Furthermore, the same year, when engineer Franquet visited, they were still looking for a good ironmaster who could oversee the workmen (see Chapter 1). In other words, problems of this kind did not arise only when recruiting the lesser ranks.

Apparently, half the workmen at the Forges would leave the country, but not until after the British Conquest. The British authorities, however, ordered seven skilled workmen to remain on the job: the founder and his keeper, as well as five forgemen,[52] all apparently from the upper forge. The workmen from the lower forge who, besides being all related to one another, had been described as "drunkards" 17 years earlier, were not kept on. The British kept only the good characters![53]

After the Conquest

The workmen from Burgundy and Franche-Comté kept on "under house arrest" by the British would finally decide to remain after the Conquest. Delorme, described in 1743 as "infirm, with a weak chest [and] like to die," would remain founder until his death in 1775, at the age of 74. As for Bellisle ("Belu"), who was unmarried, we find no trace of him after 1764. The five forgemen, as well as a number of other workmen who remained of their own free will, would leave descendants who would continue to work at the Forges for almost another 100 years. Of course, it is mainly data from the parish registers that reveal the extent to which these working families would remain entrenched after 1764; indeed, their presence is reported in various notarial deeds concerning several of them. We have in fact already seen that we would have to wait until the first official census of 1825 for a partial list of workmen at the Forges. Sixty years after the Conquest, the few families from Burgundy and Franche-Comté, as well as Champagne, were still represented, alongside several others from Canada, Ireland, Scotland and England. We know that the latter families began to arrive in the late 1760s under Pélissier, when some British moulders were hired. Other skilled workmen also joined the work force subsequently, prior to 1800, while in the early 19th century, Mathew Bell announced the hiring of moulders from Great Britain. Unlike the entries in the Catholic parish registers, the all-too-rare entries in the Anglican registers do not show the origins of the workmen recorded, but we do know the names of several of them (Table 6.7).

As John Lambert reported in 1808, they held leading positions in the main departments:

> The workpeople are chiefly French Canadians, a few English only, being employed in making models, and as foremen or principal workmen.[54]

A number of them would establish roots in the country by marrying French-Canadian women. This was the case with moulder John Slicer, whose sons John and Robert would become founder and moulder respectively.[55] Superintendent Zachary Macaulay would also hand on his position to his son Henry. It was really the English and Scottish moulders who established a tradition of moulding at the Forges, since the few moulders of French origin had chosen to leave the country after the Conquest. The old-stock and Canadian families adopted this craft, apparently, only after the British moulders had settled in.[56] As noted by John Lambert, the demand for skilled labour at the Forges after the Conquest was met largely by French-Canadian families and some English. Data from parish registers tell us that the Gilbert, Mailloux, Raymond and Rivard families were among the main new families that took root at the Forges and filled the forgeman and moulder positions; and they appear as such in the first censuses of the 19th century.

MATRIMONIAL ALLIANCES

[...] the large number of your Petitioners
are still the descendants of the above mentioned
persons, and some of them by their wives,
exercising the different trades they have learned
from their fathers, and like them,
hoped to live and die in a state of life which
seemed hereditary in their families.

The Forges workers, 1846[57]

Studies on the early iron industry in Europe, France in particular, show that ironworkers formed very closed, skilled groups.[58] The solidarity within these groups was expressed in two ways: through direct descent, and through marriage. The crafts of founder, moulder and forgeman were, on the one hand, handed down directly from father to son, whereas, on the other hand, families formed a network of alliances through intermarriage of children of fathers with the same craft. The only craft for which we have no indication of family transmission at the Forges is that of founder. Ever since Benjamin Sulte, all historians of the Forges have highlighted this practice of handing down the craft from father to son, as evidenced by the parish registers and census records in particular. The epigraph above shows that the workmen themselves were quick to recall, after 100 years of operation, that they were descended from a long line of craftsmen who had traditionally passed on their technical knowledge from generation to generation. Marie-France Fortier's work (1977 and 1981) and our own (1983 and 1986) have helped to organize the data available to us in this respect, and to establish observable links among the old-stock Forges families. Subsequently, Peter Bischoff (1989) demonstrated how the moulder's craft was passed down within families. He also showed that matrimonial alliances linked the moulders' families, and it is in this light that the words of the workmen themselves in their

1846 petition saying that they were the descendants of the first workmen, "some of them by their wives," must be interpreted. He also noted that family relationships were formed when the moulders from the Forges began to migrate to Montreal in the mid-19th century. These relationships would act as a migration network for the moulders' families, who would first move to other ironworks in the St Maurice Valley, then mainly to Montreal foundries, and to those in Quebec City. On the basis of their family trees and the 1829 employee roll, Bischoff concluded that the moulders' families "were united by a complex network of family relationships around the Terreau families."[59]

Using the same sources, we for our part have established the chain of marriages linking the forgemen's families (see Appendix 17). Most of these workmen were descended from the hammermen and finers from Burgundy and Franche-Comté who were kept on by the British. Several sons of these families would later become moulders. As with the moulders, we were not surprised to find that, aside from the Robichon and Michelin families, the forgemen were all interrelated. Intermarriages among the Terreau, Gilbert[60] and Tassé families, some of which were contracted over two successive generations, formed a nucleus. The links between the Tassé and Terreau families doubtless go back the farthest, because they officially began with the second marriage, in 1772, of the original Terreau (Joseph) and the widow of the original Tassé (Jacques). This marriage between a widower and a widow whose sons were already following in their fathers' footsteps has all the appearance of a symbolic alliance between two well-established families. It is also interesting to see how the Robichon and Michelin families later became part of the chain initiated by the first three families in the last quarter of the 18th century; the former would ally with the Terreau family upon the marriage of Nicolas Robichon (here again a second marriage), at virtually the same time as the latter would seal their alliance with the Tassé family. The only outsider in the group, forgeman John Abbott, would also be linked to the chain, albeit more indirectly and belatedly, through marriages between the Sawyer (of German origin) and Raymond families, the latter being allied to the three families forming the nucleus of the chain. The family relationships binding the workmen at the Forges were thus primarily woven through matrimonial alliances among the various craft families. The industrial community was not characterized, apparently, by marriages between close blood relations. Although this topic has never been the subject of systematic study, the absence in the parish registers of any record of consanguinity dispensations—usually granted in the event of marriages between close relatives—tends to confirm this.

One has, however, to recognize that the web of relationships between the first forgemen of the French era, both through direct descent and through intermarriage, was already extensive in the forges, the lower forge in particular. The two finers, Godard *père* and *fils*, were also related to hammerman Chaillé and finer Dautel through the marriages of their sisters Marie-Anne and Anne, contracted in 1737 and 1739 respectively, shortly after their arrival at the Forges.[61] At the upper forge, Marchand and Michelin were cousins.[62] Furthermore, it is known that master founder Delorme and Marchand, master hammerman at the upper forge, were brothers-in-law; in fact, they would twice marry two sisters, from two different families, in successive marriages.[63] The family relationship between the master founder and this master hammerman probably had something to do with the fact that in 1760 the upper forge team was kept on rather than the lower forge team, which was dominated by a doubtless more close-knit clan (Godard-Chaillé-Dautel), already reputed in 1743 to be more difficult to control (all drunkards). In noting the marriages of Delorme and Marchand, Benjamin Sulte would point out that Marchand "was with Delorme the most noteworthy workman at the Forges."[64]

As for the families that stayed on after the Conquest, alliances began to take shape during the 1770s and 1780s, when the workmen's sons and daughters married. The regularity of subsequent intermarriages seems to represent a matrimonial strategy acknowledged by the workmen 100 years later; and the remarriage of widow Tassé with widower Joseph Terreau mentioned above could be seen as the first outcome of this strategy implemented by the forgemen. Significantly, these marriages among the old-stock families took place just when the company was beginning to hire British workers for its skilled work force. The belated alliance of the Robichon and Michelin families with the other old-stock families could also be seen in this light, since it occurred later, coinciding with a second drive to recruit British workers. We can only point out this coincidence here, since a number of marriages between English-speaking workmen and the daughters of French-speaking workers could, on the contrary, be interpreted as alliances that facilitated the assimilation of foreign workmen.[65]

THE DISPERSAL
OF SKILLED WORKMEN

We have already pointed out that the sons of workmen at the Forges were not always assured of a job once they reached working age. More sons were produced than the company could employ, and this led to emigration by young men aged 20–29, as may be seen from the imbalance in the composition of the population by sex that we observed above concerning this group. From the start, sons trained in their father's craft had to seek jobs elsewhere. The phenomenon is regrettably not well documented, but the marriages of two workmen born at the Forges confirm it. In 1840 and 1845 respectively, two moulders, cousins Louis and Pierre Imbleau, direct descendants of Burgundian workman Luc Imbleau, were said to be "domiciled at Vergennes" at the time of their marriage in Trois-Rivières to two young women from the St Maurice Forges.[66] Located on Lake Champlain, in the state of Vermont, Vergennes had an ironworks that had already been mentioned by Lord Selkirk in 1804 and Lieutenant Baddeley in 1828.[67] It is quite possible that a hiring network existed between the St Maurice Forges and certain American ironworks. Indeed, Lieutenant Baddeley would turn to Edward Grieves, Mathew Bell's son-in-law and agent in Trois-Rivières, for all his information concerning American ironworks near the Canada–U.S. border.[68] Migration of workmen was not in one direction only, since American workers would also be employed at the Batiscan Iron Works in the early 19th century.[69] Canadian ironworks had long been known in the United States and the American entrepreneur Peter Hasenclever had expressed an interest in purchasing the St Maurice Forges after the Conquest (see Chapter 1).

Peter Bischoff has taken pains to demonstrate that some moulders had begun to emigrate from the Forges during the first half of the 19th century. Based on the case of one of the workmen established at Vergennes in the 1840s, Louis Imbleau, he has shown how the moulders from the Forges were gradually drawn to the foundries of Montreal through their family contacts. Louis, the son of moulder Claude Imbleau, with whom he likely apprenticed, practised his father's craft at Vergennes, Vermont, then again at the Forges, at Trois-Rivières, and at the Radnor Forges, before finally settling in Montreal in 1866, where he would belong to the Iron Molders International Union until his death in 1890.[70] If Imbleau was at Vergennes in the 1840s, this was no doubt partly because of the surplus of manpower at the Forges discussed earlier.[71] This was perhaps also the case with his cousin Pierre, and probably other workmen before them. According to Bischoff, an initial migration network developed around Robert Slicer, a former moulder at the Batiscan Iron Works. Following the closure of that ironworks in 1814, Slicer worked in Montreal before being employed at the St Maurice Forges. On returning to Montreal around 1836–37, he took his son-in-law and his sons with him. Later, in the 1840s, other moulders related to the Slicer, Terreau and Mailloux families would in turn migrate to Montreal, return for a time to the Forges, and then go back once again to Montreal.[72] It was at this time, during which the number of foundries rose substantially, especially in Lower Canada (see Chapter 6), that the migration of other families would begin to Quebec (Terreau family) and Joliette (Imbleau family), where they would open their own foundries.[73] Trois-Rivières, where Mathew Bell was already operating a foundry at the turn of the 19th century,[74] would also remain a natural destination for workmen from the Forges. A number of them would found their own establishments

▼

there. As we saw in Chapter 6, Louis Dupuis, a former moulder from the Forges, set up a foundry in Trois-Rivières in 1843, and was one of the founders of the L'Islet Forges in 1856. In 1861, Dupuis employed 24 workmen at his Trois-Rivières foundry.[75]

Nevertheless, if Louis Imbleau was in Trois-Rivières in 1852, then at the Radnor Forges in 1861 along with 13 other moulders from the Forges, this was no doubt partly because of the unpleasant atmosphere at St Maurice and because of the changes that the company's first private owners had been implementing since 1846. In legal proceedings between John Porter & Company and James Ferrier, it was emphasized that Ferrier's irresponsible administration had led to the departure of several workmen between 1847 and 1851.[76] Similarly, the closure of the establishment between 1858 and 1862 led to the exodus, primarily to Montreal, of moulders from the Forges, and the same was true of the moulders from the Radnor Forges after that establishment went bankrupt in 1866. In the 1871 census, Bischoff identified 39 moulders originally from the Forges who had settled in Montreal, where they made up a quarter of the French-Canadian moulders living there. A large number of them were at that time employed by John McDougall & Company, the car wheel foundry which took virtually all the pig iron produced at the St Maurice Forges. The brief foray into making car wheels by chill casting at the St Maurice Forges, between 1854 and 1858, had introduced them to this new technique.[77] The Day, Terreau & Deblois foundry also employed a large number of these moulders from the Forges. It is, moreover, worth noting that, as Bischoff shows, all these families continued to stick close together and intermarry "despite the flourishing marriage market in Montreal." Bischoff points out that between 1861 and 1881, 5 out of 17 marriages were contracted between families of moulders

originally from the Forges. Like some ethnic communities, the industrial community of the St Maurice Forges was reproduced in microcosm in the working-class neighbourhood of St Antoine, along the Lachine Canal. The close family ties among the workmen helped feed a grapevine that "functioned to some extent as a job placement system."

Once part of a much larger working world, the moulders from the Forges wasted no time becoming active in the trade unions. They would first join the Iron Molders Union of America and then, abandoning Local 21 of this British-dominated union for a while, a number of them would become representatives of the moulders' craft within the *Grande association des corps de métier de Montréal*, largely consisting of French-Canadian workmen from 26 trades. In the 1870s, they rejoined Local 21 again, taking part in the major labour struggles of the period alongside workers of every origin. The influence of this union would also be felt in Trois-Rivières, and in Quebec City, where in 1880 a Terreau would be treasurer of Local 176.[78] Forced to leave the establishment where they had been born and trained, the moulders from the Forges set off to establish the family tradition of their craft elsewhere, thus prolonging in an urban environment the history of the first industrial community in Canada.

HANDING DOWN THE CRAFT

Conditions of apprenticeship at the Forges are not well documented. It is only after the fact that we can observe that the craft was indeed passed on by apprenticeship within families when we see that sons succeeded their fathers in the same craft. Everything indicates that it was the family that had the most control over transmission of technical knowledge. Apparently the first French workmen were little inclined to train apprentices who did not belong to their families, and since a good number of them did not marry until after they arrived at the Forges, their sons were still too young in the 1740s and 1750s to work alongside their fathers, or to succeed them in their jobs. As we have seen, in 1748 the Minister suggested to Bigot that workmen be trained on site because it was difficult to recruit any in France, showing how little control the directors of the Forges had over the assimilation of outsiders as apprentices into the body of workmen. Nevertheless, an attempt was made from 1740 onwards to set up apprenticeships—with the forgemen in particular—no doubt with a view to averting work stoppages resulting from the illness of the incumbent workmen.[79] That year, the upper forge hands, Marchand, Michelin, Terreau and Ambleton, undertook to train the soldier Pierre Vilard, *dit* Saint-Mexant, as a finer within one year, for the sum of 200 *livres*, to be split among them. But according to the testimony of lower forge hammerman Chaillé two years later, Vilard was apprenticed for only one month, although his masters were quite pleased with him.[80] Soldiers would also be employed as day labourers or helpers. In 1750, Jonquière added two companies of soldiers in Trois-Rivières, in order to be able to provide the establishment with day labourers. Bigot reported that among them there were "even two or three journeymen blacksmiths" who worked at the Forges but who, as he added, "had nothing in common

with furnacemen."[81] Passing through the Forges two years later, engineer Franquet, who would replace engineer Chaussegros de Léry in New France,[82] stressed the acute labour problems at that time, when both the *habitants* and the soldiers who were to lay in supplies of raw materials had to be dragged kicking and screaming to work:

> [...] workmen who have to be drawn from the country or the Trois-Rivières garrison at the height of the work, the former resist going there on the pretext that they have to till their land, violence is sometimes used to force them, whence it happens that they prefer to abandon the township to go and settle elsewhere rather than submit to what is demanded of them. Then they resort to the soldiers, but they, feeling the need there is for them, come only for high rates, which they are refused, so the work languishes.

Franquet also deplored the independence of the skilled workers, who "are generally paid exorbitant rates owing to their scarcity [and] are all given lodging, heated and conveyed at the King's expense."

Emphasizing the need to have new workmen sent from France, Franquet observed that the forge workmen, "on the pretext that the term of their indenture has expired lay down the law for the work."

It is only later, in the late 18th century and above all the first quarter of the 19th century, under Mathew Bell, that we find apprenticeship contracts with master workmen at the Forges. But apparently these apprentices would not become a part of the regular work force at the Forges, since few of them genuinely settled there. Rather, it would seem that these workmen remained at the St Maurice Forges only for the duration of their apprenticeships, since for a number of years, the Forges were the only establishment in Canada, aside from the Batiscan Iron Works, that could train ironworkers. The fact that several apprenticeship indentures were signed with the Forge superintendent rather than with a designated master craftsman suggests that the establishment had become a training site. Deeds of indenture found[83] stipulate an apprenticeship of two to four years, during which time the apprentices were given room and board by the master, and paid a monthly wage of 15–20 shillings.

In the early 19th century, workmen from the St Maurice Forges were taken on at the Batiscan Iron Works. The deed of indenture for one of them, a descendant of an old-stock Burgundian family, may be indicative of the control exercised by these workmen over their working conditions at the St Maurice Forges. The terms of Pierre Terreau's indenture appear to guarantee him conditions that have all the earmarks of privileges, as engineer Franquet had implied 50 years earlier, when he said of the Forges workmen that they "lay down the law for the work." Under the terms of this deed of indenture, dated 1801, hammerman Terreau, who was to be accommodated at the Batiscan Iron Works with his family, demanded exclusive use of a forge fire 12 hours out of 24, because he was to be paid a piecework rate (per thousandweight of iron). And it was stipulated in the deed that his children could be trained as moulders or hammermen. Indeed, Terreau undertook to teach forging himself.[84]

The passing down of the craft within families is therefore a verifiable fact, but this does not mean it always went smoothly. A review of the records of the families of workmen at the Forges reveals that the eldest son was first in line to succeed his father or to work alongside him; he was often named after his father. Once the eldest son was suited, replacing the father upon his retirement could be a source of conflict, not to mention the father's own reluctance to be replaced. A rare notarial deed, dated 1820, tells us of the terms and conditions of succession to a job. Through this deed, Nicolas Robichon and his two sons, Nicolas and André, stipulated that a life annuity of $5 per month would be paid to the father for bequeathing to his son André his position as hammerman at the St Maurice Forges. André, who held the position, had to pay $3 per month, and Nicolas, who was already on the job, $2 per month. By signing this agreement, Robichon *père* renounced proceedings he had instituted against his sons. This agreement may well be highly indicative of the nature of the working environment at the Forges, but regrettably it is the only documented case of its kind. It could be a unique case, specific to the Robichon family, but such a procedure is similar to the *inter vivos* transactions current among the *habitants*, under the terms of which the father was guaranteed means of subsistence through clauses placing an obligation on his sons. This gift of a skilled position in exchange for a life annuity is rather eloquent testimony to the fact that at the time hiring was still a family affair at the Forges.

INDISCIPLINE, TROUBLE AND TENSIONS

We have seen in Chapter 5 that work stoppages were frequent, particularly at the forges. Many of these stoppages were caused by mechanical failure, but some of them were attributable to illness or indiscipline. We have also seen that several forgemen were considered drunkards during the French era. All these clues suggest that the Forges were a work environment where sturdy, uncouth men worked side by side and in confrontation, doing a hard job in extremely difficult, even dangerous conditions.

The Turbulent Early Years

It should be said at the outset that the atmosphere of chaos that prevailed on the post during construction of the Forges in 1736 and 1737, and the lack of co-ordination in hiring in particular, did nothing to institute sound corporate management.

The establishment started out with a serious management problem that would undermine the ironmasters' authority all through the French era. During the early years, the authorities would resort to ordinances issued by the Intendant to force the workmen to obey the clerks and ironmasters.[85] Since the skilled workmen had arrived too soon in an establishment that was as yet only a construction site, an attempt was made to keep them occupied, but without success, and from the start there were fears that some of them might desert. This fear of desertion by workmen hired at vast expense in France was no doubt warranted also by the surplus of forgemen, since Vézin had not managed to install all the planned equipment in the single forge he had built. In 1737, the master founder himself fled to Montreal with the apparent intention of going to New England, but a detachment of soldiers was sent to bring him back to St Maurice. The publication of the ordinance in September that year, prohibiting workers from leaving the Forges without permission, or the colony without obtaining leave from the Intendant, subject to a heavy fine of 200 *livres*, should be interpreted in light of that incident. Anyone caught deserting to the British colonies would be fined 500 *livres* and be subject to corporal punishment.[86]

Debauchery and Drunkenness

Forges allowed taverns to prosper.
Ironworkers were reckoned to be drunkards.[87]

Visiting the Forges in 1752, engineer Franquet was subjected to a ritual practice as he left the various workshops:

> [...] in each department of the forges [the blast furnace and the two forges], the workmen observed the old ceremony of brushing a stranger's boots; in return they expect some money to buy liquor to drink to the visitor's health.

Franquet implies that this ritual—which shows that the Forges hands drank a great deal, especially at work—was common practice in France in the ironmaking world.[88] The sustained physical effort made by the workmen and the intense heat around the furnace and chaferies could not fail to provoke an unquenchable thirst. Woronoff, quoted in the epigraph, puts the consumption of wine in French ironworks of the period at 3 litres a day per man. He also points out that "many circumstances—feast days, success at work, reward from a boss or reward from a client—[were] a pretext for drinking." He cites an ironmaster from Franche-Comté who wrote that the workers were "daily [...] besotted with wine."[89] The author of a memorial on the St Maurice Forges echoes this, reporting the viewpoint of ironmaster Vézin, who said "drink excites debauchery and they drink daily."[90] Since most forgemen at St Maurice came from the wine-growing regions of Burgundy and Franche-Comté, it is hardly

Order to workmen, labourers and other employees to obey Sieurs Olivier, Simonet and others charged with their orders [...].

12 February 1739

surprising that they were great wine drinkers. We saw earlier that complaints were made about this, and that, for this reason alone, forgemen from the Ardennes would have been preferred, who were less skilled but also less inclined to inebriation. Drinking on the job was moreover implicitly recognized: Allowances of wine were granted to the workmen, and this form of bonus was provided for in their contracts.[91] Contradicting himself somewhat, ironmaster Vézin even considered drinking to be an incentive to work. To justify his largesse towards his workmen, he explained:

> That it was impossible during those years not to make a great expenditure on spirits and wine to engage workers for work, in particular in 1737 and 1738 [...] and that this expenditure was indispensable for engaging workmen to achieve more in their work, which is not difficult for him to prove.[92]

Even Cugnet, the ironmaster's chief detractor, recognized the necessity of distributing this "wine which the forgemen cannot do without and we are obliged to give them."[93] He proposed that Trois-Rivières merchants be forbidden to sell retail drinks to Forges workmen, but that they be allowed to "sell it to them in casks, that is, hogsheads, half hogsheads or barrels to be taken by them to St Maurice."[94] After the bankruptcy of Cugnet et Compagnie, Intendant Hocquart would for his part instruct the trustee Estèbe to limit the distribution of wine and spirits to the workmen, specifying that he "may nevertheless, if the workmen perform their duties well, occasionally provide them with the means of making merry," and adding that "the important point is that they be kept in complete subordination." But this is where the entire problem had lain: At a time when his authority was being undermined from all sides, ironmaster Vézin had been inclined to too much complacency towards his workmen, who had been prone to drink "extraordinary quantities of wine and spirits," leading to drunkenness. Ordinances would then be published to counter this drunkenness—a source of debauchery and disorder—and to attempt to bring these workmen back to order.[95]

Thus, in 1740, an ordinance would "prohibit the keeping of a tavern at the Forges." It was specified in the ordinance that workmen "are selling wine and spirits, which leads to disorder" and it is wished to "prevent trade in spirits with the savages," failing which the offenders will be fined 100 *livres* and undergo corporal punishment. But the problem would persist, since 10 years later another ordinance had to be issued to prevent the sale of alcohol.[96] Despite all these cases of abuse reported during the French era, there were apparently no work accidents or dismissals associated with drinking by workmen. We saw in Chapter 5 that the productivity of the "worst drunkards," the forgemen at the lower forge, does not appear to have been affected, even though the new British authorities did not retain their services.

After the French era, there was no more talk of a chronic drunkenness problem, but the workmen continued to be supplied with alcohol. Barely three weeks after receiving the order from the British to retain the seven main workmen, Forges director Courval received a barrel of **tafia** for them.[97] For the first year (1760–61) of production after the Conquest, in a Forges statement of account, the provision of 98 gallons of rum for the workmen was recorded. And, shortly thereafter, workers of British origin would have a monthly supply of whisky included in their indentures. In 1784, the indebtedness of a moulder from drinking too much rum suggests that consumption of alcohol continued to be current.[98] Later, under Mathew Bell, the environment was apparently "made more salubrious." In the best paternalistic spirit, Bell would describe the Forges in 1827 as:

> [...] my quiet peaceable village (where a man can scarcely take an extra glass of grog without my permission or knowledge).[99]

Much later, an incident reported by Dollard Dubé appears to indicate that bonuses in alcohol—granted with more moderation than before—were still standard practice. Robert McDougall, it is related, one evening gave some "strong drink" to the furnacemen; if one zealous fellow had not saved the run-out at the last minute, there could have been an accident.[100] Subsequently, the excesses of certain workmen would figure in stories or legends, and this could mean that drunkenness had by then become the exception. However this may be, the work of these men would remain just as difficult and dehydrating, so they would no doubt continue to be just as thirsty as before.

Violence and Tragedy

While we cannot link them directly to excessive drinking, a number of tragic events marked the establishment's early years, which were particularly tumultuous. They remind us of the very unhealthy atmosphere of that time at the St Maurice Forges. Until the upper forge was actually put into operation in October 1739, "idle" workmen had to be kept, who were difficult to discipline.[101] Earlier, in February, a brawl had broken out between finers Ambleton and Terreau; locksmith Beaupré, who had intervened, sustained severe head injuries. Ambleton was sentenced under the terms of an ordinance and police regulation imposing a 10 *livres* fine on brawlers (payable to the Forges chapel) and requiring payment of the expenses and wages of injured parties who had been prevented from working. Four days later, Dautel and Marchand refused to change chaferies so work could be carried out on the wheelrace of the other forge chafery. Their disobedience, "accompanied by seditious talk and whispering" earned them a stay in the Trois-Rivières jail, along with a fine of 5 *livres* per day, plus another of 3 *livres*.[102] When the upper forge opened in October of the same year, the forgemen were finally split into two teams.

The same month, ironmasters Vézin and Simonet left the colony to see to business in France; Simonet's son Jean-Baptiste then acted as ironmaster. Advantage was taken of their absence to reduce their responsibilities. While the plotting was going on at the top, disorder reigned among the workmen. Tensions remained strong, particularly at the lower forge, where a tragedy occurred on 19 October between 11 o'clock and midday, just a few days after the ironmasters departed. Locksmith Beaupré, the very man who had intervened between Ambleton and Terreau, was attacked at the lower forge dam, then struck fatally with a stick by helper Jean Brissard, *dit* Saint-Jean, a soldier from Cournoyer's company. The motive of the crime is not known. Condemned to hang in absentia, the murderer, who had fled, was never found. Three days earlier, Beaupré's wife had tragically given birth to a stillborn child, François, a common occurence at the time. Midwife Marguerite Banliac, who had come that day to "lay out the child," witnessed his father's murder.[103] Tragic as they are, such events point to the tensions and violence that were common in the work environment, but did not apparently affect the pace of work, since the upper and lower forges were running on 19, 20 and 21 October. A nine-day work stoppage began on 22 October at both forges, but this was attributable to a shortage of charcoal.[104]

Another murder occurred six years later, revealing this time the tensions between a father and his son, moulders Louis and Étienne Cantenet. The murder took place at the upper forge dam, where Étienne Cantenet mortally wounded Pierre Guyon, a helper.[105] The tragedy occurred on 19 September 1745, at 8 pm, when Cantenet *fils* arrived home. His father was there with some friends, Pierre Guyon among them, apparently carousing and singing. Étienne, annoyed, launched himself at his father and grabbed him by the throat. The father's friends intervened and with the father set about thrashing the son. Guyon even struck him with an iron bar. The brawl spilled over outside, onto the dam, in front of neighbours, who were drawn by the noise. That is when Cantenet "fell violently and like a roaring lion upon the said Pierre Guyon even despite those who wished to stop him [...] he threw the said Guyon to the ground, and struck him several blows; the said Cantenet was able to give no more [...]." Guyon, crying murder, was fatally stabbed in the face and stomach. He died the next day at the Hôtel Dieu in Trois-Rivières. Cantenet managed to flee without trace, and he too was condemned to hang in absentia.[106]

These two murders, which point to the tensions that pervaded the working world, must be placed against the backdrop of a particularly unstable, troubled period, when the establishment was being set up and getting under way. As we have seen, the company at that time was experiencing serious labour shortages, among both apprentices and unskilled workmen. The employment of Jean Brissard, who killed locksmith Beaupré in 1739, and of Pierre Vilard, a witness to that murder, both of them soldiers in Cournoyer's company, attests to the fact that the Forges had to resort very early on to the services of the Trois-Rivières garrison to attempt to meet their manpower needs.[107] The integration of these "outsiders" into a working world centred on a family craft tradition no doubt caused tension. Not to mention the fact that family clans were still a source of conflict, and had not yet led to the tight-knit community that would truly take shape only after 1760. The establishment's first two decades were therefore marked by indiscipline, disorder and tension, within a working-class community that had yet to find its team spirit. Early in the year of the second murder, 1745, an ordinance aimed at workmen at the Forges had denounced disorder, absences and scandals, thus sharply highlighting the discipline problems arising in that work environment.[108] So that is the context in which the second murder occurred, the scenario of which we are better acquainted with. This event shows that the family circle itself was not free from tensions. Starting with a quarrel between a father and his son, it would lead to the murder of an outsider. Étienne Cantenet's attack on his father tells us a great deal about the tense relationships between these two rare moulders at the Forges during that period, one of whom was the other's apprentice.[109] Furthermore, the Robichon succession, which we talked about earlier, also points to tensions within the same family, although these did not lead to such extreme violence.

Criminality

There is every reason to believe that as soon as the process of passing down the craft within the family was established and matrimonial alliances had begun, the homogeneity of the community would tend to be conducive to the birth of a team spirit that would lead to a degree of social peace. The subsequent history of the industrial community yields no other cases of such violent assault. A preliminary study of criminality at the Forges between 1790 and 1876 provides no evidence that the working population was violent. The judicial data extant bear no witness to the involvement of workmen from the Forges in any serious assault leading to severe injury, or even death.[110] At most, a number of inhabitants of the Forges would be accused of common assault arising from arguments or brawls; very few of them committed breaches of the peace or assaulted peace officers on duty. One incident involving a former workman at the Forges is, nevertheless, worthy of mention. Antoine Michelin, aged 60, forgeman, was arrested on 30 June 1871 in Trois-Rivières for "vagrancy without means of subsistence," and was then released; on 12 July 1872, this "former forgeman" at the Forges would be imprisoned for "senile dementia."[111] The case is particularly pathetic, since it shows the tragic end of a descendant of one of the original forgemen, finer Pierre Michelin. And at the same time it is a reminder that the abandonment of bar iron production at the Forges caused the deskilling of these forgemen, who had handed down a formerly prestigious craft for more than 125 years. The vagrancy of this workman from an old-stock Forges family already foreshadows the end of the St Maurice Forges, which came about 10 years later.

There was no great incidence of theft or fraud either. There were some cases of petty theft of tools,[112] cast or wrought iron parts, or even produce from the garden of the Grande Maison over the years. These cases were, however, few and far between, and the lack of repeat offences by the individuals involved appears to demonstrate that the community at the Forges was essentially law abiding.

Desertions and Fraud

Most of the offences committed by workmen involved desertion from work. Statements of account from the French period reveal the debts of certain workmen, described as "deserters" or "runaways."[113] In the 19th century, in Mathew Bell's time and in the years following the sale of the Forges in 1846, permanent labourers were accused of desertion. These offences were considered cases of fraud, this being the charge levelled against employees who deserted work once they had received wage advances or were on contract. Temporary desertion—by colliers and miners working in the woods, in particular—considered to be disobedience, was sometimes also punished.[114] A number of charges concerned somewhat comical situations. For instance, a complaint was lodged against a carter from the Forges for driving a sleigh whose harness did not have enough bells, and speeding by young carters from the Forges—on Sundays!—would be denounced.[115]

A number of directors of the Forges would also commit more serious fraud. Simonet *fils*—the ironmaster who, following the 1741 bankruptcy, would continue to direct the workmen during Estèbe's temporary stewardship—was suspected by Estèbe of misappropriating 1,720 pounds of iron and stealing the sum of 660 *livres* from his cabinet. It was decided not to prosecute him, because of the disruption this would have caused among the workmen under his influence, but above all because he was the stepson of Madame Hertel de Cournoyer, who had married his father Jacques in Trois-Rivières in 1738.[116] The affair was therefore hushed up, and Simonet quietly sent back to France.[117] We have also seen that when Christophe Pélissier fled with the Americans in 1776, he made off with £2,000 of the proceeds from the Forges.

SOCIAL LIFE

Christenings

Christenings, weddings and funerals were important events in the social life of the community. The Forges register for the years 1740–64 allows us to identify individuals who acted as witnesses at such events. Christenings, the majority of the events recorded in this register, show us that workmen were present when anyone's children, without distinction of rank within the workers' hierarchy, were baptized. Delorme, the master founder, would thus attend the christening of a finer's child as he would a carter's or helper's child, despite an absence of any clear family relationship. The records show that the directors of the Forges and their foremen and clerks also joined in socially. Ironmaster Cressé, merchant Perreault and, in particular, foreman Champagne and clerk Milot would many times attend the christenings of their workmen's children, and workers would attend those of their employers' children. Director Jean Urbain Martel de Belleville attended the christening of hammerman Chaillé's son in 1744—the child was even given the names of "Claude" and "Urbain"—and the christening of carter Lacombe's twins in 1748, in the company of inspector Rouville and ironmaster Cressé.[118]

Several baptismal records also indicate the presence of workmen from the Forges at the christening of children of "Algonquin savages," who were doubtless passing through the establishment and may perhaps have performed some work there.[119] Estèbe's accounts for 1742–43 in fact show payments made to "savages" for cartage and working on forest roads.[120]

After the French period, it should be remembered, intermarriage would make the community increasingly homogeneous and close-knit. Furthermore, there is every indication that the British workmen were successfully assimilated, even though a good number of them held management positions throughout the plant. According to Laterrière, who spent five years (1775–79) at the Forges as inspector and director, it was a peaceful hamlet where everyone got along together. Indeed, the working population was very well supervised, thanks to the paternalistic way in which the community was run:

> It was the rule that no workman took anyone home with him without coming to the office to advise them and request permission; so that nothing indecent and no accidents happened without our being aware of it; we were even informed of their balls, dances and festivities.[121]

Laterrière claims to have spent five happy years among "good people" who had not lost their taste for celebrating. It should be pointed out, however, that Laterrière found love there with his boss Pélissier's young wife, Marie-Catherine Delzène—quite an unusual love story![122] At that time, the directors began to hold dinners, balls and gaming evenings at the Grande Maison, when they would play host to visitors from outside, local notables and officers from the Trois-Rivières garrison, sometimes to the accompaniment of the regimental band. In 1775, the American invaders were also greeted with open arms, with the consequences we have seen for Pélissier, who had to go into exile, and for Laterrière, who later, in 1779, paid for his collaboration with a prison sentence of more than three years.[123] In the meantime, following Pélissier's forced withdrawal with his American friends, Laterrière spent some happy years with Marie-Catherine, who gave him a daughter, Dorothée, in 1778.[124] He would relate that he had conducted memorable campaigns there, in an atmosphere of good

relations with the workmen. His attachment to the establishment would prompt him to spend three years of his imprisonment at Quebec building a working model of the Forges (a "machine") and another of the fortifications of Quebec.[125]

In his time, Mathew Bell would continue, as we have seen, to maintain paternalistic relations with the workmen. He also gave receptions and organized hunts, which were denounced by Kimber, the member of the legislature, in 1832, at a time when Bell was under attack from the citizens of Trois-Rivières (see Chapter 2).

Religious Observance and Feast Days

Religious sacraments, rites and services were important events in the social calendar. In the French period, the community was served by a chaplain who was given board and lodging at company expense at the Grande Maison. A chapel accommodating a dozen people had been set up in the house, as well, where the faithful received the sacraments, and the various rites of the Catholic Church were celebrated. Certain milestones in the working year were marked by a celebration. Father Augustin would thus be given a special honorarium of 15 *sols* on 2 July 1743 for celebrating a "mass for the success of the furnace."[126] The feast days of St Éloi (1 December) and the Translation,[127] for the forgemen, and St Thibault (8 July), for the colliers, were also solemnly celebrated, sometimes even to excess, if we are to believe Bishop Pontbriand of Quebec who issued an ordinance, entered in the Forges register in July 1755, denouncing the "scandalous excesses, far from sanctifying the days," and threatening to prohibit any "special ceremonies" on these feast days.[128] The workmen were merely perpetuating the traditional celebrations current in their home regions of

▼

France, which would moreover long continue to be part of the working man's customs in this country.[129] By giving bonuses to the workmen on these feast days,[130] the company to some extent encouraged these practices, although they sometimes led to disruption and lengthy production stoppages. We have identified work stoppages lasting two days in 1739 and three days in 1740 in both forges[131] (see Appendix 10). Work also stopped at Christmas (three days), New Year's (one day), Epiphany (6 January, one day), and Candlemas (2 February, one day). The weekly work schedule from 1739 to 1741 does not mention any other feast days during the year when the workmen's patron saints were celebrated.

The tiny chapel in the Grande Maison soon became too small. At Sunday Mass, more than 100 people had to squeeze into the corridor leading to the small room.[132] A larger church—measuring 13 m x 9.7 m—would subsequently be built, as a 1760 inventory attests.[133]

Our data concerning the religious observance of the workmen at the Forges in later years are more fragmentary. The Catholic clergy apparently remained very close to the community, which would be managed a Protestant elite for the rest of its history. In opposing a marriage, Vicar General Saint-Onge would moreover remind his flock of his authority. In 1772, during the early years of the Pélissier administration, Jacques Arnaux, a Catholic filler, whom the Vicar General had refused to allow to marry a widow from the Forges, committed the affront of having his union blessed by a Protestant minister. The Vicar General apparently managed to separate the spouses and refused them the sacraments. Having been reviled by the Vicar General and threatened with damnation if he were not remarried by a Catholic priest, Arnaux asked Pélissier to check with the Lieutenant Governor whether his marriage was valid. We know of this incident from a letter from Pélissier to his partner George Allsopp, but regrettably we know nothing of the sequel. Pélissier then suggested to Allsopp that this case be used to try to get Saint-Onge to stop insulting the Protestant clergy.[134]

First-hand accounts garnered by Dollard Dubé from the last workmen lead us to believe that, in his time, Mathew Bell had a priest come to see to the good behaviour of the workmen. Indeed, Bell described them in 1827 as a "peaceable quiet race of people."[135] Nevertheless, the report from a priest sent by the parish priest at Trois-Rivières to organize a retreat lasting several days for the workmen during the 1840s suggests that their religious observance was not so regular:

> Living three leagues from the church, burdened by continual, extraordinary work, they found themselves in urgent need of religious succour [...] The masters of the Forges for their part were so good as to grant the workmen half an hour longer to give them the opportunity to attend the religious rites. I said Mass early in the morning and in the evening I conducted a service at half past seven. I then heard the men's confessions until midnight, and sometimes even until one or two o'clock in the morning; I heard the women's confessions during the daytime.

And he added in admiring tones:

> [...] when it was granted to me upon my return to see so many people, of whom a large number for several years had been far removed from the sacraments, come to Holy Communion with admirable reverence and faith, my heart, I may say, overflowed with joy.

Following a week-long retreat, where he saw almost all of them receive communion and even "pledge themselves to temperance," the priest was escorted back to Trois-Rivières in a cavalcade of close to 40 carts by workmen wishing to thank the Vicar General for having given them the opportunity to make a retreat.[136] The docility shown by this community, which had been so turbulent 100 years before, is something of a surprise. A traveller visiting the Forges in 1847 praised the "courtesy" of the people he met. In a letter published in the Trois-Rivières *Gazette*, he protested against the plan of the owners at that time to fence in the site of the Forges.[137] The following year, fences were mentioned again when engineer Nicolas-Edmond Lacroix, who lived at the Forges, complained of the noise made by the workmen "(celebrating) *La Guignolée*, at the risk of breaking down the fences."[138] To judge by these accounts, Mathew Bell's long reign helped make the industrial community a village that was more closed-in than before, but which for all that was not always peaceful.

LIVING CONDITIONS AND STANDARDS OF LIVING

Pay

We encounter the same limitations regarding the data available on workmen's pay as in studying the other aspects of the establishment's history. The data on wages are partial and episodic and do not allow us to create statistical series that would show how wages evolved over 150 years (Table 8.10). As with other aspects, the first years of operation are the best documented, as well as the four years following the Conquest (1760–64), which show few differences with respect to wages. Subsequently, we have to go by deeds of indenture, primarily for outside workmen, on eyewitness accounts of travellers and on sparse information drawn from the correspondence of various administrations. This material does, however, tell us about the many methods of payment that were current in an ironworks that, when all is said and done, did not change much in this regard throughout most of its history.

The workmen's wages and method of payment varied according to their position in the organization. Except for the final years of production, we find no single form of monthly or annual wages applied to the work force as a whole. At each level in the organization, the method of payment was geared to the workmen's specific tasks. Such work conditions are quite revealing of the autonomy enjoyed by the workmen, who had full control over the progress of their work and their productivity. This differential pay complicated more than a little the way these workmen were managed, and prompted a number of French experts in particular to say that the ironmasters were merely "their workmen's bankers."[139] A look at the list of account books still being used in the 1850s (see Appendix 5) shows that each category of workmen had its own accounting system.

These various methods of payment applied to both outside and inside workmen, although this distinction tended to blur in the 19th century, in the case of carters, colliers and miners. Daily, monthly or annual wages, payment by the piece or by the load, by the job or by weight, or even in kind, as well as bonuses, were all used as methods of payment. And, as far as the inside workmen were concerned, two or three of these methods of payment together were sometimes combined in the course of a single year.

[...] working day and night, without even excepting holidays and Sundays according to the rules established at the said post of the Forges.

Table 8.10

KNOWN WAGES OF WORKMEN, BY DEPARTMENT, ST MAURICE FORGES, 1740–1874

DEPART-MENT	WORKMAN	1740 (in livres)	Method	1742 (in livres)	Method	1764 (in l s d)	Method	1804-06 (in £)	Method	1819 (in £)	Method	1827 (in £)	Method	1856-57 (in £)	Method	1874 (in $)	Method
INSIDE WORKMEN																	
Blast furnace	Founder	58-06-08	month	58-06-08	month	02-02-00	thousand-weight			160-00-00	year (1825)	05-00-00	month			28	month
		00-25-00	thousand-weight	00-25-00	thousand-weight	58-08-00	month					10-00-00	month				
						05-00-00	stove										
	Keeper			30-00-00	month	45-00-00	month			00-04-00	day (room)					22	month
				45-00-00	month					00-05-00	day (room)						
	Filler			35-00-00	month	40-00-00	month			03-15-00	month (room)						
						00-30-00	day										
	Charger			45-00-00	month			40-00-00	year								
								05-14-00	month								
	Moulder	100-00-00	month	30-00-00	thousand-weight			40-00-00	year	06-10-00	month	05-00-00	month				
				40-00-00	month							10-00-00	month				
	Moulder's assistant	25-00-00	month														
	Sand moulder																
	Helper					40-00-00	month										
Forges	Hammerman	87-10-00	month	75-00-00	month	75-00-00	month	00-03-00	cwt of iron	00-35-00	thousand-weight (1801)	05-00-00	month				
								07-00-00	month	07-10-00	annual bonus	10-00-00	month				
								08-00-00	month								
	Finer	60-00-00	month	58-06-08	month	22-00-00	thousand-weight	00-03-00	cwt of iron			05-00-00	month				
				12-00-00	thousand-weight	58-08-00	month	07-00-00	month			10-00-00	month				
								08-00-00	month								
	Charger																
	Helper	33-06-08	month	30-00-00	month	40-00-00	month										
Smithies	Blacksmith									04-06-00	month (room)	05-00-00	month				
												10-00-00	month				
	Edge-tool maker																
Workshops	Carpenter	83-06-08	month			100-00-00	month			04-06-00	month (room)						
	Carpenter's assistant	25-00-00	month			60-00-00	month										
	Joiner			01-10-00	day	90-00-00	month			1612	month (room)						
				03-10-00	day												
	Sawyer																
	Saddler																
	Wheelwright			1,000-00-00	year												
				350-00-00	year												
	Mason			01-10-00	day												
				02-10-00	day												
OUTSIDE WORKMEN																	
Charcoal pits	Collier			00-09-00	pipe	01-05-00	cord (dressed, feuilled and charred)		day	00-02-06	binne (1806)			binne	00-02-06		
				00-10-00	pipe	02-10-00	binne	00-02-00	binne (1806)	01-00-00	month (board)						
	Pit setter			20-00-00	100 cords	00-07-00	cord	00-00-03	cord (1806)	00-03-00	cord			00-00-04	cord		
				25-00-00	100 cords									4½d			
	Feuiller			12-00-00	100 cords	00-07-00	cord	00-00-02	cord (1806)	00-02-00	cord			00-00-04	cord		
	Garde-feu					40-00-00	month			02-00-00	month			00-02-00	day		
Mines	Miner													00-04-06	780/lb clean	0.3	barrel
	Ore washer													00-04-06	1000/lb dirty		
Forest	Woodcutter			01-10-00	day	02-00-00	cord	01-08-00	cord					00-03-00	cord		
				02-00-00	day	02-10-00	cord										
				01-00-00	cord									00-01-03	day		
														00-02-06	day		
Vehicles and scows	Carter			30-00-00	month	42-00-00	month	02-15-00	month (1805) (room, board, horses and cart supplied)	03-00-00	month (room, cart supplied)						
				01-10-00	day	06-08-00	day (board, team of horses plus fodder)	00-05-00	6 cwt of ore	03-10-00	month (room, cart supplied)						
						03-10-00	day										
						01-10-00	pipe ore										
						00-04-00	pipe charcoal										
	Scowman					02-00-00	day										
	Bateauman																
	Ferryman																
	Road maker					40-00-00	month										
Day labourers					01-10-00	day	02-00-00	day			03-00-00	month	00-02-00	day	$16 or 04-00-00	month (sawing)	0.70	day
														00-02-06	day (ware cleaning)		

Notes on currency: Figures for 1740–64 are in livres: 1 livre (l) = 20 sols and 1 sol (s) = 12 deniers (d). Figures for 1804–57 are in pounds sterling: £1 = 20 shillings (s) and 1 shilling = 12 pence (d). 24 livres = £1. Figures for 1874 are in dollars and cents.

Sources: NAC, MG 1, C¹¹A, vol. 111, fols. 58–59v, "Frais annuels de l'exploitation," appended to "Régie de l'exploitation," 24 October 1740 (estimated remuneration) and fols. 354–444, Estèbe, "Estat général de la dépense,"1741–42; NAC, MG 21, B²¹² (21681), microfilm A-615, fols. 50–88, état de dépense, 1760–64 and MG19, E, 1.1, Diary of Lord Selkirk, 1804; Not. Rec. ANQ-M and ANQ-TR, 1801–25; "Lt. Baddeley's Report," APT, vol. 5, no. 3 (1973), pp. 5–33; excerpts from account books of 1856–58; B. J. Harrington, "Notes on the Iron Ores of Canada and Their Development," Geological Survey of Canada, Report of Progress for 1873–74 (Montreal: Dawson Brothers, 1874).

It is difficult to assess a substantial part of the overall compensation—of the inside workmen in particular—namely, room and heat, as well as the bonuses in kind they were occasionally given. The workmen were given room and board by the company, although their food rations were debited from their accounts. The requirements of industrial production governed the duration and pace of their work. In France, a contemporary ironmaster estimated that such fringe benefits accounted for a quarter of the workmen's monthly income.[140] These additional costs, added to the already high wages of the skilled workmen, drew much comment, and were even denounced, in the turbulent context of the establishment of the Forges. The colonial authorities seemed to have a poor understanding of the realities of the ironmaking world. Such comments show through in the historiography of the Forges, which sometimes tends to set apart the conditions of workmen there, without relating them to general conditions in the iron industry.[141] The wages of workmen during the French regime may seem abnormally high, but these conditions would not last, apparently. They would be subject to the ups and downs of the economy. We cannot always measure them accurately, but we shall see that wages fell steadily in the 19th century. In 1827, Mathew Bell, specifically noting the firm's negative economic situation, stated that his workmen were paid "very low wages" and expressed his fear of wage competition from potential new establishments in the region.[142] Similarly, during the final years of operation, Dr Harrington described wages as "very low" at the Forges, compared with those at other establishments in Canada.[143]

Methods of Payment

Payment in Kind

The issue of how workmen were paid was highly controversial during the initial years of operation. When engaging skilled workmen in France in 1737, ironmaster Simonet had stipulated in their contracts that they would be paid cash each month, following the practice current at that time in ironworks in France.[144] But this practice was not applied to the letter since, very early on, the workmen began to be paid in kind. Vézin, who would subsequently complain about this method of payment, had been—perhaps unintentionally—behind this practice. Anxious to shield the workmen and the company from the greed of the merchants of Trois-Rivières, Vézin had asked for a store to be kept at the Forges. Not to mention that the travelling by Forges workmen to and from Trois-Rivières for supplies wasted valuable company time. In a memorial written following the company's bankruptcy, it was stated that "Sr Olivier was always in a position to give the workmen more than half their wages in cash," implying that he had not actually done so.[145] Following a review of the ironmaster's accounts from 1736 to 1739, Cugnet would note that Vézin "had not given merchandise to the workmen at less than 35% profit."[146] Criticized for his shoddy bookkeeping, Vézin would claim that he had not been employed as a merchant, and that responsibility for selling victuals and merchandise lay with his clerk Cressé.[147] In 1739, responsibility for the store was taken from him and entrusted to Perrault and Cressé. Vézin would subsequently denounce, with Simonet, the fact that advances and payment in victuals and other goods had led to overspending and indebtedness among the workmen, who had not hesitated to retaliate by demanding wage increases.[148]

Suspected of enriching themselves at the expense of the workmen, the ironmasters' associates would defend themselves. Cugnet saw behind Vézin's criticisms the machinations of the merchants of Trois-Rivières, who had "a great interest in spreading such talk to retain for themselves the trade from St Maurice that they had considered the foundation of their fortunes,"[149] and feared losing a lucrative trade to a store at the Forges. When the Forges reverted to the Crown following the bankruptcy, the store was kept, with the recommendation that the workmen be paid cash so they could purchase provisions there. It was further pointed out that the goods should be sold at cost, so that the workmen would not be tempted to go and purchase their provisions for less in Trois-Rivières or elsewhere and, even more important, it was demanded that no more alcohol be sold on credit. It was also noted that, on the pretext of going for supplies, "the workmen waste at least two days going to Trois-Rivières, and they waste more when they are obliged to go further afield."[150] It is not certain that these recommendations were really followed, and the availability of merchandise on site was probably not the only cause of "overspending" by the work force.

Payment in kind sowed disorder and discontent, from which ultimately the workmen were the only ones to gain. This atmosphere may not have been the only factor that drove them to become either drunkards or heavy spenders. Apparently they already had a certain propensity for extravagance and luxury. In the memorial cited above, reference is made to some of the conditions demanded by the workmen recruited in France.

> Sr Simonet brought to Canada all his workers dressed in broadcloth and hats trimmed with fine silver. He even obliged the Company to give them one hat trimmed each year aside from their wages. This was one of the conditions of their engagement; the introduction of luxury among them should in no way be attributed to the store. It was far more inspired by visits to towns, since the merchants to increase their trade would do their utmost to encourage them to spend.[151]

The workmen—so went the constant plaint—always played on their scarcity in the colony to keep their wages high and conditions advantageous. This did not escape Pehr Kalm, who observed in 1742 that they must be paid "large sums." He also referred to "many officers and overseers," mentioning ironically that, being maintained at the King's expense, they "appear to be in very affluent circumstances." In 1752, engineer Franquet noted that the workmen were paid "exorbitant rates" and were "all given room, heat and transport at the King's expense." Emphasizing also the affluence of the directors and clerks, he observed that the sale of victuals and merchandise was flourishing.[152] This period, marked by state supervision and the protection of the King of France, was a veritable golden age for the workmen and administrators at the Forges. In such conditions, it is not surprising that the skilled workmen tended to settle in at the Forges, even when, according to Vézin, "their indentures have been broken and cancelled."[153] The fact that they were kept on since it was not possible to replace them helped to raise their demands and encourage their independence. By the end of the French regime, any attempts there might have been at disciplining their "extraordinary" spending habits had proven unsuccessful. To the point that, after the Conquest, Governor Haldimand would still be able to remark upon the extravagance of the workmen from the Forges:

> Money does not stay long in their pockets, barely have the men received their wages than they take them to the merchants, who soon send to Quebec for new merchandise; I am persuaded that at year's end there do not remain 100 francs in any individual's purse; they like in general to enjoy themselves, and rarely think of the morrow.[154]

Laterrière's account, cited above, seems to confirm that the workmen continued to live the good life after the Conquest.[155] The Grande Maison store was always well stocked with provisions and goods "to sustain all these people."[156] Subsequent inventories, as well as eyewitness accounts of administrators and clerks, also confirm that the company store was kept running until the Forges closed down.[157] According to clerk Timothy Lamb, victualling was given special attention in the mid-19th century:

> The collecting of the provisions for the cattle and drawing the supplies for the people on the post are some of the chief objects of attention.[158]

These accounts do not specify whether the workmen were still paid in kind. It appears, however, that this practice was maintained, and that it even continued to lead to indebtedness. We have seen that at the sale of the Forges in 1846 the generous-spirited Mathew Bell burnt his account books in front of witnesses to erase his workmen's debts. The last workmen interviewed by Dollard Dubé for their part said they had been paid in **scrip** or *bons* exchangeable at the store, under the McDougalls administration, which "paid wages 50% in cash, 50% in scrip, thus ensuring that half the wages were spent at the store."[159] The Forges masters were well aware of the profit they could make from supplying the necessities for some 400 people—not to mention the seasonal employees—and this "captive" population thus easily became for them a second source of profit.[160]

Inside Workmen's Wages

Ironworkers

During the French period, monthly wages were the most widespread method of payment among inside workers, both skilled workmen and their assistants, and craftsmen in iron and wood (Table 8.10). Nevertheless, ironworkers—the founder and master moulder at the blast furnace, as well as the hammermen and finers at the forges, for instance—generally received wages only during lay-off periods. This was in fact a retainer paid to them outside the campaign, that is, for about five months over the winter for furnacemen and for three or four months for forgemen. They were also paid this retainer when the blast furnace or forges were idle, either owing to equipment failure or for lack of raw materials. During the campaign, these master workmen were paid by weight of metal produced: the founder and moulder by thousandweight of pig iron and the forgemen by thousandweight of bar iron. Since these workmen had total control over the pace and intensity of their work, their pay was determined by their productivity. The forgemen do not, however, appear to have been systematically paid by thousandweight of iron. A 1738 document proposing construction of a second forge—the eventual upper forge—suggests that separating the workmen into two different forges would create "friendly competition" that could be encouraged by paying the men per thousandweight of iron. So there is reason to believe that this method of payment had not yet been instituted for the forgemen at that time.[161] Furthermore, the accounts for Estèbe's trusteeship in 1741–42 show that only a small part of the work of the forgemen from a single forge was paid at 12 *livres* per thousandweight of iron, whereas most of their working time was paid in wages.[162] A more detailed analysis reveals that the three months when Lalouette, Marchand and Terreau worked for 12 *livres* per thousandweight were a little less remunerative for them than for their comrades who were laid off. And it may be that, as a result of this bad experience, they opted for a fixed wage for the rest of the year. The converse occurred in the case of founder Delorme, who spent four and a half months smelting close to 480,000 pounds of cast iron at 25 *sols* per thousandweight, earning almost twice what he was paid for his five and a half months out of work. Subsequent memorials are always explicit on this method of payment in regard to the founder, but not on whether the forgemen continued to be paid by the thousandweight of iron. Rather, their wages were assessed at a set amount per year, corresponding to a monthly wage of 75 *livres* for the hammerman and 58 *livres* 6 *sols* 8 *deniers* for the finers.[163] Furthermore, other documents establish that payment by weight of iron produced was current. Franquet, in 1752, reported that the workmen were paid "some per hundredweight of iron, others at fixed wages for the year, etc., others at different rates for the winter and summer months." A statement of expenditures for 1756 mentions 250 thousandweight of bar iron paid at the higher rate of 22 *livres* per thousandweight, whereas the rate remained the same for the founder (25 *sols* per thousandweight of cast iron). In a 1760 estimate of production costs for the military administration, Courval set "the making of 70 thousandweight of bar iron" at the same cost of 22 *livres* per thousandweight.[164] Later, the eyewitness accounts of Lord Selkirk in 1804 and John Lambert in 1808 would confirm the fact that the forgemen were paid by weight of iron produced:

> The workmen are paid by the quantity of work they perform.[165]

The records for subsequent periods refer to the monthly earnings of the forgemen, but there are grounds for believing that, as long as bar iron was produced in quantity, they continued to be paid by weight of iron produced. Passing through the Forges in 1827, Lieutenant Baddeley would write that the "mechanics," or ironworkers, received £5–10 a month, but he did not specify whether this was a fixed wage or payment per hundredweight. He also noted that 45–50 hundredweight of iron were produced there each week. These figures correspond roughly to the output referred to by Lord Selkirk in 1804, when the forgemen were paid 3 shillings a hundredweight.[166] While this is not documented, it is reasonable to believe that payment of the forgemen by weight of metal produced was current as long as bar iron production was maintained, that is, until the late 1850s.[167]

The ironworkers' assistants were generally paid a fixed monthly wage. During the French period (1742–43), moulder's assistants, as well as the helper—who was part of their team—were paid a certain fraction of the total output of castings: Of the 2,400 *livres* budgeted for the 60 thousandweight of cast iron (at 40 *livres* per thousandweight), the master moulder received a third, each of the two assistants a quarter, and the helper a sixth.[168] Before the advent of the first moulders, the founder was paid 5 *livres* per stove cast. The 1756 statement of account shows the same sum, without identifying the workman receiving it. But Courval's 1764 estimate showed that this work was now being performed by the moulders. In this proposal, Courval, who favoured increasing the output of castings to 400 thousandweight, refers to the employment of four moulders at 1,000 *livres* each annually, without specifying how they were to be paid.[169] As with the forgemen, we lose all trace of payment by weight subsequently, without being able to assert categorically that this practice was abandoned.[170]

The few deeds of indenture available from Mathew Bell's time reveal that moulders were paid annual wages.[171] The same was true for moulder's apprentices, who would be hired for three-to-five year terms, at wages varying between 15 shillings and £1 for the first three years, plus room and board.[172]

Smiths and Artisans

The smiths (blacksmiths, farriers, and edge-tool makers) employed during the French period were also paid by the job. Granted land at the Forges to practise their craft, they apparently enjoyed special status. For instance, Bouvet's grant in 1740 authorized him to take commissions from surrounding *habitants* whenever he had time to spare from his work at the Forges.[173] In 1741–42, forgemen Bouvet and Marineau were paid upon presentation of "work bills" for which they received varying amounts. They were moreover paid by the job for making axes (4 *livres* per axe) and picks (3 *livres* 10 *sols* per pick).[174] The memorial of 1742 provided for the hiring of a farrier at 1,000 *livres* per year. The 1756 statement of account set the wages for an edge-tool maker and his journeyman at 1,440 *livres* for the two of them. After the Conquest, especially in the 19th century, there would be several blacksmiths among the regular employees of the Forges, as shown in particular in the censuses and rolls available to us. Contracts from 1810–20 provide for the engagement of forgemen, with lodging provided by the company, for a one-year period at a monthly wage of £4–5. At the same period, apprentice forgemen were also engaged, on three-year contracts, at 15 shillings a month the first year and £1 the second and third years, plus room and board.[175]

Estèbe's accounts (1741–42) also show that woodworkers, essential for making repairs and carrying out maintenance, were paid by the day, at rates depending on the type of work. For instance, joiner Bériau and his two journeymen were paid sums running from 1 *livre* 10 *sols* per day for "minor work" to 3 *livres* for repairing bellows or 3 *livres* 10 *sols* for making the flasks for moulding cooking pots. Carpenter Bellisle, responsible for building and maintaining the plant machinery, was also taken on at 1,000 *livres* a year, and the wheelwright was paid a similar sum, while his assistant received 350 *livres* a year. So these master craftsmen were paid as much as the master founder. The masons were also paid by the day, at rates from 1 *livre* for mudwalling a house, up to 2 *livres* 10 *sols* for the annual essential task of relining the furnace, "making and placing the inwalls and hearth of the furnace," which could take close to 40 days.

The few annual deeds of indenture extant for joiners and carpenters from the late 18th century, and especially from 1810–20, set their monthly wages at £4–5, with lodging provided for the workmen and their families "according to the custom of the post," with the added stipulation that a stove would be provided and garden ground made available.[176] Most deeds of indenture from the period specify that all categories of workmen are responsible, for instance, for not being absent while on duty, for working at night, on Sundays and feast days when necessary, and for making themselves available to the Forges should they be needed to help fight a fire. Fines of up to one month's wages could be imposed for absence, or for non-compliance with clauses of the contract. Furthermore, a workman not wishing to renew his annual contract was obliged to give the employer three months' notice, subject to having to work an additional month.

Carters and Bateaumen

Water transportation—by scow or boat on the St Maurice River—and land transportation—by cart and sleigh—constantly required the hiring of scowmen, bateaumen and carters. Different methods of payment were used: monthly wages, and payment by the day, load or job. Documents from the French period clearly indicate this, and subsequent data show a degree of continuity in how transportation was managed. The regular use of wage-earning carters was doubtless a sound way of ensuring a regular supply of the raw materials required to operate the plant. And there was no doubt a wish to guard in this way against manpower supply problems and the seasonal variations in cost inherent in using independent carters. The accounts for 1741–42 reveal that most of the ore and charcoal was carried by these carters who, using the company's horses, were paid 30 *livres* a month. We have worked our way back to this conclusion. Only the volume of raw materials hauled by certain carters paid per *pipe* of charcoal or ore is recorded in the accounts, and the volume of charcoal hauled by the three carters who were paid by the *pipe* makes up only 30% of the coal produced that year. Similarly, the nine carters paid by the *pipe* carried only 120 (8%) of the 1,500-odd *pipes* of ore required each year.[177] Table 8.11 shows that the average number of days worked by carters paid by the month was much higher than for employees paid by the day.

In fact, according to an estimate based on Estèbe's trusteeship in 1741–42, the cost of using carters paid by the month was virtually the same as, or even slightly lower than, the cost at 4 *sols* per *pipe* of charcoal (Table 8.12). The estimated daily cost of 3 *livres* 5 *sols* includes 1 *livre* for the carter, 1 *livre* for each of the two horses (including amortization of the purchase price) and 5 *sols* for the cost of the wagon box, wheelbase and maintenance. Estèbe's accounts show that he paid a little less (3 *livres*) for carters using a two-horse team, who were paid by the day.

Payment of 2 *livres* per day was the most frequent (Table 8.13), especially for scowmen. This amount included their food, since 10 *sols*, representing the cost of a daily ration, were sometimes deducted, with the notation that the man was given board. Provision of a ration could also explain the variation of 10 *sols* observed in the price paid to carters, but this variation could also be due to the fact that a single horse was used. The difference of 1 *livre* could also be explained by the fact that the company provided fodder for the horses, as the author of the 1743 memorial points out.

Table 8.11

METHODS OF PAYMENT FOR CARTERS, ST MAURICE FORGES, 1741–42 [178]

Method of payment	Number of carters	Total days' employment	Average days per man
Monthly wages	12	1,880	156.6
Daily wages	37	333	9.0
Per pipe of charcoal*	3	338	112.7

* Number of days estimated on the basis of cartage of 17 *pipes* per day, the bulk of it (93%) by only two carters working an average of 157.7 days each. Not included in this table is transportation by the job or by trip, which cannot be converted into days' work and accounts for only 16% of costs.

Table 8.12

DAILY COST OF A CARTER, BY THREE METHODS OF PAYMENT, ST MAURICE FORGES, 1742*

Means of transport	Method of payment	Unit of cost	Cost
Two-horse team	Monthly wages	Day	3 l 5 s
Two-horse team	Pipe of coal	Day	3 l 8 s
Two-horse team	Daily wages	Day	3 l 0 s

* From "Mémoire concernant [...]" (see note 2) and "Estat général [...]" (see note 9).

Table 8.13

DAILY TRANSPORTATION COSTS ACCORDING TO ESTÈBE, ST MAURICE FORGES, 1741–42*

Goods carried	Means of transport	Wages per day (in *livres-sols-deniers*)	Remarks
Oats, flour and hay	Scow	2-00-00	
Iron	Scow	2-00-00	From the Forges to Trois-Rivières
Stone	Scow	2-00-00	From La Gabelle to the Forges
		1-10-00	From La Gabelle to the Forges
		2-00-00	From the riverside quarry
		1-10-00	From the riverside quarry
		1-10-00	Board provided. From La Gabelle to the Forges
Victuals	Scow	2-00-00	From Trois-Rivières
Firewood	Cart	2-00-00	
		2-10-00	Two-horse team
		3-00-00	Firewood and hammer helves
		3-10-00	
Oats, flour and hay	Cart	2-05-00	
Hammer helves	Cart	3-00-00	Two-horse team
Manure	Cart	2-10-00	

* From "Estat général [...]," Estèbe, 1741–42.

The estimate in the 1743 memorial mentions 14 wage-earning carters, including the one assigned to the Grande Maison. Estèbe's statement of account for 1741–42, for its part, reports a complement of 12, all paid a monthly wage of 30 *livres* and employed for one to 10 months: two were employed for 10 months, two for eight months, and one for seven months. These were likely workmen assigned to meet the direct needs of the plant and the Grande Maison. The other regular carters were employed for periods of less than five months, sometimes starting in the winter months—doubtless to haul the ore by sleigh—sometimes starting in the springtime, when stone and flux for the blast furnace were carted (see Annual Supply Maintenance Schedule, Chapter 2). The wages paid to these regular carters in 1741–42 make up 42% of the company's transportation costs, with the other costs split between haulage paid by the day (17%) and by the job or load (41%) by carters and scowmen from the neighbourhood of the Forges.[179] In 1750, Intendant Bigot would write that "the wood is ever more distant, so transportation costs rise."[180] And the monthly wages of the carters would increase at the end of the French period. The accounts for 1756 suggest that each of the eight carters received monthly wages of approximately 40 *livres*, while the 1760–61 accounts refer to monthly wages of 42 *livres*. In his 1764 recommendation, Courval would still allow a monthly wage of 30 *livres* for the six carters to be employed, but his estimate of the haulage cost by the load (*barrique*) of ore showed a substantial increase, as we shall see below.

Until the 19th century, the company apparently continued to employ regular wage-earning carters alongside carters paid by the job. For lack of detailed accounts, we must go by the number of horses inventoried on the post for a better understanding of how transportation was managed during the period. According to the tally of horses in 1784 (22), the number of carters living at the Forges had remained substantially the same as during the French period. There is, however, reason to believe that the company continued to employ carters on a seasonal basis, now providing them with the team of horses and paying them wages.

A series of group indentures spread over the period from 1805 to 1820 shows a certain consistency in the number of carters employed full-time by the company. These contracts point to the renewed annual engagement of 9, 11, 13 and 15 carters who, paid monthly wages of £3 with lodging provided by the company,[181] would be taken on as carters and day labourers.[182] No doubt owing to the importance assumed by this "department," it was at this time that we find a "master carter." On the 1829 roll, Jean Vadeboncoeur was designated as being in charge of the other 13 carters. Later, Hamilton Rickaby, John Porter & Company's clerk during the 1850s, stated that the master carter had to report to him on his men's activities; the clerk recorded the time the men worked in two account books, the "Carters a/c book" and the "Carters a/c (Little)."[183] Rickaby's comment suggests that they were paid wages. In a statement of account for the same administration (1852–58), the monthly cost of a carter, along with his two horses and cart, was actually estimated at $26 a month.[184]

In criticizing Jeffrey Brock's poor management during those years, Rickaby also referred to the fact that the price of hauling ore depended on the point in the season:

> [...] owing to the prices given by Mr Brock in the early part of the season being lower than the carters wanted, the carting was not carried on vigorously and as the season closed higher prices had to be paid to get the mine in.[185]

And Timothy Lamb would observe likewise that:

> collecting and drawing home of the ore was sometimes done out of proper seasons which increased the costs.[186]

These accounts suggest that, as far as transportation of ore at least was concerned, carters were being paid by the load.

As we saw earlier, the category of carter disappears from the inventory of occupations listed in censuses from 1861 onwards; this does not necessarily mean that the company no longer employed carters after that date, since there were clearly some carters among the day labourers. The 30 horses reported in the 1871 census indicate that there were still some carters at the Forges that year. And the last eyewitness accounts gathered by Dollard Dubé refer to daily wages of $1 for carters; these documents also show that transportation by the job or load was still current, at the rate of 10¢ per cord for wood, and 10¢ per thousand-weight of flux.[187]

Table 8.14

UNITS OF MEASURE FOR PAYMENT BY LOAD, ST MAURICE FORGES, 1741–42[188]

Merchandise	Unit of measure	Means of transport	Unit price (livres-sols-deniers)	Distance
Oats	minot	scow	00-04-00	
Hay	100 bales	scow	04-00-00	Between Trois-Rivières and the Forges
		cart	08-00-00	Between Machiche and the Forges
Wood	cord	cart	00-08-00	
Iron	thousand-weight	scow	02-00-00	Between the Forges and Trois-Rivières
		sleigh	01-00-00	
Stone	toise	scow	12-00-00	Between the Quarry and the Forges
Ore	pipe		01-10-00	
Coal	pipe	wagon	00-04-00	

Outside Workmen's Wages

Carriers Paid by the Job

Used on a case-by-case basis for the carters, payment by the job or by the load was generally applied to water transportation. Regardless of the merchandise carried—hay, flour, stone or iron—Estèbe in 1741–42 paid the scowmen 2 *livres* a day for their services. Payment by the job was sometimes assessed on a per-trip basis, but most frequently by volume, based on the unit of measure specific to each type of merchandise (Table 8.14).

Forges wares were shipped by water from Trois-Rivières to Quebec at the rate of 8 *livres* per tonne. In the 1743 memorial, the cost of shipping was estimated "at 6 *livres* per thousandweight from St Maurice to Trois-Rivières and from Trois-Rivières to Quebec or Montreal," which amounts to 12 *livres* per tonne; yet Estèbe had paid some scowmen 4 *livres* per tonne for shipping from St Maurice to Trois-Rivières.

The few clues from the 1756 accounts and the 1764 estimate do not show any price changes, at least with respect to haulage of ore.[189] As we saw above, it seems that payment by the load was maintained later on, to judge by the clerks from the 1850s, who spoke of variations in transportation costs over a season. Rickaby in particular said costs were lower in November, in that interval between the times when coal was transported and ore was transported.[190]

Forest Workers

Most of the men at the Forges were forest workers employed at different times of year. The woodcutters were taken on as early as Michaelmas (29 September), cutting hardwood until winter began and, during wintertime, cutting softwood until March. The colliers, pit setters and feuillers were employed from May until All Saints Day (1 November).[191] "Road makers" were also employed during the summer. The woodcutters were the most numerous: Estèbe employed 216 of them in 1741–42, to deliver 11,282 cords of wood for charcoal; they were paid 1 *livre* per cord.[192] In 1756, they were paid 1 *livre* 10 *sols* a cord, and 2 *livres* in 1760. In the early 19th century, they would receive 1s 8d per cord, the equivalent of 40 *sols* (2 *livres*). In 1833, the same price was still being paid. Legal proceedings instituted against a woodcutter for fraud that year reveal that the output of these workmen was measured using a tally stick, "a small piece of wood stamped with the figures reflecting the quantity of wood the bearer may have cut." This tally stick was authenticated by the Forges foreman, Joseph Michelin.[193] A statement of account for 1857 shows that 3s (75¢) was being paid for a cord of wood, or a little more than double the price 20 years earlier.[194] Workmen assigned to sawing "small wood" by circular saw for the blast furnace were paid 1s 3d (28¢) a day, equivalent to monthly wages of

close to $7 (allowing for six working days per week); these wages correspond to what was paid by logging operators elsewhere in the region during the same period.[195]

A comparative analysis of the complements of woodcutters from 1742 and 1857 could be indicative of some concentration of wood cutting in the 19th century. For one thing, there is substantial variability in deliveries of cords of wood by suppliers on the 1742 list. It establishes that the 216 woodcutters each delivered from 3 to 250 cords, an average of 52 cords each, although only 26 of them delivered more than 100 cords each. Second, the 14 woodcutters on the 1857 roll delivered an average of 187 cords each. Obviously then, there were major producers during both periods, but there were also many small producers in 1742 and none in 1857.

After the blast furnace was rebuilt in 1854, doubling its capacity, more wood was needed to make charcoal. Before undertaking this work for John Porter & Company, Timothy Lamb, who had been employed by Mathew Bell for 11 years, reported on the firm's needs:

> I also know from long experience, that the Forges consume from 18 to 20,000 cords of wood annually (and which will now be much increased), and which has been hitherto charred in the woods [...][196]

The lots of the fief of St Étienne would henceforth be cleared by settlers, exposing the Forges' wood reserves to the risk of fire. Lamb went on to say that the company had lost 2,000 cords of wood in this way in the spring of 1852. And to counter these risks, he recommended carting the wood to the Forges for coaling—despite the higher transportation costs involved. Without putting a number on it, John Porter & Company stated that they still employed a large number of woodcutters in the winter.[197] Nor do we have any clues that would enable us to establish whether the increased wood requirements led to more woodcutters.

Dollard Dubé's investigation revealed that in the 1870s a dozen *bûcheux* (woodcutters) were employed; the best of them produced four or five cords a day and were paid 40–45¢ a cord.[198] In 1874, Dr Harrington noted that at St Maurice the wages of miners and day labourers were particularly low compared with those at similar establishments elsewhere in Canada, where wages varied from $1.25 to $1.40 a day:

> At the St. Maurice Forges wages were very low, an ordinary labourer getting in some cases as low as 70 cents a day and boarding himself.[199]

So the standard of living of these forest workers remained rather low, even though wages had doubled over the previous 20 years. René Hardy and Normand Séguin have found comparable rates specified in deeds of indenture for forest workers elsewhere in the St Maurice Valley in the 1870s.[200]

Colliers, Feuillers and Pit Setters

The colliers, feuillers and pit setters worked together on the same charcoal-making process. This common effort was at one time expressed in lump-sum turnkey coaling contracts with the master colliers, who were then responsible for the four operations—banking, dressing, feuilling and charring. Contracts of this type were signed in 1740, and Estèbe's accounts for 1741–42 refer to payments of 9 and 10 *sols* per *pipe* of coal. This was clearly the preferred method of payment for coaling, and the company would even have liked the colliers to receive the wood cut by the woodcutters directly, so as to avoid fraud in deliveries. In their view, lump-sum turnkey contracting for charcoal did not just preclude fraud, but yielded higher quality charcoal:

> [...] because they themselves pay for the dressing and preparation of their furnaces, it is in their interest to be economical with the wood, to ensure themselves that their furnaces are well dressed, feuilled and banked, and to control the fire there so that the wood is not reduced to breeze and produces all the charcoal it must yield, it is even in their interest to ensure that the wood is cut to the right length and well corded, so that the cord of wood for which they would pay for the dressing and other work would produce for them two and a half *pipes* of charcoal [...][201]

Cases of fraud were apparently common among woodcutters and colliers in France at that time.[202] Nevertheless, a number of colliers would refuse to make charcoal on a turnkey basis and preferred to be paid only for the charring, at 3–4 *sols* per *pipe*, according to the practice current during the very first years of operation; at that time, pit setters were paid 20–25 *livres* for 100 dressed cords, and the feuillers were paid 12 *livres* for 100 feuilled (leafed) cords. This was the division of labour that was ultimately adopted, as evidenced by the accounts for 1856 and 1857 that show these workmen were paid successively 2s 6d per wagonload and 4 1/2d and 4d a cord for "charring wood," "dressing wood" and "feuilling wood at the Ventes." Payment based on productivity would therefore be maintained, but as a few rare indentures show, monthly wages would also sometimes be paid.[203]

In the coaling season, a wage-earning *garde-feu* or pit warder was always hired to monitor the charcoal pits. He was paid 40 *livres* a month during the French regime and £2 a month in the early 19th century. During the last 20 years of operation, when pit coaling was abandoned in favour of kiln charcoal, the *chargeurs de kiles* were paid from 75¢ to $1 a day, and the *gardien de kiles* or kiln warder, the successor to the *garde-feu,* $35 a month, with lodging provided.[204]

Finally, the road makers, another category of forest workers, received 1–2 *livres* a day during the French period. They were responsible for opening up and maintaining the entire network of roads and bridges, linking the establishment to the ore mines, the charcoal pits, the river and the town of Trois-Rivières (see Chapter 2). More than 50 men were employed in this work, for varying lengths of time.[205] In 1761, two road makers were employed at 40 *livres* a month, 2 *livres* less than the wages paid to the carters.[206]

In the 1850s, a clerk would be responsible for supervising the work of the road makers who, at that time, apparently handled both the upkeep of winter roads and the haulage of ore by sleigh.[207] Some stretches of the many roads made for the Forges since their founding wound up on private land after the Forges were sold in 1846. Under the terms of the deed of sale, the purchasers were obliged to transfer lots from the fief of St Étienne to colonists wishing to settle there. In other words, the owners of the Forges, who still needed the ore found on that land, now had to transport it across private property. Arrangements were made with some of these private landowners who supplied ore, but others would hold out, doubtless with a view to obtaining better prices:

> Another difficulty I have also had [...], that is, in every instance where I have been able to arrange with one party for taking ore off their Lands, another party on the line of road required to bring the same to the Forges, have closed the roads against us, which roads were made by the Forges, but which have now become the property of the persons owning the Lots.[208]

To the author of this 1852 account, clerk Timothy Lamb, this was a new constraint that he had clearly not had to deal with during the years he had worked for Mathew Bell. These new conditions would help push up the costs that the last private entrepreneurs of the Forges had to pay to maintain production.

Miners and Quarrymen

The cost of ore was usually calculated "delivered at the furnace," that is, mined, washed and hauled. In 1735, Vézin put it at 3 *livres* (60 *sols*) per *pipe*. The memorial of 1743 mentions that it had cost up to 50 *sols* to obtain a *pipe* of ore, that is, 8 *sols* to mine it, 40 *sols* to haul it and 2 *sols* to wash it. So the miners likely received 10 *sols* per *pipe*, since they washed it at the mine (Plate 2.7). When it was realized that the ore would have to be extracted from mines that were less easy to work, located farther and farther away, the maximum cost of a *pipe* of ore delivered at the furnace was set at 4 *livres* (80 *sols*). But later accounts show a larger increase in costs. In 1756, 1,200 man-days were paid for at 40 *sols* to collect 1,000 *pipes* of ore, and 60 *sols* per *pipe* were paid to haul it. In 1764, Courval, basing his estimate on the costs of the last years of operation under the French regime, anticipated a cost of 30 *sols* per *pipe* to mine it and 60 *sols* per *pipe* to haul it, for a total cost of 90 *sols* per *pipe* delivered at the blast furnace.[209]

As we can see, both payment by the load and wages were used concurrently for these workers too, and this would always be so. The 1829 employee roll includes two miners and two ore washers living at the Forges, who were likely paid wages. Statements of account for 1857 refer to deliveries for which the ore-supplying settlers mentioned earlier were paid. That year, 29 such men delivered 1,272,553 lbs. (577 t) of iron ore. They received 4s 6d (37.5¢) for each 780-lb

(354-kg) *barrique* of clean ore or each thousandweight (454 kg) of dirty ore delivered.[210] Twenty years later, Dr Harrington, still observing the low wages of employees at the Forges, would report that they were then being paid 30¢ a *barrique*, or 7¢ less than in 1857. And Dollard Dubé's survey of the last workmen at the Forges would show that the ore was still being collected by employees of the Forges and farmers from the surrounding area. The going rate per *barrique* was "cheap" at 15¢, and paid in scrip exchangeable at the store, or in cash if the suppliers were from the outside. Dubé would also tell us that the miners worked in teams, consisting of a prospector, five or six ore miners, and three or four washers. Three such teams were employed by the Forges in 1872. Dubé would also refer to six day labourers who were "woods washers" or "bog washers," whom he differentiated from the ore washers at the washery on the post.[211]

The quarrymen, employed to quarry stone for flux or construction, were paid by the day or by the *toise* of stone. Estèbe's accounts show that it was often the masons themselves who carried out this work for 30–40 *sols* per day. In 1740, Cugnet said he had contracted with "two different bands of quarrymen to quarry stone at 7 *livres* 10 *sols* per *toise*." He also estimated the cost of stone at 40 *livres* per *toise* "delivered at the furnace."[212] In 1764, Courval for his part would evaluate the cost of quarrying stone at 40–48 *livres* per *toise*, but did not mention how the workmen were paid. As to later periods, we have but little information. Dubé's investigation reveals that the workmen were paid 28¢ per thousandweight to "raise the stone,"[213] or virtually the same as the price paid for a *barrique* of ore according to Harrington (see above).

Physical Conditions

The Workmen's Accommodations

A review of estate inventories from the French period and the first part of the 19th century allows us a closer look at the physical conditions in which Forges workmen lived.[214] Ethnologist Luce Vermette noted some evolution from one period to the other, and pointed out advances in the workmen's standard of living, which in the craftsmen's dwellings translated into a better-appointed interior with, for instance, furniture and accessories of greater value.

In the 18th century, the workmen's lodgings were most often limited to a single room, known as the *"chambre" or "salle,"* similar to the dwellings at French ironworks of the period.[215] This was a multipurpose area, where all the daily activities took place—meals, housework, entertainment and sleep. The generally rudimentary furnishings included a table, a few chairs, a dough box, beds, a dresser, a cupboard, and a buffet or a chest. Lighting was usually by candle and some workmen had lamps for when they went outside.[216] Life centred around the hearth, which was used for cooking and for heating the room. By 1760, all the dwellings had heating stoves made at the Forges,[217] but it was not until the 19th century that these stoves would also be used for cooking. In the houses—but not the huts—the common room often had a "closet," a small room sometimes located in the attic,[218] which could be used as a bedroom and where a chest and some furniture were sometimes placed.

The interiors of 19th-century lodgings are not much different. The few inventories that refer to them suggest that life was still being lived in the main room, with a separate bedroom reminiscent of the 18th century "closet." This room may be mentioned apart from the main room to emphasize that it was for the parents' exclusive use. According to the *Journal des Trois-Rivières*, workmen's houses generally still had only a single room in 1865, "the parents' room being curtained off."[219] This separation of the rooms, with a view to giving the parents some degree of privacy, was a sign of changing mores;[220] the mention of curtains at the windows was another such sign. Another room or annex was mentioned in the inventories, namely the dairy room (*laiterie*), which also served as a larder for beef, salt pork and other provisions.[221]

So household activities were still concentrated in the main room. Until the early 19th century, the hearth was still commonly used for cooking. The stove, always located near the hearth, was used for heating but, in the following decades, would begin to be used for cooking. The references are not, however, explicit on this point; rather, it is the presence of such utensils as kettles, cooking pots and saucepans, as well as references to double box stoves, which indicates this new use of the stove.[222] The interior of the dwellings was also better appointed than before, and the furnishings were more numerous and varied. Luce Vermette points out that this increase in the number of objects to be listed would prompt the notary to alter the way in which he drew up his inventories. Abandoning the 18th-century technique of moving from area to area—which she actually reconstructed—the notary would take inventory of furniture and other items by type and by group.[223] We often find more than one table, armchairs and a variety of other chairs, several cupboards, the dough box, and toilet accessories, as well as, albeit more rarely, several suitcases and even clocks. The walls are decorated with more mirrors than before, of different sizes, as well as "frames," shelf racks and pictures. Colour is also a new element. During the French period, any colour was primarily to be found in the bedding, while the 19th-century inventories refer to several items of furniture painted in contrasting colours, with red predominating. Wood is varnished. As in the French period, lighting is by candle, which some people make themselves, and lanterns are available. There are, however, few oil lamps.

When the workmen were given lodging, they were also provided with firewood. In 1742, 1,500 cords of wood were supplied for heating the Grande Maison and workmen's lodgings.[224] And according to the memorial, wood was also provided to the seasonally employed carters, who probably lived in shanties or huts for some time, particularly in fall and winter.[225] In 1804, Lord Selkirk would talk of 2,000 cords of firewood.[226]

Food

The need to feed the workmen's families on an isolated post, out in the woods, was a primary concern from the outset. We shall see in Chapter 9 that the establishment would end up with its own grist mill and farm. Furthermore, the need to feed ironworkers brought its own special requirements. Although out in the country, these workmen did not live in tune with the rhythms of country life and had no time to see to feeding themselves. For its workmen to be able to keep up the pace of shift work that was required of them, the company provided them with daily rations, which were debited from their accounts. It also supplied rations to the day labourers and carters employed on a casual or seasonal basis, deducting them from their wages. Another special feature was the physical effort entailed by the workmen's hard labour, which had to be sustained by a good diet. As Denis Woronoff points out, "to withstand the rigours of heavy iron-making crafts, the workmen needed food with high energy value." Bread and meat were the staples of their diet. In France, forgemen of the period consumed their three pounds of whole-meal bread per day and up to seven pounds of beef per week, not to mention their three litres of wine per day. These workmen appear to have consumed spectacular amounts daily—twice a Napoleonic soldier's ration.[227] The ration of St Maurice workmen, or the "*habitant*'s ration," was comparable: two pounds of bread and one pound of beef or a half pound of salt pork a day. This ration was given to each family member. Vézin pointed out that every workman, "together with his wife and each member of his family, eats his half pound of salt pork or pound of beef a day."[228]

Wine was not included in the ration but, as we saw earlier, the workmen did not lack for it. A number of employees, including the foreman, received bonuses of two jugs of spirits per month, and others, such as the baker, a jug of tafia per month.[229] The *habitant*'s ration was equivalent to virtually double the ration of a soldier in New France—at least as far as meat is concerned—which consisted of one and a half pounds of bread, four ounces of peas and four ounces of salt pork.[230]

The bread eaten at that time was made with wholemeal flour, and bread making during the French period took 3,000 *minots* of wheat a year, ground in the neighbouring seigneuries until the Forges set up their own grist mill. A baker was always employed full time to work in the bakery built close by the Grande Maison.[231] He was also responsible for preparing and distributing victuals, as specified in records from the 19th century, where he is sometimes called a baker and sometimes a butcher.[232] Furthermore, since there are some references to several *minots* of flour in the estate inventories of the workmen, some of them probably made their own bread.[233]

So the workmen ate a great deal of beef and pork. Apparently pork was more common, doubtless because it was easy to preserve in brine. Indeed, company pork was salted in the cellars of the Grande Maison.[234] In the 19th century, consumption of meat would still be substantial, and a number of workmen would have a good supply of pork and beef in their dairies. In 1865, the *Journal des Trois-Rivières* mentioned that the Forges workmen ate meat twice a day, fresh in the summer and frozen in the winter.[235] The workmen always owned at least one pig and, as we shall see below, the herd would become particularly large in the 19th century. They would also, though more rarely, have an ox, as well as a dairy cow to provide them with milk for making butter and sometimes cheese, primarily in the 19th century.[236] They also ate a lot of eggs and poultry, which were found in large quantities at the Forges. Small game was also on the menu, since the workmen were apparently very fond of hunting. In fact, ordinances had to be issued during the French period to prohibit hunting on the Forges land, for fear of forest fires. And the periodic renewal of these bans suggest that the workmen would not stop hunting in the Forges woods, which at that time were rich in passenger pigeons, pigeons and partridge.[237] Indeed, they would continue to hunt in the 19th century, as evidenced by inventories of workmen, and of certain clerks, listing rifles.[238]

Although they lived alongside a river, the workmen were apparently less fond of fishing than of hunting. Nevertheless, especially during the French regime, fish hooks were sold at the company store. A number of references, plus the remains of bones found, suggest that people ate fish.[239] According to clerk Hamilton Rickaby, fish was sold at the store in the 1850s.[240]

Peas were apparently the only vegetable eaten during the French era, although in the 19th century, the workmen's kitchen gardens would also supply beans, potatoes, cabbage, corn and even a few pumpkins. Fruit was rare, except for wild berries, as we shall see in Chapter 9. The 19th century saw the appearance of sugar, more particularly local maple sugar. A number of workmen had large quantities of it. Indeed, there would be four maple groves with sugar shacks on Forges land in Mathew Bell's time, which he would rent to "sugar makers."[241] The company store also stocked molasses in the 1850s.[242]

As to drink, we have seen that alcohol, wine in particular, abounded at the Forges, and this would also hold true in the 19th century, although less reference would be made to its harmful effects during that period. Creek water was also drunk, and we shall see in Chapter 9 that, unlike wine, it sometimes had beneficial effects attributed to it. The presence of cows and dairies also indicates that milk was consumed; it was widely used for cooking, especially in the 18th century.[243] A number of references to utensils imply that drinking tea and coffee was not very common in the French era, but would become more widespread in the 19th century.[244]

Health

The workmen at the Forges certainly had the look of sturdy fellows. In view of the high-risk environment they worked in, it is quite astonishing to observe that, all in all, very few cases of illness and particularly of work accidents were reported. The air was mostly unbreathable in the plant, and the temperature variations were huge, especially in the winter. Nevertheless, during the French regime, several work stoppages occurred because of illness and it was feared that the illness or disability of the main workmen could compromise the operation. We therefore cannot fail to be astonished at the longevity of most of these workmen, who lived at a time when life expectancy was no more than 40 years. The case of Delorme, the founder, who was one of the ironworkers kept on after the Conquest, is interesting. In 1743 he was said to be "infirm, with a weak chest [and] like to die"; but he was not to die until 1785, at the age of 74. Pierre Bellu, his keeper, who had been described as "incapable of taking his place in case of illness," would die at 95. Finer Pierre Michelin, who was also said to be "infirm and almost always sick," would die at 72. Hammerman Pierre Marchand and finer Luc Imbleau died at 72 and 60 years of age respectively.[245]

As for the managers, they were apparently harder hit by illness. Francheville died at 40, and Vézin said that he himself was afflicted by "a long and dangerous malady" in 1740.[246] In those years, surgeon Alavoine was paid by the Compagnie des Forges to attend to the staff at the Grande Maison and the workmen; the latter had, however, to pay for his services, as well as for any medicines. But the company did pay the expenses of two sick carters, deemed insolvent in 1743.[247] Little information is available subsequently. In the 19th century, workmen consulted surgeons from Trois-Rivières and even Montreal, to whom they became indebted. The estate inventory of surgeon François Rieutard of Trois-Rivières, who died in 1819, shows that some 50 workmen from the Forges had accounts outstanding, mostly for amounts not exceeding 50 *livres* for the most part.[248] The workmen who witnessed the last 20 years of the Forges would relate to Dollard Dubé that illness was rare at the Forges but that, when necessary, physicians from Trois-Rivières were sent for. Dr Beauchemin, who married Onésime Héroux's widow and lived at the Forges farm, also provided care for the working population.[249] All in all, then, the health of workmen at the Forges appears rather good. And, if we are to go by Benjamin Sulte, it was thanks to the iron-rich water of the creek from which the cyclops of St Maurice drank!

Plate 9.1
The Forges on the St Maurice River, by Mary Millicent Chaplin, 1842.
NATIONAL ARCHIVES OF CANADA, MAP DIVISION, C-820.

IX

The Village

ESTABLISHMENT OR VILLAGE?

The establishment has always been generally known as the "Forges," although its exact name varies in period sources and in the historiography: "les Forges de Saint-Maurice," "les Forges Saint-Maurice," "les Vieilles Forges" and "les Forges du Saint-Maurice" in French, and the "St Maurice Works," "St Maurice Ironworks" and "St Maurice Forges" in English. Benjamin Sulte, and later Dollard Dubé and Albert Tessier, used the name *Les Forges Saint-Maurice* as the title of their books. The "Forges *du* Saint-Maurice" was chosen as the French name of the national historic site, as it refers to the St Maurice River, which is commonly called "*le* Saint-Maurice." The company founded by Francheville was originally called the "Compagnie des Forges *de* Saint-Maurice," since the name was based on that of the seigneury of St Maurice.

The Forges were never granted the legal status of a village or even a parish. In most cases they were referred to as a post or establishment, and regarded as such, because of their specific industrial character, which was their distinguishing feature throughout their history. The size, physical layout, and population of the Forges gave them the appearance of "a tolerably large village," in the words of Surveyor General Joseph Bouchette, in 1815, but the place was never officially designated as or considered to be a village. At the same time, Mathew Bell spoke of his "quiet peaceable village,"[1] for which he would have liked recognition as a "Provincial establishment" rather than "a Local or District one" (see Chapter 2). But, over a hundred years after the founding of the Forges, just before they were to be put up for sale, it was still being deplored, by commissioner of inquiry Étienne Parent, this time, that a village, or even a town, had never grown up around the establishment:

> It might be expected that the trade of the mines would have created, in forty-five years, at least, a manufacturing village, if not a town [...] In the immediate vicinity of the Forges it appears to the Committee to be proper to reserve the necessary extent of Land for establishing a Village or Town, and to lay out this Land in Town Lots, which should be granted to fit applicants, but in no case to the Tenant of the Forges.[2]

Some people hoped that the existence of the Forges would spur the founding of other iron-processing industries there, and that these industries would attract a larger working population. As early as 1735, the colonial authorities had promoted Vézin's ironworks project by suggesting to the Minister that "the proposed establishment could give rise to others of the same kind."[3] Yet, 60 years later, the establishment had still not engendered the expected industrial development, at least not in its immediate vicinity. In 1798, the Executive Council had expressed the wish that the seigneury of St Maurice would become the

"cradle of a large population" and that other craftsmen would be drawn to live there.[4] The Forges, which had only 200 inhabitants at the time, saw their population double thereafter, but this still did not lead to the creation of the manufacturing village that Étienne Parent would have expected to find. At the same time, however, the intensified manufacture of castings led Mathew Bell to set up a foundry in Trois-Rivières. The Batiscan Iron Works was also founded that same year, following in the footsteps of the St Maurice Forges, from which it recruited many of its workers.

The Forges were for a long time a Crown-owned enterprise, and that is no doubt the main reason why they remained a post or establishment. Since the lessees merely had the use of facilities and resources provided to them by the Crown, they had no incentive to launch new industries by reinvesting their profits. They were required only to keep the plant in good working order, but many of them failed to do even that. A significant seasonal work force was employed on the post, which meant that the continual movement of workers no doubt helped created the impression that the establishment was merely a place of work: the "foundry," as it was increasingly referred to in the 19th century. Whatever the case, the Forges still constituted, at least in fact if not in status, a true industrial village.

So far we have chiefly used the term "post" to refer to the Forges. In this chapter, however, we will be using the word "village" more, in order to focus on the side of the Forges that gave it the appearance of an organized community: its dwellings and services.

THE FORGES IN PICTURES

[...] Their situation is admirable,
and I would be pleased to be ordered
to re-establish their operation.

Haldimand, 1762[5]

The first general views of the Forges, produced in the 1840s (Plate 9.1), show the post as it existed over a hundred years after its founding, at the time when the population was at its peak of 425 inhabitants. In addition to the main shops of the blast furnace and forges (upper and lower), there was a grist mill, a sawmill and a charcoal mill (see plan, Plate 2.16), which were powered by no fewer than eight waterwheels distributed along the creek over a distance of 610 m. Some 100 workers lived on the post with their families, while another 250 others came to work there or worked in the vicinity on a seasonal basis. Yet these views convey no impression of the bustle, work and energy usually associated with an industrial plant. They do not conjure up the roar of the water as it raced down the hill, with its 10 cubic feet per second (0.3 m³/s) turning the creaking waterwheels, or the din of the two 250-kg hammers striking the iron close to 100 times per minute! Nor do we feel the suffocating heat or black dust of the shops. Instead, these pictures of the Forges tend to give the impression of a small rural village where the smoke from the shops mixes with that from the chimneys of the workmen's cottages (Plate 9.2). This impression is not necessarily the fruit of the artist's idealized vision, or of our own nostalgia a century and a half later. Some contemporary observers, such as John Lambert in 1808 or Surveyor General Joseph Bouchette in 1815, were very much taken by the site, particularly the beauty of the valley (Plate 9.3). Lambert even found the place "truly romantic."[6]

This representation, although it masked noisy, dirty industrial activity, is a reminder of a time when, in a rural setting, industry was powered by streams and rivers driving the rudimentary machinery of its workshops. Contrary to the spectacular facilities of modern factories, the Forges' hydraulic works do not immediately reveal the know-how of their creators, the ironmasters and master craftsmen of the 18th century. Still, their art is no less present in the choice of site, in the assessment of the flow of the creek, in the siting of the dams and shops along the creek, and in the craftsmanship of the furnaces, wheels and gear mechanisms.

At a time when theory rarely had the upper hand over practice, especially in the area of water management, the ironmasters were essentially hands-on workers whose vast knowledge and experience was difficult to sum up in terms of general principles. But they were effective, nevertheless.

All the ups and downs surrounding the foundation of the Forges shed some light, sometimes with invaluable details, on a number of aspects of what could be referred to as "the art of setting up an ironworks in the 18th century." So although the pleasant views of the Forges are far too short on technical detail, a second look may discern a hidden energy and know-how.

Plate 9.2
The Forges near Three Rivers, 1845, by Captain Pigott.
ARCHIVES DU SÉMINAIRE DES TROIS-RIVIÈRES, DRAWER 258, NO. 48.

DISCOVERING THE POST

When visitors take the Trois-Rivières road (now Boulevard des Forges) to get to the Forges, they first see the site from the top of a hill, from the southwest; the hill curves westward, less steeply than it used to. Most of the general views of the site, including the only known photograph of the whole post (Plate 4.9), are from the top of the hill. Coming over the "big hill," one had the best view of the site in its entirety, taking in most of the buildings and dwellings on the plateau. But it is difficult from this distance to make out the deep gully where the creek rushes down a 30-m slope and empties into the St Maurice River. The only sign that there is a creek is the blast furnace, with its great wheel and long headrace. To get a better view of how the workers' housing was laid out on either side of the gully and to discover the creek down below—the establishment's energy source, where the plant with its water-powered machinery was concentrated—one must move closer to the blast furnace.

Other views of the site are from the east, across the St Maurice River, as if one were discovering the establishment while making one's way up the river (Plate 9.5), and it is from this perspective that one has the best view of the gully dividing the plateau in two. In the foreground is the creek emptying into the river, and above, the lower forge with its imposing chimney. Higher up the gully, we can see a mill and the chimney stack of the upper forge. So someone approaching the site from the big hill sees primarily the village, that is, the dwellings, with the blast furnace embodying the site's industrial purpose in the middle. Someone approaching from the river mainly sees the rest of the ironworks, at the foot of the 37-m terrace of the Grande Maison.

None of these views explicitly shows the five millponds located upstream from the workshops, that is, the washery pond, the blast furnace pond, the upper forge pond, the grist mill pond and the lower forge pond. Using a topographic map, a plan of archaeological digs (Plate 9.10), archival documents and the photograph dating from the 1860s, an artist has recreated a bird's-eye view of the site (Plate 9.6). The recreation shows the arrangement of the various millponds at the different points along the creek, making it easier to associate each of them with its shop. A large-scale model of the village of the Forges in 1845 has also been built using a similar technique[7] (Plates 9.7a, 9.7b and page 375).

Plate 9.3
View of the St Maurice Forges, by Joseph Bouchette, Jr., 1831.
NATIONAL ARCHIVES OF CANADA, MAP DIVISION, C-4356.

BUILDING THE VILLAGE

The site was not originally as well cleared around the St Maurice Creek. When Francheville had the first forge built in 1733, at the same place where the lower forge now stands, the site was still in the wilderness, in the middle of the woods (Plate 1.2). In the summer of 1736, when the new forges were being built, extensive clearing had to be done, as Vézin himself reported:

> The Company had to have lodgings built for all the workmen because the establishment is in a place where no trees had yet been cut down, and so large areas had to be cleared [...] The work of clearing was more costly than expected because of the large number of enormous trees on the land, and because of all the roads that had to be built to carry the mine, flux stone, charcoal and other materials.[8]

Plate 9.4
St Maurice Forges, circa 1870, seen from the big hill, as they appeared to visitors arriving from Trois-Rivières.
PHOTOGRAPH BY JOHN HENDERSON, CIRCA 1870, NATIONAL ARCHIVES OF CANADA, PA-135-001.

The purpose was not only to build a *plant*, but also to found a *village* and create, in the middle of the woods, a network of roads leading to the post.[9] So the Forges were built from scratch, much as Quebec mining towns were in the 20th century.

The establishment erected by Francheville consisted of just four buildings (see Chapter 1), built on a plateau just over an acre in area, on the edge of the St Maurice. The plan drawn by Vézin in 1735 shows the four buildings set in a line just to the north of the creek (Plate 9.8). It also shows the road to Trois-Rivières running north around the site and down to the river.[10] At that time, then, there was no road to the works along the creek, in the gully; the creek road between the blast furnace and the lower forge was not built until the new forges were installed.[11] Only the forgemen's house was kept when Vézin began building the establishment, in the summer of 1736. The new forges occupied a much larger area. Francheville's site was rearranged to accommodate the lower forge and adjoining buildings. The upper forge and blast furnace—and later, the mills—were built higher up in the gully. On the plateau, most of the dwellings and utility buildings, as well as the various smithies, were built on either side of the gully. With the 19th-century boom, the village further expanded to cover more than 400 acres.

APPEARANCE OF THE VILLAGE

The layout of the establishment built by Vézin remained essentially unchanged until the end of the 18th century. When the first inventory of the Forges was drawn up, in 1741, there were 30 or so buildings, including 14 dwellings (see Appendix 15). By the turn of the century, under Monro and Bell, the resident population had grown, doubling the number of dwellings to 29, and various other new buildings increased the occupied surface area. The size of the establishment did not change thereafter.

In the fall of 1735, Vézin submitted a detailed project with costings, but made no mention of an overall plan. What are mentioned—in a letter, dated the following year, accompanying a report on the status of the work—are plans that the ironmaster Simonet was supposed to submit to the Minister of Marine to keep him abreast of the progress being made.[12] A plan of the Grande Maison was apparently also drawn up.[13] These plans have not been found, and so we do not know what was in them exactly. Considering the way that work on the Forges was undertaken, these plans—if they did put forward a comprehensive development scheme—would probably not have been a definitive blueprint for the managing and carrying out of the work. The many reports and memorials dealing with the construction of the establishment never mention the existence of any such plans. And the atmosphere of haste that reigned over the work site certainly did not give the impression that it was a well-planned undertaking. Speed was the watchword, and as we saw in Chapter 4, the dams and shops appear to have been designed on the spot, with the result that execution of the work was not always well coordinated. The primary documents on how these structures were built show that many different decision makers were involved in the project, including Intendant Hocquart himself. Even the location of the blast furnace was chosen without much forethought. The only drawing by ironmaster Vézin that has been found is a sketch of the slope of the gully, based on six level measurements that he recorded at the bottom of a map (Plates 2.3 and 3.2). The ironmaster chiefly oversaw the building of the dams and workshops, and if his remarks are to be believed, the construction of the dwellings, and even the Grande Maison, was a real nuisance to him. As a result, only the layout of the dams and workshops follows an organizing principle consistent with the need to channel and reuse the power of the creek. But as we have seen, even in this case, engineer Chaussegros de Léry had to step in and rectify the system.

The gully was therefore the central axis along which the dwellings were erected, one by one, as needed, without following any overall plan, as engineer Franquet noted when he visited the site in 1752:

> The dwellings [...] stand higgledy-piggledy, with no symmetry, and no relationship between them. Everyone has his own separate dwelling, so that there is a motley collection of houses, as well as sheds and shelters [...][14]

The few buildings dotted across a wide area no doubt underscored, at the time, the haphazard appearance of the establishment, as can be seen on a plan from 1738 (Plate 9.9). It shows, on a very small scale, 16 buildings spread out randomly on either side of the gully, with no discernible pattern. Murray's plan of 1760, drawn on a similar scale, shows over 30 buildings scattered about (Plate 2.17). A hundred years later, when the land was more densely occupied, the lack of symmetry was less noticeable, and in views of the village, it is easier to distinguish residential areas from those set aside for storage and work (Plates 9.1 to 9.6).

For a good overall view of the village, it has to be seen from the creek that runs down through the gully to the river. As we saw in Chapter 3, this is the heart of the establishment. It is the reason for its existence; everything else depends on it. The creek, harnessed initially by three dams and then by five, was the driving force of the plant. Built along the course of the creek, one after another at different locations down the gradient, the blast furnace, upper forge and lower forge were altered somewhat over the years, but they stood always on the same sites, which were chosen in relation to the dams. The creek corridor therefore accommodated only these main shops with their sheds and iron stores, and three water mills, which were successively incorporated into other levels of the grade: one at the foot of the blast furnace, at the tailrace of the great wheel, another just downstream from the upper forge, and the last one near the lower forge. There is no known trace of habitation in the gully itself. Only the two houses for the lower forge workmen, situated at the bottom of the gradient on the terrace giving onto the river, were adjacent to the gully. The remainder of the dwellings were spread out on the plateau, on either side of the gully, to the south and north.

While there were few dwellings originally, the occupied areas always remained the same, becoming more densely inhabited in the 19th century. These areas were connected by roads, which now help us to determine their boundaries. All the dwellings and buildings were located within a perimeter of just over 240 m (Plate 9.11).

Plate 9.6
Bird's-eye view of the St Maurice Forges,
after the photograph by Henderson (circa 1870),
the plan of the ruins and various pictorial sources.
RECONSTRUCTED BY ILLUSTRATOR BERNARD DUCHESNE.

First there was the big hill, which was the route to the site when coming from Trois-Rivières. At the foot of the hill, a large slag heap was the first sign of the industrial activity of the Forges; this was the "scum" from the furnace that is mentioned in some documents[15] and shown in a wash drawing from the 1840s (Plate 9.12). Archaeological digs at the site have revealed that the slag was used to surface the sandy roads of the village, giving them a blackish appearance.[16] Even today, vitrified slag can be found scattered around the site. Also at the foot of the big hill, there was a three-way intersection. Straight ahead, towards the washery pond and the Forges farm to the west, was the *chemin du roi* (King's Highway), which later continued on to the village of St Étienne des Grès. The right-hand fork led to the centre of the village by the *chemin de l'empellement* (sluice gate road). And farther to the right was the *rangée du meunier* (miller's row). At the intersection, on either side of the *chemin du roi*, was a first group of dwellings, some of which lined the millpond of the upper forge. Behind these houses, set back at the foot of the hill, were the charcoal barns. The *rangée du meunier*—so called because of the tenement house found there, as well as the wheelwright's shop—led to the edge of the gully, on the south side, where a path ran down into the gully near the upper forge.

The *chemin de l'empellement*, running north, went to the blast furnace complex on the edge of the gully. Another residential area was situated opposite the furnace, on the west side of the road, and included two tenement houses for furnacemen. The *chemin de l'empellement* was later extended northward to the site of the kilns and charcoal sheds. Right behind the blast furnace complex was a double intersection on the *chemin de l'empellement*. The *chemin du gros marteau* (hammer road) led to the gully, the workshops and the river, and from there, southward along the river to the dock. The *chemin de la grande maison* (road to the big house), which was the densest residential area, ran along the north side of the gully. Ordinary houses stood next to craftsmen's shops and two tenement houses built perpendicular to the road, and there were also various utility buildings, one of which was used as a chapel at one point. The road, as its name indicates, led to the Grande Maison, located at the edge of the plateau and overlooking the St Maurice River. There a turn to the left led to the bakery and to the stables at the far end. About halfway along the road from the intersection was a turnoff to the north on the *chemin de la pointe à la Hache* (axe point road). There, near some houses, stood more charcoal barns and utility buildings, and then the road left the site, heading north toward the point. Archaeological digs have unearthed part of the corduroying of this road at the exit from the village[17] (Plate 9.13).

LIVING QUARTERS AND WORKPLACE

Despite the apparent lack of any discernible pattern of land use, the layout of the dwellings around the gully nevertheless obeys certain principles that were common in ironworks in France at the time, that is, proximity to the place of work, individuality of dwellings and lodgings, and consolidation of the living quarters of the workers from the same department.[18]

Proximity to the workplace was dictated by the round-the-clock system of shift work that was integral to plant operations. In some parts of France, such as Ariège, ironworkers even "took turns sleeping" at work on a straw-covered "perch" secured to a wall.[19] At St Maurice, similar considerations led to the construction of the founder's first living quarters—a "room" in the blast furnace complex, where the casting house joined the bellows shed. Another room, "where the moulder lives," was also located near the bellows shed.[20] These workers were indeed living on the job, sleeping in a room separated by a mere partition from the huge bellows powered continuously by the grinding gears for six to eight months of the year and breathing the fumes of burning charcoal and smelting iron issuing from the furnace. Likewise, the workers at the lower forge lived in a house adjoining their shop, but separate from it. In 1740, there had been plans to provide the same living arrangements for the men at the upper forge, to make it easier for them to work in shifts:

> If there is not enough water to keep four forgemen busy, we will have two work at a time, and while they are working, we will make the most suitable arrangements for their bed and board at the upper forge, if they so desire.[21]

Even when he was off his shift, the founder had to be ready at any time to step in to adjust the set of the furnace or put out a fire. The forgemen had to stay nearby because of their two six-hour shifts each day, not to mention the fires that could break out at any time in the chafery chimney or in the shop, the walls of which were covered in charcoal dust.

WORKERS' DWELLINGS

The various types of dwellings generally reflected the hierarchical social structure of the work force.[22] From the outset, care was taken to provide the skilled workers and craftsmen with "houses." Some of these houses were single-family dwellings, but most of them consisted of more than one unit and were home to more than one family,[23] although generally each family had its own living quarters.

A house had all the features of a permanent dwelling. In most cases, it was a *pièce sur pièce* log construction with foundations and a stone chimney having one or two hearths, placed back to back in adjoining dwellings. The windows in these plank-covered houses were usually glazed (mullioned and transomed), but in the 18th century, they were sometimes of oilcloth. The houses usually had two or three rooms (bedrooms and small rooms or closets), with perhaps a cellar and attic.[24]

Fellow craftsmen generally shared a house, and sometimes even rooms, but workmen of differing occupations could also be put together. According to his 1735 plan, Vézin intended to have the founder and his four workmen live together,[25] and the practice of lodging workers together continued thereafter. Sources from the late 18th and early 19th century mention the "moulders' house," "forgemen's house," "carters' house," "quarrymen's house" and "carters' and labourers' house," each of which consisted of several lodgings.[26] Some of the houses for workmen of

Plate 9.7a
Upper forge (detail of model of the St Maurice Forges),
at the Forges du Saint-Maurice National Historic Site.
PHOTOGRAPH BY PARKS CANADA/JEAN AUDET,
130/MQ/PR-7/SPO-00116, 1989.

Plate 9.7b
Dwellings near the blast furnace millpond (detail of model of the St Maurice
Forges), at the Forges du Saint-Maurice National Historic Site.
PHOTOGRAPH BY PARKS CANADA/JEAN AUDET, 130/MQ/PR-7/SPO-00077, 1989.

the same trade were semi-detached or row houses, in which more than one family lived, and some were tenement houses (*corps de logis*), the long buildings consisting of a number of dwellings seen in the first views of the Forges in the 1840s. The inventory of 1807 (see Appendix 4) specifically mentions a building of this type measuring 30.5 m by 6 m "serving as several lodgings for the m. founder, keeper and moulders" that had been built under Davison and Lees (1787–93). In the same inventory, a distinction seems to be made between these tenements and the row houses or semi-detached houses, as in such descriptions, for example, as the "five houses in a single building," 24.4 m by 7 m, where the carters lived, or the "two houses in a single building," 12 m by 6 m, where two moulders lived. These "houses in a single building," of post-and-groove log construction, consisted of independent units side by side under the same

gable roof. Some houses shared a chimney, which was built into the party wall, while others each had a single chimney.[27] Most often the houses were semi-detached, two by two, with a central double chimney, but there could be up to five joined together. Obviously this means that units were added as they were needed for new workmen; the existing buildings were extended as much as space would allow. The same thing occurred at French ironworks in the late 18th century:

> When the entrepreneur wanted to house additional workers (after setting up more workshops), he put up a new building; sometimes, if there was enough room, all he had to do was extend a building by the number of dwellings needed. Although the ironworkers' houses were clearly inspired by rural houses, their identical design and side-by-side construction differentiate them.[28]

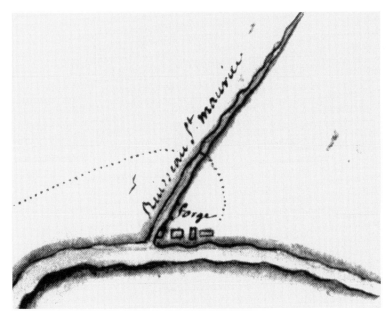

Plate 9.8
Sketch of the Forges road built by Francheville in 1732–33.
OLIVIER DE VÉZIN, PLAN OF THE MINES AT TROIS-RIVIÈRES (DETAIL), 1735, FRANCE,
BIBLIOTHÈQUE NATIONALE, PARIS, CARTES ET PLANS, PORTEFEUILLE 127, DIVISION 8, PIÈCE 50.

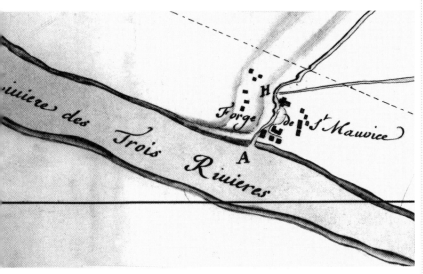

Plate 9.9
Plan of the Forges in 1738. The blast furnace, and lower down, the lower forge,
can be seen along the creek. The upper forge was not built until the following year.
"MR DE LÉRY'S PLAN OF THE FORGES AT TROIS-RIVIÈRES" (DETAIL), CIRCA 1738,
NATIONAL ARCHIVES OF CANADA, MAP DIVISION, C-8347.

The tenement houses were built in rows with a view to accommodating several families at once. Their design followed an overall plan for a symmetrical arrangement of the dwellings. They formed rather massive buildings with hip roofs (Plate 2.1, to the left of the blast furnace). Archaeological excavations have uncovered two types of tenement houses. The first, very typical, was marked the "foreman's house" on the plan drawn by

surveyor J.-P. Bureau in 1845 (Plate 7.2, No. 33). According to the remains found, this post-and-groove log building consisted of eight dwelling units on either side of a lengthwise partition. The units were in two groups of four, each around a chimney with four fireplaces. Each unit was independent, having its own door.[29] The other type was similar to the houses in a single building described above, but was laid out more symmetrically. Archaeologists have unearthed a long building with four living units side by side under the same roof, with double chimneys built into the party walls.[30]

These tenement houses probably date from the 1780s, after the ironworks had been left in a pitiful state under Alexandre Dumas (1778–83). The new lessee, Conrad Gugy, began the overhaul that would be continued by Davison and Lees, then Monro and Bell. Indeed, the inventory of 1807 (see Appendix 4) lists the new buildings that had been put up by these two administrations.[31] The overhaul coincided with the period in which the village population doubled. The tenements housed the families of the new workers, whose employment contracts stipulated that they would be provided with free lodgings. This proviso explains the presence of other tenements on Bureau's plan (Plates 2.16 and 7.2) and Pigott's water colour of 1845 (Plate 9.2). In the photograph dating from the 1860s (Plate 9.4), two tenement houses can clearly be seen on the site, facing each other across the gully. On the north side, at right angles to the gully, is the foreman's house with a bell turret, which the Forges workmen called "bell row," indicating that there was a row of dwellings, and on the south side, there is another building along the gully, which the workers called "miller's row."[32]

Aside from the houses, there were also more modest dwellings, which are not generally counted in the inventories. The inventories from the French regime list shanties and huts in which the day or seasonal labourers lived. These dwellings are not mentioned afterwards, but that does not mean they no longer existed. Small houses that were not company property were occasionally listed, but considered separately.[33] An inventory from 1785 thus mentioned "several little houses on the Post," and later, in 1807, "fourteen little houses built by labourers and workmen associated with the place and living there" were mentioned.[34] The building of these little houses suggests that the 1739 policy of granting land to craftsmen and carters was finally being implemented.[35] A good example is the case of edge-tool maker Pierre Bouvet, to whom "a grant of a square arpent above Marineau's" was to be made in 1740.[36] The presence of small houses later on would have been the result of the extension or adaptation of this policy, which allowed employees to settle at the Forges on their own property.

Plate 9.10
Overall plan of the Forges, with location of ruins excavated.
DRAWING BY LOUIS LAVOIE, PARKS CANADA, QUEBEC, 81-25G1-4.

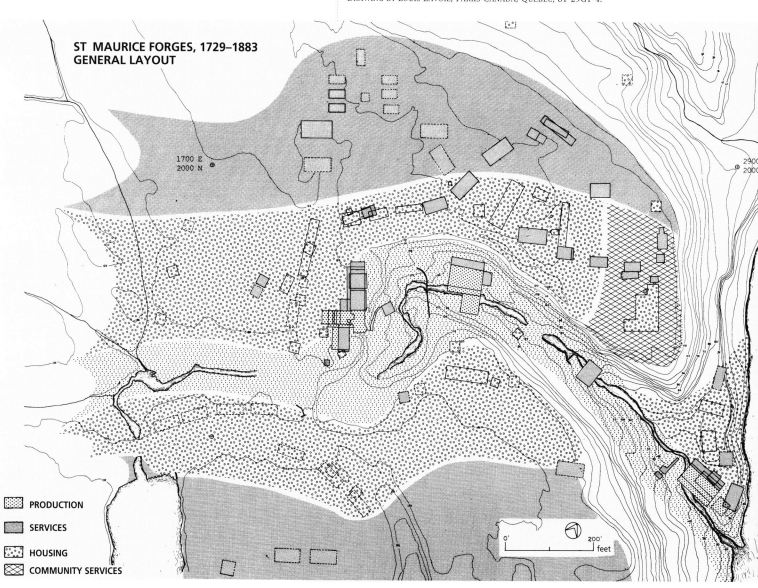

**ST MAURICE FORGES, 1729–1883
GENERAL LAYOUT**

1700 E
2000 N

2900
2000

PRODUCTION

SERVICES

HOUSING

COMMUNITY SERVICES

0' 200' feet

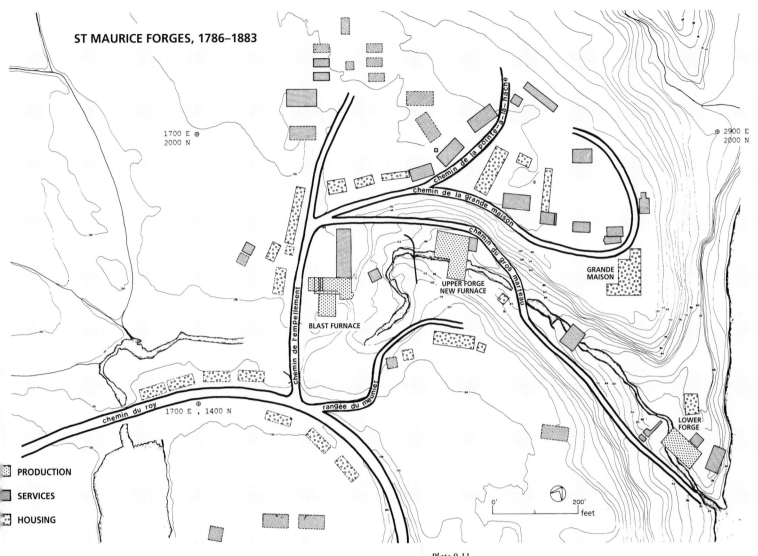

ST MAURICE FORGES, 1786–1883

1700 E ⊕
2000 N

⊕ 2900 E
2000 N

chemin de la pointe-à-la-hache

chemin de la grande maison

chemin du gros marteau

chemin de l'empellement

GRANDE
MAISON

UPPER FORGE
NEW FURNACE

BLAST FURNACE

chemin du roy 1700 E , 1400 N

rangée du meunier

LOWER
FORGE

▨ PRODUCTION

▨ SERVICES

▨ HOUSING

0'   200'
└──────────┘ feet

Plate 9.11
Roads at the Forges.
DRAWING BY FRANÇOIS PELLERIN, PARKS CANADA,
QUEBEC, 81-25G1-5.

The shanties, and later the small houses, may only have been temporary accommodations, occupied by seasonal workers—chiefly carters, colliers and other labourers.[37] The shanties listed in the inventories from the French regime were simple log constructions with neither foundations nor wooden floors, having "earthen" chimneys. It is hard to imagine families braving the Canadian winter in such dwellings, which offered little protection against the cold.

The craftsman's shop was the last type of dwelling. It was a building with two or three rooms, one of which was used as a workshop. These were the homes of farriers, carpenters, joiners and wheelwrights. Some craftsmen shared a shop, as did the carpenter and the wheelwright in 1741, or the blacksmith and joiner in 1807.

THE GRANDE MAISON

A house was needed for the masters,
the chaplain, the clerk and the goods [...]
We reckoned that a stone building [...]
would save us from the fires [...] that are
all too frequent in this country, especially
in the middle of the woods
and in the vicinity of a forge.

François-Étienne Cugnet[38]

In the 18th century, an ironworks generally included a "house" serving as the ironmaster's residence. It also housed the clerks, servants and occasionally a few workmen. At the St Maurice Forges, the "Grande Maison," known as the "master's house" or the "iron-master's house," was also called the "main house" or even the "inspector's house."[39] The building served a number of purposes, however, for it was also used as a warehouse, storehouse, company store and even place of worship. In his 1735 plan, Vézin had intended that the ironmaster should have only a house of logs "lathed and roughcast inside and out," and covered with planks and shingles. He estimated the cost of construction at 4,000 *livres*.[40] He was to get much more, but he had to spend five times as much. Once it was built, amid much controversy, the Grande Maison at the Forges would be one of the most immense private residences in New France.[41]

Plate 9.12
St Maurice Forges, circa 1840 (anonymous wash drawing). At the foot of the big hill can be seen a slag heap in front of the long blast furnace headrace. The Grande Maison is on the far right.
PARKS CANADA, CULTURAL RESOURCE MANAGEMENT, QUEBEC.

Plate 9.13
Remains of corduroy road
unearthed at the Forges.
PHOTOGRAPH BY PARKS CANADA,
25G-82R9X-3, 1982.

Due to the high cost of construction, the huge mansion was never built in its entirety (Plate 9.14). Construction was begun in the spring of 1737, and ended in the fall of 1738. According to the original plan, the Grande Maison should have had two rear wings, one on either side, with a courtyard between them, but in the end, only the south wing was built. The original construction was never significantly altered as long as the Forges were in operation. It was not always well maintained and required major repairs at times, including at the beginning of Monro and Bell's tenure and after a fire in 1863, when John McDougall had just taken possession of the establishment.[42]

The style of the house is similar in some ways to the architecture of Burgundy, where the first ironworkers and ironmasters came from. The pitch of the roof, the framing on purlins, and the many supporting walls that compartmentalize the interior space, are typical. The existing remains of the building reveal the five separate cellars created by the thick limestone inside walls that rise up to the ridge of the roof (Plate 9.15).

Situated at the edge of the northern plateau, overlooking the St Maurice River, this imposing house made a strong impression even when it was being built. It seemed to be out of proportion to the rather modest ironworks that surrounded it. In 1739, the King's engineer, Chaussegros de Léry, was the first to criticize it, deeming it too costly:

> I see that they have built a house to live in, too fine by far, which cost them a great deal [...].

The building was clearly designed to prevent the spread of fire, and this was the reason Cugnet cited to justify the construction costs (see the epigraph to this section). The main block, 16.5 m deep, extended over a length of 32 m on the river side. The south wing, which is close to 8 m in width, extended the building to a depth of 24 m. The cellars were used for storage of supplies and wares. The ground floor had approximately 10 rooms, while under the roof, with its six chimneys and 11 dormer windows, were two huge attics, one above the other (Table 9.1).

A stone supporting wall divided the ground floor along its entire length. The apartments of the ironmasters and managers looked out over the St Maurice, while the rooms used as offices, the company store and apartments for the clerks and foremen gave onto the village. During the French regime, as well as during Mathew Bell's tenure, one of the rooms of the house was made into a small chapel.[44]

According to director Cugnet, it cost 20,000 *livres* to build the Grande Maison, whereas according to ironmaster Vézin, who was opposed to the project, it cost four times that amount.[45] A large number of masons, quarrymen, carpenters, joiners and labourers were put to work on the project, which was perceived from the outset as being extravagant. From that point on, this imposing building, which dominated the humble dwellings of the workers, would symbolize the prestige and lavish lifestyle of some of the Forges masters. In their time, Vézin, Christophe Pélissier, Pierre de Sales Laterrière and Mathew Bell played host to important guests, and they gave parties and receptions that remained etched in the collective memory of the people of the Trois-Rivières region[46] (Plate 9.16).

When the company ceased operations, the building was gradually demolished by the local inhabitants, who were authorized to use the cut stone for their own purposes. Photographs from the early 20th century show that the main part of the building had already been completely demolished, only the cellars remaining intact[47] (Plates 9.17a and 9.17b).

Table 9.1	INSIDE THE GRANDE MAISON[43]	
Divisions	**According to 1741 inventory**	**According to 1786 inventory**
Ground floor	one room	Room of R. Brydon, chief clerk
	two bedrooms	storeroom
	chaplain's room	large room
	chapel	five small rooms
	two small rooms	
Wing		
Annex	kitchen with brick oven	kitchen with bed
First attic	attic	attic
	four panelled bedrooms	one bedroom (one bed)
Second attic	attic	
Staircases	two to the attics, one to the cellar	
Cellars	four walls with four doors	cellar

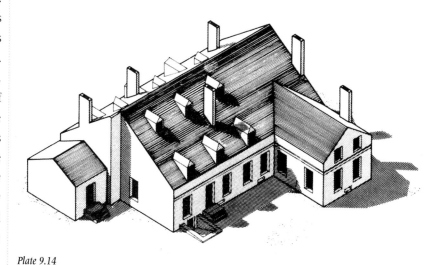

Plate 9.14
The Grande Maison, unfinished.
JEAN BÉLISLE, "LA GRANDE MAISON DES FORGES DU SAINT-MAURICE: TÉMOIN DE L'INTÉGRATION DES FONCTIONS: ÉTUDE STRUCTURALE," MANUSCRIPT REPORT NO. 272 (OTTAWA: PARKS CANADA, 1977).

Plate 9.15
Remains of the cellars
of the Grande Maison.
QUEBEC, MINISTÈRE
DE LA CULTURE ET
DES COMMUNICATIONS,
MICHEL GAUMOND
COLLECTION, 1967.

UTILITY BUILDINGS

Other Workshops

At different times, there were a variety of workshops where moulds were made and where items manufactured at the blast furnace and two forges were finished and assembled. Directly related to the industrial production of the Forges, the finishing shops were different from the farrier's or blacksmith and edge-tool maker's shops, or smithies, which made the tools and hardware needed to run the ironworks. Yet there is reason to believe that the division between these types of work, performed in different shops, was not always so strict, and that smiths sometimes worked on finishing certain ironwares. Remains found may be those of a double forge mentioned in documents from the French regime, where finishing work of this kind could have been done.[48] The trimming and sanding of castings

Plate 9.16
Yard of the
Grande Maison,
late 19th century.
COLLECTION OF
DAVID MACDOUGALL,
CIRCA 1895 (DETAIL).

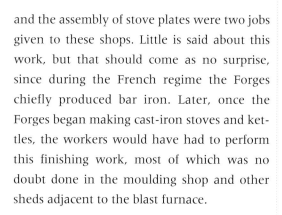

Plate 9.17a
The Grande Maison, derelict.
HENRY RICHARD S. BUNNETT, *THE HOUSE
OF THE FORGE*, OIL ON CANVAS, 1886,
MCCORD MUSEUM OF CANADIAN HISTORY, M-737.

Plate 9.17b
Demolition of the Grande Maison at the end of the 19th century.
COLLECTION OF ERIC SPRENGER.

and the assembly of stove plates were two jobs given to these shops. Little is said about this work, but that should come as no surprise, since during the French regime the Forges chiefly produced bar iron. Later, once the Forges began making cast-iron stoves and kettles, the workers would have had to perform this finishing work, most of which was no doubt done in the moulding shop and other sheds adjacent to the blast furnace.

As more and more casting was done, beginning in the last quarter of the 18th century, the Forges had to build shops for making moulds and finishing fine castings. The 1807 inventory mentions a "pattern shed," that is, a workshop for making wooden casting moulds. Remains dating from the early part of the 19th century indicate that finishing was done in another workshop identified as a "stove shed." The fact that a "ware dresser" is listed on the employee roll of 1829 and the 1831 census indicates that there was a certain degree of specialization in the finishing work at the time. The accounts of 1857 also mention the payment of a worker to do "ware cleaning."[49] Another shop, built after 1845, was also involved in making moulds. These shops had storage areas, as well.[50]

Shops

We saw earlier that craftsmen's shops were built into some dwellings. All the ancillary work required to keep the ironworks running smoothly, including the maintenance of tools and machinery, was performed in these shops. Two shops were inventoried in the 18th century, one used for ironworking and the other for woodworking. The smithy listed in Estèbe's inventory of 1741 seems to have been used for all kinds of ironwork done by the same craftsman, since it served as an edge-tool maker's shop,[51] locksmith's shop and farrier's shop. The woodworking shop was just as multipurpose, since carpentry, joinery and wheelwright's work were done there.[52]

In the 19th century, the woodworking shops tended to specialize. The inventory of 1807 lists two smithies, one blacksmith and joiner's shop, and one carpenter's shop. The remains of smithies that have been unearthed by archaeologists reveal that the blacksmiths were still just as versatile; they performed as farriers, but also made parts—such as strap hinges and pintles for stove doors—needed for finishing Forges products.[53] As for woodwork, the fact that two separate shops for the joiner and the carpenter are mentioned suggests that the needs of the establishment, which had become more densely populated since the early 19th century, had increased. Three or four woodworkers lived at the Forges henceforth. Later, from 1845 on, the wheelwright, too, had his own workshop, adjoining his house.[54]

Mills

According to a statement of account prepared by director Cugnet in 1741, it took 16,000 thin boards and 5,700 heavy planks to build the Forges (from 1736 to 1739).[55] Although this wood was not sawed in a mill on the site, it was not long before one was built at the Forges. A "sawmill with its utensils" was listed in an inventory of 1760, along with a saw. The "building on posts, 40 feet long by 20 feet wide, together with its works," which is the next item in the inventory, could well be the structure housing the mill with its wheels and gear mechanisms. A building of the same size and type, "covered only," with "its works" was listed, along with two saws, in an 1807 inventory. This may have been the mill machinery, which was covered only by a rudimentary roof[56] (Plates 9.18a and 9.18b). Some indications from the 1785 estimate suggest that it was installed near the lower forge dam.[57] Furthermore, water colours from the 1840s show a building standing beside the creek, which seems to correspond to the saw and charcoal mill of the Bureau plan of 1845 (Plate 7.2, No. 5). A new mill "containing two runs with one saw each" was built in the 1850s, probably in the same area; it was equipped with circular saws. The charcoal mill, which was in the same building as the sawmill, according to the 1845 plan, had already been mentioned 15 years earlier by Lieutenant Baddeley, who situated it at the same spot. Charcoal was ground in this mill to make a powder used by the moulders.[58]

Plate 9.18a
Sawmill in Baie-Jolie with the saw run from Ernest Marchand's
sawmill at the Forges after they closed.
PHOTOGRAPH BY PARKS CANADA/MICHEL BÉDARD, 1977.

In its initial years of operation, the Compagnie des Forges de Saint-Maurice had the 3,000 or so bushels of wheat it procured annually from the Government of Montreal ground at Sieur Tonnancourt's mill, at Pointe du Lac, and at the Jesuits' mill, at Cap de la Madeleine.[59] This practice no doubt continued throughout the French regime. In 1743, in an effort to reduce the cost of grinding and hauling the flour, which amounted to close to 1,300 *livres*, a proposal was put forward, in a memorial, to establish a grist mill near what would later become the washery pond, "two arpents above the furnace." It was also thought that part of the cost of the mill might be paid by attracting the local *habitants* to come and grind their wheat there, especially when the Tonnancourt mill could not operate because of low water.[60] The grist mill was not built until the 1770s, however, and we do not know where it was located, or what kind of equipment it had. In 1786, notary Papineau drew up an inventory that included a "grist mill,

the works, the bin, a pair of millstones and fittings,"[61] and the 1807 inventory mentions a grist mill equipped with a pair of millstones that had been built by Monro and Bell. There was a miller at the Forges at that time.[62] This recent mill was apparently a building on pilings, located near the blast furnace wheelrace. Dollard Dubé made a drawing of it, and archaeological digs downstream from the tailrace of the great wheel at the blast furnace have unearthed a race that apparently served to convey water to the spot where the building stood on pilings, in order, it seems, to drive an undershot wheel.[63] This building can be seen in the water colour by Pigott dating from 1845 (Plate 9.2), and even more clearly in the photograph from the 1860s (Plate 9.4). Another grist mill was apparently built by Onésime Héroux in a huge brick shed near the creek. This building, located just downstream from the upper forge (Plate 9.19), was also used as a workshop and craftsman's shop.[64]

Sheds and Stores

We have already seen that the furnace and forges had their own storage areas for iron wares. A store had also been built at Trois-Rivières, in the first few years of operation.[65] More ironwares were kept at the Grande Maison, the cellars and attics of which were largely devoted to storing victuals and goods for the company store. Around the 1870s, it seems that another building, called the "blue store," was used as well. It was located on the *chemin de l'empellement*, near the blast furnace complex.[66] Large quantities of wood, commodities, and all sorts of other goods were kept in warehouses called "sheds"and "barns," which varied in number and location over time.[67]

Plate 9.18b
Saw run from Ernest Marchand's sawmill
at the Forges after they closed.
PHOTOGRAPH BY PARKS CANADA/MICHEL BÉDARD, 1977.

Every effort was made to lay in stocks of charcoal that would last at least a year, and this required large storage areas. The blast furnace and each of the forges had its own charcoal shed. A charger was employed to make up the furnace charges, cutting the pieces of charcoal to the right size and assembling the right quantities.[68] Additional storage areas had to be created for the stockpiles of charcoal. From the outset, Vézin had recommended the construction of "sheds of little expense with removable roofs" at the charcoal pits for storing the charcoal that was to be hauled by sled in winter.[69] It is possible that some of the charcoal may have been stored at the pits. During the French regime no mention was ever made of charcoal sheds other than those adjacent to the shops, on the post itself. It was not until the 19th century that additional charcoal barns were listed on an inventory. In 1807, there were two large ones that had been built under Monro

Plate 9.19
Héroux's mill with millpond downstream from the upper forge, circa 1940.
ARCHIVES DU SÉMINAIRE DES TROIS-RIVIÈRES, ALBERT TESSIER COLLECTION.

and Bell. The 1845 plan prepared by surveyor Bureau (Plate 2.16) shows 10 of them, a few of which are represented in period drawings and in the 1860 photograph of the site (Plate 9.4).

Stables and Outbuildings

The ironworks had a good number of stables and outbuildings where the horses belonging to the company and its workers, as well as those of casual carters, could be sheltered and tended. We know that Francheville's establishment had a stable down by the river. A large building, measuring 50 m by 26 m, was built as early as 1737. It was later altered on several occasions. Fodder was stored in a hayloft. In the 1807 inventory, four smaller stables were listed, the largest measuring 15 m in length. Archaeological digs have unearthed large stables dating from the 1820s. Bureau's plan indicates these large stables, and period illustrations, show the building in the background to the north, near the *chemin de la pointe à la Hache*. The 1845 plan also indicates some small stables standing together on the south plateau and at the foot of the big hill (Plate 2.16); they were probably used for the workers' horses. There was also a cowshed near the large stables.

In the inventory of 1807, and on the plan of 1845, barns are indicated along the *chemin de la pointe à la Hache*. Two other large barns (one measuring 30 m and the other 18 m) were built in the 1850s.

Bakery, Icehouse and Wells

A bakery was built, early on, to the north of the Grande Maison. This simple shack, made of stakes during the French regime, was renovated several times before taking on its final form as a building of respectable size in the 19th century. The remains of the bakery that have been found reveal two foundations, the larger of which formed a rectangle 10 m by 7.3 m.[70] At the north end of the building, the fieldstone foundation of the bread oven can be seen. The first bakery, listed in the inventory of 1741, had a brick oven, "surrounded by masonry," with a capacity of 6 bushels, and two large kneading troughs.[71]

The inventories of 1746 and 1748 mention an icehouse that seems to have been later converted into a shed. A rectangular pit unearthed in an archaeological dig near the kitchen of the Grande Maison could be an icehouse. It was apparently abandoned around 1770, and then put back into use in 1785. Archaeologists have discovered two similar pits on the south and north terraces, near dwellings.[72]

At the request of director Cugnet, the possibility of digging a well near the Grande Maison had been considered in 1740.[73] Although no such well is mentioned later, and although no trace of it has been unearthed, it would be rather surprising if a well had not been dug at this location, since the water table is easy to reach. The only well mentioned in the 1860s is that of Onésime Héroux, which was located close to his farm.

It is also possible that, since it was easy to draw water from the creek, the inhabitants did not feel the need to dig a well at the site. We saw in Chapter 2 that a "water cart" was mentioned in 1786. Dollard Dubé noted that in the McDougalls' time, an employee had the job of drawing water from the washery pond to fill up a "tun" installed on a cart and distributing it to the households.[74] And Benjamin Sulte, who lived at the Forges for a few years at that time, reported that the inhabitants drank the ferruginous water of the creek, to which he attributed certain virtues:

> [...] in it I found the taste of iron that gives the blood of the people of the Forges the vigour, health, ardour and spring in the step that one does not usually associate with working people. The gait, cheerfulness and bearing of the women and men of this small group of families, the spryness of their movements, all come from the Forges creek, the forest air, the elevation of the land.[75]

Chapel

One room in the Grande Maison served as a place of worship for about 10 years. Later, around 1750, a chapel was built nearby. The inventory of 1760 lists a "church" of *pièce sur pièce* log construction measuring 13 m by 9.7 m. After that date, the building was no longer used as a chapel, and towards the end of the 18th century it served as a shed, while, once again, another room of the Grande Maison, in the lower attic, was used as a chapel. In the 1850s, a second chapel was built, this time at the edge of the site, west of the washery pond, before the Forges farm[76] (Plate 9.20).

The first chapel in the Grande Maison was particularly well appointed: The church plate was estimated to be worth 1,352 *livres*. At the time, the company maintained a chaplain at its expense, who lived at the Grande Maison.[77]

The archival records make no specific reference to a cemetery at the Forges. Napoléon Caron, referring to Sulte, claims that there was a cemetery "along the St Maurice," at the site of what was to become a huge garden.[78] Sulte appears to have based his hypothesis that a cemetery existed from the outset at the Forges on death certificates from 1745. Two such records, concerning the young children of Joseph Aubry, contain the following specific information: "buried at the request of the parents in the Trois-Rivières cemetery" and "buried at Trois-Rivières with the permission of Clément Lefebvre [the chaplain]."[79] The fact that the parents asked for their children to be buried in the Trois-Rivières cemetery suggests that the dead were usually interred on the post itself, or else nearby. Nevertheless, this is the only statement supporting the hypothesis. In the case of two other deaths, in 1747 and 1748, it was specified that "the body was in due course interred in the cemetery of this parish."[80] This suggests, in contrast, that the dead were buried in the cemetery of the Parish

Plate 9.20
Chapel at the Forges, early 20th century.
PARKS CANADA, CULTURAL RESOURCE MANAGEMENT, QUEBEC.

▼

of l'Immaculée Conception de Trois-Rivières. Archaeological excavations have not revealed any traces of a cemetery, but the site of the Forges has not been dug in its entirety. Only digs in very specific areas may turn up new findings. The existence of a cemetery in the 19th century has not been confirmed either, and it appears that in the second half of the century, the dead were buried either at Trois-Rivières or at St Étienne des Grès after 1859, the year when the parish was raised to official canonical status.[81]

School

An enumeration done in 1838, which put the population at 393, found that only 14 people at the Forges knew how to read and write.[82] Luce Vermette points out that even though Mathew Bell was president of the Education Society of the Town of Three Rivers at the time, he does not seem to have provided a school for the children of the workers of his establishment. She also notes the exceptional case of the founder John Slicer, who sent his sons William and Henry to a school in Bath, England![83]

The 1871 and 1881 censuses were the first ones that recorded a schoolmistress.[84] The last workers interviewed by Dollard Dubé told him there was a school "in a small house on the east side, near the dam, to the right of a large building with five apartments." This was most likely the residential area at the foot of the big hill, next to the blast furnace pond. According to Dubé, the school had 25 to 30 pupils in the 1860s.[85] In 1871, the census taker counted 64 "schoolgoers"; in 1881, he counted 60. At the time, most of the adults at the Forges were illiterate.[86] The schoolmistress boarded at the Héroux farm,[87] where the post office was also located, at the expense of the people of the Forges.[88]

CROPS, GARDENS AND PASTURELAND

Before an industrial community could be built from the ground up in the midst of the St Maurice forest, the means to feed the community had to be planned and provided. The colonial authorities examined this question before setting up the establishment. There were plans to cultivate the surrounding land to produce enough food to feed the workers and horses, but the sandy soil of the St Maurice seigneury did not lend itself to extensive farming. This fact was subsequently confirmed when, in the 19th century, the monopoly exercised by the masters of the Forges was contested by settlers who wanted to move onto the land.

Before embarking on his tour of observation, in the fall of 1735, Vézin was given instructions that noted three disadvantages to establishing the Forges at the location chosen by Francheville, in the fief of St Maurice. It was pointed out that since the seigneury was "not settled," it would be difficult, because of the poor quality of the soil, to attract *habitants* from within four leagues to go there. The cost of bringing food in from outside was also high. The other site that had been considered, on the Batiscan River, offered a distinct advantage from this standpoint. The Batiscan seigneury was already settled, and "the wheat, peas, salt pork and other victuals required for the subsistence of all the workers who would be employed at the forges" could be found there.[89] The fact that people were already living there was also seen as an advantage, since it would be possible to find workers for the Forges nearby, and labour costs would be lower than at St Maurice. After weighing the pros and cons of the two sites, Vézin still recommended that the establishment be built at St Maurice. In his "Observations," he noted laconically, regarding the question of victuals, that supplies should:

[...] be procured from places where it would be cheaper and the necessary provisions made in due season. Food is plentiful and cheap in Canada.[90]

Food supplies would therefore come from outside, chiefly from the neighbouring seigneuries of Pointe du Lac and Yamachiche. And having to rely on the Quebec and Montreal markets was not without its disadvantages, especially in years when food was scarce[91] (1737 and 1738), since purchases could not always be made so cheaply. Throughout the French regime, there are records of purchases of food, hay, oats, wheat and flour from local *habitants*,[92] but there is no evidence to suggest that the company had taken up farming at the time. Later, although the lessees would be authorized under the terms of their lease to cultivate the land for their own needs, it seems that they did not take advantage of these provisions. The 1784 census mentions 120 arpents of cleared land, but gives no particulars on harvests. Hay, however, was apparently harvested at the time, and perhaps even earlier.[93]

It was not until the 19th-century censuses that the existence of a farm at the Forges was recorded. In 1815, Surveyor General Bouchette had mentioned that only a small part of the fief of St Maurice was cultivated, without being any more specific. The employee roll of 1829 includes a farmer, William Hooper, who had been listed in the census of 1825 and was listed again in the fief of St Maurice in 1831.[94] Under Hooper's name, the census taker reported 60 arpents of land under cultivation, plus a harvest of 50 bushels of peas and 3,000 bushels of potatoes for the preceding year[95]; he also listed a servant working for the farmer. The development of farming probably began when Bell brought new families into the village, chiefly after 1810. In 1851, the census taker recorded 140 arpents being farmed, and then 160 arpents in 1861. In 1861, the farm at the Forges was appraised at £750 ($3,000). It was the farm, incidentally, that interested Onésime Héroux the most when he decided to buy the Forges in 1862[96] and he kept it when he sold the Forges to John McDougall in 1863. Héroux died in 1865, and three months later his widow married Dr Beauchemin, who continued to run the farm.[97] By 1871, there were 220 arpents under cultivation, chiefly oats (1,000 bushels), potatoes (1,000 bushels), hay (500 bales) and corn (16 bushels).[98] We saw in Chapter 2 that the amounts of hay and oats harvested represented only a tiny fraction of the quantities consumed each year at the Forges. The 1871 census also tells us that workers grew potatoes, rye, buckwheat and corn beside the Beauchemin farm.[99]

The documents do not specify the farm's precise location. In his short report of 1861, H. R. Symmes, who appraised the farm at $3,000, stated that the "farm is good." The Forges were not in operation at the time of his visit, and he said that "the fencing upon the farm had almost entirely disappeared."[100] According to Dollard Dubé, Héroux's farm and house were on the road to St Étienne, west of the washery pond, and set back a little[101] (Plate 9.21). As we saw earlier, the McDougalls also had a farm at the L'Islet Forges. The photograph of the site taken several years later does not show that part, but fenced-off areas can be seen, and the large fields visible on the edges of the establishment give every appearance of being cultivated (Plate 9.4).

The workers' kitchen gardens are evidence that they grew some of their own fruits and vegetables. The families at the Forges must always have kept small vegetable patches, and as we shall see later, raised some livestock, as well. As in the case of the farm, however, little mention is made of this until the 19th century. Some references from the late 18th century are fairly explicit, however. The first dates from 1771.[102] Writing about the years he spent as inspector and then as director at the Forges, from 1775 to 1779, Pierre de Sales Laterrière, turned physician, mentions "good, beautiful gardens" for the workmen.[103] In 1808, John Lambert wrote of "the habitations of the superintendent and workpeople belonging to the establishment, with their little gardens and plantations."[104] Twenty years later, Mathew Bell spoke of the workers' little gardens and his own next to the Grande Maison.[105] A gardener, Joseph Mendes, is moreover listed on the employee roll of 1829, which suggests that the land surrounding the Grande Maison may have been planted with flowers. Sulte, who lived at the Forges for a while, spoke of parterres and flower beds. In the 1871 census, under the headings "gardens and orchards," the

Plate 9.21
Farm next to the Forges, early 20th century.
PHOTOGRAPH BY PINSONNEAULT, TROIS-RIVIÈRES,
PARKS CANADA, CULTURAL RESOURCE MANAGEMENT, QUEBEC.

census taker noted one arpent under the names Robert and David McDougall. David McDougall was a flower lover who apparently cultivated a variety of yellow rose.[106]

In the Henderson photograph (Plate 9.4), behind the Grande Maison and beside the new house the McDougalls had built, fenced-off areas that are clearly gardens and plantations are visible. Other enclosed areas near some workers' houses can also be seen. Napoléon Caron, in 1889, like the last workmen questioned by Dollard Dubé, spoke of beautiful big gardens.[107] We have seen that potatoes, buckwheat, rye and corn were grown at the Forges in 1871. Traces of barley, wheat and pumpkin have been found in 18th-century macroremains from the site. Many traces of wild plants have been found as well, including raspberry and medicinal plants such as sarsaparilla, hawthorn, strawberry, stonecrop and sumac.[108]

Table 9.2

LIVESTOCK AT THE ST MAURICE FORGES, 1741–1871 [112]

	1741	1746	1760	1764	1784	1804	1831		1851		1861		1871		
							Co.	Workers	Co.	Workers	Co.	Workers	Co.	Workers	Dr Beauchemin
Horses	24	30	6	6–7	22	30–40	55	22	14	13	1	19	33	7	9
Pigs		3			4		100	22		6		6	15	43	20
Cattle		1 c*			25		60	83	8	40	8		17	8	27
					23 c				7 c	31 c	4 c	10 c	8 c	5 c	19 c
Sheep					0		60		13	1		3	24	17	63
Total	24	34	6	6–7	51	30–40	275	127	35	60	9	38	89	75	119

*c: cow(s); included in total number of cattle, where applicable.

Domestic Animals

Workers at the Forges always kept livestock and poultry, and starting in the late 18th century, the company itself owned animals. Estate inventories dating from the French regime indicate that many workmen owned poultry, pigs, cows and a few oxen.[109] The problem of setting aside pastureland soon arose. As early as 1740, Sieur Cugnet suggested building a pen. That year, a nearby pond had been drained to create a "meadow." But an ordinance had to be issued to prohibit "all forgemen and residents of the Forges from letting their animals wander into the reserves, woods, copses and other lands belonging to the Forges."[110] This was an attempt to protect new growth in the areas cleared to make charcoal in order to renew the resources. But these instructions were scarcely obeyed. Five years later, another, more explicit ordinance had to be issued:

> Workmen, carters and other residents shall be permitted to raise as many cows and sheep as they wish. But they must ensure that they do not go beyond the land behind the stables, on pain of a fine of 10 *livres* to owners allowing them to get into the reserves beyond the big hill. [...] Belleville and Cressé will issue a special regulation to hire a herdsman at the owners' expense.[111]

Murray's map of 1760 (Plate 2.17) shows cleared areas around the etablishment that must have been used for pasture and hayfields. The number of livestock increased later (Table 9.2).

The 1831 census was the first to give precise figures on the type and number of livestock. The census taker counted 402 animals on the site, including 55 horses belonging to the company and 22 to the workers. In addition to its horses, Mathew Bell's company had 60 cattle, 60 sheep and 100 pigs. The other 105 animals belonged to the workers. Among the resident employees, the skilled workers owned the most livestock, including half of the horses and pigs, and over half of the cattle. Nevertheless, almost all the workers had at least one horse, one head of cattle and one pig. Only the company owned sheep.

The Forges were idle at the time the census was taken in 1861, so the census figures indicate a drop in the number of livestock owned by the company in comparison with 1851 (Table 9.2). We also know that under John Porter & Company, in the 1850s, the company still had a herd and farm implements that would later be appraised in connection with a legal proceeding.[113] According to the 1871 figures, there were still livestock on the site, and although there were fewer workers than before, they kept more pigs. The McDougalls and Dr Beauchemin, who had been running Onésime Héroux's farm since 1865,[114] owned most of the animals. The produce from this farm no doubt accounted for a significant part of the establishment's provisions.

MAINTENANCE OF THE ESTABLISHMENT

Throughout the 150 years that the establishment was in operation, the workshops, dwellings and other buildings were repaired, converted, demolished or replaced. While regular maintenance and repair work is normal in an industrial setting, it seems that it was not as frequent as would be expected, if we are to believe the descriptions and inventories we have. There appears to have been a tendency to let the buildings become quite dilapidated before undertaking any repairs. This no doubt explains why some observers reported that the workers lived in shanties. Their impressions depended on the time of their visit. The establishment was in its worst state of repair when the lease approached its expiry date, and many complete overhauls of workshops, dwellings and other buildings were carried out by new lessees or owners. In 1829, after major repairs were made under Mathew Bell, a visiting doctor commented favourably on the dwellings:

> Mr. Bell's workmen appeared contented and comfortable; they occupied good cottages, with a small plot of garden attached to each.[115]

Yet some 15 years later, Étienne Parent's report to the government gave a quite different impression:

> It is stated that the only residents of St Maurice are the common workmen who live in shanties or small log houses the construction of which is not intended for permanent residence.

> The Committee learns that one or two clerks of the Lessee reside upon the spot, and are properly accommodated in the only buildings which are entitled to the name of houses.[116]

It is worth noting here that the above report was written at a time when Mathew Bell's monopoly was being contested, and in the end the sale of the Forges was recommended. The new masters would put no greater effort into the upkeep of the place. And, as we have seen, James Ferrier was taken to court by John Porter & Company for damage caused to their establishment.

THE OLD FORGES

In its century and a half of existence, the first industrial village in Canada was never more than a post or industrial establishment. Although it looked like a village of respectable size, according to visitors, and offered services somewhat akin to a village, the Forges remained until the end a village unto itself. Its strictly industrial vocation and population of skilled workers always defined and limited its social character. A century after their founding, people were surprised that a real village or even a town had not sprung up around the Forges. But there is nothing in the history of the Forges to indicate that anyone had ever envisioned the establishment's developing in that way. Even the famous Mathew Bell, who wished to obtain recognition as a "Provincial establishment," only saw that as official confirmation of a status it already enjoyed. The Forges had been built on the model of an 18th-century industrial establishment, to be worked for maximum profit with no thought for expansion and growth. The industrial world it parallelled was changing and expanding rapidly, so sooner or later, the St Maurice Forges were bound to become the old Forges and slip into the realm of legend and history.

Site of the lower forge, with chafery chimney intact, on the west bank of the St Maurice.
PHOTOGRAPH BY PARKS CANADA/JACQUES BEARDSELL,130/00/ PR-7/SPO-00058, 1985.

Conclusion

The St Maurice Forges constitute one of the most important legacies of the French regime. In the words of Albert Tessier, the final French page of the history of the Forges, like that of Canada itself, reflects little glory on the mother country.[1] The French officials who were its masters clearly neglected to develop the establishment's full potential, preferring to run it for their personal profit. Nor did the wars of the closing years of French rule bring much benefit to this industry, which had been founded during the regime's most prosperous decade.[2] Yet the potential of the Forges was real, as the British were quick to realize; in the immediate aftermath of the Conquest, they saw to it that the ironworks was kept running by ordering the key craftsmen to stay on the job. Subsequent history was not only to confirm the viability of this first Canadian industrial enterprise, but also show how much it contributed to supplying the country's demand for manufactures.

As evinced by the conditions under which the Forges were founded, exploitation of Canadian iron resources was to all intents and purposes a state-run business. In fact, the Forges were to remain Crown property for most of their history. This ubiquitous government presence was not due solely to the colonial situation, wherein, given the small size of New France's economy and level of wealth of its citizens, private capital was simply inadequate to the task of establishing such an industry. Even in the mother country the government subsidized and controlled the development of the iron and steel industry, often intervening directly. Of course, it has to be said that this was not just any industry; it was capital-intensive, and its output was of eminent strategic importance. It produced not only numerous items of everyday use, but also tools and manufacturing equipment, as well as arms and munitions. Such an industry could not expect to operate without the state, which could at any time become its leading client, seeking to exercise some degree of oversight. Strategic concerns over development and defence of the country were not the only grounds for state interest in this industry. Government action was needed to create the conditions for growth of an industry so demanding of resources and land. To start with, mining was within the royal prerogative, so that even when mining rights were granted, access to water, mineral and timber resources were privileges emanating from the King of France. Without resources, there could be no

Blast furnace interpretation centre, with remains
of the last furnace and surrounding shops.
PHOTOGRAPH BY PARKS CANADA/JACQUES BEARDSELL,
130/00/ PR-7/SPO-00006, 1985.

development, and without ready access to these resources, there could be no sustainable development. Resources were thus the bedrock of the Forges' history, and state "protection" of these resources, in the form of the privilege of free access, was to have a direct bearing on their longevity. With this privilege, the enterprise had the advantage of low production costs for 125 years (1738–1863); when this privilege was withdrawn, it lasted less than 20 years (1863–83), allowing for shutdowns.

Under the French regime, the vast land reserve allocated to the Forges was as good as a guarantee of an inexhaustible supply of raw materials. The problem, then, was less one of rights to resources than of collecting them, since there was not always enough manpower available at the right time. After the Conquest, more restricted access to mineral and timber resources became a central concern of the masters of the Forges. Territorial protection and expansion became a real obsession for Mathew Bell, who sought to retain control over the huge reserve of the French period, to the point where, as the Trois-Rivières population swelled, people began to see his insatiable thirst for land as indicative of monopolistic ambitions. The sale of the Forges in 1846 was a direct

View of remains of the founder's quarters in the blast furnace interpretation centre.
PHOTOGRAPH BY PARKS CANADA/JACQUES BEARDSELL, 130/PE/ PR-7/SPO-00018, 1985.

Reconstructed Grande Maison with modern interior providing exhibition space.
PHOTOGRAPH BY PARKS CANADA/JACQUES BEARDSELL,
130/IN/ PR-7/SPO-00037, 1989.

consequence of this land issue. Thereafter, though the government continued to grant land concessions to the first sets of proprietors, the history of the enterprise would never be the same. For the first time since the days of Cugnet et Compagnie, the masters of the Forges would have to deal with chronic indebtedness. Thenceforth, they had to meet the real costs of their plant and raw materials. Forced to pay the true costs of production for the first time—and in a highly competitive market at that—they could not take into the 20th century an enterprise still rooted in the 18th.

The longevity of the Forges was therefore largely attributable to government patronage. Even after their demise, the state championed their cause again, with its involvement in commemorating the site, culminating in the creation of the national historic site in 1973.

Apart from its lifespan, the enterprise's conditions of operation have been the chief object of our attention. Precisely because it operated for nearly a century and a half we have been able to study the enduring factors that made this possible. The successive masters of the Forges were impelled to recognize that development of an industry like this meant allowing for certain internal constraints, one of which was the imperative of a strict schedule. The historiography of the Forges places considerable emphasis on the fact that the enterprise was badly run during the French regime,

Model of the St Maurice Forges as they were mid-19th century, on exhibit at the Grande Maison.
PHOTOGRAPH BY PARKS CANADA/JACQUES BEARDSELL, 130/IN/ PR-7/SPO-00065, 1990.

Mural showing master craftsmen at the St Maurice Forges.
Located at the visitor's centre at the national historic site, it depicts the human side of the first industrial community in Canada.
MURAL BY BERNARD DUCHESNE, PHOTOGRAPH BY PARKS CANADA/ JEAN AUDET, 130/PE/ PR-7/SPO-00041, 1991.

particularly by Cugnet et Compagnie. Taking a longer view, however, it has to be admitted that, construction costs aside, the main setbacks the enterprise faced arose from a failure to abide by the operating schedule. Any delays in laying in raw materials, merchandise and victuals—and hence in hiring the workers who would use or need these things—resulted in higher costs and work stoppages in a plant designed to work non-stop. Whatever the case, the charges of mismanagement levelled at Vézin were just as valid over a century later against someone like Jeffrey Brock of Weston Hunt & Company, commissioned by John Porter & Company to manage the business.

On every front, the operation was characterized by a number of factors that tended, regardless of the period, to keep the business running. On the resource front, only ore and limestone were not renewable, so that control over the lands where they could be obtained had to be maintained. The same principle applied to timber stands, with the added proviso that good forestry practices were required to renew the resource. Yet, as we have seen, it was thanks to the huge timber reserves allocated to them that the operators were able, in the final analysis, to compensate for their lack of planning in this regard. The energy source for driving the machinery was water power, the crucial factor, key to which was firstly the choice of a water source, and then, once this was known to be reliable, the system of dams and ponds through which the energy potential of the creek could be harnessed at every drop in level down the gully. On the labour front, skilled crafts were essentially handed down from father to son, a tradition endorsed by the British order of 1760 to keep the French ironworkers on the job.

The key factor in capital renewal was the growth of the market, spurred by demographic expansion and development of the country. We have seen how the Forges successively supplied the state, contributed to the development and defence of the colony and met the demands of industry. Yet we have to remember that government protection played a major role in keeping the enterprise viable and that it operated for more than a hundred years shielded from national and local competition. It can nonetheless be said, with little exaggeration, that, given the factors that underpinned the Forges' survival until their sale in 1846, the enterprise flourished. State protection, stability of the skilled labour force and weak competition were undoubtedly the key factors in their survival. Once in private hands, in a rapidly changing industrial and economic environment, it would prove difficult to reproduce on the various operational fronts the strategic resource and capital conditions that would have given the Forges a new lease on life.

Over and above the undeniable contribution of the St Maurice Forges to the industrial and economic history of Canada, there is another legacy worthy of mention, namely the knowledge they have brought to the history of technology. A particular focus of our work has been to describe the plant and technology as revealed by archaeological work and perusal of archival sources. The study of the Forges has afforded us a remarkable opportunity in this regard, often to a considerable level of detail. The history of the Forges has thus advanced the history of both technology and science in Canada. In describing the smelting process in the blast furnace and fining in the forges we gain great insight into the current state of knowledge in the embryonic field of metallurgical engineering then known as "the art of ironfounding." We see how the craftsmen's empirical knowledge, what the ironmasters called the "mysterious industry," was to dominate smelting and ironworking until the late 19th century. The Forges were a veritable training school for ironworkers, as attested by the apprenticeship contracts of the late 18th and early 19th centuries; this vocation was consolidated by the spawning of new establishments in the vicinity. Workers from the Forges would spread out to other ironworks and foundries throughout the St Maurice region (Batiscan, L'Islet, Radnor and Trois-Rivières) and to those of Quebec City, Montreal and US border settlements. Migration of its master craftsmen gave the Forges an influence whose ramifications still largely remain to be explored.

The Forges' reach was also extended especially by the McDougalls, who were the first to develop an industrial strategy for the iron industry in the St Maurice Valley and beyond. Because they bought the ironworks at both St Maurice and L'Islet and acquired shares in other enterprises in the region (Grondin and Trois-Rivières) and farther afield (St Francis), they appear as the only true industrialists to preside over the fortunes of the Forges.

Lastly, the Forges have made a significant contribution to the history of hydraulic engineering in Canada. The engineering experiments of Chaussegros de Léry, together with his notes, drawings and calculations, constitute, in our view, the first chapter in the history of this applied science in this country. These documents, with the explanations of ironmasters Vézin and Simonet and of their detractor Cugnet, provide us with an astonishingly accurate and detailed picture of the hydromechanical system powering an 18th-century industrial plant.

Study of the industrial community at the Forges sheds new light on a milieu that was to mark the social history of the St Maurice region until the dawning of the 20th century, even later if we accept the premise that the Radnor Forges represent an extension of what the St Maurice Forges had accomplished. This attention to the population of the Forges, especially recent study on the migration of the moulders from St Maurice, has shown that the establishment never really was the closed community it is represented as in legend and in much of the historiography. It was, in fact, quite an open community, though this did not prevent the village from developing its own demographic and social characteristics. It has also been established that the turmoil and instability of the early years soon gave way to a cohesive and homogeneous community, cemented by the handing down of crafts from father to son and intermarriage among the original families, supplemented by the arrival of new families after the Conquest.

The St Maurice Forges, as painted by Lucius Richard O'Brien shortly before they closed. Inset, tapping at the blast furnace.
GEORGE MUNRO GRANT, ED., *PICTURESQUE CANADA: THE COUNTRY AS IT WAS AND IS* (TORONTO: BELDEN BROS., 1882), VOL. 1, P. 97.

Mention must also be made of the invaluable contribution that study of the Forges has made to historical archaeology, and to industrial archaeology in particular. Here again the Forges have served as a major training ground for many archaeologists and material culture specialists. These people have reconstructed the complex, multidimensional fabric of Canada's first industrial village. Their work, often highly painstaking, makes it possible for us today to reconstitute, in many cases in intricate detail, the physical components of an early ironworks. Their digs have brought to light the actual layout of the site and revealed hitherto unknown aspects which forced us to reinterpret the documentary history. Without the contribution of archaeology, it would have been impossible to verify many of the descriptions in the archival record.

Finally, the study of the Forges owes much to work done in a multidisciplinary framework, supported by a number of enlightened managers at Parks Canada, and undertaken by engineers, architects, ethnologists, specialists in material culture, interpretation, and museology, (including interpretive guides who took an active part in digs and did historical research and are now managing the national historic site), and by illustrators. It is largely due to the contributions of these varied specialists and to the melding of their respective efforts that it has been possible to produce this fresh, expanded account of the history of Canada's first industrial community, the cradle of the country's iron and steel industry.

Notes

INTRODUCTION

1. Quoted in Marc Vallières, *Des mines et des hommes, histoire de l'industrie minérale québécoise, des origines au début des années 1980* (Quebec City: Les Publications du Québec, 1988), p. 23.

2. Ibid., pp. 16–19.

3. Colbert to Frontenac, 24 June 1672, quoted in Mgr Albert Tessier, *Les Forges Saint-Maurice* (Montreal and Quebec City: Les Éditions du Boréal Express, 1974 [1952]), p. 37 (hereafter cited as Tessier 1974), and requoted in Réal Boissonnault, "La structure chronologique des Forges du Saint-Maurice des débuts à 1846," typescript (Quebec City: Parks Canada, 1980) (hereafter cited as Boissonnault 1980), pp. 13–14.

4. Bertrand Gille, *Les origines de la grande industrie métallurgique en France* (Paris: Éditions Domat, 1947) (hereafter cited as Gille 1947), pp. 34–35.

5. Ibid.

6. Ibid.

7. Boissonnault 1980, p. 15.

8. Gille 1947, pp. 33, 39.

9. Boissonnault 1980, p. 19.

10. Gille 1947, pp. 133–36.

11. Project to build two milldams, a furnace, a forge and a slitting mill, in the vicinity of the Etchemin River on the south shore of Quebec, and employing some thirty skilled workers and upwards of 150 others. NAC, MG1, C11A, vol. 111-2, fols. 195–200v, "Mémoire des minnes et des lieux propres à bâtir des forges à fer que moy P. Hameau maître de forge ay vues et découvertes, depuis [...] 5e de juillet 1687 tant dans l'Acadie qu'au Canada [...] jusques au 19e décembre 1688 [1689][...]."

12. Boissonnault 1980, pp. 20–24.

13. A survey conducted in France in 1789 counted 363 blast furnaces and 640 forges; Denis Woronoff, *L'industrie sidérurgique en France pendant la Révolution et l'Empire* (Paris: Éditions de l'École des hautes études en sciences sociales, 1984) (hereafter cited as Woronoff 1984), pp. 500–501.

14. Bertrand Gille, "Le moyen âge en Occident (Ve siècle –1350)," in M. Daumas, ed., *Histoire générale des techniques*, vol. 1: *Les origines de la civilisation technologique* (Paris: Presses Universitaires de France, 1962), pp. 513–14.

15. Denis Woronoff, "Le monde ouvrier de la sidérurgie ancienne: note sur l'exemple français," *Le Mouvement social*, No. 97, 1976 (hereafter cited as Woronoff 1976), pp. 113–14.

16. Woronoff 1984, pp. 288–89.

17. Gille 1947, pp. 151–52.

18. H. Clare Pentland, *Labour and Capital in Canada, 1650–1860* (Toronto: James Lorimer & Company, Publishers, 1981), pp. 24–25.

19. Gille 1947; Hardach, citing a report from the early 19th century, mentions the particular arrogance of Franche-Comté workers, which grew after the French Revolution; Dr Gerd H. Hardach, *Der soziale Status des Arbeiters in der Früindustrialisierung. Eine Untersuchung über die Arbeitnehmer in der französischen eisenschaffenden Industrie zwischen 1800 und 1870.* Schriften zur Wirtschafts- und Sozialgeschichte (Berlin: Duncker & Humblot, 1969), pp. 81–82.

20. Gille 1947, p. 156.

21. Woronoff 1976, p. 118.

22. Cameron Nish, *François-Étienne Cugnet. Entrepreneur et entreprises en Nouvelle-France* (Montreal: Fides, 1975), p. 113.

23. Boissonnault 1980, p. 118.

24. Édouard Montpetit in 1925 praising the works of Lionel Groulx, during the first Canadian History Week. Quoted in Tessier 1974, p. 18.

CHAPTER 1

NOTE ON SERIES C11A

In this work, all references to series C11A of the French colonial archives are to the microfilmed original documents on file at the National Archives of Canada (NAC). The microfilm numbers correspond to volume numbers. This consistent reference to the folios of the original documents is meant to simplify once and for all the manner in which those documents are referred to. This applies in particular to volumes 110, 111 and 112 of series C11A, devoted exclusively to the St Maurice Forges. Previously, the many reports written by Parks Canada researchers referred for the most part to transcriptions of the original documents, which had not yet become available on microfilm. The transcriptions, variously made by NAC and by the Quebec Department of Cultural Affairs (Michel Gaumond), some of which are printed and assembled in book form, do not reproduce the original pagination of the manuscripts, complicating precise reference to the original documents. We are indebted to archivist André Desrosiers for cross-referencing the NAC transcriptions and the original documents, and to Marcelle Cinq-Mars for cross-referencing the transcriptions of Michel Gaumond and the originals.

1. Except for specific points, the chronological structure of this chapter is based on the works of Réal Boissonnault and Michel Bédard. To avoid having too many, references to those authors are explicit only when pertaining to their own interpretations or references. Réal Boissonnault, "La structure chronologique des Forges du Saint-Maurice des débuts à 1846", typescript (Quebec City: Parks Canada, 1980) (hereafter cited as Boissonnault 1980); Michel Bédard, "La privatisation des Forges du Saint-Maurice 1846–1883: adaptation, spécialisation et fermeture," manuscript on file (Quebec City: Parks Canada, 1986) (hereafter cited as Bédard 1986).

2. NAC, MG 1, C11A, vol. 51, fols. 100–101, 25 October 1729, Beauharnois and Hocquart supporting Francheville's petition.

3. NAC, MG 1, C11A, vol. 110, fols. 260–61v, letter to Maurepas, signed Francheville, n.d. [1729].

4. Beauharnois and Hocquart supporting Francheville's petition, 25 October 1929 (see full reference at note 2).

5. NAC, MG 1, F3, vol. 11-2, fols. 429–37, "brevet qui permet au Sr Poulin de Francheville d'ouvrir de fouiller et d'exploiter pendant vingt ans des mines de fer en Canada," 25 March 1730; appended is an extract of the Council of State register for 4 April 1730.

6. NAC, MG 1, C11A, vol. 110, fol. 105, "Traité de société [...]," 5 February 1737.

7. 9,244 *livres*, 9 *sols*, 5 *deniers*, 30 December 1732; Cameron Nish, *François-Étienne Cugnet. Entrepreneur et entreprises en Nouvelle-France* (Montreal: Fides, 1975) (hereafter cited as Nish 1975), p.44; NAC, MG 1, C11A, vol. 57, signed Francheville, n. p., n.d. [1732]. "[...] [the establishment] is only two leagues by land from Trois-Rivières and Sieur Francheville has had a cart road made there [...]," NAC, MG 1, C11A, vol. 110, fol. 290, Beauharnois and Hocquart to Maurepas, 28 September 1734.

8. Workers of unspecified trade hired for one year in 1733 probably worked as miners and colliers. During that year Francheville had 600 *barriques* of ore mined and 400 *barriques* of charcoal made in anticipation of starting operations in the fall, ANQ-Q, Not. Rec. Jacques Pinguet, 18 July 1733; NAC, MG 1, C11A, vol. 110, fols. 262–63, Francheville to Maurepas, 22 October 1733; Marie-France Fortier, "La structuration sociale du village industriel des Forges du Saint-Maurice; étude quantitative et qualitative," Manuscript Report No. 259 (Ottawa: Parks Canada, 1977, (hereafter cited as Fortier 1977), pp. 5–7, 178–81, 216; Nish 1975, p. 46.

9. ANQ-Q, Not. Rec. Jacques Pinguet, 16 January 1733; Boissonnault 1980, pp. 29-32; Nish 1975, pp. 44–45.

10. He was buried in Montreal on 30 November, "in the crypt of the St Amable chapel," Hubert Charbonneau and Jacques Légaré, eds., *Répertoire des actes de baptême, mariage, sépulture, et des recensements du Québec ancien*, Université de Montréal, Département de démographie) (Montreal: Les Presses de l'Université de Montréal, 1984, vol. 24).

11. NAC, MG 1, C11A, vol. 57, fol. 111, Beauharnois and Hocquart to the Minister of Marine, 15 October 1732; vol. 63, fols. 190–192; "Observations faites par moy...," Olivier de Vézin, 17 October 1735; vol. 110, fols. 284–92v, Beauharnois and Hocquart to the Minister, 28 September 1734; vol. 112, fols. 46v–47v, 54v–55, "Inventaire des Forges 1741," Estèbe; chart of the mines of Trois-Rivières (referred to in a letter from Beauharnois and Hocquart dated 26 October 1735), Bibliothèque Nationale, Paris, Cartes et plans, portefeuille 127, division 8, pièce 5 d; AJTR, Not. Rec. Polet, "Vente - Les héritiers du deffunt Lafond," 16 July 1736.

12. Anvil weighing 450 French pounds (220.2 kg) recorded by Estèbe in 1741, from Francheville's forge. Two hammers, one of forged iron and the other of cast iron, also recorded in 1741, were said to be "for the tilt hammer that the late Sr Francheville had had sent from France." These hammers were thus smaller and lighter than those used later, which weighed as much as 450 French pounds. NAC, MG 1, C^{II}A, vol. 112, fols. 46v–47v, "Inventaire des Forges 1741," Estèbe.

13. NAC, MG 1, C^{II}A, vol. 110, fol. 290, Beauharnois and Hocquart to Maurepas, 28 September 1734.

14. Boissonnault 1980, pp. 36–37.

15. Boissonnault 1980, p. 44.

16. Ibid., pp. 47–48.

17. Name of the company formed on 16 October 1736 and officially registered at Quebec on 11 February 1737 by François-Étienne Cugnet, Thomas-Jacques Taschereau, Pierre-François Olivier de Vézin, Jacques Simonet and Ignace Gamelin. Cugnet, who acted as director and treasurer of the company, was to sign as "Cugnet et Compagnie" all notes, contracts, undertakings, etc., made on behalf of the Compagnie des Forges de Saint-Maurice. NAC, MG 1, C^{II}A, vol. 110, fols. 112 and 114.

18. NAC, MG 1, C^{II}A, vol. 64-3 fols. 494–95, Minister to Beauharnois and Hocquart, 14 March 1736; vol. 110, fols. 320–22, "Obligation des intéressés aux forges de Saint-Maurice envers le roi," Not. Rec. Jacques-Nicolas Pinguet, 18 October 1736; vol. 110, fols. 318–19v, Minister to Beauharnois and Hocquart, 15 May 1736.

19. NAC, MG 1, C^{II}A, vol. 110, fols. 96–103, "Société entre les intéressés en l'établissement des Forges de Saint-Maurice," 16 October 1736; vol. 110, fols. 104–118v, "Traité de société […], 5 February 1737; vol. 67, fols. 208–208v, Hocquart to the Minister, 1 June 1737; Boissonnault 1980, pp. 50–52.

20. NAC, MG 1, C^{II}A, vol. 110, fols. 96–103, "Société entre les intéressés en l'établissement des Forges de Saint-Maurice," 16 October 1736; the company consisted of 20 *sols* or shares.

21. The fief was first reunited to the King's domain before being regranted to the company. Benjamin Sulte, *Les Forges Saint-Maurice* in *Mélanges historiques: études éparses et inédites,* (Montréal: G. Ducharme, 1920), vol. 6, p. 56, deed of concession of 12 September 1737 (hereafter cited as Sulte 1920); Allan Greer, "Le territoire des Forges du Saint-Maurice, 1730–1862," Manuscript Report No. 220, (Ottawa: Parks Canada, 1975), p.7. The deed indicates clearly that wood was cut only on the land granted to the company: "In order to obtain the quantity of wood necessary to operate the forges […]."

22. "The promptness with which this establishment was begun greatly increased its cost and the rigours of the Canadian climate made it necessary to have buildings that were much more solid, better sealed and consequently of much greater expense than those made for ironworks in France, it is essential to shelter the movements and wheelraces from the excessive cold of the country," NAC, MG 1, C^{II}A, vol. 110, fol. 389, "Mémoire du Sieur Olivier de Vézain sur les Forges de Saint-Maurice, à Monseigneur le Comte de Maurepas Ministre et Secrétaire d'état de la Marine," 28 December 1739.

23. Boissonnault 1980, pp. 79, 88.

24. NAC, MG 1, C^{II}A, vol. 110, fols. 144–49, 17 October 1735, signed Hocquart and Vézin; vol. 112, fols. 2–27, Hocquart to the Minister, Quebec, 20 October 1741.

25. This was an estimate drawn up in 1740, and not a record of specific expenditures. NAC, MG 1, C^{II}A, vol. 110, fols. 154–56, n.s., n.d. but clearly from 1740, since it was attached to a statement of expenditures for the ironworks on which the last date entered was March 1740. This was a list of expenditures recorded against the "account of Mr Cugnet" and the "account of Mr Olivier" since 1735.

26. The cost of building a 500-ton *flûte*, estimated at 87,793 *livres* in 1731, was finally 208,208 *livres* 10 years later. Incidentally, the cost of such a vessel was close to the cost of constructing and operating the Forges for three years, as estimated by Vézin in 1735; the ship *Le Canada* built in the same period (1739–42) cost about the same (218,000 *livres*). The cost of work on the fortifications of Quebec had been estimated by Chaussegros de Léry in 1720 at roughly the same amount (529,000 *livres*) as was actually invested in the Forges between 1735 and 1741. In 1745, Chaussegros de Léry again estimated the work at close to 400,000 *livres*, but the cost passed the million mark by 1751. Jacques Mathieu, *La construction navale royale à Québec, 1739–1759,* (Quebec City: La Société historique de Québec, Cahiers d'histoire n° 23, 1971), pp. 68, 101. Estimates of 1720 and 1745 by Chaussegros de Léry in André Charbonneau, Yvon Desloges, Marc Lafrance, *Quebec, the Fortified City: From the 17th to the 19th Century* (Ottawa: Parks Canada, 1982), p. 305.

27. "Maladministration, creditors going bankrupt, treasurers going missing, such was, in short, the financial life of the French iron industry in the 18th century: and all of these facts prove a lack of financial education on the part of industrialists and a distinct shortage of capital." Hocquart also pointed to a lack of capital when writing to Maurepas in 1741: "The objection that arises is to know why this establishment which is now come to perfection has not turned a profit: I have touched on the reasons in this dispatch, but the solidest reason, and the real one, in my estimation, is the lack of money." Bertrand Gille, *Les origines de la grande industrie métallurgique en France* (Paris: Éditions Donat, 1947) (hereafter cited as Gille 1947), p. 143; NAC, MG 1, C^{II}A, vol. 112, fol. 27, Hocquart to the Minister, Quebec, 20 October 1741.

28. Maurepas had submitted Vézin's project to the French Council of Commerce, which had deemed it realistic and desirable. Bertrand Gille points out that in France at that time the state "remained and would remain for a long time yet the principal source of credit for the nascent heavy industry," and that requests for advances and subsidies were transmitted to the Council of Commerce. The Council members must therefore have had a good idea of the cost of constructing and operating a plant such as the Forges. Gille, 1947, p. 133.

29. NAC, MG 1, C^{II}A, vol. 112, fol. 25v, Hocquart to the Minister, Quebec, 20 October 1741.

30. Nish 1975, p. 96.

31. NAC, MG 1, C^{II}A, vol. 112, fols. 13–13v, Hocquart to the Minister, Quebec, 20 October 1741.

32. Boissonnault 1980, p.133.

33. Cugnet gave no details of this arrangement, which enabled him to pay his creditors "with even more ease and less risk"; NAC, MG 1, C^{II}A, vol. 112-2, fols. 316–17v, Cugnet to Minister Rouillé, Quebec, 28 October 1750.

34. Boissonnault 1980, p. 141.

35. "An ironworks remains advantageous to it [the colony] even if it is not to the interested parties," NAC, MG 1, C^{II}A, vol. 112-1, fols. 61–68v, "Cession et abandon du Sr Cugnet de son interest dans les forges de St Maurice 4 octobre 1741."

36. Boissonnault 1980, p. 86.

37. He ended up staying 10 months, from 1 November 1741 to 20 August 1742; Boissonnault 1980, p. 98.

38. See Chapter 8.

39. Boissonnault 1980, p. 138.

40. Boissonnault 1980, p.136. The cumulative profit from 1741 to 1747 was 72,000 *livres*.

41. Because of the small amount of cast iron produced during the 1741 campaign (the blast furnace was not blown in until 4 July, and taken out of blast in August), the two forges were able to operate for only a month and a half between 1 October 1741 and 25 April 1742, the date on which the blast furnace was blown in. The forgemen were able to work only 110,000 pounds of cast iron into 70,000 pounds of wrought iron. NAC, MG 1, C^{II}A, vol. 112, fol. 369, "Mémoire concernant les Forges du Saint-Maurice," n.p., n.d. (but describing operations of 1741–42).

42. An error in transcribing the figure for expenditures suggests a surplus to Boissonnault; the figure should read 5,208 *livres* 9 *sols* 0 *deniers* in expenditures and not 60,168 *livres* 10 *sols* 11 *deniers*, Boissonnault 1980, p. 97, and Réal Boissonnault, *Les Forges du Saint-Maurice, 1729–1883, 150 years of occupation and operation*, Les Forges du Saint-Maurice National Historic Park series, Booklet No. 1, (Quebec City: Parks Canada 1983), p. 27. NAC, MG 1, C^{II}A, vol. 111, fols. 278–305, "Estat général de la dépense faite pour l'exploitation des forges de St Maurice depuis le 1er octobre 1741 jusqu'au premier aoust 1742," signed Estèbe, Quebec, 2 October 1742. In the retrospective accounts that Estèbe produced in 1746, 1748 and 1750, the figures for the year 1741–42 were changed. They showed first no balance, and then a deficit of over 8,000 *livres* after revision of the account in April 1744; NAC, MG 1, C^{II}A, vol. 112, fols. 267–69v, signed Hocquart, 14 October 1746; vol. 112, fol. 295, signed Estèbe and Bigot, 8 August 1748, and fol. 323, signed Estèbe and Bigot, 21 September 1750.

43. In the statement of account for 1 October 1741 to 1 January 1746 signed by Hocquart in October 1746, the Intendant gives figures where "revenues […] offset expenditures," and he bases the "good revenue or profit" on the Forges wares in stock at that date. Other statements of account produced on three subsequent occasions by Estèbe and Bigot showed in fact that this period resulted in a cumulative operating deficit of 53,340 *livres* 10 *sols* 9 *deniers*, that is, the difference between revenues and expenditures. According to the last statement of account available, for 1752, only the years 1747, 1751 and 1752 were profitable. But for the overall calculation of revenues and expenditures, two items were taken into account: the established value of the Forges in 1744 (expenditure) and the value of the Forges after inventory on the date of the account (revenue). Stocks kept at Quebec and Montreal (see tables in Chapter 6) were also taken into account. NAC, MG 1, C^{II}A, vol. 112, fols. 267–69v (1746), fol. 295 (1748), fol. 323 (1750), fol. 331 (1752).

44. *Dictionary of Canadian Biography*, s.v. "Martel de Belleville, Jean-Urbain," vol. III (1741–1770), pp. 432–33.

45. Jean Latuilière was placed in charge of the Forges in 1750, succeeding Martel de Belleville. His name appears in a statement of account for 1750 to 1751; NAC, MG 1, C¹¹A, vol. 112, fol. 331. Latuilière, a Quebec merchant, moved to Bordeaux in 1757; *Dictionary of Canadian Biography*, s.v. "Lamalétie, Jean-André," vol. IV (1771–1800), pp. 433–34.

46. Hertel de Rouville had been lieutenant general for civil and criminal affairs in the royal jurisdiction of Trois-Rivières since 1745. In 1747 Hocquart had delegated him to look into disputes among workers at the Forges; Boissonnault 1980, p. 126; *Dictionary of Canadian Biography*, s.v. "Hertel de Rouville, René-Ovide," vol. IV (1771–1800), pp. 343–47.

47. NAC, MG 1, C¹¹A, vol. 77, fols. 337–41, Hocquart to the Minister, 28 June 1742. See Chapter 8.

48. With the exception of the Macaulays, father and son (1796–1844), none who followed had an ironmaking background.

49. Tessier 1974 (see full reference at note 111), p. 115 quoted in Boissonnault 1980, p. 135.

50. Peter Kalm, *Travels into North America*, trans. J.R. Forster (Warrington: William Eyres, 1771), vol. 3, p. 89.

51. NAC, MG 4, C², fol. 210e, "Des forges de St Maurice," Franquet, 1752.

52. Franquet said of Mathieu Molérac that he was "an intelligent man, capable of running a workshop such as this, his family has managed forges from time immemorial and five or six of his relatives are today scattered about the Kingdom, running them with success as proven by the Certificates of respected people who are the proprietors of those forges […]." NAC, MG 4, C2, fol. 145, Franquet to Rouillé, Louisbourg, 10 October 1754.

53. Boissonnault 1980, pp. 146–47.

54. NAC, MG 11, CO ⁴², vol. 1-1, fols. 159–65, 21 June 1764, "Memorial of John Marteilhe of Quebec to the Lords Commissioners of Trade & Plantation."

55. Trudel identified this Courval as François Poulin de Courval, but he was actually Claude-Joseph, son of Claude, known as Cressé; this note at the bottom of his memorial of 1764 put us on the right track: "The Estimate was conveyed to me by Mr Cressé the Father, and based on the prices that were paid at the forges when he was Director," Marcel Trudel, "Les Forges Saint-Maurice sous le régime militaire (1760–1764)," *RHAF*, vol. V, No. 2 (September 1951) (hereafter cited as Trudel 1951) p. 164; NAC, MG21, B²¹⁻², fols. 139–44, microfiche A-615, Haldimand Papers, memorial from Courval, 20 September 1764. See Poulin family tree at Appendix 16.

56. NAC, MG 1, C¹¹A, vol. 112, fols. 340–42v, "Forges de Saint-Maurice, Inventaire," Hertel de Rouville, 8 September 1760. Also 300 cords of wood in the "vente du nord," a quantity that could provide only 990 *pipes* of charcoal, based on the figures given by Courval in the same period; NAC, MG 21, B ²¹⁻², fols. 139–44, microfiche A-615, Haldimand Papers, memorial from Courval, 20 September 1764. The 1,000 *pipes* of ore in stock corresponded to half the annual requirement, and this stock level was not unusual at that time of year, since ore was transported mainly in winter. See tables on raw materials required for the annual operations of the Forges in Chapter 2.

57. "Delorme, Robichon, Marchand, Humblot, Torrant, Michelin, Belie." "To M Courval at the Forges," signed J Bruyère, 1 October 1760; RCA, 1918, pp. 85–86. The census conducted in September 1760 did not include, in Trois-Rivières and the surrounding area, the names of the seven workers ordered to remain at the Forges (except possibly Belu, the keeper); however, births involving the families of some of those workers in 1760 were recorded in the register of the parish of L'Immaculée Conception-de-Trois-Rivières; RPA, 1918, p. 158. In a 1765 petition by those workers, signed by hammerman Pierre Marchand, it was said "that after the Conquest of this country, they all wanted to cross to France, since they did not expect to find any more work in this province; but his Excellency General Amherst […] detained them all and gave them orders to continue working at the forges just as they had always done under the French"; NAC, RG 1, E 15 A, vol. 4, "Petition of Pierre Marchand, Mr Blacksmith at Trois-Rivières, 17 sept 1765," with, appended, "Workmen at the Forge Three Rivers, accnt against the Government" (mentioning the workers' trades).

58. Ibid. The spelling of the names has been corrected.

59. Governor Haldimand paid for the expenditures of the Trois-Rivières government between May 1762 and March 1763 out of the proceeds of the sale of 45,000 pounds of iron (22 t); Trudel 1951, p. 163.

60. Trudel 1951, pp. 159–85.

61. Courval was also compensated for that "idleness." Boissonnault 1980, pp.164–65; NAC, RG 1, E 15 A, vol. 4, "État de ce qui est dû aux ouvriers des forges St. Maurice depuis le 1ᵉʳ octobre 1764 jusqu'au 1ᵉʳ aoust 1765 temps où ils ont reçu l'ordre de se retirer, joint à une pétition signée Pierre Marchand," 17 September 1765.

62. Boissonnault 1980, p. 162.

63. Trudel 1951, pp. 178–79.

64. NAC, MG 21, B ²¹⁻², fols. 139–44, microfiche A-615, Haldimand Papers, memorial from Courval, 20 September 1764.

65. Boissonnault 1980, p. 164.

66. The workers had been ordered to leave the post without otherwise being instructed that they had been laid off.

67. Boissonnault 1980, p. 168.

68. This was the territory of the seigneury as augmented in 1737 by the inclusion of the fief St Étienne and the land to the rear (see Appendix 2). That same year, Pélissier was conceded a strip of land by the Jesuits measuring 20 arpents by 2 leagues (1,148 ha), part of the seigneury of Cap de la Madeleine, on the east bank of the St Maurice River, opposite the Forges; Greer 1975, p. 13.

69. An inventory of the ironworks had been drawn up by Pélissier in March of that year. Boissonnault 1980, p. 169.

70. Fortier 1977, pp. 121–26; *Dictionary of Canadian Biography*, s.v. "Dunn, Thomas," vol. V, pp. 287–93.

71. Boissonnault 1980, pp. 167–76. In 1771 and 1772, Pélissier bought out his partners in return for the delivery at Quebec of 90 tons of "gueusets or pigs marked 3 Rivers" for each share (1/9) purchased; ANQ-Q, Not. Rec. Saillant, No. 2152, 4 April 1771, and No. 2298, 26 June 1772, and NAC, MG 11, Q ³⁸, p.125 (noted by Simon Courcy, "Gueuses et gueusets fabriqués aux Forges du Saint-Maurice," documentary record and inventory of archaeological artifacts prepared as part of the Grande Maison development project, (Quebec City: Canadian Parks Service, 1989), pp. 9–11). *Dictionary of Canadian Biography*, s.v. "Price, Benjamin," vol. III; "Pélissier, Christophe," "Saint-Martin, Jean Dumas," vol. IV; "Allsopp, George," "Dunn, Thomas," "Dumas, Alexandre," "Watson, Brook," vol. V.

72. Greer 1975, p.13. Based on 1 league = 84 arpents, and 1 arpent = 180 *pieds* (French feet) or 191.83 English feet; from Marcel Trudel, *Les débuts du régime seigneurial au Canada*, (Montreal: Fides, 1974).

73. In January 1768, Pélissier was still working to get the Forges back on their feet; Boissonnault 1980, pp. 169–70.

74. Boissonnault 1980, p. 170. When four partners sold their shares to Pélissier on 4 April 1771, it was said that "each of them has spent to date a sum of six hundred and fifteen pounds Halifax currency." In actual fact, therefore, those nine shares represented an investment of £5,535 currency; ANQ-Q, Not. Rec. Saillant, No. 2152, 4 April 1771.

75. Fortier 1977, pp. 18, 23.

76. ANQ-Q, Not. Rec. Saillant, No. 2152, 4 April 1771, and No. 2298, 26 June 1772; NAC, MG 11, Q ³⁸, p. 125.

77. Boissonnault 1980, pp. 175–76.

78. Laterrière had employed this prisoner as a woodcutter at the Forges, and had also recommended him to the merchant Alexandre Dumas; NAC, MG 21, B 185–1, pp. 168–71, deposition of John Oakes (the American prisoner) before G. de Tonnancour, Trois-Rivières, 24 February 1779; Pierre de Sales Laterrière, *Mémoires de Pierre de Sales Laterrière et de ses traverses* (Quebec City: Imprimerie de l'Événement, 1873), pp. 109–23.

79. Fortier 1977, pp. 92–93. In 1771, he also acquired the fief and seigneury of Dumontier. Sulte tells us that Gugy took refuge at the Forges in 1776 when the retreating Americans looted his house; Sulte 1920, p. 171.

80. It should be mentioned in his defence that Gugy was not involved in formulating the Council's recommendations favouring him! Boissonnault 1980, pp. 178–80.

81. Boissonnault 1980, p. 179.

82. See announcement reproduced in Chapter 6, Table 6.5.

83. *Dictionary of Canadian Biography*, s.v. "Bell, Mathew," vol. VII, pp.64–69.

84. This petition was aimed at countering Hugh Finlay's, seeking a lifetime lease on the Forges. Despite Governor Dorchester's support, Finlay did not pursue the matter and lost interest in the Forges in 1794; Boissonnault 1980, pp. 186–87.

85. They were protégés of the Duke of Northumberland; *Dictionary of Canadian Biography*, s.v. "Bell, Mathew," vol. VII, pp. 64–69.

86. It was said that Mathew Bell burned them when he left in 1846. Marcelle Caron, "Analyse comparative des quatre versions de l'enquête de Dollard Dubé sur les Forges Saint-Maurice," manuscript on file (Quebec City: Parks Canada, 1982) p. 85. See quotation at the end of this part.

87. Based on Boissonnault 1980, Appendix A, p. 265.

88. The Batiscan Iron Works Company had been founded in 1798, and the ironworks closed early in 1814. Claire-Andrée Fortin and Benoit Gauthier, "Aperçu de l'histoire des Forges Saint-Tite et Batiscan et préliminaires à une analyse de l'évolution du secteur sidérurgique mauricien, 1793–1910," research report submitted to the Regional Branch of the Quebec Department of Cultural Affairs, Centre de recherches en études québécoises (Trois-Rivières: Université du Québec à Trois-Rivières, December 1985), pp. 4–7.

89. The 1801 lease was for a term of five years, but stood unchanged until 1810, without official renewal, because of the government's refusal to accept the outcome of the public auction of the lease for £60 a year.

90. The other bidders were John Mure and Michel Berthelot d'Artigny, the latter representing a company formed by Pierre de Sales Laterrière. Thomas Dunn dismissed the company, which made a single bid of £16, as being made up of people without means. Boissonnault 1980, p. 209.

91. Member of the House of Assembly for St Maurice from 1800 to 1804, and for Trois-Rivières from 1809 to 1814, Bell was an avowed supporter of the English party; John Lees had been the member for Trois-Rivières before him and was an honorary member of the Executive Council; Bell became a member of the Legislative Council on 30 April 1823; *Dictionary of Canadian Biography*, s.v. "Bell, Mathew" vol. VII, pp. 64–69.

92. On a visit to the Forges in 1828, Lieutenant Baddeley noted in his report that the extent of the Forges lands was 120 square miles (31,079.86 ha): "The property belonging to Government and leased with the iron works [...] contains about 120 square miles." He did not include the fiefs of St Maurice and St Étienne. "Lieutenant Baddeley's (Rl Engineers) report on the Saint Maurice iron works, near Three Rivers, Lower Canada, jany 24th 1828," in *APT*, vol. V, No. 3 (1973) (hereafter cited as Baddeley 1828), p. 12.

93. In 1796, Barthélémy Gugy applied for a 10-square-mile township adjacent to the Forges lands. The survey showed that it lay within the boundaries of the Forges reserve. In 1800, Moses Hart claimed part of the Forges territory but his claim was rejected; Louis Gugy also made a claim in 1806. Some outside claims for portions of the Forges lands stemmed from the lack of an official survey of the St Maurice and adjoining seigneuries; Monro and Bell requested a survey on several occasions so as to be exempt from paying stumpage dues and ore-mining fees on what they considered to be their own land. Boissonnault 1980, pp. 191, 203, 205–206.

94. Boissonnault 1980, p.196. Greer 1975, p. 14, says that this was probably the seigneury of Pointe du Lac.

95. Greer 1975, p. 17.

96. Greer 1975, p. 197; this concession had already been granted to Pélissier in 1767.

97. The land between the Forges and the Gatineau fief, and the land of Caxton Township, along with the islands in the St Maurice River; Boissonnault 1980, p. 203.

98. Boissonnault 1980, p. 205.

99. Ibid., p. 212.

100. Ibid., p. 222.

101. Pierre-Benjamin Dumoulin, member for Trois-Rivières from 1827 to 1832, and for Yamaska from 1851 to 1854; Bédard 1976, p. 95; *Répertoire des parlementaires québécois 1867–1978*, Legislative Library of Quebec, political records section, (Quebec City: Quebec National Assembly, 1980), p. 190.

102. Boissonnault 1980, pp. 222–25.

103. *Dictionary of Canadian Biography*, s.v. "Bell, Mathew," vol. VII, pp. 64–69.

104. The lease was issued on 25 November 1834. Boissonnault 1980, pp. 238–39.

105. Boissonnault 1980, p. 245.

106. He received this information from John Lees, the former lessee of the Forges and partner of Alexander Davison from 1787 to 1792; Boissonnault 1980, pp.214–15.

107. In 1829, Bell stated that the activity of the Forges put £20,000 annually into circulation in the surrounding region. Boissonnault 1980, pp. 225–26, 258.

108. Based on the 1831 census record of ironworkers (founders, moulders, forgemen) at Trois-Rivières, the Trois-Rivières foundry employed an estimated 16 workers; Roch Samson, "Les ouvriers des Forges du Saint-Maurice: aspects démographiques (1762–1851)," Microfiche Report No. 119 (Quebec City: Parks Canada, 1983), p. 165 (Appendix 8); NAC, RG 4, A 1, vol. 225, p. 84, list of workers at the St Maurice Forges in 1829.

109. Two cupola furnaces formed the equipment of this foundry, in which old iron mixed with pigs from the Forges was melted down. Baddeley 1828, p. 12.

110. According to an analysis that will be discussed in Chapter 7, a list dating from 1829 included 37 workers who were not natives of the Forges, most of whom had settled there during the Bell regime; of that number, 22 arrived between 1814 and 1825, after the 21-year lease had been granted in 1810. Added to the 51 natives of the Forges, those 37 workers brought to 88 the total number of workers living on the post in 1829. NAC, RG 4, A 1, vol. 225, p. 84 (1829); RG 4, A1, S, vol. 86, 26692–95 (1805); RG 4, B 15, vol. 18, p. 8824.

111. Accounts of workers recorded by Dollard Dubé in 1933 quoted in Mgr. Albert Tessier, *Les Forges Saint-Maurice*, (Montreal and Quebec City: Les Éditions du Boréal Express, 1974 [1952]), p. 180.

112. ANQ-TR, court papers, deposition of 3 January 1848, in Luce Vermette, *La vie domestique aux Forges du Saint-Maurice*, History and Archaeology No. 58, Ottawa, Parks Canada, 1982, pp. 222–23.

113. *Le Constitutionnel*, 6 August 1869.

114. The other bidders were Hart, Boutillier and Judah. Bédard 1986, pp.19–20; Michel Bédard "La structure chronologique des Forges du Saint-Maurice (1846–1883)," typescript (Quebec: Parks Canada, 1979) (hereafter cited as Bédard 1979), p. 347.

115. Bédard 1986, p. 20.

116. The 36,209 acres (14,653 ha) had been valued at 6 shillings an acre, for a total value of £10,862. The overall price that Stuart paid therefore saved him £4,962; Bédard 1979, p. 348.

117. Stuart sold no more than one-fifth of the available lots, thus remaining well in control of his raw materials, which was not the case for his successor, after the expiry of the five-year privilege granted to Stuart. Michel Bédard, "Le contexte de fermeture des Forges du Saint-Maurice (1846–1883)," manuscript on file (Quebec City: Parks Canada, 1980) (hereafter cited as Bédard 1980), pp. 15–17.

118. Bédard 1986, pp. 20–25.

119. Ibid., p. 24.

120. Ibid.

121. Stuart was apparently advised by a French engineer (possibly Nicolas-Edmond Lacroix, according to Bédard) on how to increase plant productivity at the Forges, in particular the blast furnace. He replaced the bellows with a hot-air furnace, likely coupled to an air compressor. Compared with a "cold blast," a "hot blast" could boost the output of the blast furnace by up to 50%, not to mention the savings in fuel and flux. Bédard 1980, pp. 20–22, and Bédard 1986, pp. 79–83.

122. Bédard 1980, p. 27.

123. He had advanced Stuart £16,947 13s 10d for Forges operations. The £5,000 claimed corresponded to the separate loan made on 30 October 1847; ibid., p. 31.

124. Ibid., p. 32.

125. APJQ, Superior Court, docket no. 614, John Porter et al. v. James Ferrier, 3 May 1853.

126. Eldest son of Andrew Stuart (1785–1840). In 1885 he was named Chief Justice of the Superior Court for the Province of Quebec. He was also knighted by Queen Victoria on 9 May 1887. Bédard 1986, p. 29.

127. A justice of the peace for the district of Montreal, from whom Henry Stuart had borrowed £1,500 in anticipation of the sale at auction of the fiefs St Maurice and St Étienne. Bédard 1986, pp. 20–21.

128. It was Stuart and Porter who arrived at this figure, starting from the initial price charged to Henry Stuart (£11,475). They subtracted from that sum the principal and interest already paid by Henry Stuart, as well as the excess charged for the area of the fiefs, arriving at the sum of £7,526 12s 08d, which is what they believed they owed the government. Bédard 1986 p. 69.

129. JLAPC, 1852–53, vol. 11, app. CCC, p. 28, letter from Timothy Lamb, 31 August 1852.

130. According to the total interest owed by Henry Stuart as shown on a statement produced by the Crown Lands Department on 13 March 1852. Bédard 1986, pp. 69–70.

131. The company consisted of Weston Hunt and Jeffrey Brock. Brock was to be paid the £400 to manage the Forges on site.

132. "[...] the said Defendants [Weston Hunt & Co.] undertook to furnish the necessary capital to work the same [Forges] limited by the said parties to a sum of Seven Thousand Five Hundred pounds to take the management of the said Forges and of the business thereat, and to render an account thereof **annually**" [our emphasis]. APJQ, Superior Court, docket no. 2238, J. Porter et al. v. Weston Hunt et al., 10 November 1856.

133. Before coming to Canada, Hunter had supposedly been chief engineer at one of the largest ironworks in Britain. However, that was not the impression of the superintendent of the Forges at the time, as his recollection of the explosion of 1854 makes plain: "The furnace blew up two days after I had come there from the incompetency of the Engineer in every respect he was a common blacksmith and he could not even speak English correctly. He was selected, I believe, by Mr Hunt's correspondent in England"; ANQ-Q, Superior Court, District of Quebec, docket no. 2238, John Porter et al v. Weston Hunt et al., testimony of William Henderson, 14–17 December 1861. Bédard 1986, p. 33.

134. JLAPC, 1852–53, vol. 11, app. CCC, A 1852 letter from William Hunter to Messrs Stuart and Porter, 24 August 1852.

135. Henderson, at Weston Hunt's request, audited the accounts of the Forges in the winter of 1853–54, and was employed as manager by Weston Hunt & Company at the end of the summer of 1854. After attempting in vain to mediate between the two companies, he resigned and was brought back a short time later by Andrew Stuart to conduct an inventory of the Forges. It was after determining the value of the ironworks and its potential return (about £7,000 by his calculation) that Henderson saw no risk in coming to the aid of Porter and Stuart by providing a bond worth £8,287 9s 0d secured by a mortgage on some of his land to Weston Hunt & Company. On the same date, 5 September 1854, when the parties agreed to dissolve their partnership in his presence, William Henderson was appointed manager of the Forges, which he would remain until 5 May 1856. He arrived at the Forges on 14 October 1854, two days before the blast furnace exploded. According to Henderson's testimony on 14 December 1861 before the Superior Court of the District of Quebec, quoted in Bédard 1986, pp. 36–37.

136. This is the first mention of a bank's involvement in the affairs of the Forges.

137. Henderson would acknowledge paying them during later testimony; APJQ, Superior Court, docket no. 2238, J. Porter et al. v. Weston Hunt et al., testimony of William Henderson, 14 December 1861.

138. According to an anonymous article in the newspaper *Le Constitutionnel*, published on 13 June 1870; quoted in Bédard 1986, p. 50.

139. The minutes of the notary W. D. Campbell for the year 1867, which record the agreement, were lost; Bédard 1986, pp. 49–50.

140. Ibid., pp. 53–54.

141. The testimony of Hamilton R. Rickaby confirmed that there was indeed a kiln, referred to as a "pyrolygneous acid machine," installed before 1854. Bédard 1986, p. 91.

142. See Chapter 8.

143. For more details, see Chapter 7.

144. MER, Service de la concession des terres, St Maurice Township, general file No. 25203/1936, excerpt from two letters from H. R. Symmes to P. M. Vankoughnah, Crown Lands Commissioner, 30 September 1861.

145. The deed of sale did not appear to include movable property. Rather, it shows that Héroux was involved in a lawsuit with John Porter & Company over property that had been seized. He specified in the deed that he was dropping his claims to this property, and the purchaser, McDougall, agreed to withdraw his suits and to pay some of Héroux's costs. We also know that John McDougall acquired the movables of John Porter & Company in December 1862 for $1,000. ANQ-TR, Not. Rec. Petrus Hubert, No. 4575, 27 April 1863, "Vente des Forges St Maurice," Onésime Héroux to John McDougall, and No. 4492, 15 December 1862.

146. David J. McDougall, "The St Francis Forges and the Grantham Iron Works," unpublished manuscript, Appendix VII; Bédard 1986, p. 127, Table 5.

147. ANQ-Q, Not. Rec. W. D. Campbell, 28 February 1860, No. 498.

148. His son William put up a third of the investment needed to buy the two works, according to a notarized statement by John McDougall, 11 June 1863; Bédard 1986, p. 129.

149. In addition to founding a family firm to manage the St Maurice Forges and the L'Islet Forges, John McDougall & Sons also set the terms governing the management of the store in Trois-Rivières, though this involved only the father and two of his sons, John Jr and James, who were in charge of running the store. Bédard 1986, p. 131.

150. Timothy Lamb, who had served under Mathew Bell and later under John Porter & Company, had recommended to the latter that charcoal be made on the post itself. A kiln was in fact in use during the time of John Porter & Company.

151. Bédard 1986, Table 8, p. 178.

152. Bédard also puts forward another hypothesis. Since 1865, John McDougall and John McDougall & Sons had supplied the overwhelming bulk of their production of pig iron (over 90%) to John McDougall & Company of Montreal, which manufactured wheels for rail cars. In 1874 John McDougall & Company bought the St Francis Forges at St Pie de Guire to secure its supply of pig iron. Although we know that John McDougall & Company was still doing business with John McDougall & Sons in 1875 and 1876, it is possible that it had reduced its orders for cast iron from Trois-Rivières, thus causing financial difficulties for John McDougall & Sons which borrowed $80,500 from the Quebec Bank at that same time. Bédard 1986, p. 184.

153. The L'Islet Forges closed for good. The St Maurice Forges were at a standstill, but George McDougall undertook repairs that he interrupted in 1879 to settle a disagreement with his cousin, Alexander Mills McDougall; AJTR, Superior Court, District of Trois-Rivières, 1879, docket no. 230, Napoléon Dufresne v. Hyacinthe Grondin, testimony of Alexander McDougall, 17 January 1881, quoted in Michel Bédard, "Le territoire des Forges du Saint-Maurice, 1863–1884," Manuscript Report No. 220, (Ottawa: Parks Canada, 1976), p. 12; Bédard 1986, p. 140.

154. Bédard 1979, p. 363.

155. An employee of the Forges would say later that George McDougall had realized that he could operate both furnaces with the same number of men. Bédard 1986, p. 163.

156. A turbine caisson was found on the site of the new furnace during archaeological digs. Another caisson of the same type was also found in the furnace wheelrace. This second turbine was probably installed around the same time, between 1880 and 1883. See Chapter 3.

157. Bédard 1986, p. 66.

158. Bartlett, Wurtele, Donald, and Bellemare pointed each in turn to the depletion of raw materials as the immediate cause of the closing of the Forges, without taking the trouble to support their conclusions with explicit references. James Herbert Bartlett, *The Manufacture, Consumption and Production of Iron, Steel, and Coal, in the Dominion of Canada* (Montreal: Dawson Brothers Publishers, 1885); J.E. Bellemare, "Les Vieilles Forges Saint-Maurice et les Forges Radnor," *Bulletin des recherches historiques*, vol. 24, No. 9 (September 1918), pp. 257–69; W.J.A. Donald, *The Canadian Iron and Steel Industry, a Study in the Economic History of a Protected Industry* (Boston & New York: Houghton Mifflin Company, 1915); Mgr Albert Tessier, *Les Forges Saint-Maurice* (Montreal and Quebec City: Les Éditions du Boréal Express 1974, [1952]); F.C. Wurtele, "Historical Record of the St. Maurice Forges, The Oldest Active Blast-Furnace on the Continent of America," *Proceedings and Transactions of the Royal Society of Canada for the Year 1886*, vol. 4, Montreal (1887), section 2. Other authors ignored the more immediate causes and blamed the economic situation. Tessier (1952, 1974) was the first to emphasize the consequences of the recession in the last quarter of the 19th century. McDougall (1971) pointed to the drop in prices for cast- and wrought-iron goods between 1874 and 1892. And finally, Faucher (1973) found that the immediate causes of the closing of the Forges still escaped historians ("[...] questions remain about the causes of their demise"). He tried to explain the closing by trends in the iron industry and the marked shift to mineral coal which was to make the older charcoal process of ironmaking obsolete. Bédard 1986, pp. 1–4.

159. Campbell had taken over a mortgage granted well before James McDougall's in 1863 to a certain Dr George Taylor. Campbell agreed in 1866 to take back the mortgage so that Taylor could be reimbursed, on condition that he be given precedence over the $20,000 mortgage that James McDougall had held since 1863. He and William McDougall reached a private agreement on those terms, which Campbell had registered on 15 February 1866. Bédard 1986, p. 143.

160. George McDougall's main competitor, the Radnor Forges, were served by a railway line as of 1880, which gave them an undeniable edge over the St Maurice Forges; Bédard 1986, p. 190.

CHAPTER 2

1. In this regard, merchant iron produced in the colony fetched a better price than iron imported from Europe. See Chapter 6.

2. Our emphasis. NAC, MG 1, C^{11}A, vol. 110, fol. 290, Beauharnois and Hocquart to the Minister, 28 September 1734.

3. We will see in Chapter 5 that this rationality came into play in the worker control process.

4. With regard to the founders, Bouchu wrote: "It is unusual to see a founder from one province used to certain types of mine succeed in another province with different types of mine." *Encyclopédie ou Dictionnaire raisonné des Sciences, des Arts et des Métiers*, s.v. "Forges, (Grosses-)" by M. Bouchu, iron-master at Veuxsaules, near Château-vilain, vol. 7, 1757, p. 136.

5. David S. Landes, *The Unbound Prometheus: Technological Change and Industrial Development in Western Europe from 1750 to the Present* (Cambridge: Cambridge University Press, 1969), p. 174.

6. The royal warrant granted to Francheville and subsequently to Cugnet et Compagnie called for payment of compensation to the owners of neighbouring seigneuries only when operations affected land under cultivation.

7. See especially N. Séguin, *La conquête du sol au 19e siècle* (Sillery: Les éditions du Boréal Express, 1977) and R. Hardy and N. Séguin, *Forêt et Société en Mauricie* (Montreal: Boréal Express, 1984) (hereafter cited as Hardy and Séguin 1984).

8. "The commencement of the present lease is dated 1st Jany 1810 and it terminates 31st March 1831. By it for this period the lessees have entire and absolute control over all mineral and ore etc. Productions of the soil with the exception only of Gold and Silver." "Lieutenant Baddeley's (Rl Engineers) report on the Saint Maurice Iron Works, near Three Rivers, Lower Canada, jany 24th 1828," in *APT*, vol. V, no. 3 (1973) (hereafter cited as Baddeley 1828), p. 12.

9. We will see that no care was taken to cut wood in a systematic, regulated way, which would have enabled the species harvested to renew themselves periodically (every 15–20 years). Concern for sound forest management, while sometimes expressed, does not appear in demands for land expansion. See Marcel Moussette, "L'histoire écologique des Forges du Saint-Maurice," Manuscript Report No. 333 (Ottawa: Parks Canada, 1978) (hereafter cited as Moussette 1978).

10. Peter Kalm, *Travels into North America*, trans. J. R. Foster (Warrington: William Eyres, 1771), vol. 3 (hereafter cited as Kalm 1771), p. 88.

11. Joseph Bouchette, *A Topographical Description of the Province of Lower Canada, with Remarks upon Upper Canada and on the Relative Connexion of both Provinces with the United States of America* (London: printed for the author, and published by W. Faden, Geographer to His Majesty and the Prince Regent, Charing-Cross, 1815) (hereafter cited as Bouchette 1815), p. 304.

12. Excerpt from a debate in the Legislative Assembly; *Quebec Gazette*, Monday, 24 December 1832. But Bell wasted no time in responding: "Had Mr Kimber taken the trouble to ascertain the fact, by visiting the Iron Works this Fall, he must have closed his eyes and ears on a considerable part of the road leading from the Town to the Forges, not to have seen the piles of wood recently cut, and not to have heard the sound of many an axe, for the woodcutters were at work on the very side of the road." Mathew Bell to Colonel Craig, 31 December 1832, JLAPC, 1844–45, vol. 4, app. O.

13. In 1796, Monro and Bell had obtained permission to have a survey carried out by Walter Waller of Forges land which they considered exhausted of wood and ore reserves; NAC, RG 4, A1, S, vol. 63, fols. 20383–86, 5 April 1796. The survey of 1806 was reported by Joseph Bouchette; NAC, RG 1, L3L, vol. 144, fols. 70766–69.

14. NAC, RG 1, L3L, vol. 144, fols. 70766–69, Joseph Bouchette, "Return of the survey of the Seigniory of St-Maurice and the adjoining tracts thereto," 23 September 1806.

15. Francheville, the seigneur of St Maurice, thus set up his forge on his neighbour Lafond's fief. Lafond's heirs did try, in 1736, to have their rights over the site of the Forges recognized, but the following year the colonial authorities finally decided by ordinance to take their fief away from them and reunite it to the royal domain. They cited the fact that the seigneur of the fief had failed to develop it, in breach of the royal decrees of Marly, which governed the conditions of seigneurial tenure. The fief was thus annexed to the fief St Maurice, which was also reunited to the royal domain and placed at the disposal of the Forges. Lafond's heirs were no doubt trying to demonstrate possession by signing a notarial deed on July 16, 1736, selling the fief in question, on which Francheville's buildings were located, to Geneviève Trottier "residing in the fief of St Étienne." This sale took place just as the Forges were being set up by Vézin. On 28 May 1736, the colonial authorities issued an ordinance banning them from any trade with the Indians at St Maurice and in the depths of the Trois Rivières River pending better justification of their title to the property. The risk of fire in a forest to be used for the needs of the Forges was also cited. AJTR, Not. Rec. Polet, 16 July 1736, "Vente-Les héritiers de deffunt Lafond"; P-G Roy, *Inventaire des ordonnances des intendants de la Nouvelle-France conservées aux Archives provinciales de Québec* (Beauceville: L'Éclaireur limitée, 1919) (hereafter cited as Roy 1919), vol. 2, pp. 203, 221–22; MG 8, A 6, vol. 13, fols. 389–92 (1736), vol. 14, fols 107–33 (1737).

16. Roy 1919, pp. 221-22.

17. Marcel Trudel, "Les Forges Saint-Maurice sous le régime militaire, (1760– 1764)," *RHAF*, vol. V, no. 2, September 1951, pp. 178–79.

18. JLAPC, 1844–45, vol. 4, app. O, cited in the report to the Executive Council signed by Étienne Parent, 15 September 1843.

19. JLAPC, 1844–45, vol. 4, app. O, Mathew Bell to D. Daly, Provincial Secretary, St Maurice, 26 September 1844.

20. Claire-Andrée Fortin and Benoit Gauthier, "Les entreprises sidérurgiques mauriciennes au XIXe siècle: approvisionnement en matières premières, biographies d'entrepreneurs, organisation et financement des entreprises," René Hardy, ed., research report submitted to the Regional Branch of the Quebec Department of Cultural Affairs, Centre de recherches en études québécoises (Trois-Rivières: Université du Québec à Trois-Rivières, November 1986), pp. 27–28.

21. According to the royal warrant he obtained in 1730, Francheville had the right to set up his forge at the most convenient locations on his own and neighbouring seigneuries. NAC, MG 1, F³, vol. 11–2, fols. 429–37, "brevet qui permet au Sr Poulin de Francheville d'ouvrir, fouiller et exploiter pendant 20 ans des mines de fer en Canada," 25 March 1730.

22. There was not always really a decrease, since the deeds of sale could entitle the Forges to mine ore or cut wood on the land sold.

23. In his petition of 1819, seeking authorization to take resources on the land belonging to the Cap de la Madeleine seigneury, Bell cited the near exhaustion of the resources on the fiefs of St Maurice and St Étienne, their excessive distance from the Forges, or their inaccessibility. Visiting the Forges in 1827, Lieutenant Baddeley confirmed this: "The wood is now procured by permission from the eastern side of the St. Maurice river on the Seigniory of Cap de la Magdeleine it being considered as exhausted or nearly so on the Seignory of the St. Maurice." NAC, RG 4, A 1, S, vol. 178, pp. 17–19, Mathew Bell to the Duke of Richmond, 6 March 1819; Baddeley 1828, p. 10.

24. Bell, quite rightly, was afraid of forest fires. In 1832, he wrote of being exposed to the carelessness of trespassers on Forges land and said that he had lost several hundred cords of wood. NAC, RG 4, A 1S, vol. 219, no. 69, M. Bell to A. Cochrane, Civil Secretary, Quebec, 6 September 1825; Benoît Gauthier, "La criminalité aux Forges du Saint-Maurice," (preliminary report), June 1982; JLAPC 1844–45, vol. 4, app. O, Bell to Craig, 31 December 1832.

25. The seigneury, previously part of the Jesuit estates, had been confiscated by the government in 1800 following dissolution of the Society of Jesus by the Pope in 1773. Bell had no doubt put his political influence to good use in obtaining permission to exploit this area now reverted to the Crown.

26. The McDougalls operated the L'Islet Forges, in the parish of Mont Carmel, until 1878.

27. In late 1854, the year the Radnor Forges were established, Auguste Larue and Company held 50,000 arpents (171 km²) of land in this area and in Radnor Township. At the same time (1852), John Porter & Company, owners of the St Maurice Forges, had 62 km² of territory. Benoît Gauthier, "Les sites sidérurigiques en Mauricie (Radnor, Saint-Tite, L'Islet)," manuscript, April 1983, p. 42; Fortin and Gauthier 1986, p. 16.

28. Sugar bush on Mount Carmel, in the seigneury of Cap de la Madeleine. Allan Greer, "Le territoire des Forges du Saint-Maurice, 1730–1862," Manuscript Report No. 220 (Ottawa: Parks Canada, 1975), p. 25; Hardy and Séguin, 1984, p. 25.

29. This was the consensus view embodied in the newspaper *L'Ère nouvelle*, describing the harmful effects of Bell's land monopoly. The article, published on 1 March 1854, rejoiced in the fact that the "barrier has now been removed" following the loss of land privileges after the sale of the Forges.

30. Hardy and Séguin, 1984, pp. 30–35.

31. For example, around 1890, the Baptist family controlled 7,381 km² of forest in the St. Maurice Valley; Hardy and Séguin 1984, pp. 37–39.

32. The Canada Iron Furnace Company obtained its land, which was already part of a forest concession to the Laurentide Pulp Co., through a special act of the Legislative Assembly. Fortin and Gauthier 1986, pp. 21–24; Claire-Andrée Fortin and Benoît Gauthier, *Description des techniques et analyse du déclin de la sidérurgie mauricienne, 1846–1910*, René Hardy ed., research report submitted to the Regional Directorate of the Quebec Department of Cultural Affairs, Centre de recherches en études québécoises: Trois-Rivières: Université du Québec à Trois-Rivières, November 1988 (hereafter cited as Fortin and Gauthier 1988), pp. 129–130.

33. Deeds of indenture in the early 19th century obliged workers (joiners, carpenters, wheelwrights, etc.) to be available to fight forest fires; see ANQ-TR, Not. Rec. Jos Badeaux, indenture of J.-B. Gagon, carpenter and wheelwright, and Louis Pépin, joiner, 12 December 1814; ANQ-M, Not. Rec. N. B. Doucet, indenture of Louis Pépin, joiner, 28 November 1810, No. 2736.

34. "[...] considered as exhausted or nearly so"; Baddeley 1828, pp. 9–10.

35. He dealt first of all with the water resources, evaluating the energy potential of the St Maurice Creek. This was a vital resource to which Chapter 3 is devoted. NAC, MG 1, C¹¹A, vol. 63, fols. 189-95v, "Observations faites par moy [...]," Olivier de Vézin, 17 October 1735.

36. Moussette 1978, pp. 49–50.

37. NAC, MG 6, C¹¹A, vol. 110, fols. 211 and 216, "Procès verbal de la vente des forges et mines de Saint-Maurice, Jean Eustache Lanouiller de Boiscler, conseiller du Roy et Grand voyer de la Nouvelle-France," 26 September 1740.

38. NAC, MG 6, C1, E-1E, vol. 122, fols. 103–5, letter from Minister Maurepas (Port of Rochefort), Marly, 15 February 1735; NAC, MG 6, C 1, E-1E, vol. 124, fols. 1–6, letter from Minister Maurepas (Port of Rochefort), Versailles, 10 January 1736,.

39. NAC, MG 1, C¹¹A, vol. 110, fol. 285, "Messieurs de Beauharnois et Hocquart sur les Forges de St Maurice," letter to Minister Maurepas, 28 September 1734.

40. Hunt wrote in 1870: "The results of several analyses of the ores of this vicinity, made by me in 1852, are given in Geology of Canada [1863] p. 511 [...]"; Geological Survey of Canada, *Report of Progress for 1866–1869*, report of Dr Sterry Hunt, p. 257. According to Moussette, these analyses were carried out by William Edmund Logan in 1853 and subsequently compiled by Thomas Sterry Hunt; Moussette 1978, p. 37.

41. Founded in 1842, the Geological Survey of Canada was first directed by William Edmund Logan, who also had contacts with the St Maurice Forges. Marc Vallières, *Des mines et des hommes, histoire de l'industrie minérale québécoise, des origines au début des années 1980* (Quebec City: Les Publications du Québec, 1988), p. 63.

42. In his report of 1874, Harrington described limonite as follows: "This ore, which in some of its forms is often called brown hematite, consists essentially of peroxide of iron combined with water, the theoretical proportions being 85.6 of the former to 14.4 of the latter. The term limonite is generally made to include bog ores [...]"; Moussette 1978, p. 39; Dr B. J. Harrington, "Notes on the Iron Ores of Canada and Their Development,"Geological Survey of Canada, *Report of Progress for 1873–74* (Montreal: Dawson Brothers, 1874) (hereafter cited as Harrington 1874), p. 228.

43. McGill University Archives, Logan Papers, accession 1207/11, item 94, John Porter & Co. to W.E. Logan, Provincial Geologist, St. Maurice Forges, 11 January 1855, signed W. Henderson, Manager. See Appendix 6.

44. "No ore is more easily reduced than bog ore; for not only is it porous and readily permeable by reducing gases, but the organic matter undoubtedly aids in its reduction." Harrington 1874, p. 235, citing *Geology of Canada*, 1863, p. 683.

45. Ibid., pp. 236–37.

46. In 1828, Baddeley made the same remark on the inconvenience of using bog ore for the manufacture of plate iron, wire or steel, saying that "for the first it may be too brittle owing to the presence of the phosphoric acid which is always found in metal obtained from this ore." Baddeley 1828, p. 10.

47. NAC, MG 1, C¹¹A, vol. 112, fol. 136, "Mémoire concernant les Forges de St Maurice," n.s., n.d. (around 1743, since it is based on Estèbe's trusteeship of 1742).

48. Experts from Rochefort who examined the file on Francheville's tenure in 1734 were already stressing the fact that the distance between the mines and the forge was too great. In 1750, when construction of a second blast furnace was under consideration in order to supply a cannon foundry, it was suggested that it be set up close to the mines rather than on the Forges site. NAC, MG 1, C^{11}A, vol. 110, fol. 303, Maurepas to Beauharnois and Hocquart, 10 May 1735. See Chapter 5.

49. NAC, RG 4, A1, vol. 18, fol. 6425–27, François-Joseph Cugnet to Governor Guy Carleton, 26 November 1768 (concerning mining rights on the seigneury of Sieur Joseph Godefroy de Tonnancour (Pointe du Lac seigneury).

50. A. B. McCullough, *Money and Exchange in Canada to 1900*. (Toronto and Charlottetown: Dundurn Press Ltd., 1984) (hereafter cited as McCullough 1984), p. 70., Table 5, Ordinance of 1764.

51. Several years earlier, in 1764, the estimated cost of mining 3,000 *barriques* of ore was 2,250 *livres* (or £93.15, the £ being worth 24 *livres*); transportation costs were estimated at 4,500 *livres* (£187.10); NAC, MG 21, B21-2, fols. 139–44, microfiche A-615, Haldimand Papers, memorial from Courval, 20 September 1764.

52. Cugnet put forward the argument that a clause in the seigneurial grant obliged the seigneur to "*notify His Majesty or his governors and intendants of mines, ore or minerals found in his concession;* that His Most Christian Majesty did not thereby claim ownership of them; in granting the seigneury, he also granted what lay under the surface in the subsoil; that the mine belonged to the seigneurs to whom the grant had been made on the principle that *whoever owns the soil owns what is on and under it.* That it is the law in France that no seigneur may open or exploit any mine found on his land without the permission of His Majesty, to whom a tenth of it belongs in the event that it is exploited. And that in consequence of this right of tithe, this clause is inserted in all land-grant deeds for seigneuries in Canada." (our emphasis). NAC, RG 4, A1, vol. 18, fols. 6425–27.

53. APJQ, Superior Court, docket no. 2238, J. Porter et al. v. Weston Hunt et al., exhibit 18, "Recapitulation of inventory," August 1854; ASTR, N3-H30, St Maurice Forges, excerpt from an account book 1856–58 (8 p. s.); McCullough 1984, pp. 156–57. See Chapter 8.

54. McGill University Archives, Logan Papers, accession 1207/11, item 94, "John Porter & Co to W.E. Logan, Provincial Geologist, St Maurice Forges," 11 January 1855, signed by W. Henderson, Manager.

55. Geological Survey of Canada, *Report of Progress for the Year 1852–3* (Quebec City: Lovell & Lamoureux 1854), pp. 43–47.

56. Moussette 1978, p. 44.

57. Bouchette 1815, pp. 303-4. This should in fact read "the falls of Grès." For a description of the fief of St. Étienne, see Bouchette 1815, p. 302.

58. In his estimate of annual expenditures, produced at the same time, he doubled this price to one *livre* per *pipe*. MG 1, C^{11}A, vol. 63, fol. 192, "Observations faites par moy [...]" and vol. 111, fol. 168-72, "Projet des dépenses à faire [...] signé Hocquart et de Vézin," 17 October 1735.

59. NAC, MG 1, C^{11}A, vol. 111, fol. 8, memorial from Cugnet, 25 September 1740.

60. The inventory of 1807 mentions "An old shed for dressing stone"; AJTR, Not. Rec. Jos Badeaux, 1 April 1807, Inventory of the St Maurice Forges post.

61. "The smelting furnace and the quarry may be said to be one and the same thing for without the quarry the other could not be carried on. This quarry must now have been opened for upwards of ninety years, and during the last thirty nine years of that period about thirty to fifty Toises of stone p. annum have been drawn from it: so tenacious was I of this stone, that I believe not a single piece of it that could be made use of in the building of a furnace has ever otherwise been employed, or ever brought away from my Establishment with the exception of three to four pieces that may have brought down at different times to show as a curiosity to friends and others [...]." Mathew Bell feared at that time that the government would make other use of the Gabelle quarry following the visit of Lieutenant Baddeley of the Royal Engineers. Following this intervention by Bell, it was recommended that the quarry, which was considered too deep, not be used, but it was also requested that a special clause be inserted in the next Forges lease, authorizing its use by the government. It was Bell's understanding that this stone was being considered for the monuments to Wolfe and Montcalm in Quebec City, which is confirmed in Baddeley's report. NAC, RG4, A1, vol. S-191, nos. 17–17a, Mathew Bell to A. W. Cochran, Quebec, 4 December 1827; Baddeley 1828, pp. 16-20.

62. It was located near the Gabelle Falls, where the dam of the same name is located today; Moussette 1978, p. 48.

63. This observation recalls the fact that, in the spring of 1737, when construction began on the Grande Maison, it was impossible to quarry stone because of the spring floods, and it had to be brought in from Quebec at great expense.

64. Moussette 1978, p. 50.

65. Baddeley 1828, pp. 16–20.

66. Harrington, in 1874, associated it with the Trenton Formation, Clark and Globensky with the Leroy Age. Harrington 1874, p. 248; Moussette 1978, p. 49.

67. Geological Survey of Canada, *Report of Progress for the Year 1852–3* (Quebec: Lovell & Lamoureux, 1854), p. 63.

68. McGill University Archives, Logan Papers, accession 1207/11, item 94, letter to W.E. Logan, Provincial Geologist, St Maurice Forges, 11 January 1855, signed W. Henderson, Manager.

69. Harrington speaks of 45 charges of 45 lbs. (20 kg) of limestone every 24 hours. Harrington 1874, p. 247.

70. NAC, MG 1, C^{11}A, vol. 112, fol. 139, "Mémoire concernant les Forges de St Maurice," n.s, n.d. (around 1743, since it is based on Estèbe's trusteeship of 1742).

71. Eight *toises* were equal to 155 long tons or 157.4 t. See Note 93 for further details on calculating a *toise* of limestone.

72. André Bérubé, "L'évolution des techniques sidérurgiques aux Forges du Saint-Maurice, 1: La préparation des matières premières." Manuscript Report No. 305 (Ottawa: Parks Canada, 1978) (hereafter cited as Bérubé 1978), p. 75.

73. NAC, MG 1, C^{11}A, vol. 112, fols.114–15, "Mémoire sur les Forges de Saint-Maurice," n.s, n.d. (1746, based on a reference to a report of Cressé's, dated 13 September of that year).

74. Hardy and Séguin 1984, pp. 12–16.

75. The spread of the indirect reduction process in France between 1480 and 1540 resulted in a timber supply crisis at the forges. In 1540, there were over 460 ironworks in France, 400 of which had been built in the previous 50 years. The ensuing deforestation led to the first royal ordinances on rationalizing timber felling and marked the introduction of modern silviculture methods. Jean-François Belhoste, "Une sylviculture pour les forges, XVIᵉ - XIXᵉ siècles" in Denis Woronoff, ed., *Forges et forêts: recherches sur la consommation proto-industrielle de bois* (Paris: Éditions de l'École des hautes études en sciences sociales, 1990) (hereafter cited as Woronoff 1990), pp. 219–61.

76. This attitude on the part of the Forges operators was not confined to Canadian entrepreneurs. In the United States, ironworks also devoured the forest. Citing travellers in the late 18th century, Theodore W. Kury reports that, in New Jersey, a blast furnace could deplete nearly 20,000 acres of forest (8,094 ha) in 12 to 15 years; Theodore W. Kury, "The Iron Plantation: Agent in the Formation of the Cultural Landscape," paper presented at the Symposium on the Industrial Archaeology of the American Iron Industry, 13th Annual Conference, Society for Industrial Archaeology, Boston, Massachusetts, June 16, 1984 (hereafter cited as Kury 1984).

77. André Lafond and Gilles Ladouceur, *Description des groupements forestiers du Québec*, cited in Moussette 1978, p. 59.

78. Cited in Kalm 1771, p. 88.

79. APJQ, Superior Court, docket no. 2238, J. Porter et al. v. Weston Hunt et al., exhibit 5 and exhibit D "List of books kept," deposition of Hamilton Rickaby, 27 January 1860.

80. Harrington 1874, pp. 247–56.

81. JLAPC, 1852–53, vol. 11, app. CCC, pp. 25–26, J. Brock, Forges manager, 1 September 1852.

82. Kury 1984.

83. JLAPC, 1852–53, vol. 11, app. CCC, p. 17, Étienne Parent to the Commissioner of Crown Lands, Quebec, 20 September 1852.

84. The weight of a *pipe* of the ore used by Francheville was provided by Vézin; NAC, MG 1, C^{11}A, vol. 63, fol. 192v, "Observations faites par moy [...]," Olivier de Vézin, 17 October 1735.

85. See reference to Table 2.9 at note 87.

86. However, we have seen that at the time of Harrington's visit a campaign could last as long as 13 months. The amounts of ore and iron used annually would increase proportionally.

87. NAC, MG 1, C^{11}A, vol. 111–2, "Projet des dépenses à faire [...]," Hocquart and Olivier de Vézin, 17 October 1735, p. 110; NAC, MG 1, C^{11}A, vol. 111, fols. 58–59, "frais annuels de l'exploitation," 24 October 1740; NAC, MG 1, C^{11}A, vol. 111, fol. 135, "Mémoire concernant les Forges de St Maurice," n.s, n.d. (probably around 1743, since it is based on Estèbe's trusteeship of 1742); NAC, MG21, B21-2, fols. 139–44, microfiche A-615, Haldimand Papers, memorial from Courval, 20 September 1764; Baddeley 1828; Geological Survey of Canada, *Report of Progress for 1866–1869*, report of Dr Sterry Hunt, p. 257; Harrington 1874, p. 247.

88. McGill University Archives, Logan Papers, accession 1207/11, item 94, John Porter & Co. to W. E. Logan, Provincial Geologist, St. Maurice Forges, 11 January 1855, signed W. Henderson, Manager. See Appendix 6.

89. APJQ, Superior Court, docket no. 2238, J. Porter et al. v. Weston Hunt et al., exhibit 5, deposition of Hamilton Rickaby, 27 January 1860.

90. "To separate earthen matter from the gangue, which is not fusible, use is made of the property that multiple silicates melt at the temperature of the blast furnace, while simple silicates do not. So a flux is added to the ore, variable according to the nature of the gangue and likely to form fusible products with the earthen matter, or slag, contained in the gangue, which is eliminated," Pierre Léon, *Les techniques métallurgiques dauphinoises au dix-huitième siècle.* (Paris: Hermann, 1961), pp. 35–36, cited in Bérubé 1978, p. 21.

91. Limestone was also used to make lime, as its name would indicate. An inventory of 1807 lists a "limestone furnace" at the Forges. ANQ-TR, Not. Rec. Joseph Badeaux, 1 April 1807.

92. Bérubé 1978, p. 20.

93. The weight of one cubic foot of sandstone is calculated as follows:

 - Weight of an English cubic foot = density of the sandstone × the weight of one cubic foot of water (62.4 lbs), eg. 2.66 (density of Sillery sandstone) × 62.4 = 166 lbs.

 - The cubic *toise* = 216 French cubic feet × 1.21 = 261.4 English feet.

 - The weight of 261.4 English cubic feet × 166 lbs = 43,385.76 lbs. = 19.37 long tons.

 - One cubic *toise* weighs 19.37 long tons or 19.7 t *(rounded off to 20 t)*.

 A statement of expenditure, likely from 1740, mentions "8 *toises* of limestone for flux" or "6 *toises* of limestone for the crucible and furnace lining," as well as "80 *barriques* of lime." NAC, MG 1, C¹¹A, vol. 110, fols. 152–53v, "État des fonds nécessaires à l'exploitation, Suite des dépenses du fourneau," n.s., n.d. [1740].

94. The stone is estimated at 48 *livres* per *toise*. He also allowed two more *toises* of "limestone" at 40 *livres* per *toise*, likely to make lime. NAC, MG 21, B21–2, fol. 139–44, microfiche A-615, Haldimand Papers, memorial from Courval, 20 September 1764.

95. JLAPC, 1852–53, vol. 11, app. CCC, p. 26, J. Brock, 1 September 1852.

96. Allan Greer, "Le territoire des Forges du Saint-Maurice, 1730–1862," Manuscript Report No. 220 (Ottawa: Parks Canada, 1976), pp.40,104 and 106.

97. Bérubé cites a notarial deed of 1874, involving the purchase by William McDougall of all the limestone on the land of Charles Gélinas for 25 *piastres*. Bérubé 1978, p. 21.

98. With a hot blast, Hunt speaks of a charge of 500 lbs. of ore and 25 lbs. of limestone. For a cold blast, Harrington reported 600 lbs. of ore to 45 lbs. of limestone. The same amount of charcoal, 16 bushels, was used in both cases.

99. NAC, MG 1, C¹¹A, vol. 111, fols. 58–59, annual operating costs, 24 October 1740, Kalm 1771, p. 88 (quoted in Bérubé 1978, p. 21).

100. Baddeley 1828, p. 12.

101. H.S. Osborne, *The Metallurgy of Iron and Steel, Theoretical and Practical: in All its Branches; with Special Reference to American Materials and Processes* (Philadelphia: Henry Carey Baird and Londers, Trubner & Co, 1869), p. 151 (quoted in Bérubé 1978, p. 22).

102. NAC, MG 1, C¹¹A, vol. 112, fol. 140, "Mémoire concernant les Forges de St Maurice," n.s., n.d. (around 1743, since it is based on Estèbe's trusteeship of 1742).

103. NAC, MG 1, C¹¹A, vol. 110, fols. 152–153v, n.s., n.d. [1740], "État des fonds nécessaires à l'exploitation, Suite des dépenses du fourneau."

104. Based on the explanation provided by Lieutenant Baddeley in 1828 (cited in Moussette 1978, pp. 52–53).

105. Ibid., p. 54.

106. McGill University Archives, Logan Papers, accession 1207/11, item 94, John Porter & Co. to W.E. Logan, Provincial Geologist, St. Maurice Forges, 11 January 1855, signed W. Henderson, Manager. See Appendix 8.

107. ASTR, Forges Papers, St Maurice Forges, December 1857, fol. 278.

108. Cited in Moussette 1978, p. 54.

109. NAC, MG 1, C¹¹A, vol. 112, fol. 156, "Mémoire concernant les Forges de St Maurice," n.s., n.d. (around 1743, since it is based on Estèbe's trusteeship of 1742).

110. On this subject, see Woronoff 1990.

111. Cited in Moussette 1978, p. 75.

112. JLAPC, 1852–53, vol. 11, app. CCC, pp. 25–26. J. Brock, Forges manager, 1 September 1852.

113. ASTR, Forges Papers, St Maurice Forges, December 1857, fol. 242.

114. This estimate is less than half that given by Moussette (22,000 acres or 8,903 ha) in his 1978 study. Aware that his estimate was overly generous, Moussette tempered it with comparable data from Scotland (14,000 acres), which is in the same order as Brock's evaluation of 1852 for a cut of 18,000 to 20,000 cords. In fact, Moussette's overestimate results from confusion with regard to the diameter (3–4 inches) necessary for a tree to be used for charcoal. He mixed up circumference (6–12 inches) and diameter to project a regeneration time of 40 rather than 20 years. Moussette 1978, pp. 75–76.

115. From 1767 to 1861, Forges entrepreneurs had at their disposal an area three to 14 times as large (see Plate 2.1a).

116. In 1870, the McDougalls had six kilns at the Forges; *Le Constitutionnel*, 13 June 1870.

117. He specified, however, that to produce 100 bushels of charcoal, 3.5 cords of wood were required for the pit charcoal process and two cords for the kiln process. Baddeley mentions the same proportions in 1828; Baddeley 1828; Harrington 1874, p. 247.

118. Three workers worked 5 nights, 16 days, and 11 days respectively, cutting wood on the circular saw at the Forges mill; ASTR, Forges Papers, St Maurice Forges, April 1857, fol. 164.

119. See references at Table 2.9.

120. Height of 12 feet (3.65 m) reported by Lord Selkirk in 1804 and 30 feet (9 m) reported by Harrington in 1874. Thomas Douglas, Earl of Serkirk, *Lord Selkirk's Diary 1803–1804*, edited by Patrick C.T. White (Toronto: The Champlain Society, 1958), p. 230; Harrington 1874, p. 247.

121. See Serge Benoit, "La consommation de combustible végétal et l'évolution des systèmes techniques," in Woronoff 1990, pp. 87–150.

122. André Bérubé, "Rapport préliminaire sur l'évolution des techniques sidérurgiques aux Forges du Saint-Maurice, 1729–1883," Manuscript Report No. 221, (Ottawa: Parks Canada, 1976) (hereafter cited as Bérubé 1976), p. 72.

123. Hunt reported production figures for the month of December 1868. He counted a total weight of 163.25 tons for the month, or 5.27 tons (5.35 t) per day. Hunt added later that average daily production was in the order of 8 tons. Geological Survey of Canada, *Report of Progress for 1866–69*, Report of Dr. Sterry Hunt, p. 257.

124. Bérubé 1976, p. 75.

125. Fortin and Gauthier 1988, p. 148.

126. Data on Allevard taken from Serge Benoit, "La consommation de combustible végétal et l'évolution des systèmes techniques," in Woronoff 1990, pp. 130–31.

127. In Burgundy, in the Dijon area, where a number of the skilled workers at the St Maurice Forges originated, Grignon reported an average yield of 2,370 kg of charcoal for 1,000 kg of cast iron in 1778, with square kilns of the same type as that used at the Forges; ibid., p. 91.

128. NAC, MG 1, C¹¹A, vol. 110, fol. 185 (January–February 1739). Notes and calculations attributed to Chaussegros de Léry. A *pipe* of charcoal weighed 176 pounds or 86.14 kg.

129. Calculated on the basis of Baddeley's figures of 1828 giving an estimate of 1,344 t of charcoal for an eight-month blast.

130. NAC, MG 1, C¹¹A, vol. 63, fol. 192v, "Observations faites par moy […]," Olivier de Vézin, 17 October 1735.

131. Cameron Nish, *François-Étienne Cugnet. Entrepreneur et entreprises en Nouvelle-France* (Montreal: Fides, 1975) p. 36.

132. "The employment in which a number of workers have been engaged since the spring to establish a forge has led us to believe that the springs in this township could be mineral springs. I spoke about this to Sr Poulin, the Undertaker of the said forge, and told him that he had only to put a pinch of powdered gall into a glass of this water, which would change on the spot if the water contained minerals and iron […] which leads me to say that the discovery of this ferriginous water has so far been a hidden treasure for Canada's invalids, whose conditions are often resistant to ordinary medicine and that this water has only been discovered by pure chance […]"; NAC, MG1, C¹¹A, vol. 58, fols. 99–103, "Description des eaux du Cap de La Madeleine," Michel Sarrazin, 10 October 1732.

133. Benjamin Sulte, "Les Forges Saint-Maurice" in *Mélanges historiques : études éparses et inédites* (Montreal: G. Ducharme, 1920) vol. 6, p. 104.

134. There were also some mineral deposits in the dry land at the foot of hills. Baddeley 1828; Bérubé 1978, p. 6.

135. Ibid.

136. Courtivron and Bouchu, "Art des Forges et fourneaux à fer," in *Description des Arts et Métiers, faites et approuvés par Messieurs de l'Académie Royale des Sciences* (Paris: Dessaint and Saillant, 1761), cited in Bérubé 1978, p 6.

137. Bérubé 1978, p 7.

138. […] asked Sieur Jutras *père* whether he knew of other mines, and he told us that he had seen a number of small ones during his research for the late Sieur Francheville […]," NAC, MG 1, C¹¹A, vol. 110, fols. 217–18, "Procès verbal de la vente des forges et mines de Saint-Maurice, Jean Eustache Lanouiller de Boiscler, conseiller du Roy et Grand Voyer de la Nouvelle-France," 26 September 1740.

139. Beauharnois and Hocquart, in 1735 (cited in Moussette 1978, p. 39).

140. Bérubé 1978, pp. 9, 14 and 15.

141. See Chapter 8.

142. Bérubé 1978, p. 18.

143. Ibid, pp. 18–19.

144. Ibid, p. 16.

145. André Lepage, "Étude du travail et de la production aux Forges du Saint-Maurice à deux moments de l'histoire de l'entreprise," typescript (Quebec City: Parks Canada, June 1984) (hereafter cited as Lepage 1984), Table 3.6.

146. Bérubé 1978, pp. 23–24.

147. McGraw-Hill Dictionary of Scientific and Technical Terms, Daniel N Lapedes, Editor in Chief, (New York: McGraw-Hill Book Company, 1976), p. 223.

148. APJQ, Superior Court, docket no. 2238, J. Porter et al. v. Weston Hunt et al., exhibit 5, deposition of Hamilton Rickaby, 27 January 1860.

149. Michel Bédard, "La privatisation des Forges du Saint-Maurice 1846-83: adaptation, spécialisation et fermeture," manuscript on file (Quebec City: Parks Canada, 1986) (hereafter cited as Bédard 1986), p. 90. "Vente" meant a tract of forest that had just been cut.

150. NAC, MG 1, C¹¹A, vol. 110, fols. 209–218. "Procès verbal de la vente des forges et mines de St- Maurice, Jean Eustache Lanouiller de Boiscler, conseiller du Roy et Grand voyer de la Nouvelle-France," 26 September 1740.

151. Marie-France Fortier, "La structuration sociale du village industriel des Forges du Saint- Maurice: étude quantitative et qualitative." Manuscript Report No. 259 (Ottawa: Parks Canada, 1977), p. 13; Bérubé 1978, p. 28.

152. NAC, MG 1, C¹¹A, vol. 112, fols. 150-51, "Mémoire concernant les Forges de St Maurice," n.s., n.d. (around 1743, since it is based on Estèbe's trusteeship of 1742).

153. NAC, MG 1, C¹¹A, vol. 111, fol. 287, "Estat général de la dépense […]," signed Estèbe, 2 October 1742.

154. Bérubé 1978, pp. 30–31.

155. There is no explicit description in the archival record on the Forges of the charring process, or pit-setting. However, Bérubé, from whom this description is taken, drew on Duhamel DuMonceau's *L'Art du charbonnier* (1761), as well as a detailed description of pit-setting by the last collier at the Hopewell Furnace in Pennsylvania. Bérubé 1978, pp. 25–33, 86–104.

156. Bérubé 1978, p. 33.

157. In 1742, a collier was killed when a tree fell on his hut, cited in Luce Vermette, *Domestic Life at the Forges du Saint-Maurice*, History and Archaeology No. 58 (Ottawa: Parks Canada, 1982), p. 53.

158. Bérubé 1978, p. 33.

159. NAC, MG 1, C¹¹A, vol. 111, fol. 83v, memorial signed Cugnet and Simonet, 18 March 1740.

160. Joseph Risi 1942, pp. 26–27, cited in Bérubé 1978, p. 57.

161. *Le Constitutionnel*, 13 June 1870.

162. *Geology of Canada* 1863, p. 683 (cited in Harrington 187, p. 247).

163. Dollard Dubé located the kilns where they are shown in the photograph at Plate 2.13.

164. Dubé's dimensions correspond to those provided by Harrington in 1874. Dollard Dubé, *Les vieilles forges il y a 60 ans*, coll. Pages Trifluviennes, series A, no. 4, (Trois Rivières: Les éditions du Bien Public, 1933), p. 44.

165. Quoted in Marcelle Caron, "Analyse comparative des quatres versions de l'enquête de Dollard Dubé sur les Forges de Saint-Maurice," manuscript on file (Quebec City: Parks Canada, 1982), p. 22. The gist of Dubé's description is given in Bérubé 1978, pp. 36–39.

166. Ibid, p. 39.

167. Ibid., p. 21.

168. NAC, MG 1, C¹¹A, vol. 110, fols. 284–92, Beauharnois and Hocquart to the Minister, 28 September 1734.

169. Bouchette 1815, p. 303.

170. Baddeley 1828, p. 9.

171. Ibid.

172. NAC, MG 1, C¹¹A, vol. 111, fol. 79, memorial signed Cugnet and Simonet, 18 March 1740 (cited in Moussette 1978, p. 33).

173. Moussette 1978, p. 33.

174. NAC, MG 1, C¹¹A, vol. 112, fol. 359v, "Inventaire des Forges," Estèbe, 22 November 1741. The boats would return lighter and were possibly towed from the bank by horses on a towpath, although we have no confirmation of this.

175. NAC, MG 1, C¹¹A, vol. 63, fol. 193, "Observations faites par moy […]," Olivier de Vézin, 17 October 1735.

176. NAC, MG 1, C¹¹A, vol. 111, fol. 79, memorial signed Cugnet and Simonet, 18 March 1740. Estèbe's accounts for 1741–42 show that stone was also carted in October and November; NAC, MG 1, C¹¹A, vol. 111, fols. 278–305v, "Estat général de la dépense […]," signed Estèbe, 2 October 1742.

177. NAC, MG 1, C¹¹A, vol. 110, fol. 247–52, "Plan de la régie qu'il convient d'établir aux forges de Saint Maurice en Canada […]," Olivier de Vézin, n.d. [1741] (cited in Moussette 1978, p. 33). The bateaux recommended by Vézin were larger than those previously used. He was actually describing double bateaux. In a memorial that followed this plan, to save money on the cost of transportation, the ironmaster recommended a "flat-bottomed boat that could carry 12 to 15 thousandweight of iron which three men could easily navigate after the high water on this river in the spring." NAC, MG 1, C¹¹A, vol. 110, fols. 254–59, "Mémoire pour les resgles des forges de Saint-Maurice en Canada, qu'il conviendrait de suivre pour les conduire à leur degré de perfection." Olivier de Vézin, n.d. [1741].

178. Ibid.; Moussette 1978, p. 33.

179. Lepage 1984, p. 36, Table 2.2.

180. Bérubé 1976, p. 28.

181. John Lambert, *Travels through Canada and the United States in the Years 1806, 1807, 1808, to which are added biographical notices and anecdotes of some leading characters in the United States* (London: Baldwin, Cradock and Joy; Edinburgh: W. Blackwood; and Dublin: J. Cumming), 1816, p. 488.

182. Baddeley 1828, p. 9.

183. NAC, MG 1, C¹¹A, vol. 110, fols. 231–36, "Mémoire instructif des observations à faire par le sieur Olivier dans la visite des mines de fer de Saint-Maurice et lieux en dépendant," n.s., 1 September 1735.

184. "Extraordinary and unforeseen expenses
For a Ferry boat attempted to bring over the Charcoals from the other Side of the River St. Maurice & save the expense of Battoes & Battoemen, to five men employed for 12 days in cutting & squarring proper wood & constructing the sd. boat
180:-:-
For smith's work about the same
162:
For a cable of 90 fathoms long, six inches thick & the expenses of 3 men sent in a Course to Quebec on purpose to fetch the sd. cable
606:-:-"
NAC, MG 13, War Office 34, vol. 6, fols. 136–40, "A General Account of the Monies received for and expended at the forges St. Maurice in the Government of Three Rivers & produce thereof from the 1st of October 1760 to the last day of December 1761. Copied after the Accounts closed with the Inspector of the sd. forges during that Interval of time; Trois Rivieres, Decr the 20th 1761, J. Bruyères, Secy," 1760–1761.

185. NAC, MG 21, B21–1, fol. 21681, p. 66–69 (cited in Moussette 1978, p. 33).

186. ASTR, Forges Papers, a few pages taken from an account book dating from July 1856 to April 1858.

187. Caron 1982, pp. 132, 142.

188. "Jusqu'aux Forges par eau," *Le Constitutionnel*, 13 June 1870.

189. Moussette 1978, p. 33.

190. Estèbe's accounts for 1741–42, cited above, report on the payment of Léveillé, master of the *Le Manon*, for the transportation of iron and munitions from Trois-Rivières to Quebec. "The wares are chiefly taken to Three Rivers in *bateaux*, in the summer season, and those destined for Montreal and Quebec shipped from Three Rivers shipped in schooners and sometimes in steamers." APJQ, Superior Court, docket no. 2238, J. Porter et al. v. Weston Hunt et al., exhibit 5, deposition of Hamilton Rickaby, 27 January 1860; Michel Bédard, "Tarification, commercialisation et vente des produits des Forges du Saint-Maurice," typescript (Quebec City: Parks Canada, 1982), p. 37.

191. *Dictionary of Canadian Biography*, s.v. "Bell, Mathew," vol. VII, pp. 64–69. ANQ–TR, Not. Rec. A. Badeaux, 16 April 1796, sale by Firmin Comeau to David Monro and Mathew Bell.

192. NAC, RG 42, 1, vol. 476, fol. 61, Register of Shipping, Montreal, Quebec (cited in M. Bédard, 1986, p. 168).

193. NAC, MG 1, C¹¹A, vol. 112, fol. 57, "Inventaire des Forges," Estèbe, 1741.

194. Jean Hamelin and Jean Provencher, "La vie de relations sur le Saint-Laurent entre Québec et Montréal au milieu du XVIIIᵉ siècle," *Cahiers de Géographie du Québec*, 11ᵉ année (1967) no. 23 (cited in Moussette 1978, p. 106).

195. NAC, MG 1, C¹¹A, vol. 110, fols. 247–52, "Plan de la régie qu'il convient d'établir aux forges de Saint Maurice en Canada […]," Olivier de Vézin, n.d. [1741].

196. Lepage 1984, Tables 3.1 and 3.2.

197. Louis Franquet, *Voyages et Mémoires sur le Canada, (1752–1753)* (Montreal: Éditions Élysée, 1974), pp. 18–19, 22.

198. Caron 1982, pp. 26–28.

199. Roads maintained by Forges workers in 1741–42, according to Lepage 1984, Table 3.2, pp. 67–68; and according to the accounts of interviews conducted by Dollard Dubé in 1933; Caron 1982, pp. 26–28, 100–104.

200. "For two roads made, one above the forges to go to the Intended ferry, & the other below the forges to go to the foot of the Rapids in order to embark the iron for Trois Rivieres Magazine"; NAC, MG 13, War Office 34, vol. 6, fols. 136–40, "A General Account of the Monies received for and expended at the forges St Maurice in the Government of Three Rivers & produce thereof from the 1st of October 1760 to the last day of December 1761. Copied after the Accounts closed with the Inspector of the sd. forges during that Interval of time; Trois Rivieres, Decr the 20th 1761, J. Bruyères, Secy," 1760–61.

201. Since the road from Pointe à la Hache appears on Murray's map, it may be supposed that it existed during the French regime. Similarly, the desire to reduce the cost of transporting charcoal from the east bank of the river by bateau suggests that charcoal was already being made on that side by the end of the French era. In 1750, Bigot spoke of the need to make charcoal on the Cap de la Madeleine seigneury, "whence it is carried by bateaux and on the ice when the river is frozen"; Bérubé 1978, p. 41.

202. Moussette 1978, p. 109.

203. Bédard 1986, p. 162.

204. NAC, MG 1, C¹¹A, vol. 110, fols. 231–36, "Mémoire instructif des observations à faire par le sieur Olivier dans la visite des mines de fer de Saint-Maurice et lieux en dépendant," n.s., 1 September 1735.

205. Moussette 1978, p. 109.

206. APJQ, Superior Court, docket no. 2238, J. Porter et al. v. Weston Hunt et al., exhibit 5, deposition of Hamilton Rickaby, 27 January 1860.

207. Ibid.

208. "The business of the year began at the St Michel, as the wood cutting was the first operation." Ibid.

209. This comment by Hamilton Rickaby would seem to indicate that coal, ore and provisions were carted by the same carters. He added later that transportation costs were lower at that time of year. Ibid.

210. ANQ–M, Not. Rec. N.B. Doucet, 16 August 1806, No. 1050, "contrat entre Jean Sauer, demeurant aux Forges St. Maurice, et Monro & Bell, représenté par Zacharie Macaulay, demeurant aux Forges St. Maurice;" John Munro and Henry Voyer, of the Forges, were present as witnesses.

211. According to Bérubé 1976, pp. 27–28; Bérubé 1978, pp. 9–14; Serge Saint-Pierre, "Les charretiers aux Forges du Saint-Maurice," typescript (Quebec City: Parks Canada, 1977), pp. 7–15 (hereafter cited at Saint-Pierre 1977).

212. NAC, MG 1, C¹¹A, vol. 111, fol. 84, "Mémoire des articles arrestez entre nous soussignés François Étienne Cugnet and Jacques Simonet […]," 18 March 1740; (cited in Bérubé 1978, p. 10.)

213. Bérubé 1978, p. 11.

214. Probably LeProust, the Forges agent at Trois-Rivières at that time.

215. "Heard Mr Proust Captain of militia at Three Rivers respecting the construction of the slays made use of for the carrying ore to the Forges which he describes as follows: frame six to eight feet long in the runner, the breadth four inches in the front tapering to three inches in the back, the depth of the runner nine inches clear from the bottom of the sled, the shafts are (?) fix to it by an iron bolt as used in trucks. That those sleds prevent cahots may cost about half a [?] & will last for 20 years." ASQ, manuscript no. 157, pp. 89–90, "Committee of Merchants, Montreal 26, 30 December of 1786."

216. Denis Woronoff, *L'industrie sidérurgique en France pendant la Révolution et l'Empire* (Paris: Éditions de l'École des hautes études en sciences sociales, 1984), p. 229.

217. Moussette 1978, p. 66.

218. Bédard 1986, p. 166.

219. Bérubé 1978, pp. 12–14.

220. Ibid.

221. Ibid; Saint-Pierre 1977, p. 13.

222. Based on a weight of 88 French pounds per *pipe* of charcoal (1 French pound = 489.41 gr), according to Chaussegros de Léry; NAC, MG 1, C^{11}A, vol. 110, fol. 185 (January–February 1739), Notes and calculations attributed to Chaussegros de Léry.

223. Cited in Bérubé 1978, p. 113. According to Harrington, a *minot* or bushel = 2.250 cu. in. or 37 litres. Harrington 1874, p. 247.

224. NAC, MG 1, C^{11}A, vol. 112, fol. 157v, "Mémoire concernant les Forges de St. Maurice," n.s., n.d. (around 1743, since it is based on Estèbe's trusteeship of 1742).

225. Ibid.

226. Based on Moussette 1978, p. 98, modified. In 1871, 9 of the 16 workers' horses belonged to Dr Beauchemin's farm.

227. Based on 60 bales of hay per horse per month, as in the French era. Forty thousand bales could feed 55 horses for a year.

228. APJQ, Superior Court, docket no. 2238, J. Porter et al. v. Weston Hunt et al., exhibit 5, deposition of Hamilton Rickaby, 27 January 1860.

229. A 16-lb. (7.3 kg) bale of hay, the equivalent of a bale from the French regime. René Hardy and Benoit Gauthier, "La sidérurgie en Mauricie au 19e siècle: les villages industriels et leurs populations," research report submitted to the Regional Branch of the Quebec Department of Cultural Affairs, Centre de recherches en études québécoises (Trois-Rivières: Université du Québec à Trois-Rivières, May 1989), p. 166, Table XV.

230. APJQ, Superior Court, docket no. 2238, J. Porter et al. v. Weston Hunt et al., exhibit 5, deposition of Hamilton Rickaby, 27 January 1860.

231. Lepage 1984, Tables 3.10 and 3.11.

232. Saint-Pierre 1977, pp. 29–30.

233. Serge Saint-Pierre, "La technologie artisanale aux Forges du Saint-Maurice, 1729–1883," Manuscript Report No. 307 (Ottawa: Parks Canada, 1976), pp. 4–5.

234. NAC, MG 11, Colonial Office 5, vol. 59, fols. 307–13, "Inventaire des Fers, Fontes, Mines, Bois pour charbon, Maisons, Bâtiments, Ustenciles, appartenant aux Forges St. Maurice," Hertel de Rouville, 8 September 1760.

235. Saint-Pierre 1977, p. 29.

CHAPTER 3

1. Cited in M. Carlier, *Hydraulique générale et appliquée* (Eyrolles, 1980), p. 4.

2. We are referring to the French foot in this chapter: 1 foot = 1,065 English feet = 32.484 cm; NAC, MG 1, C^{11}A, vol. 63, fol. 190, "Observations faites par moy [...]," Olivier de Vézin, 17 October 1735.

3. Joseph Bouchette, *A Topographical Description of the Province of Lower Canada, with Remarks upon Upper Canada, and on the Relative Connexion of Both Provinces with the United States of America.* (London: printed for the author and published by W. Faden, Geographer to his Majesty and the Prince Regent, Charing-Cross, 1815), p. 304.

4. NAC, MG 1, C^{11}A, vol. 111, fols. 246-77, 8e article, memorial from Vézin and Simonet to Monsieur Delaporte Lalane, 10 June 1741.

5. Called a *roue à la jantille* or *roue à la gentille*, the *jantille* being defined as "each of the wooden boards attached to the circumference of a bucket wheel that formed the sides of each bucket," Réjean L'Heureux, *Vocabulaire du moulin traditionnel au Québec des origines à nos jours*, Documents lexicaux et ethnographiques, "Langue française au Québec," 3e section (Quebec City: Les Presses de l'Université Laval, 1982) (hereafter cited as L'Heureux 1982), p. 155. This well-documented work is an indispensable reference tool for the study of mills and hydraulic mechanisms in Quebec.

6. NAC, MG 1, C^{11}A, vol. 111, fols. 246-77, 8e article, memorial from Vézin and Simonet to Monsieur Delaporte Lalane, 10 June 1741.

7. Maurice Daumas, *Histoire générale des techniques*, tome III, "L'expansion du machinisme," (Paris: PUF, 1968), pp. 15-16.

8. The energy capacity of a stream was usually estimated in terms of mill wheels, no doubt because assessments of watercourses were usually done by a millwright, who was also responsible for building milldams, wheelraces and wheels that could be used in various industries. In the fall of 1735, Vézin was accompanied by such an expert, Jean Costé. He was known as a millstone dresser [*amoulangeur*]; Paillé, who developed the blast furnace mechanisms, was known as a mill carpenter [*charpentier de moulins*]. L'Heureux 1982, p. 359, "Bazil Costé, amoulangeur," from Quebec, in 1742, and pp. 355 and 358, Léonard and Charles Paillé, father and son, "charpentier de moulins" from Montreal, in 1708 and 1715.

9. NAC, MG 1, C^{11}A, vol. 57, pp. 114-116, Francheville to the Minister [October 1732].

10. Vézin said he had kept some space; NAC, MG 1, C^{11}A, vol. 112, fol. 92v, memorial from Olivier de Vézin [1742]; see the plan for a slitting mill and finery, NAC, MG 1, C^{11}A, vol. 57, p. 111, Beauharnois and Hocquart to the Minister, 15 October 1732.

11. NAC, MG 1, C^{11}A, vol. 110, fol. 287, Beauharnois and Hocquart to Minister Maurepas, 28 September 1734.

12. NAC, MG 1, C^{11}A, vol. 63, fol. 189v, "Observations faites par moy [...]," Olivier de Vézin, 17 October 1735.

13. Michel Bédard, André Bérubé, Claire Mousseau, Marcel Moussette and Pierre Nadon, "Le ruisseau des Forges du Saint-Maurice," Manuscript Report No. 302 (Ottawa: Parks Canada, 1978) (hereafter cited as Bédard et al. 1978) pp. 9-10. Water levels were based on sea level.

14. MER, Service de la concession des terres, St. Maurice Township, general file no. 25203/1936, excerpt from a letter from H.R. Symmes to P.M. Vankoughnah, Crown Lands Commissioner, 30 September 1861.

15. NAC, MG 1, C^{11}A, vol. 110, fol. 287, Beauharnois and Hocquart to the Minister, 28 September 1734.

16. NAC, MG 1, C^{11}A, vol. 57, p. 111, Beauharnois and Hocquart to the Minister, 15 October 1732.

17. NAC, MG 1, C^{11}A, vol. 110, fols. 304-8, Beauharnois and Hocquart to the Minister, 26 October 1735, and vol. 110, fol. 73, "Sur les forges Saint-Maurice administrées par le sieur Olivier," Cugnet to Minister Maurepas, 17 October 1741.

18. Bertrand Gille, *Les origines de la grande industrie métallurgique en France* (Paris: Éditions Domat, 1947).

19. NAC, MG 1, C^{11}A, vol. 63, fol. 189, "Observations faites par moy [...]," Olivier de Vézin, 17 October 1735.

20. See note 16.

21. These waterwheels would not necessarily all be running at the same time, as grist mills only operated intermittently. This was likely true for other mills as well, such as the slitting mill and plate mill.

22. According to Sieur Demeloize, as reported by Cugnet; NAC, MG 1, C^{11}A, vol. 110, fol. 73, "Sur les forges Saint-Maurice administrées par le sieur Olivier," Cugnet to Minister Maurepas, 17 October 1741.

23. Bédard et al. 1978, p. 10.

24. On 30 October 1735, Vézin wrote: "[...] a perfect establishment, of which he is sending a plan, as well as a plan of the stream whose slope is 60 feet 9 3/4 inches, as surveyed exactly with that of the Three Rivers, in one of which the stream is discharged [...]." This should in fact read 65 feet, after checking the addition of the figures, as attested by his cross-section and his document of 17 October 1735, quoted in the epigraph to this chapter; NAC, MG 1, C^{11}A, vol. 110, fols. 309-310v, Olivier de Vézin to the Minister of Marine, 30 October 1735.

25. NAC, MG 1, C^{11}A, vol. 63, fol. 190, "Observations faites par moy [...]," Olivier de Vézin, 17 October 1735.

26. Later in the Forges' history, there was a periodic fear that the creek would dry up or not have enough water. For example, when the McDougalls acquired the establishment in 1863, the seller, Onésime Héroux, who kept part of the site, agreed not to cut down the trees around the washery pond to prevent it from drying up. Furthermore, the buyers were assured of "[...] the right to bring water from the big Yamachiche bog across the seller's land to increase the said water power [...]"; ANQ-TR, Not. Rec. Petrus Hubert, No. 4575, 27 April 1863, "Vente des Forges St Maurice," Onésime Héroux to John McDougall.

27. NAC, MG 1, C^{11}A, vol. 63, fols. 189-90, "Observations faites par moy [...]," Olivier de Vézin, 17 October 1735.

28. NAC, MG 1, C^{11}A, vol. 112, fol. 156, "Mémoire concernant les Forges de St Maurice," n.s., n.d. (probably 1743, as it is based on Estèbe's trusteeship of 1742). St Maurice Creek sometimes went by the name "St Étienne." See the 1845 map in Plate 2.15.

29. Achille Fontaine, *Génie d'hier et d'aujourdhui aux Forges du Saint-Maurice*, text presented to guides - summer 1978, manuscript on file, Engineering and Architecture, Environment Canada, Parks, (May 1978).

30. NAC, MG 1, C^{11}E, vol. 10, p. 80v-82; cited in Marcel Moussette, "L'histoire écologique des Forges du Saint- Maurice," Manuscript Report No. 333 (Ottawa: Parks Canada, 1978) (hereafter cited as Moussette 1978), pp. 30-31.

31. Moussette 1978, pp. 31-32; Bédard et al. 1978, pp. 54-55.

32. The 25 horsepower of the sawmill and the 35 horsepower of the grist mill were also referred to in the 1871 census; see Michel Bédard, "Les moulins à farine et à scie aux Forges du Saint-Maurice," Manuscript Report No. 301 (Ottawa: Parks Canada, 1978) (hereafter cited as Bédard 1978), pp. 53-54.

33. NAC, MG 1, C^{11}A, vol. 63, fols. 190-91, "Observations faites par moy [...]," Olivier de Vézin, 17 October 1735.

34. We know that a model of Francheville's bloomery was sent to France to be studied by forge experts, who did study it, as we have their observations through Minister Maurepas in 1734; we also know that the forge's hydraulic mechanisms were built by Costé, who accompanied Vézin to the site in 1735. NAC, MG 1, C^{11}A, vol. 110, fols. 284-92 (model of the forge), Beauharnois and Hocquart to Minister Maurepas, 28 September 1734, fol. 297-300, the Minister's reply to Beauharnois and Hocquart. With regard to Costé, see the references at note 17 above.

35. NAC, MG 1, C^{11}A, vol. 111, fol. 249v, 8e article, memorial from Vézin and Simonet to Monsieur Delaporte Lalane, St Maurice Forges, 10 June 1741.

36. At the time of the Intendant's visit, Vézin had already modified his plan of the forge by removing three waterwheels. It was originally supposed to have six wheels, and Francheville's flume, which fed two wheels, would assuredly not have been able to drive six. NAC, MG 1, C¹¹A, vol. 111, fols. 246-77, 8ᵉ article, memorial from Vézin and Simonet to Monsieur Delaporte Lalane, 10 June 1741.

37. In 1740, Cugnet denounced Vézin, writing that he "[...] was never prepared to clearly admit that he was mistaken about the quantity of water, although he was forced to agree that there was not enough to keep two hearths going in the same forge where he initially planned to establish six movements [...]"; NAC, MG 1, C¹¹A, vol. 111, fols. 7-7v, Cugnet, Quebec, 25 September 1740 [the most detailed memorial about water problems]. Cugnet repeated the same argument in his memorial of 1741. The same year, Vézin briefly explained the changes to his plan as follows: "[...] having himself studied the stream to see whether it had enough water to furnish the Establishment with two *renardière* chaferies according to the last plan that the said Sieur Olivier had made [...]"; NAC, MG 1, C¹¹A, vol. 111, fols. 246-277, 8ᵉ article, memorial from Vézin and Simonet to Monsieur Delaporte Lalane, 10 June 1741; NAC, MG 1, C¹¹A, vol. 110, fol. 70v, "Sur les forges Saint-Maurice administrées par le sieur Olivier," Cugnet to Minister Maurepas, 17 October 1741.

38. Vézin had to clear a large amount of clay and sand (3,000 *toises*) to make a solid base for the milldam and forge. NAC, MG 1, C¹¹A, vol. 110, fol. 345v, "État des ouvrages qui ont été faits cette année à Saint-Maurice pour l'établissement des forges de fer," Olivier de Vézin, Quebec, 19 October 1736.

39. NAC, MG 1, C¹¹A, vol. 111, fols. 246-77, 8ᵉ article, memorial from Vézin and Simonet to Monsieur Delaporte Lalane, 10 June 1741.

40. Quoted by Jean-François Belhoste and Hubert Maheux regarding the blast furnace at La Poitevinière; France, Ministère de la Culture, "Les forges du Pays de Châteaubriant," Cahiers de l'Inventaire 3, Inventaire général des Monuments et Richesses artistiques de la France, Pays de Loire, Département de Loire-Atlantique, 1984, p. 155.

41. In 1737, the measurements of the milldam built by Vézin were given as 74 feet long by 18 feet high by 4 to 5 feet thick (French measure); NAC, MG 1, C¹¹A, vol. 110, fol. 370, 30 October 1737.

42. NAC, MG 1, C¹¹A, vol. 111, fols. 246-77, 7ᵉ article, memorial from Vézin and Simonet to Monsieur Delaporte Lalane, 10 June 1741.

43. See Denis Woronoff, *L'industrie sidérurgique en France pendant la Révolution et l'Empire* (Paris: Éditions de l'École des hautes études en sciences sociales, 1984), p. 288, and several plans in France, Ministère de la Culture, "Les forges du Pays de Châteaubriant," Cahiers de l'Inventaire 3, Inventaire général des Monuments et Richesses artistiques de la France, Pays de Loire, Département de Loire-Atlantique, 1984.

44. Bédard et al. 1978, pp. 55, 74-75.

45. NAC, MG 1, C¹¹A, vol. 111, fol. 12, Cugnet, Quebec, 25 September 1740.

46. NAC, MG 1, C¹¹A, vol. 72, fol. 239-43, Chaussegros de Léry, October 1739.

47. NAC, MG 1, C¹¹A, vol. 110, fol. 237, n.s., n.d., "Projet pour parvenir à faire une petite forge [...]," [1738].

48. See André Charbonneau, Yvon Desloges and Marc Lafrance, *Quebec, The Fortified City: From the 17th to the 19th Century* (Ottawa: Parks Canada, 1982).

49. NAC, MG 1, C¹¹A, vol. 72, fols. 239-43, Chaussegros de Léry, October 1739.

50. NAC, MG 1, C¹¹A, vol. 110, fols. 175-208, (n.s., 1739), Notes and calculations on the hydraulic mechanisms of the existing forge and the planned forge (upper forge); these notes were probably written by Chaussegros de Léry or one of his clerks.

51. Ibid.

52. NAC, MG 1, C¹¹A, vol. 72, fols. 239-43 (microfilm F-72), October 1739. In this letter to Minister Maurepas, Chaussegros de Léry wrote that the second forge (the upper forge) had been in operation since 10 October 1739.

53. We know that not only could Vézin not have driven six wheels at the same time, he had trouble driving three of them. According to the accounts of Sieur Demeloize and the forgemen he met, the water level in the millpond dropped before their eyes. The wheels turned and drove the bellows and hammer, but at too high a price in terms of water use. Léry not only solved the problem of where to site the forges but also that of the output of the first forge.

54. This is Galileo's method (1643) of calculating speed. The formula is: $V = \div kh$; the same definition appeared in Diderot's *Encyclopédie* in 1757: "The speeds are to each other as the square roots of the heads or in subduplicate ratio of the heads"; M. Carlier, *Hydraulique générale et appliquée*, (Eyrolles, 1980), p. 5; *Encyclopédie ou Dictionnaire raisonné des Sciences, des Arts et des Métiers*, vol. 7 (1757), p. 120.

55. Léry never wrote in² or in³.

56. The effort needed for the hammer wheel was approximately 7,416 inches, and approximately 2,296 inches for the chafery wheels.

57. The result was 38.62 but Léry rounded it up to 39.

58. The wheelrace was the walled pit in which the waterwheels were mounted. It formed an extension of the forge itself.

59. NAC, MG 1, C¹¹A, vol. 111, fols. 246-77, 29ᵉ article, 4°, memorial from Vézin and Simonet to Monsieur Delaporte Lalane, 10 June 1741.

60. NAC, MG 1, C¹¹A, vol. 112, fol. 91-95v, "À Monseigneur le Comte de Maurepas, [...]," Vézin, 1742.

61. NAC, MG 1, C¹¹A, vol. 112, fol. 46, Inventaire des Forges 1741, Estèbe, 22 November 1741.

62. AJTR, Not. Rec. J. Badeaux, 1 April 1807, inventory of the St Maurice Forges post.

63. Pierre Beaudet, "Vestiges des bâtiments et ouvrages à la forge basse, Forges Saint-Maurice," Manuscript Report No. 315 (Ottawa: Parks Canada, 1979).

64. This measurement of flow of 14 pints per minute established by hydraulic engineer Edmé Mariotte in 1686 gives us a highly inadequate measurement of the flow; Augustin-Charles d'Aviler, *Dictionnaire d'architecture civile et hydraulique, et des Arts qui en dépendent [...]*, (Paris), Charles-Antoine Jombert, 1755, s.v. "dépense d'eau," p. 145. Roch Samson and Achille Fontaine dealt with this problem in the following article: "La mise en opération des Forges du Saint-Maurice (1736-1741): Une étude pluridisciplinaire," *APT*, vol. 18, nos. 1-2, 1986, pp. 15-31. Marcel Moussette hypothesized that the 240 inches of water may have corresponded to the sluice aperture, converting this into cubic inches per minute and horsepower. He obtained a flow of 390.4 cubic feet per minute, which he converted into 54.9 horsepower. However, Moussette had some doubts about the site and measurement conditions. The flow was more than twice what Inspector Symmes had observed in 1861 (20 horsepower). Unfortunately, the creek's current configuration does not allow us to reproduce conditions at the time. Moussette 1978, pp. 26-27.

65. Bédard et al. 1978, p. 32.

66. Bédard 1978, pp. 29-31.

67. Dollard Dubé told a story about a cache of cannonballs which was disturbed on the site of the grist mill to install a driving wheel. Marcelle Caron, "Analyse comparative des quatre versions de l'enquête de Dollard Dubé sur les Forges du Saint-Maurice," manuscript on file (Quebec City: Parks Canada, 1982) (hereafter cited as Caron 1982), p. 76. See also Chapter 9.

68. This turbine is associated with a dam built in the 1920s by Ernest Marchand, who owned the site at the time. The turbine drove a centrifuge, butter churn and washer. Bédard et al. 1978, pp. 92-93.

69. Claire-Andrée Fortin and Benoit Gauthier, "Description des techniques et analyse du déclin de la sidérurgie mauricienne, 1846-1910," research report submitted to the Regional Branch of the Quebec Department of Cultural Affairs, Centre de recherches en études québécoises (Trois-Rivières: Université du Québec à Trois-Rivières, February 1988), p. 146.

70. André Bérubé, "Rapport préliminaire sur l'évolution des techniques sidérurgiques aux Forges du Saint-Maurice, 1729-1883," Manuscript Report No. 221 (Ottawa: Parks Canada, 1976), p. 74.

71. Ibid., p. 78.

72. Ibid., pp. 72-74.

73. ANQ-TR, Not. Rec. Petrus Hubert, No. 4575, 27 April 1863, "Vente des Forges St Maurice," Onésime Héroux to John McDougall.

74. The geologist described a blast furnace with two tuyeres like the one at St Maurice that is driven by a 24-foot waterwheel. Dr B. J. Harrington, "Notes on the Iron Ores of Canada and Their Development," Geological Survey of Canada, *Report of Progress for 1873-74* (Montreal: Dawson Brothers, 1874), p. 247.

75. Caron 1982, p. 59.

76. "Incendie aux Forges," *Le Journal des Trois-Rivières*, (21 February 1881), p. 2.

CHAPTER 4

1. NAC, MG 1, C¹¹A, vol. 110, fol. 303, Maurepas to Beauharnois and Hocquart, 10 May 1735.

2. Ibid., fol. 301.

3. Ibid., fols. 297–300, Maurepas to Beauharnois and Hocquart, 19 April 1735.

4. Réal Boissonnault, "La structure chronologique des Forges du Saint-Maurice des débuts à 1846," typescript (Quebec City: Parks Canada, 1980) (hereafter cited as Boissonnault 1980), p. 34.

5. NAC, MG 1, C¹¹A, vol. 111, fol. 246, memorial from Vézin and Simonet to Monsieur Delaporte Lalane, 10 June 1741. Bertrand Gille mentions the Sionne ironworks in Champagne, which were still in operation in 1788 and were owned at that time by the widow Michel, who had three other operations; Bertrand Gille, *Les origines de la grande industrie métallurgique en France* (Paris: Éditions Domat, 1947) p. 177.

6. NAC, MG 1, C¹¹A, vol. 63, fols. 190–92, "Observations faites par moy [...]," Olivier de Vézin, 17 October 1735.

7. NAC, MG 1, C¹¹A, vol. 111, fols. 168–72, Hocquart and Vézin, 17 October 1735.

8. Boissonnault 1980, p. 49.

9. NAC, MG 1, C¹¹A, vol. 110, fols. 144–49, estimate signed Hocquart and Vézin, 17 October 1735.

10. Boissonnault 1980, pp. 41–42.

11. Ibid., p. 46.

12. NAC, MG 1, C¹¹A, vol. 110, fol. 304, 26 October 1735. The experts were Grand Voyer Boisclair, Jesuit Brother Leclerc, and Jean Costé, millwright.

13. NAC, MG 1, C¹¹A, vol. 110, fols. 69–71, Cugnet to Maurepas, 17 October 1741.

14. NAC, MG 1, C¹¹A, vol. 111, fol. 248v, 8ᶜ, memorial from Vézin and Simonet to Monsieur Delaporte Lalane, 10 June 1741.

15. NAC, MG 1, C¹¹A, vol. 110, fols. 69–71, Cugnet to Maurepas, 17 October 1741.

16. NAC, MG 1, C¹¹A, vol. 111, fol. 246v, memorial from Vézin and Simonet to Monsieur Delaporte Lalane, 10 June 1741.

17. NAC, MG 1, C¹¹A, vol. 110, fol. 68, Cugnet to Maurepas, 17 October 1741.

18. NAC, MG 1, C¹¹A, vol. 72, fols. 239–43, October 1739. In this letter to Maurepas, Chaussegros de Léry states that the second forge (the upper forge) had been operating since 10 October 1739. This information is confirmed by Vézin and Simonet, who left for France that fall, eight days after the upper forge began operating. Tabulations for the two forges, however, did not begin until 15 October (see Chapter 5); NAC, MG 1, C¹¹A, vol. 111, fols. 255v–56, memorial from Vézin and Simonet to Monsieur Delaporte Lalane, 10 June 1741.

19. Boissonnault 1980, p. 46.

20. Ibid., p. 79.

21. NAC, MG 1, C¹¹A, vol. 110, fol. 339, Beauharnois and Hocquart to Maurepas, 19 October 1736.

22. Boissonnault 1980, p. 49.

23. Vézin's plan shows four buildings, but only three were described in his "Observations," which did not include a description of the blacksmith's shop mentioned in the deed of 16 July 1736. "Plan des mines de Trois-Rivières" (referred to in a letter from Beauharnois and Hocquart dated 26 October 1736), Bibliothèque Nationale, Paris, Cartes et plans, portefeuille 127, division 8, pièce 5 d; AJTR, Not. Rec. Polet, 16 July 1736, "Vente - Les héritiers du deffunt Lafond"; NAC, MG 1, C¹¹A, vol. 111, fol. 170v, Hocquart and Olivier de Vézin, 17 October 1735.

24. When the Forges were sold in 1846, the village encompassed 165 ha; Michel Bédard, "La privatisation des Forges du Saint-Maurice 1846-1883: adaptation, spécialisation et fermeture," manuscript on file (Quebec City: Canadian Parks Service, 1986) (hereafter cited as Bédard 1986), p. 30.

25. NAC, MG 1, C¹¹A, vol. 111, fol. 246v, memorial from Vézin and Simonet to Monsieur Delaporte Lalane, 10 June 1741.

26. NAC, MG 1, C¹¹A, vol. 110, fol. 312, Maurepas to Beauharnois and Hocquart, 14 March 1736.

27. NAC, MG 1, C¹¹A, vol. 110, fol. 339, Beauharnois and Hocquart to Maurepas, 19 October 1736: "The workers naturally took advantage of the undue haste with which they were engaged to advance the work in accordance with your intentions."

28. NAC, MG 1, C¹¹A, vol. 111, fols. 246v–47v, memorial from Vézin and Simonet to Monsieur Delaporte Lalane, 10 June 1741.

29. Ibid., fol. 250. He was from the commune of Labergement-Foigny, in the bailiwick of Dijon; "Carte de la province de Bourgogne au XVIIIᶜ siècle," *Annales de Bourgogne*, tome LV, année1983, fascicule III, juillet-décembre.

30. NAC, MG 1, C¹¹A, vol. 110, fols. 318–19v, Maurepas to Beauharnois and Hocquart, 15 May 1736.

31. NAC, MG 1, C¹¹A, vol. 110, fols. 339–42v, Beauharnois and Hocquart to Maurepas, 19 October 1736.

32. NAC, MG 1, C¹¹A, vol. 110, fols. 345–46, Olivier de Vézin, 19 October 1736 and fols. 339–42v, Beauharnois and Hocquart to Maurepas, 19 October 1736 (reference to plans).

33. "The merchant Sr Huguet has informed me that he has just received the implements and other utensils for an ironworks to be established in Canada from Sr le Blanc, ironmaster at Clavière, to have them sent with dispatch to Bordeaux, even overland, since this must be effected with haste. It would be absolutely impossible to use this route, since the utensils weigh 12 thousandweight [...]." Nantes, 15 March 1737, excerpt from a letter from M Dionis (to unknown) in the Colonies and other locations; MG 2, B3, vol. 380, fols. 32–33.

34. NAC, MG 1, C¹¹A, vol. 111, fol. 250v, memorial from Vézin and Simonet to Monsieur Delaporte Lalane, 10 June 1741. Simonet brought back [...] fifty people, men, women and children, who have already cost the company nearly 18,000 francs [...]."

35. Ibid., article 16.

36. Ibid., fol. 253. Our estimate of a production of about 250 thousandweight of pig iron (122 t) for a six-week period (42 days) is based on a maximum output of six thousandweight (3 t) per 24 hours established by Chaussegros de Léry in 1739. NAC, MG 1, C¹¹A, vol. 72, fols. 239–43, October 1739.

37. NAC, MG 8, D2, Trois-Rivières Notarial Records, "procès-verbal du premier feu mis au fourneau le 20 août 1738 entre 11 et 12 h"; ANQ-Q, NF-25, no. 1177, "procès-verbal concernant le fourneau des forges," 7 October 1738. This *procès-verbal* establishes the date of the initial payment of advances to the King; Jean-Noël Fauteux, *Essai sur l'industrie au Canada sous le régime français* (Quebec City: Ls-A Proulx, The King's Printer, 1927), vol. 1, p. 87.

38. NAC, MG 1, C¹¹A, vol. 111, fols. 252v–53, memorial from Vézin and Simonet to Monsieur Delaporte Lalane, 10 June 1741.

39. NAC, MG 1, C¹¹A, vol. 110, fol. 73, Cugnet to Maurepas, 17 October 1741.

40. Roch Samson and Achille Fontaine, "La mise en opération des Forges du Saint-Maurice (1736–1741). Une étude pluri-disciplinaire." *APT*, vol. 18, nos. 1-2, 1986, pp. 15–31.

41. This was either Léonard or Charles Paillé, father and son, millwrights from Montreal; Réjean L'Heureux, *Vocabulaire du moulin traditionnel au Québec des origines à nos jours*, Documents lexicaux et ethnographiques, "Langue française au Québec," 3ᶜ section (Quebec City: Les Presses de l'Université Laval, 1982), p. 355.

42. In the fall of 1739, he explained to Maurepas that, in 1737, when he had been forced to raise the foundation sills of the forge by 2 feet to escape the spring flooding of the St Maurice, he had had to reduce the diameter of the waterwheels by 2 feet, to 8 from 10 feet. The leverage of the wheels was reduced accordingly, so that more water was needed to make up for the lack of height. In addition, the resistance of the gear mechanism called for larger diameter wheels. In these circumstances, Vézin had to give up on the idea of using three wheels at once. NAC, MG 1, C¹¹A, vol. 110, fol. 391, "Mémoire du Sieur Olivier de Vézin sur les Forges du Saint-Maurice," Versailles, 28 December 1739.

43. According to Léry. Another reference gives the date as 15 October: NAC, MG 1, C¹¹A, vol. 72, fol. 239–43, Chaussegros de Léry, October, and vol. 111, fols. 31–33, Hocquart to Maurepas, Quebec, 24 October 1740.

44. NAC, MG 1, C¹¹A, vol. 111, fols. 255v–56, memorial from Vézin and Simonet to Monsieur Delaporte Lalane, 10 June 1741, and vol. 111, fols. 64–65v, Olivier de Vézin to Maurepas, Quebec, 12 October 1740.

45. The assistance of soldiers from the Trois-Rivières garrison was required to make the repair. Boissonnault 1980, p. 62.

46. Ibid., 10 June 1741, fols. 268–69.

47. Boissonnault 1980, p. 60. Such an output in a little over 10 months equals about a thousandweight of bar iron per day, if idle time on Sundays and feast days, as well as the inevitable work stoppages for mechanical breakdowns, are taken into account. In his memorial of 28 December 1739 to Maurepas, Vézin claimed to have produced roughly 300 thousandweight of bar iron between October 1738 and October 1739. Chaussegros de Léry put the output of each chafery at a thousandweight of iron per day. NAC, MG 1, C¹¹A, vol. 72, fols. 239–48, October 1739.

48. Ibid.

49. Boissonnault 1980, p. 62.

50. NAC, MG 1, C¹¹A, vol. 110, fol. 391, "Mémoire du Sieur Olivier de Vézin sur les Forges du Saint-Maurice," Versailles, 28 December 1739.

51. Boissonnault 1980, p. 62.

52. NAC, MG 1, C¹¹A, vol. 111, fols. 246–77, memorial signed Cugnet and Simonet fils, 18 March 1740.

53. See the detailed figures in Appendix 10. Cugnet stated in 1741 that the cracks had formed during construction of a furnace retaining wall on the creek side. The excavation for the foundations for this wall had caused the wall of the furnace to slip a bit, so that it had to be reinforced, which caused cracks. NAC, MG 1, C¹¹A, vol. 110, fols. 71-72, Cugnet to Maurepas, 17 October 1741.

54. NAC, MG 1, C¹¹A, vol. 111, fols. 31–33, Hocquart to Maurepas, 24 October 1740.

55. Ibid.

56. Ibid. Cugnet had used revenue from the King's domain to finance Forges operations.

57. NAC, MG 1, C¹¹A, vol. 112, fols. 28–59v, Inventaire des Forges 1741, Estèbe, 22 November 1741, signed Ignace Gamelin, Simonet *fils* and Estèbe.

58. NAC, MG 1, C¹¹A, vol. 112, fol. 93 and fol. 95v, Olivier de Vézin to Maurepas, n.d. (probably October 1741).

59. NAC, MG 1, C¹¹A, vol. 112, fols. 25v–26, Hocquart to Maurepas, 24 October 1740.

60. In his memorial of 25 September 1740, Cugnet reported on the complaints of a blacksmith from Ste Anne des Grondines, who said that workmen at the Forges "were slacking off" by producing "flawy" iron "that was not shingled enough." He also took the opportunity to report complaints from Quebec blacksmiths and even clients in France. However, Vézin vigorously defended his workers. In a memorial to Maurepas in 1742, Hocquart vaunted the excellent quality of Forges iron. NAC, MG 1, C¹¹A, vol. 111, fols. 2–30, memorial from Cugnet on the St Maurice Forges, 25 September 1740; Boissonnault 1980, p. 100.

61. See Chapter 9. Louis Franquet, *Voyages et Mémoires sur le Canada (1752–1753)* (Montreal: Éditions Élysée, 1974) p. 20.

62. NAC, MG 1, C¹¹A, vol. 112, fol. 92v, Olivier de Vézin to Maurepas, n.d. (probably October 1741): "which would have been easily accomplished, if the said Olivier had not wanted to have a third forge between the first two, which would have allowed and would still allow them to be spaced out [...]."

63. NAC, MG 1, C¹¹A, vol. 72, fols. 239–43, October 1739.

64. More specifically, the northeastern side, since it faced not quite 90⁰ north. For readability, we have named the four sides in accordance with the four points of the compass.

65. In 1741, Estèbe described the pillar rising from the gully to support the shaft of the great wheel as 21.5 feet (7 m). He gave the same height for the "southeastern transverse wall at the tailrace." This height brings us approximately to the level of the furnace crucible, under which Estèbe measured a 7-foot (2.3 m) foundation. NAC, MG 1, C¹¹A, vol. 112, fols. 38v–42, "Inventaire des Forges 1741," Estèbe."

66. NAC, MG 1, C¹¹A, vol. 110, fol. 71v, memorial from Cugnet on the St Maurice Forges, 17 October 1741.

67. Parks Canada engineers had to solidify it to ensure that archaeologists could work in safety. Achille Fontaine, "Étude des mécanismes hydrauliques du haut fourneau. Forges du Saint-Maurice," December 1980, pp. 15–16, internal document, Engineering and Architecture, Environment Canada, Parks. This buttress probably dates from the time when the blast furnace was totally rebuilt (1854) to double its original size. See the part of this chapter entitled "Changes to the Blast Furnace."

68. NAC, MG 1, C¹¹A, vol. 111, fols. 247v–48v, memorial from Vézin and Simonet to Monsieur Delaporte Lalane, 10 June 1741.

69. Pierre Drouin and Alain Rainville, "L'organisation spatiale aux Forges du Saint-Maurice: évolution et principes," typescript (Quebec City: Parks Canada, 1980) p. 124.

70. The charcoal house extended 3.3 m beyond the north side of the furnace and was supported for 3.3 m of its width by the casting house, which was the same width as the west side of the furnace. It is easy to picture a door between the two bays in this location, as can be seen later in the engraving by Lucius O'Brien, published in 1882 in *Picturesque Canada* (Plate 4.2). André Bérubé, "Rapport préliminaire sur l'évolution des techniques sidérurgiques aux Forges du Saint-Maurice, 1729-1883," Manuscript Report No. 221, (Ottawa: Parks Canada, 1976) (hereafter cited as Bérubé 1976), p. 40.

71. Monique Barriault, "Le moulage au haut fourneau des Forges du Saint-Maurice: les grandes périodes de moulage et les bâtiments correspondants," manuscript on file (Quebec City: Canadian Parks Service, 1979), n.p.

72. In 1736, Vézin said he had built a furnace 28 feet high measuring 26 feet (8.4 m) square at the base, tapering to 21 feet (6.8 m). These measurements are quite close to those in his spending estimates of 17 October 1735, where he said that the height of 28 feet (9 m) included the foundations. In 1874, Harrington gave the height as 30 English feet (9.1 m), without specifying whether this included the foundations. In 1933, based on the data from Dollard Dubé's interviews with the last workers at the Forges, the architect Ernest Denoncourt drew a blast furnace 30 feet (9.1 m) wide on each side and 36 feet (11 m) high, not including about 9 feet (2.7 m) of foundation. The architect's cross section does not show the blast furnace as a pyramid, which contradicts later drawings and photos after the Forges were closed. NAC, MG 1, C¹¹A, vol. 110, fols. 144-49v, Olivier de Vézin, "Projet des dépenses [...]," 17 October 1735; Dr B.J. Harrington, "Notes on the Iron Ores of Canada and Their Development," in the Geological Survey of Canada, *Report of Progress for 1873-74* (Montreal: Dawson Brothers, 1874) (hereafter cited as Harrington 1874), pp. 192–259. Ernest L. Denoncourt, "Relevé du Haut Fourneau, Forges du Saint-Maurice" (cross section), 1933, reproduced in Marcelle Caron, "Analyse comparative des quatre versions de l'enquête de Dollard Dubé sur les Forges Saint-Maurice," typescript (Quebec City: Parks Canada, 1982) (hereafter cited as Caron 1982), pp. 119–21.

73. These dimensions, provided by Vézin himself shortly after construction of the furnace in the fall of 1736, differ from those given by Estèbe in his inventory of 1741. We think that this discrepancy results from the method used to measure the furnace. NAC, MG 1, C¹¹A, vol. 110, fols. 345–46, Olivier de Vézin, 19 October 1736.

74. In the foundations of the last blast furnace at the Forges, archaeologists have located a void about 30 cm high, with an area of 1.2 m by 1.5 m, under the crucible; P. Nadon to L. Gohier (memorandum of 10 May 1978). The present ruins of the furnace also show the vents in the stack across from the crucible, obviously to allow moisture in the masonry to escape under the crucible. Richard Cox, "Maçonnerie de la salle des soufflets et emplacement des engrenages," typescript (Quebec City: Parks Canada, 1976); Pierre Nadon, "Recherches archéologiques aux Forges," typescript (Ottawa: Parks Canada, April 1975).

75. *Encyclopédie ou Dictionnaire raisonné des Sciences, des Arts et des Métiers,* vol. 7, 1757, s.v. "Forges, (Grosses-)" by M. Bouchu, ironmaster at Veuxsaules, near Châteauvilain (hereafter cited as *Encyclopédie 1757*), p. 153; *A Diderot Pictorial Encyclopedia of Trades and Industry,* edited by Charles Coulston Gillespie (New York: Dover Publications Inc, 1959), vol. 1, Plate 87, The Blast Furnace II.

76. Concerning the design of furnaces, particularly in Champagne and Burgundy, Courtivron and Bouchu write: "In the end, there are no established proportions for the bottom, top, middle and the position of the tuyere. This is where the mysterious industry of the founders comes into play." M. le Marquis de Courtivron and M. Bouchu, Correspondant de l'Académie des Sciences, "Art des forges et fourneaux à fer," in *Descriptions des Arts et Métiers faites et approuvées par Messieurs de l'Académie Royale des Sciences* (Paris: Dessaint et Saillant, 1761), s.v. "des fourneaux," (hereafter cited as Courtivron and Bouchu 1761), p. 45.

77. "There is perhaps some question as to why the interior cavity has this double funnel shape, and it would be difficult to get the workers to give a reason. However, it would appear that prior to the adoption of this shape, a number of others were tried and found wanting, and it would appear that this shape works very well. It can be seen that with this shape, where the upper opening is narrower than the furnace towards the boshes, it is clear that the heat from the fire dissipates less; that the inwalls reflect towards the ore some of the heat that would rise if the interior cavity was the same width everywhere. Nevertheless, it seems that it would be even better to have a round circumference to the shape of the inwalls; the heat would have the same effect everywhere and would not wear down the inwalls more in one area than another. If the funnel forming the crucible and the boshes is upside-down, another reason can be found: width is required near the boshes to contain the charcoal and the ore, which continually feed the fire; [...]." Courtivron and Bouchu 1761, p. 8.

78. It is easier to understand the "engineering" of the boshes by comparing the belly of the blast furnace to that of the low shaft furnace, which had vertical inwalls that did not hold the heavy charges of ore in check above the fire. In 1735, Vézin himself, in one of his rare technical papers, lauded the superiority of a bosh furnace in criticizing the direct ore reduction process in a simple chafery at Francheville's forge. "The mine, which was barely heated before it reached the tuyere and could not melt unless it was directly on the blast, fell to the bottom of the receiver, *since it had no boshes to retain it* in the area of most intense heat [...]." He later went on to say: "The role of the mine is to produce a cast iron to which the founder gives the quality which it should have, by observing it carefully as it is worked in the receiver, which is, properly, the reservoir or repository of the iron with its proportions, where it develops its qualities during the consumption of six or seven charges of mine, charcoal, limestone and clay, which cannot melt unless they are *halted by the boshes* in the area where the heat is most intense, which is above the tuyere, during which time the iron comes to nature" (our emphasis). NAC, MG 1, C¹¹A, vol. 63, fols. 190-91v, "Observations faites par moy [...]," Olivier de Vézin, 17 October 1735.

79. Metric measure: 4.6 m × 2.4 m × 2.1 m.

80. In his "Projet des dépenses [...]" of 1735, Olivier de Vézin allowed for "10,000 [pounds of] fired brick for the inwalls," NAC, MG 1, C¹¹A, vol. 110, fols. 170-72. Quebec, 17 October 1735, signed Olivier de Vézin and Hocquart. It would appear, however, that sandstone was used, as this 1740 memorial suggests: "[...] he will have to quarry sandstone for the chimneys and for the inwalls of the furnace [...]," NAC, MG 1, C¹¹A, vol. 111, fol. 79, Cugnet and Simonet, 18 March 1740. Somewhat later, an unsigned and undated document in which the year 1742 is mentioned provides details on the materials used. NAC, MG 1, C¹¹A, vol. 112, fols. 120v-21, n.s., n.d., "Mémoire sur les Forges de St. Maurice." In 1828, Lieutenant Baddeley, describing the Gabelle quarry, wrote: "A very valuable fire stone is also found in the same bank, but about one quarter of a mile lower down the river, of this the furnaces are formed, and it is found to stand unaltered the longest campaign." "Lieutenant Baddeley's (Rl Engineers) report on the Saint Maurice iron works, near Three Rivers, Lower Canada, jany 24th 1828," in *APT*, vol. V, no. 3 (1973), p. 12. In 1855, W. Henderson, then manager of the Forges for John Porter & Company sent two cases of samples of bog ore to the Paris Exhibition. One case also contained "two pieces of what we call here Fire stone, as it is used in building the furnace and stands the fire so well that a furnace usually lasts from 3 to 4 years. It is also an excellent building material, and is found only at one place viz the Gabelle above referred to. I am not aware of its existing elsewhere in this part of the Province. It forms the face of the steep & high bank of the River and in quantity

appears inexhaustable [...]." McGill University Archives, Logan Papers, accession 1207/11, item 94, John Porter & Company to W. E. Logan, Provincial Geologist, St Maurice Forges, signed W. Henderson, Manager, 11 January 1855.

81. In the 1740 memorial cited in the preceding note, under the item "Repairs to the furnace," the first two sentences are: "It is important to have the furnace in working condition by next April; the stone and lime are already at the site to begin the lining. The stone for the receiver is dressed." The entire blast furnace was to be rebuilt several times, as indicated by a number of sources, from which the following excerpts are taken. In 1744, a source notes: "This furnace is getting old, and can last only another 2 or 3 years at most; preparations will have to be made in advance and the materials assembled to make a new one and, in the meantime, some repairs will have to be made to the old one [...]." NAC, MG 1, C¹¹A, vol. 112, fols. 242v–43v, Hocquart to Maurepas, 18 October 1744. In 1857, in a report on recent repairs, we find the following:

"Entire blast furnace, for stone and work 300.00

This furnace was rebuilt a second time because of an explosion." MER, Service de la concession des terres, St Maurice Township, general file no. 25203/1936. Estimate made 4 September 1857 by Édouard Normand, accompanied by Sieur Thelesphore Lemay, master joiner and contractor, and Sieur Édouard Parent, master mason and contractor. Other references: RG 1, L3L, vol. 155, p. 76225–29 (1769); AJTR, Not. Rec. J. Badeaux, 1 April 1807, "Inventaire du poste des Forges St Maurice"; NAC, RG1, E1, vol. 64, p. 412 (1845).

82. "It is always a good idea to repeat that, in Sweden, furnaces are perfectly round inside, and that, in France, many are square, and that the least defective there have eight unequal sides. We are forced to believe, as M. de Réaumur states, that, having been left completely to the workers, the shape that was used was the one they found easiest to make." And Bouchu writes elsewhere: "The interior of furnaces in most of Champagne and Burgundy is an elongated square, although they differ according to the founders, who do not want to build something just like their neighbours, and who, in similar mines, argue the quality of the mine." Courtivron and Bouchu 1761, troisième partie, article VII, p. 62 and première partie, article VII, p. 44.

83. Vézin wrote in 1741, describing his troubles in blowing in the blast furnace in 1737: "[...] the chimney and receiver of this furnace are well made, in accordance with the proportions and degree of heat suitable for the mines of this country [...]," NAC, MG 1, C¹¹A, vol. 111, fol. 253, memorial from Vézin and Simonet to Monsieur Delaporte Lalane, 10 June 1741.

84. Bouchu, in the *Encyclopédie*, uses the word *ouvrage* to mean the crucible and the boshes; in his *Art des Forges* [...], he describes the *ouvrage* as "[...] the entire lower part from the inwalls [...]," and later he distinguishes the boshes from the *ouvrage*, which refers to the part below the boshes and includes the crucible. *Encyclopédie* 1757, vol. 7, p. 150; Courtivron and Bouchu 1761, article V, p. 59 and article VI, p. 61.

85. The description of the crucible is taken from Bouchu's article in the *Encyclopédie*, which gives the following dimensions for the crucible of a 21-foot-high blast furnace (the height of the blast furnace at St Maurice, according to Vézin in a 1736 document): length $3^1/_2$ to 4 feet; width 13 inches; height 12 to 13 inches (French measure). *Encyclopédie* 1757, p. 159. See note 70, Chapter 2, for the exact dimensions of the forge crucible stones around 1740.

86. See reference in Chapter 2 to the use of two grey stones (probably one on top of the other) for the bottom. According to Bouchu, the bottom could also be of sand or a mixture of sand and firestone, or sandstone or any other refractory stone.

87. An old anvil was often used to make the dam, and when it did not have a slanted side, *gentilshommes* (two parallel iron bars) were laid against it, on which the slag freed from the crucible ran out. See Courtivron and Bouchu 1761, section "des fourneaux," Plate I, Figure 5.

88. The original bellows were made by the mill carpenter Paillé. In 1741, Vézin wrote: "[...] the bellows were as good as if they had been made by a French bellowsmaker. They had only one defect, just like the bellows of the said forge, which stemmed from the quality of the wood, which should have cured for another year before being used. But this was not possible, since we had no other, and they are so solid that they still exist today." NAC, MG 1, C¹¹A, vol. 111, fol. 252, memorial from Vézin and Simonet to Monsieur Delaporte Lalane, 10 June 1741.

89. Some sources mention the purchase of hides (caribou and others), tallow and oil for the bellows. The hides were likely used for the bellows valves and the oil to lubricate the sides of the box and the bag, which were fitted together very precisely. In the inventory of 1741, in the section on "Ustanciles à maréchal," (farrier's tools), there is mention of "one cowhide bellows," which is only a single bellows for a farrier's shop; NAC, MG 1, C¹¹A, vol. 110, fol. 167, "État des divers achats faits et autres payements pour l'exploitation des Forges de Saint-Maurice depuis le 27 avril 1742 jusqu'au 1er octobre 1743"; vol. 110, fol. 225v, "Mémoire sur les Forges de Saint-Maurice"; vol. 111, fol. 298, "État général de la dépense [...]," 1741–42; vol. 112, fols. 169-169v, "Mémoire sur les Forges de Saint-Maurice," 1742–43; NAC, MG 1, C¹¹A, vol. 112, fol. 47v, "Inventaire des Forges 1741," Estèbe.

90. *Encyclopédie* 1757, vol. 7, p. 154.

91. J. Nef, "La civilisation industrielle," s.v. "*Industrie*," in *Encyclopaedia Universalis* (1968), 5th edition, 1973, vol. 8, p. 969.

92. JLAPC, 1852–53, vol. 11, app. CCC, p. 9, "Memorial from Andrew Stuart and John Porter to the provincial secretary, 23 June 1852."

93. Ibid., p. 17. Letter from Étienne Parent to the Honourable John Rolph, 20 September 1852.

94. However, replacing the bellows with a compressor did not necessarily mean that the great wheel had to be replaced (see Chapter 3).

95. Bédard 1986, p. 24.

96. André Bérubé, "Technological changes at Les Forges du Saint-Maurice, Quebec, 1729–1883," *CIM Bulletin*, May 1983, vol. 76, no. 853; "Les changements à l'intérieur de la filière technique des Forges du Saint-Maurice entre 1729 et 1883," Congrès de l'Institut d'histoire de l'Amérique française, October 1980.

97. According to Bérubé 1976, p. 72, the savings on fuel thus realized would be in the order of 25%.

98. In September 1852, Andrew Stuart and John Porter had sent Étienne Parent Hunter's report, in which he estimated the cost of the necessary repairs at £3,600 to £4,000. When the Executive Council re-evaluated Stuart and Porter's contractual obligations pursuant to the sale of the Forges, one of the conditions the Council set was that the £4,000 earmarked for improvements be spent within two years. Bédard 1986, pp. 42–47.

99. JLAPC, 1852–53, vol. 11, app. CCC, p. 27, William Hunter, engineer, 24 August 1852.

100. Bédard 1986, pp. 67–69.

101. The managers at that time, Weston Hunt and Jeffrey Brock, claimed, after the first explosion, that there had been "a supposed defect in the bottom of the said furnace," and installed a "new bottom." The workers had warned them of the lack of vents in the furnace masonry; "[...] That the said defendants while in the management of the said Forges proceeded to pull down and rebuild the large furnace at the said Forges, and contrary to the advice and remonstrances of the workmen, at the said Forges, proceeded to rebuild the same, in a manner wholly different from that in which the same had always previously thereto been; and in ignorance and violation of the simplest and known rules in the construction of such furnaces, built the same of solid masonry, without any interval to admit of the necessary expansion of the same, when subjected, as it must be, to great heat; the consequence whereof was that for the first time since the Forges have been in operation the said furnace, on or about the first day of April (17 April, according to the coroner's inquest) one thousand eight hundred and fifty four, exploded and two men were killed thereby and others very much burnt and injured; [...] and [...] on or about the sixteenth day of October one thousand eight hundred and fifty four again exploded very seriously burning two other men, [...] it became necessary to demolish the said furnace to discover the cause [...]." AJQ, Superior Court, docket no. 2191, John Porter et al. v Weston Hunt et al., affidavit, 6 August 1855, fol. 4.

102. See previous note.

103. Harrington 1874, p. 247.

104. Caron 1982, pp. 58, 64–67.

105. His drawing does, however, give a clue to a change in the height of the last furnace. On the drawing, a smaller mass superimposed over the old charging platform represents the upper part of the belly of the new furnace. The height of 22 feet (6.7 m) shown for the platform corresponds to the platform of the first furnace; ibid., p. 65.

106. Bérubé 1976, p. 75.

107. In his manuscript notes, Dubé drew in "Plan C" a "hot air chamber" up against the south side of the furnace. Fonds Marchand, fol. 46. See the reheating chamber on Denancourt's cross-section in Plate 4.7. See also the illustration of a hot blast furnace in Plate 4.8.

108. Bérubé 1976, pp. 72–73.

109. Caron 1982, pp. 58, 121.

110. Bérubé 1976, p. 74.

111. Ibid., p. 75.

112. Bédard 1986, p. 165.

113. Ibid., p. 127.

114. AJTR, Superior Court, docket no. 281, Robert Wilson et al. v George McDougall, testimony of Philippe R. Hamilton, exhibit no. 52, 23 April 1883, fol. 4.

115. In his testimony cited in the preceeding note, Hamilton stated that "When the said new furnace was built, an engine from L'Islet was brought to the St Maurice Forges and used *to supplement the water power*, which was inadequate to operate the two furnaces" (our emphasis). A turbine caisson, possibly installed at this time, was excavated in the furnace wheelrace (see Claire Mousseau, "L'évolution fonctionnelle de la forge haute à travers la transformation des ouvrages, 1739-1883," Manuscript Report No. 398 (Ottawa: Parks Canada, 1979) (hereafter cited as Mousseau 1979). At that time, it was common practice to use mixed energy sources when possible (particularly during high water), since water power cost less than steam; during low water, steam enabled plants to continue operating. The two forms of energy thus complemented each other.

116. Ibid., fol. 6.

117. Cited in Bérubé 1976, p. 77.

118. Ibid., p. 77.

119. NAC, MG 1, C¹¹A, vol. 72, fols. 239–43, Chaussegros de Léry, October 1739. Remarks confirmed by Cugnet in 1741: "[...] he built a forge of the size required to install the 6 movements included in his plans [...]," NAC, MG 1, C¹¹A, vol. 112, fol. 69, memorial from Cugnet on the St Maurice Forges, 17 October 1741.

120. NAC, MG 1, C¹¹A, vol. 110, fols. 345–46, "État des ouvrages [...]," Olivier de Vézin, 19 October 1736 and NAC, MG 1, C¹¹A, vol. 63, fol. 190. "Observations faites par moy [...]," Olivier de Vézin, 17 October 1735. This clean-up of the site of Francheville's little forge, which was demolished, likely explains why archaeological digs on the site turned up almost no trace of the first forge. Pierre Beaudet, "Vestiges des bâtiments et ouvrages à la forge basse, Forges du Saint-Maurice," Manuscript Report No. 315 (Ottawa: Parks Canada 1979 (hereafter cited as Beaudet 1979), pp. 12–16.

121. NAC, MG 1, C¹¹A, vol. 112, fols. 39–39v, "Inventaire des Forges 1741," Estèbe.

122. NAC, MG 1, C¹¹A, vol. 111, fols. 168–172, "Projet des dépenses à faire [...]," Hocquart and Vézin, Quebec, 17 October 1735.

123. Beaudet 1979, p. 20.

124. Ibid., p. 23.

125. Ibid., pp. 28–30. We note, moreover, that, in 1742, this hearth had not yet been moved: "There can be no thought of re-establishing the lower hearth of the lower forge this year [...]"; NAC, MG 2, A6, vol. 16, p. 376, "Mémoire pour servir d'instructions aux sieurs Cressé et Martel, 30 août 1742." A memorial of the same year confirms that this chafery was out of use, since only three were operating on the site: "Five *goujats* have to be brought in from France, two for each forge and one for the third chafery, operated by sluicing [...]"; NAC, MG 1, C¹¹A, vol. 112, fols. 174v–75, "Mémoire sur les Forges du Saint-Maurice," (around 1742–43). One might also wonder whether the chafery was actually rebuilt at that time or was replaced by a hearth for a tilt hammer after the fire in the forge in 1747 (see below for more about the tilt hammer). Moreover, a note in a document of 1762 would seem to indicate that there was only one chafery in the lower forge: "[...] the lower forge, where the big cannon could have been moulded, has been out of service for four days, since the chafery shaft is broken, so that only cannons of no more than 8 pounds can be cast [...]"; they were busy remelting (fining) old ordnance left by the French after the Conquest; NAC, MG 12, B61 (W.O. 34, vol. 6), fol. 292.

126. At the St Maurice Forges, the term hammerman was reserved for the person in charge of each forge. The workers under him were called "finers." This did not mean that finers did nothing but fining and that the hammerman did nothing but hammer, since the two operations were carried out by both. The memorial of 1742 describing the character and productivity of each forgeman confirms this, stating that some finers produce less iron and others more. See description of work in Chapter 5. NAC, MG 1, C¹¹A, vol. 112, fols. 149–50. "Mémoire concernant les Forges de St. Maurice," n.s., n.d. (1742 or 1743) (hereafter cited as NAC, MG 1, C¹¹A, vol. 112, Mémoire concernant [...]).

127. Denis Woronoff, "Le monde ouvrier de la sidérurgie ancienne: note sur l'exemple français," *Le Mouvement social*, 1976, no. 97, pp. 113–14.

128. In his initial plans of 1735, Vézin provided for a "forge made up of two fineries, a chafery and a plate mill, with all their gear mechanisms and hammer and tilt hammer hurst frames." The finery-chafery set-up was completely in keeping with the traditional Walloon process, in which fining and heating were done in two distinct hearths. It is initially surprising that an ironmaster from a province (Champagne) bordering Franche-Comté would have chosen this design! However, it is possible that he was proceeding on the basis of productivity rather than fuel economy, since the Forges were surrounded by forests where hardwood was "very common," as he noted in his "Observations." In fact, even though it saved charcoal, the *renardière* process was slightly less productive, since a single hearth was used alternately for the two operations. Encyclopedist Bouchu admits as much: "It is true that blooms are made

more quickly in a finery than in a *renardière*, since both the hearth and the worker have only the one job; but is there any question of weighing abundance against economy in a manufactory?" In his view, work specialization, while more productive in this instance, did not outweigh the lower production costs of the *renardière* process. But Vézin, who had further activities in view for his forge (plate mill and tilt hammer) probably foresaw that it would be necessary to produce larger quantities of iron. In this case, the specialization of each hearth would be more efficient. Nevertheless, it is known that the constraints imposed by the flow of the creek caused him to opt at the last minute for the *renardière* process, which meant one less hearth, as he said himself in 1741: "the establishment of two *renardière* chaferies, according to the final plans of the said Sieur Olivier." NAC, MG 1, C¹¹A, vol. 111, fol. 168v, "Projet des dépenses à faire [...]," Hocquart and Olivier de Vézin, 17 October 1735; *Encyclopédie* 1757, vol. 7, p.162; NAC, MG 1, C¹¹A, vol. 111, fol. 248v, memorial from Vézin and Simonet to Monsieur Delaporte Lalane, 10 June 1741.

129. "Faisins" in French, term still used at the Forges in the late 19th century; Caron 1982, p. 70.

130. "Instead of a bed of dross, why not substitute the hearth bottom? Does not the dross itself absorb a considerable amount of iron? Crush the cinder from a *renardière* and the dross from a finery in the stamp mill if you want to be convinced. They say that the iron is fattened and softened by the slag: that is true when it is lacking, but in all cases and with molten iron always in the bottom of a *renardière*, the iron is more likely to absorb it than on the dross of a finery: has experience not shown us that with the same quality of cast iron, *renardière* iron is the best?" *Encyclopédie* 1757, vol. 7, pp. 157–58 and 161–62.

131. Encyclopedist Bouchu adds: "Many people would like the type of hearth to be the answer to making wrought iron malleable or brittle from the same cast iron. I repeat once more, the essential qualities of wrought iron stem from the type of mine; its relative qualities result from how it is worked, which can purify, rectify, diminish, enhance or alter it, but can never change its nature." Ibid., p. 162.

132. Beaudet 1979, pp. 21–30.

133. Dimensions in French feet: one foot = 32.484 cm; NAC, MG 1, C¹¹A, vol. 112, fols. 42-46, Inventaire des Forges, 22 November 1741, signed Ignace Gamelin, Simonet *fils*, Estèbe.

134. "[...] chimney cracked in several places." The chimney flues had already been judged too narrow; note the dimensions at the top. NAC, MG 1, C¹¹A, vol. 112, fol. 45v, "Inventaire des Forges 1741," Estèbe; NAC, MG 1, C¹¹A, vol. 111, fol. 12v, memorial from Cugnet, 25 September 1740.

135. "[...] to protect the workers from the great heat of the fire and better retain vapours, smoke and charcoal sparks within the hearth [...]." *Encyclopédie. Recueil de planches, sur les sciences, les arts libéraux et mécaniques, avec leur explication.* 3ᵉ livraison, 298 planches (Paris: Briasson, David, Le Breton, 1765), "Forges ou Art du fer," (hereafter cited as *Encyclopédie, Recueil de planches* 1765), sect.4, plate II.

136. *Encyclopédie* 1757, vol. 7, p. 157.

137. "[...] the pillars [...] which support the front of the chimney should be solidly built of stone blocks or, even better, iron plates cast to a suitable shape and size to be set one on top of the other with mortar [...]," *Encyclopédie, Recueil de planches* 1765, sect. 4, plate II. Estèbe, in his inventory, mentions, for one of the two chimneys of the upper forge, "Seven squares of cast iron used instead of masonry in one pillar of the said chimney, together weighing about 2,000 pounds." However, he does not mention this type of pillar for the lower forge, although the inventory also mentions "2 cast iron squares to repair the pillars of the chimneys of the forges, weighing 600 pounds." This could mean that, even if the remaining chimney is not the original one built, it is undoubtedly of 18th-century construction. At the time when Vézin built his forge, the blast furnace was not yet in operation and we learn in 1737 that he even had to line the hearth with stone, since he could not yet produce iron hearth plates to run the first trials of the chafery. So he could not at that time have made an iron pillar: "The furnace not being in working condition [...] Sieur Olivier [...] decided to have stone plates made, since he did not have any yet to make the receivers for the chaferies of the lower forge. This he did, and undertook to produce wrought iron not from cast iron but from pure ore. In a few hours, this resulted in two small iron bars, which he had the honour of sending that same year to your Excellency." We also learn that the two forge hearths were rebuilt because they were too straight; possibly iron pillars were built during these repairs. NAC, MG 1, C¹¹A, vol. 112, fols. 41v–42 and 43–43v, Inventaire des Forges 1741, Estèbe; NAC, MG 1, C¹¹A, vol. 110, fols. 387–88, "Mémoire du Sieur Olivier de Vézin sur les Forges du Saint-Maurice," Versailles, 28 December 1739; NAC, MG 1, C¹¹A, vol. 111, fol. 12v, memorial from Cugnet, 25 September 1740.

138. Our description is taken from the *Encyclopédie* 1757, vol. 7, p. 157 and the *Encyclopédie, Recueil de planches* 1765, sect. 4, plate VI.

139. The bottom is slightly sloped and is also cooled: "The void under the bottom answers to the pipe to cool it: keep the bottom sloping forward slightly towards the front and the fore spirit plate to draw the slag into this part," *Encyclopédie* 1757, vol. 7, p. 157.

140. Plugging the opening around the tuyere with "wet stone and clay is known as *faire le mureau* [...], a job carried out by the *goujats*. This plug could easily be removed to adjust or replace the tuyere; ibid.

141. Ibid., pp. 157–61.

142. Little specific information exists on the forge bellows that were installed in 1737, except that new ones had to be made in 1739; the first bellows "had been made with green wood." NAC, MG 1, C^{11}A, vol. 110, fol. 391, "Mémoire du Sieur Olivier de Vézin sur les Forges du Saint-Maurice," Versailles, 28 December 1739; in 1749, Pehr Kalm wrote, concerning the bellows of the two forges: "[...] the bellows were made of wood, and everything else, as it is in Swedish forges" Peter Kalm, *Travels into North America*, trans. J. R. Forster (Warrington: William Eyres, 1771) (hereafter cited as Kalm 1771), vol. 3, p. 87.

143. *Encyclopédie* 1757, vol. 7, pp. 147–48.

144. NAC, MG 1, C^{11}A, vol. 110, fol. 185 (n.s, 1739), notes and calculations on the hydraulic mechanisms of the existing and planned forge (upper forge); these notes were probably written by Chaussegros de Léry or one of his clerks (hereafter cited as NAC, MG 1, C^{11}A, vol. 110, Léry, notes and calculations).

145. In 1738, after the charcoal house at the blast furnace built by Charlery collapsed, Vézin reinforced the lower forge, which had also been built by Charlery. Charlery defended himself by saying that "if there was some danger, it was caused by the vibration created by the hammer, and that the building could not have been built more solidly, since space had to be left for the hearths and the studs could not be placed closer than 16, 14 or 13 feet apart, whereas they should have been set every 10 feet [...]"; NAC, MG 8, A6, vol. 14, fols. 332–34.

146. *Encyclopedie* 1757, vol. 7, p. 158.

147. Exactly 28 are shown in the *Encyclopédie, Recueil de planches* 1765, sect. 4, plate II.

148. Document included with the notes attributed to Chaussegros de Léry, discussed at length in Chapter 3. NAC, MG 1, C^{11}A, vol. 110, fols. 175–208 (n.s., 1739), attributed to Chaussegros de Léry.

149. "The drome-beam must be strong and stout to keep the entire frame steady, and long enough so that the workers can turn around it with the iron bands and dress them without touching the lesser hammer post." *Encyclopédie* 1757, vol. 7, p. 159.

150. "[...] the block is just an assembly of four large squared beams set vertically to support the anvil and absorb the excess energy from the impact of the hammer, but special means had to be used to hold the beams together [such as iron bolts, wrought iron bands and wooden frames or casings [...]," Beaudet 1979, p. 31.

151. NAC, MG 1, C^{11}A, vol. 111, fols. 15–15v, memorial from Cugnet on the St Maurice Forges, 25 September 1740. We also learn that they had neglected to install the joint (*le pied d'ecrévisse*) that fastened the main hammer post and the other timbers into which the legs were slotted. But Vézin attributed the breaking of the hurst frame to "the ill use it had had in previous years from the forgemen moving the hammer too quickly, which often happens in France." NAC, MG 1, C^{11}A, vol. 111, fols. 256v–57, memorial from Vézin and Simonet to Monsieur Delaporte Lalane, 10 June 1741.

152. NAC, MG 1, C^{11}A, vol. 110, Léry, notes and calculations, fol. 185.

153. Ibid., fols. 186–89.

154. The way in which Vézin's plans are drawn up leave no doubt that the shop adjoined the forge, in the same way as the charcoal shed mentioned just before in the section entitled *Forge composée de.* In addition, in the same document, which also includes plans for a slitting mill, Vézin provides for another shop, the same as that at the forge. While this shop was not described in more detail at the time, Rainville believes it is logical that such a shop would have been located near the forge, where forge equipment was repaired. In addition, a shop like this was located near Francheville's forge. NAC, MG 1, C^{11}A, vol. 111, fols. 169–70v, "Projet des dépenses à faire [...]," Hocquart and Olivier de Vézin, 17 October 1735; Alain Rainville, "Les bâtiments de service et les dépendances aux Forges du Saint-Maurice," Manuscript Report No. 307 (Ottawa: Parks Canada, 1977), pp. 13–17.

155. NAC, MG 1, C^{11}A, vol. 112, fols. 25v–26, Hocquart to Maurepas, 24 October 1740.

156. NAC, MG 1, C^{11}A, vol. 111, fol. 15, memorial from Cugnet on the St Maurice Forges, 25 September 1740.

157. NAC, MG 1, C^{11}A, vol. 112, fol. 47, Inventaire des Forges 1741, Estèbe, p. 45.

158. NAC, MG 1, C^{11}A, vol. 112, fols. 254–55v, Hocquart to the Minister, 16 October 1746.

159. NAC, MG 1, C^{11}E, vol. 10, fols. 80–82, La Galissonnière and Hocquart to the Minister, 24 September 1747. Shortly after the tilt hammer was installed, Hocquart increased the price of iron sold in the colony to 30 *livres* a *quintal*, obviously to pay for the newly rebuilt lower forge with its tilt hammer: "[...] this slight increase will allow me to make the Establishment better and better"; NAC, MG 1, C^{11}A, vol. 88, fol. 72, Hocquart to the Minister, 7 October 1747.

160. H. Charbonneau and J. Légaré, *Répertoire des actes de baptême, mariage, sépulture et des recensements du Québec ancien*, Université de Montréal, Département de démographie (Montreal: Les Presses de l'Université de Montréal, 1983), vol. 23, s.v. "Saint-Louis-des-Forges-de-Saint-Maurice."

161. NAC, MG 1, C^{11}A, vol. 72, fols. 239-43, letter from Chaussegros de Léry, October 1739.

162. It is possible that, in the rebuilt forge, a tilt hammer hurst frame and associated hearth took the place of the second chafery. The most easterly hearth base could well have been for the tilt hammer. The gearing for a tilt hammer and its hearth could also have been used alternately with those of the chafery and hammer.

163. NAC, MG 1, C^{11}A, vol. 112, fol. 291, Inventaire général des Forges, signed Estèbe and Bigot, 10 February 1748.

164. Kalm 1771, vol. 3, p. 87.

165. NAC, MG 1, C^{11}A, vol. 112, fols. 340–42, 8 September 1760; eight 100-pound hammer heads are also mentioned; André Bérubé 1976, p. 61.

166. ASTR, N 3 H 20, Not. Rec. J. Bte. Badeaux, 2 June 1785. In this document, there is no mention of more than two wheels at either the lower or upper forge.

167. In a document from 1742–43, there are already plans for a tilt hammer at the lower forge where it was proposed to melt pigs suitable for the tilt hammer: "It would be a good idea [...] to set up the tilt hammer in the lower forge, leaving the two hearths complete with movements and bellows available for the tilt hammer work, in order to have one always ready to make up for a malfunction of the other. Bars from nine to 14 *lignes* will be made in the chaferies at the upper forge. Bars of four to eight *lignes* will be made in the lower forge chaferies, making sure to melt 1,000- to 1,200-pound pigs for the use of this forge. Accordingly, the St Maurice Creek, not having to be overloaded with movements, since the tilt hammer takes less water than a forge hammer, will always have enough water to supply it." NAC, MG 1, C^{11}A, vol. 112, "Mémoire concernant [...]," fol. 155.

168. NAC, RG 68, vol. 274, fols. 460–61, inventory of the Forges accompanying the Pélissier syndicate lease, 9 March 1767.

169. Beaudet, pp. 48–51. The author puts forward the hypothesis of a cupola (reverberatory) furnace, whose by- products he recognized.

170. Bédard 1986, pp. 161–62.

171. This remark, by Napoléon Caron in 1889, suggests that there may have been a turbine: "A channel conveyed the water to the millwheel, and this water was used to run energy cylinders, grindstones and all the machinery that was required. All of this is completely in ruins." Napoléon Caron, *Deux voyages sur le Saint-Maurice*, (Trois-Rivières: Librairie du Sacré-Coeur, 1889), pp. 290–91.

172. Harrington 1874, p. 248.

173. NAC, MG 1, C^{11}A, vol. 110, fols. 237–38 (n.s, n.d.; probably 1738 or early 1739).

174. NAC, MG 1, C^{11}A, vol. 110, fol. 185 (January-February 1739).

175. See bottom of the "Plan of Trois-Rivières mines," Plate 2.3.

176. NAC, MG 1, C^{11}A, vol. 110, fols. 241–42, "Forges, Plan de M. De Léry," n.s., n.d. Elsewhere Léry wrote: "I have drawn up plans and stated the proportions for the second forge"; NAC, MG 1, C^{11}A, vol. 72, fols. 239–43, October 1739.

177. The charcoal shed and the iron store were not included in Chaussegros de Léry's plan. The inventory of 1741 does not mention a masonry foundation for these sheds, which were added in 1740, as it does for the lower forge. At the lower forge, the two foundations (6 feet high) on which the sheds and the north wall of the forge were built ran the entire length of the forge (80 feet), 12 feet apart. These dimensions, similar to those of the foundations of the wheelrace on the south side, suggest that these foundations, which are rather substantial for simple sheds, had initially been designed for a second wheelrace for the three other wheels included in Vézin's original plans. The only difference is the height of the foundations (6 feet for the sheds, 8 feet for the wheelrace); but it should be kept in mind that the foundations of the wheelrace were originally also 6 feet high since Vézin stated that he had increased their height by 2 feet because of the "high water" in the river; NAC, MG 1, C^{11}A, vol. 110, fol. 391, "Mémoire du Sieur Olivier de Vézin sur les Forges du Saint-Maurice," Versailles, 28 December 1739.

178. To explain the weakness of the hydraulic system at the lower forge, Vézin cited the lowness of the milldam, which did not allow larger diameter wheels to be installed, and mentioned the upper forge in support of his demonstration: "the proof lies in the second forge building, where all the movements can run almost continuously." NAC, MG 1, C^{11}A, vol. 112, fol. 92v, Olivier de Vézin to Maurepas (n.d., probably 1741).

179. NAC, MG 1, C^{11}A, vol. 112, fol. 358, Mémoire sur les Forges du Saint-Maurice (n.d., around 1742).

180. In 1740, the milldam was strengthened with a view to increasing the water level so that the two chaferies could be operated simultaneously: "strengthen it on the other end as well, so that it will be strong enough from one end to the other to withstand a similar accident, even if the water is raised to its original height, if that is necessary to provide water for the two chaferies at the upper forge when we want to operate both. When only one chafery is in operation, the milldam water level should not be any higher than at present." NAC, MG 1, C^{11}A, vol. 111, fols. 80v–81, memorial between Cugnet and Simonet, 18 March 1740.

181. NAC, MG 1, C^{11}A, vol. 112-2, fol. 85, memorial from Cugnet, Gamelin and Taschereau to the Minister, 26 October 1744. In this document, the authors add: "There is only one in the forge built by S. Olivier that we can use because we rebuilt the hearth completely in 1739. The other chafery is not of any use [...]."

182. "Five *goujats* have to be sent out from France, two for each forge and one for the third chafery run by sluicing." NAC, MG 1, C^{11}A, vol. 112, fol. 149v, "Mémoire concernant [...]."

183. NAC, MG 1, C^{11}A, vol. 111, fol. 81v, memorial between Cugnet and Simonet, 18 March 1740. "A bridge 120 feet long and 24 feet wide used to move the pigs"; NAC, MG 1, C^{11}A, vol. 112, Inventaire des Forges 1741, Estèbe, fol. 44.

184. A careful reading of the plans shows that the east chafery wheel would have turned the wrong way if the bellows had been activated by a double gear mechanism set up east of the chimney. In addition, other information tends to confirm this interpretation: "The chafery wheel is made the same as that of the other chafery, with the same diameter and width and the water flows through it in the same direction"; NAC, MG 1, C^{11}A, vol. 111, fol. 81v, memorial between Cugnet and Simonet, 18 March 1740. However, the same careful reading shows the hammer wheel turning in the opposite direction from that indicated by an interpretation of the remains of the hammer.

185. Mousseau 1979, pp. 10–31.

186. Ibid., p. 18.

187. Ibid,. p. 20.

188. Ibid., p. 22.

189. Archaeological digs have not uncovered a masonry foundation under the hurst frame similar to the one at the lower forge shown in the inventory of 1741 and the remains of the original forge. The lack of such foundations is quite surprising, since the repeated blows of the hammer must have shaken the whole structure. Ibid., p. 26.

190. "[...] a second forge above the first, also containing two chaferies in working condition. There will always be two running, one in each forge, because if anything goes wrong with one chafery in each forge, the other will always be ready to take its place." NAC, MG 1, C^{11}A, vol. 110, fol. 391, "Mémoire du Sieur Olivier de Vézin sur les Forges du Saint-Maurice," Versailles, 28 December 1739.

191. NAC, MG 1, C^{11}A, vol. 111, fols. 278–80, "Estat général de la dépense [...]," 1741–42.

192. We discuss this problem in Chapter 5. It is also possible that, in such circumstances, one chafery would have been used for fining and another for heating, the way it is done in the Walloon method.

193. NAC, MG 1, C^{11}A, vol. 112, fols. 298–99, memorial (from Bigot) on the St. Maurice Forges in Canada, 1748, accompanying a letter to the Minister of 11 October 1748.

194. Marcel Trudel, "Les Forges Saint-Maurice sous le régime militaire (1760–1764)," *RHAF*, vol. V, no. 2, September 1951, p. 173.

195. NAC, MG 21, B 21-2 (21681), microfiche A-615, fols. 147–48 (1764); NAC, RG 68, vol. 274, fols. 260–63 (1767).

196. An inventory of 1785 reports on: "Various items missing after the inventory [...] twenty-nine *quintals* of cast iron to replace the lintels and parts of the pillar of the upper forge, which Sieur Alexandre Dumas had demolished to convert the cast iron into bar iron for his own profit, according to the deposition of a number of workers"; ANQ-TR, Not. Rec. J.-B. Badeaux, No. 25, 2 June 1785. Dumas leased the Forges from 1778 to 1783; thus the second chafery was not demolished until after the foundry was set up at the time of the American invasion (1776).

197. Mgr. Albert Tessier, *Les Forges Saint-Maurice (1729-1883)* (Montreal and Quebec City: Les Éditions du Boréal Express [1952] 1974), p. 133. The author bases his statement on the notarial records of J.-B. Badeaux of Trois-Rivières on the American invasion, cited by Sulte in 1920. These records report on the collaboration of the manager of the Forges at that time, Christophe Pélissier, with the Americans. Benjamin Sulte, "Les Forges Saint-Maurice," in *Mélanges historiques: études éparses et inédites* (Montreal: G. Ducharme, 1920), vol. 6, pp. 148–49.

198. Data on digs are taken from Mousseau 1979.

199. Archaeologist Claire Mousseau noted that the wall had been rebuilt on the remains of the floor of the moulding shop. Mousseau 1979, p. 55.

200. Ibid., pp. 55, 39–40.

201. Ibid., pp. 46–49. According to an account of repairs carried out between 1853 and 1857, there actually was a 32-foot (9.75 m) red brick forge chimney (10 feet shorter than the old chimney) at the Forges; Quebec Department of Lands and Forests, M 30, B 19, "Estimation de E. Normand," 4 September 1857.

202. The chimney in the watercolour could also have belonged to a cupola. In his 1852 report already mentioned, Hunter speaks of a "cupola furnace" that needed to be rebuilt, but we saw that this could just as well have been a cupola located in the moulding shop adjoining the blast furnace. Mousseau 1979, p. 72.

203. See note 201.

204. "[...] southeast of the railcar wheel moulding shop"; AJTR, Not. Rec. Petrus Hubert, No. 4575, 27 April 1863, "Vente des Forges St-Maurice," Onésime Héroux to John McDougall; "I.W. Leaycraft [...] for Nett proceeds of 413 Railway Wheels per his account sales [...] at \$15"; ASTR, Forges Papers, St Maurice Forges, April 1858, fol. 279.

205. Two cupolas were also used for the same type of production at the Radnor Forges; Claire-Andrée Fortin and Benoît Gauthier, "Description des techniques et analyse du déclin de la sidérurgie mauricienne, 1846-1910," research report submitted to the Regional Branch of the Quebec Department of Cultural Affairs, Centre de recherches en études québécoises (Trois-Rivières: Université du Québec à Trois-Rivières, February 1988), p. 86, and p. 258 for illustration of a cupola.

206. Mousseau 1979, pp. 58–61.

207. Bédard associates this furnace with a hot blast furnace, then a kiln or a "pyrolygneous acid machine," to which Hamilton Rickaby refers later. The small size of this furnace and its connection with a chimney and annealing pits in the estimate of 1857 suggests that it was instead used as a source of heat for the annealing pits. Bédard 1986, pp. 84, 92.

208. This was at the former Turcotte and Larue foundry, which McDougall leased (see Chapter 1). After the Forges closed in 1883, it imported pig iron from the United States; AJTR, Superior Court, docket no. 281, Robert Wilson et al. v George McDougall, exhibits 25 and 53; depositions of Philippe R. Hamilton of 12 and 23 April 1883.

CHAPTER 5

1. *Encyclopédie ou Dictionnaire raisonné des Sciences, des Arts et des Métiers*, vol. 7, 1757, s.v. "Forges, (Grosses-)" by M. Bouchu, ironmaster at Veuxsaules, near Château-Vilain (hereafter cited as *Encyclopédie* 1757), p. 152.

2. "Another building on the side of said furnace, where the founder is lodged, between the casting house and the bellows shed"; NAC, MG 1, C^{11}A, vol. 112, fols. 39–39v, Inventaire des Forges 1741, Estèbe.

3. NAC, MG1, C^{11}A, vol 112, fol. 244, Hocquart to Minister Maurepas, 18 October 1744.

4. NAC, MG 1, C^{11}A, vol. 63, fol. 191v, "Observations faites par moy [...]," Olivier de Vézin, 17 October 1735.

5. Our numerous references to Bouchu's article in the *Encyclopédie* allow us to reconstruct the facilities and processes for which an incomplete description or no description is available in archival materials on the St Maurice Forges. Bouchu's descriptions are also valuable in that he came from Burgundy, the same region as Simonet and many other workers at the Forges. One worker, Louis Trotochaux, was even from the same *commune* or community as Bouchu (Vaux-Saules, *bailliage* [bailiwick] of Châtillon-sur-Seine), and at least four others were from communities in the same bailiwick, including Godard, a finer at the lower forge, and Aubry, the collier. The machinery and practices described by Bouchu were in all probability the same as those used by the ironmasters and workers from Burgundy who came to work at the St Maurice Forges.

6. "The founders are usually very mysterious about their work; this is the way they deal with questions they cannot answer: they know this or that dimension by rote; they fear creating too many of their kind." Encyclopédie 1757, vol. 7, p. 136. "The passing down of knowledge and jobs from father to son certainly created additional inertia in technical matters. There was resistance to any technological innovation, since these innovations challenged the "traditional obstinacy of workers," criticized by enlightened employers, scientists and administrators [...] Owners were, to a large degree, outside the production process, often serving only as formal co-ordinators. The construction of the hearth, the composition of the charge (ore, charcoal and flux), the length of hammering were all decisions definitely left to the workers. Furthermore, the ironmasters were in complicity with their workers." Denis Woronoff, "Le monde ouvrier de la sidérurgie ancienne: note sur l'exemple français," *Le Mouvement social*, 1976, no. 97, pp. 115–16.

7. See Franquet's comments in Chapter 1. Contained in a memorial dating from around 1743, the requirements for the iron-master's position are particularly revealing of the profession's shortcomings; one particularly significant requirement is that he be a hands-on worker: "for a manager/ironmaster (expert, also a skilled founder, hammerman and even a finer, a true worker, able to oversee all the workers at the furnace and forges, able himself to know when and how they are lacking, rectify their errors both to correct their work and show them the correct proportions and the degree of fire required to produce good pig iron and well-made wrought iron) at 2,000 *livres* per year, board not included [...]"; NAC, MG 1, C¹¹A, vol. 112, fol. 135, " Mémoire concernant les Forges de St. Maurice," n.s., n.d. (around 1743, since it is based on Estèbe's trusteeship of 1742), (hereafter cited as NAC, MG 1, C¹¹A, vol. 112, "Mémoire concernant [...]").

8. The forgemen were said to work in shifts; NAC, MG 1, C¹¹A, vol 111, 28ᵉ, memorial from Vézin and Simonet to Monsieur Delaporte Lalane, 10 June 1741.

9. ANQ-M, Not. Rec. N. B Doucet, No. 490, indenture of Antoine Buisson, keeper, to Zacharie Macaulay, 20 May 1805.

10. Initially, Vézin expected to produce one million pounds of cast iron in eight months of work; other sources confirm the non-stop campaign, the length of which could vary somewhat depending on the year. On the other hand, in 1741–42, the founder received a salary for 5 months and 13 days of work, which corresponds to the period the furnace was not in blast; that year, the campaign thus lasted less than 7 months. The founder, who was normally paid by the thousandweight, in effect received unemployment pay during the shutdown of production, like the other skilled workers. NAC, MG 1, C¹¹A, vol. 110, fol. 146, "Dépenses annuelles de l'exploitation," Hocquart and Olivier de Vézin, 17 October 1735; NAC, MG 1, C¹¹A, vol. 111, fol. 59, "Frais annuels de l'exploitation," 24 October 1740, ("A furnace operating 7 months a year will provide enough pig iron to make up to 750 thousandweight of wrought iron. In reality, the furnace produced between 22 May and 3 September 1740, 537,740 pounds of pig. Therefore, in seven months, it could produce 1,200 thousandweight of pig. It could operate for eight months."); NAC, MG 1, C¹¹A, vol. 111, fols. 278–81 "Estat général de la dépense faite pour l'exploitation des forges de Saint-Maurice depuis le 1ᵉʳ octobre 1741 jusqu'au 1ᵉʳ août 1742," Estèbe (hereafter cited as NAC, MG 1, C¹¹A, vol.111, "Estat général de la dépense [...]"). In France, there was the same constraint of having to repair the furnace annually, except that the blast furnace was not shut down in winter, when water levels were at their peak, but rather in summer, when water levels were at their lowest; Guy Thuillier, *Georges Dufaud et les débuts du grand capitalisme dans la métallurgie, en Nivernais, au XIXᵉ siècle* (Paris: S.E.V.P.E.N., 1959), p. 9; Bertrand Gille, *Les origines de la grande industrie métallurgique en France* (Paris:

Éditions Domat, 1947) (hereafter cited as Gille 1947), p. 61–64. See also Chapter 2.

11. "The furnace can operate eight months a year, or even longer, but by doing so it would produce more pig iron than the forges would use, so it only works six months a year here. By reducing the campaign to six months, we can count on continual operation during this time even should operations be interrupted for a longer period than occurred this year should the inwalls require repairs, the axle-tree or the wheels break or something else happen to the furnace." The author of the document adds that the average daily production of 4,500 pounds of pig iron would add up to 810 thousandweight in six months and "no more than this would be required to supply the forges"; NAC, MG 1, C¹¹A, vol. 112, "Mémoire concernant [...]."

12. In November and December 1741, two workers were engaged to "empty the furnace and assist in the repairs"; NAC, MG 1, C¹¹A, vol. 111, fols. 279–80, "Estat général de la dépense [...]."

13. In 1783, Alexandre Dumas, the furnace's director at the time, could not get operations at the Forges started until 9 June, because of a severe winter and a bad spring; Marcel Moussette, " L'histoire écologique des Forges du Saint-Maurice," Manuscript Report No. 221 (Ottawa: Parks Canada, 1978) (hereafter cited as Moussette 1978), p. 23.

14. The first known furnace keeper, Pierre Belu, also received a monthly allowance when the furnace was shut down. The few deeds of indenture that have been found specify the duration of employment as between "1 May to the moment when the furnace is shut down at the end of the campaign" for keepers and fillers; NAC, MG 1, C¹¹A, vol. 111, fols. 278–81, "Estat général de la dépense [...]"; ANQ-M, Not. Rec.N. B Doucet, Nos. 490 and 491, 20 May 1805, indenture of Antoine Buisson, keeper, to Zacharie Macaulay and indenture of Michel Robert and Michel Brousseau, fillers; ibid., No. 446, 26 March 1805, indenture of Joseph Houle *dit* Jean-Claude, of Trois-Rivières, furnace keeper.

15. John Lambert, *Travels through Canada and the United States of North America, in the years 1806, 1807 & 1808: To Which are Added Biographical Notices and Anecdotes of Some Leading Characters in the United States* (London: Printed for Baldwin, Cradock, and Joy; Edinburgh: for W. Blackwood and Dublin: for J. Cumming, 1816) (hereafter cited as Lambert 1808), pp. 487–88.

16. Dr. Gerd H. Hardach, *Der soziale Status des Arbeiters in der Frühindustrialisierung. Eine Untersuchung über die Arbeitnehmer in der französischen eisenschaffenden Industrie zwischen 1800 und 1870.* Schriften zur wirtschafts-und sozialgeschichte (Berlin: Duncker & Humblot, 1969) vol. 14 (hereafter cited as Hardach 1969), p. 53 (author's translation).

17. In the article in the *Encyclopédie* ("Forges, (Grosses-)," he writes, however, "with the given materials, a furnace in blast can, with 20 charges, produce five thousandweight of iron in 24 hours, and sustain a year's work [...]"; in the text accompanying Plate VII of section 2, he specifies that four charges are loaded during each shift, adding that "after two fillers have each completed a shift, comprising four charges, a ninth charge is made by both, during which time the mould for the pig is prepared [...]"; according to the second citation and the table of charges shown in Plate VII, 18 charges would be done in 24 hours, resulting in two tappings, and there would be four shifts of six hours each. *Encyclopédie* 1757, vol. 7, p. 152; ibid., *Recueil de planches, sur les sciences, les arts libéraux, et les arts méchaniques, avec leur explication.* 3ᵉ livraison, 298 planches (Paris: chez Briasson, David, Le Breton, 1765) "Forges ou Art du fer," (hereafter cited as *Encyclopédie, Recueil de planches*, 1765), sect. 2, plate V11.

18. Dr B. J. Harrington, "Notes on the Iron Ores of Canada and Their Development," Geological Survey of Canada, *Report of Progress for 1873–74* (Montreal: Dawson Brothers, 1874) (hereafter cited as Harrington 1874), p. 247. "By reducing the proportion of ore and increasing that of charcoal, thus dividing the burden more than before, by extending the duration of time before the furnace is tapped (owing to the dimensions of the new furnaces), manufacturers obtain a more complete reduction of the ore, and better yield both in terms of quantity and quality." Denis Woronoff, *L'industrie sidérurgique en France pendant la Révolution et l'Empire* (Paris: Éditions de l'École des hautes études en sciences sociales, 1984) (hereafter cited as Woronoff 1984), p. 291.

19. Marcelle Caron, "Analyse comparative des quatre versions de l'enquête de Dollard Dubé sur les Forges du Saint-Maurice," manuscript on file (Quebec City: Parks Canada, 1982) (hereafter cited as Caron 1982), p. 143.

20. Roch Samson, "Les ouvriers des Forges du Saint-Maurice: aspects démographiques (1762-1851)," Microfiche Report No. 119 (Ottawa: Parks Canada, 1983) (hereafter cited as Samson 1983), p. 91.

21. Caron 1982, p. 143.

22. *Encyclopédie* 1757, p. 152. Describing the founder's tasks, Vézin writes: "in the receiver of the furnace, which is the reservoir or recipient of the cast iron in its proportions, where it takes on its proper qualities through the consumption of six to seven charges of ore, charcoal, limestone and clay flux." We do not think that the ironmaster meant to say six to seven charges per tapping and therefore 12 to 14 charges in 24 hours, which would be too little when compared with other data, but rather that iron begins to run in the hearth once the furnace has consumed the first six to seven charges of ore after it has been blown in. In the furnace at the Rancogne ironworks, Chevalier Le Mercier observed that, after the seventh charge after blowing in, "the mine started to fall to the bottom"; NAC, MG 1, C¹¹A, vol. 63, fol. 191v "Observations faites par moy [...]," Olivier de Vézin, 17 October 1735; NAC, MG 1, C¹¹A, vol. 112, fols. 334v–35, "Mémoire dans lequel on a détaillé [...] la forge de Rancogne," Chevalier Le Mercier," 2 April 1750.

23. Charges were made in baskets. The 1741 inventory lists 13 charcoal baskets and 2 ore baskets; the 1746 inventory (fol. 261) lists 87 charcoal and ore baskets and 350 *barriques* of unwashed mine in front of the furnace" (fol. 264); the 1748 inventory lists 33 charcoal baskets (fol. 286); NAC, MG 1, C¹¹A, vol. 112, fols. 28–59v, Inventaire des Forges 1741, Estèbe; fols. 260–66, Inventaire général des Forges, Estèbe, 12 February 1746; fols. 285–82v, Inventaire général des Forges, Estèbe,10 February 1748.

24. Speaking of the work done by ironworkers at iron forges in France in the late 18th century, Woronoff writes: "The team of fillers, for example, must carry between 7 and 12 t of raw materials to the throat in 24 hours, in other words, each worker must move from two and a half to three tonnes in 25-kg baskets, which requires a hundred trips," Woronoff 1984, p. 299.

25. The founders had to share some of their trade secrets if they wanted to be relieved by other workers, but they were jealous of their privileges and very close-mouthed when it came to their techniques. Delorme, the second founder at the Forges, was said to be "always ready to take advantage of the fact that he was the only founder"; NAC, MG 1, C¹¹A, vol. 112, fols. 148–49, "Mémoire concernant [...]."

26. "[...] 4 shillings a day as furnace keeper, including work at night which will be paid at the end of each month [...]"; ANQ-M, Not. Rec. N. B Doucet, No. 490, 20 May 1805, indenture of Antoine Buisson of the St Maurice Forges, furnace keeper, to Zacharie Macaulay, and No. 446, 26 March 1805, indenture of Joseph Houle *dit* Jean-Claude, of Trois-Rivières, furnace keeper.

27. NAC, MG 1, C¹¹A, vol. 112, fol. 149, "Mémoire concernant [...]."

28. Ibid., fol. 140.

29. In documents about Estèbe's trusteeship in 1741-1742, there is no mention of *goujats* or helpers being assigned specifically to the blast furnace, although four were assigned to the two forges. The Forges were said to employ "day labourers or soldiers to act as helpers, the kind of men who are unstable and eventually have to be let go. The finers do not want anything to do with teaching them the trade." Later, in 1804, Lord Selkirk lists the workers at the blast furnace as follows: "The furnace employes 1 Charger, 2 gardes & a boy - The two forges each 4 men & 2 boys - half day - half night." The boy at the furnace was probably one of these helpers; NAC, MG 1, C¹¹A, vol. 111, fols. 278-81,"Estat général de la dépense [...],"; NAC, MG 1, C¹¹A, vol. 112, fol. 140,"Mémoire concernant [...],"; *Lord Selkirk's Diary 1803–804*, edited with an introduction by Patrick C. T. White (Toronto: The Champlain Society, 1958) (hereafter Selkirk 1804), p. 230–31.

30. This division of labour between the fillers above and chargers below was remarked on by Dollard Dubé, who relates that a loading ramp was installed after the furnace was modified in 1854. The loading method that was used before is unknown, but a hoist was probably used to winch up the charges, although the fillers may have carried the baskets up a staircase. See probable plan of the blast furnace under the French in Plate 4.5. According to Chevalier Le Mercier, who was describing operations at the Rancogne ironworks, the charger, whom he called the *arcqueur*, was responsible for filling the baskets [see also Benjamin Sulte (note 32 gives full reference), who cites Laterrière ("argueurs") at the Forges in 1775]: "the *arcqueur* fills the baskets and it is his responsibility to measure everything carefully, he is never relieved and does not sleep until he has prepared the charge, and when the fillers are ready to charge the furnace, he is woken up and must help them hoist the baskets up on their shoulders, then he begins to prepare the next charge." NAC, MG 1, C¹¹A, vol. 112, fol. 334, "Mémoire dans lequel on a détaillé [...] la forge de Rancogne," Chevalier Le Mercier, 2 April 1750.

31. "For each charge to be done with the care required, the fillers should be required to announce it to the founder or furnace keeper; for this, on one of the penthouse walls at the furnace top hangs an iron plate and a hammer, which the filler uses to strike the plate, which acts as a bell; after sounding the carillon, the filler strikes the plate as many times as required to signal to the founder of which of the four charges in the shift is being loaded into the furnace: one ring for the first charge and two, three and four rings for the other charges"; *Encyclopédie, Recueil de planches* 1765, "Forges ou Art du fer", sect.2, plates VII and II (window in the *batailles*).

32. Benjamin Sulte, *Les Forges Saint-Maurice* in *Mélanges historiques*, vol. 6 (Montreal: G. Ducharme, 1920) (hereafter cited as Sulte 1920), p. 106.

33. "De Lorme [...] is not a skilled founder and did not come here in this capacity, but as a finer at the forges [...] A good founder must be brought from France"; NAC, MG 1, C¹¹A, vol. 112, fol. 149,"Mémoire concernant [...]". Delorme was considered at the time to be an interim founder. Two years before, Vézin had brought over the founder Tortillier from Burgundy, who died a month after he arrived. Delorme, who was said in 1743 to suffer from "chest problems" and to be at risk of dying or being unable to work for perhaps a year, remained on the job until 1775; NAC, MG 1, C¹¹A, vol. 111, fol. 65, Olivier de Vézin to Minister Maurepas, Quebec, 12 October 1740; Marie-France Fortier, "La structuration sociale du village industriel des Forges du Saint-Maurice : étude quantitative et qualitative," Manuscript Report No. 259 (Ottawa: Parks Canada, 1977), p. 193.

34. This episode is more revealing of the hydraulic mechanisms in place than the blowing in of the furnace, since the problem with blowing in the furnace was attributed erroneously to the bellows not working as they should.

35. André Bérubé, "Rapport préliminaire sur l'évolution des techniques sidérurgiques aux Forges du Saint-Maurice, 1729–1883," Manuscript Report No. 221 (Ottawa: Parks Canada, 1976) (hereafter cited as Bérubé 1976), p. 45.

36. *Encyclopédie, Recueil de planches* 1765, "Forges ou Art du fer", sect.2, plate VII; Chevalier Grignon (1723–84), cited by Bouchu, was an ironmaster in Champagne, like Olivier de Vézin; Pierre Léon, *Les techniques métallurgiques dauphinoises au dix-huitième siècle* (Paris: Hermann, 1961) (hereafter cited as Léon 1961), p. 29.

37. *Encyclopédie* 1757, vol. 7, p. 150.

38. Ibid., charcoal dust.

39. NAC, MG 1, C¹¹A, vol. 111, fols. 252v–53, memorial from Vézin and Simonet to Monsieur Delaporte Lalane, 10 June 1941.

40. APJQ, Superior Court, District of Quebec, docket no. 2238, J. Porter et al. v. Weston Hunt et al., deposition of Antoine Mailloux *père*, day labourer at the St Maurice Forges, 28 January 1860.

41. Caron 1982, p. 61. See Table 2.12 in Chapter 2 for the composition of a charge as reported in 1868 and 1874.

42. Sulte also simplifies the operation, while adding a few details: "The ore mixed with charcoal and liberally sprinkled with sand is put into a box or basket. This is the charge, which is then emptied into the top of the blast furnace. The furnace is in the shape of a chimney. To empty the charge into the furnace, the workers have to climb up a staircase about 10 feet high and each time they put in four or five loads, immediately afterwards using an iron bar to strike a suspended iron plate four or five times, which makes a strange sound, drawn-out and resonant." Sulte 1920, p. 182; Caron 1982, p. 61.

43. "This is the order that must be observed when charging the furnace: when gauge XX (shown at the bottom of Plate X of this section), can be inserted in its entire length, which is 36 inches,three baskets of charcoal, half a basket of limestone and over this two baskets of charcoal, with the last one containing the smallest pieces, those that passed through the teeth of the rake, are to be emptied into the furnace; the smallest bits of charcoal are put in last to fill the spaces between the other pieces of charcoal, so arranged as to form an even surface, tilted at an angle of about 30 degrees with the tymp side, or, which amounts to the same thing, so that the surface of the charcoal is even with the top of the hearth plates on the back wall side, which is the side of the throat where the worker is standing (Fig. 1), and around seven and a half inches down on the opposite side, which is the tymp side. The slant in the charcoal layer is required because the mine charged on the back wall side is quite heavy, weighing down this part and flattening out the charcoal; too steep a slope will topple the charge, with all the charcoal going down into the lowest part. When the charge has settled, in other words, when the fire has consumed the previous charges and the mine has settled to the level of the charcoal, the rest of the limestone is put in the centre; this method of loading the furnace in two stages allows the charge to be mixed more precisely: Next, the piles of clay marl around the throat set out for drying are broken up and poured into the throat on the tuyere and fore spirit sides where the fire is hottest; lastly, ten baskets of mine are emptied into the back wall side, which the worker in Plate 1 is doing"; *Encyclopédie, Recueil de planches* 1765, "Forges ou Art du fer," sect. 2, plate VII.

44. "[...] too much mine causes the cooling of the furnace and is evidenced by black smoke, while when there is not enough mine, the flames produce white smoke; the happy medium between the two extremes is grey smoke; one filler monitors the furnace but both work together doing the charging when the time comes [...]." Observations made by Chevalier Le Mercier at the Rancogne ironworks in 1750. Le Mercier had been sent on a mission to France to study cannon founding. Cannons were to be produced at the Forges, but the project never came to fruition. NAC, MG 1, C¹¹A, vol. 112, fol. 333, " Mémoire dans lequel on a détaillé [...] la forge de Rancogne," Chevalier François Le Mercier, 2 April 1750.

45. "Lieutenant Baddeley's (Rl Engineers) report on the Saint Maurice iron works, near Three Rivers, Lower Canada (1828)," in *APT*, vol. V, no. 3 (1973), note 3 (hereafter cited as Baddeley 1828), p. 15.

46. ANQ-TR, court records, coroner's inquests no. 6 and 7 on the deaths of François Boisvert and Louis Boisclair, 17–18 April 1854; ANQ-Q, Superior Court, docket no. 2191, John Porter et al. v. Weston Hunt et al., 1855, affidavit, 6 August 1855, fol. 4.

47. "Incendie aux Forges," *Le Journal des Trois-Rivières*, 21 February 1881, p. 2; Dollard Dubé, typescript, p. 52. Fonds Cécile Marchand, Quebec City, Canadian Parks Service.

48. Ibid. pp. 54-55, under the subhead "le haut fourneau." The keepers and fillers, taking advantage of a moment of compassion by Robert McDougall, drank so much at work that they forgot to tap the furnace; an explosion was narrowly avoided.

49. Louis Franquet, *Voyages et mémoires sur le Canada (1752-1753)* (Montreal: Éditions Élysée, 1974) (hereafter cited as Franquet 1752), p. 20.

50. He adds: "The building is completely open on three sides; despite this, the heat is intense. Everyone goes outside until the time comes to wield the long rakes to recover the iron which has turned darkish but is not yet iron grey." Sulte 1920, p. 107.

51. Ibid.

52. "When the slag is almost ready to run out over the dam, the task of the founder or the person replacing him is to stir the molten iron in the hearth with a ringer, which helps the metal to become purified; it loosens the front of the furnace and allows the slag to run off." *Encyclopédie* 1757, vol. 7, p. 152.

53. "The worker of course tries to help along the fusion of the charges above by wielding his ringer and increasing the blast"; *Encyclopédie* 1757, vol. 7, p. 151.

54. Concerning the recommendations in the 18th-century memorial by Chevalier Grignon, who wished to perfect the founder's art and charging methods, Pierre Léon writes: "Despite the new spirit that infuses the work, this memorial shows the persistence of the old mentality of empiricism and practicality, in which 'the practised eye' and 'experience' play an infinitely more important role than Science based on Reason. The reforms recommended are very concrete, intended for an uneducated audience not open to innovations. They undoubtedly had to be effective by very reason of their simplicity." Léon 1961, p. 133.

55. NAC, MG 1, C¹¹A, vol. 112, fol. 335v, "Mémoire dans lequel on a détaillé [...] la forge de Rancogne," Chevalier Le Mercier, 2 April 1750.

56. Ibid. p. 316.

57. In 1743, the figure mentioned for average daily production was 4,500 French pounds (2.2 t); with two tappings a day, each pig would weigh around 2,250 pounds (1.1 t). The 1743 document cited by Bérubé 1976 mentioned a weight of 1,000 to 1,250 pounds (0.5– 0.6 t); the author states, in fact, that in order to use a tilt hammer at the lower forge, pigs of this weight were required for melting in the forge chaferies. Undoubtedly, he meant that smaller pigs must be produced for the smaller bars to be produced with the tilt hammer. NAC, MG 1, C¹¹A, vol. 112, fol. 155 "Mémoire concernant [...]."

58. Sulte is describing what he saw at the Forges around 1850: "An asbestos mask on their faces, their upper body protected with heavy leather armour, a few men watched over the boiling molten iron. Those working closer to the furnace were only wearing a simple undergarment." He adds: "This reminds me of a curious incident brought to my attention by Dr. N.-E Dionne. Around 1750, the inhabitants of the Forges had the habit in the summer of working just in a shirt without an undergarment, to stay cool in the heat, which was exacerbated by the heat from the furnace. The missionaries, who were shocked to see the rules of propriety broken, so spoke out about this. The matter did not seem to have gone any farther, since Dr. Dionne had found nothing else on the subject." Sulte 1920, p. 106.

59. Excerpt from the memoirs of Léonard Defrance (1735–1805) quoted in Maïté Pacco-Picard, *Les manufactures de fer peintes par Léonard Defrance*, coll. Musées vivants de Wallonie et de Bruxelles, no. 3, (Liège: Pierre Mardaga, 1982), p. 4.

60. NAC, MG 1, C¹¹A, vol. 110, fol. 253, "Mémoire sur les Forges de Saint-Maurice pour présenter à Monseigneur le Comte de Maurepas Ministre et Secrétaire d'Estat," Olivier de Vézin, n.d. (written during his trip to France in 1740).

61. NAC, MG 1, C¹¹A, vol. 63, fol. 192, "Observations faites par moy [...]," Olivier de Vézin, 17 October 1735.

62. Léon 1961, p. 123, note 49. It is interesting that Grignon, who was from Champagne like Vézin, also preferred grey iron and disliked white iron; this is what Léon has to say on the subject: "First of all, Grignon considered white iron to be of insufficient purity and of poor quality [...], and this prejudice was not his alone, but resulted from an insufficient understanding of the composition and properties of the different kinds of cast iron. Grey iron is undoubtedly more stable, but white iron has perhaps more possibilities, because this forge pig lends itself to steel production [...] In the 18th century, however, people were put off by its brittleness and hardness (it was more difficult to make into steel because it had a higher carbon content)."

63. Ibid., p. 124.

64. René Leboutte, *La grosse forge wallonne (du XVᵉ au XVIIIᵉ siècle)*, (Liège: Éditions du Musée de la vie wallonne, 1984), pp. 29–30.

65. NAC, MG 1, C¹¹A, vol. 111, fol. 80, "Mémoire des articles arrêtés [...]" between Cugnet and Jacques Simonet, 18 March 1740.

66. *Encyclopédie* 1757, vol. 7, p. 154.

67. The printed text states that "10 per cent. was white and 10 per cent. mottled iron." We believe, however, that the second 10% was a typographical error, and should read 90% mottled iron, which would correspond better to the Forges' production at the time, most of which (mottled) was used for pig iron and only a small part was bar iron from white iron. The 1871 contract stated in fact that, out of the 2,000 to 2,500 tons to be produced for Montreal, 10% was to be white iron. Harrington 1874, p. 247 (p. 300 in the French version, which repeats the error). AJM, Not. Rec. W.F. Lighthall, 15 February 1871, No. 5293.

68. "Though it is generally stated that the wrought-iron made from bog ores is cold-short, such is not always the case, and bar iron produced in an old-fashioned hearth-finery was seen at the St. Maurice Forges which was not all cold-short, and which, on analysis, shewed only traces of phosphorus." Harrington 1874, p. 235.

69. The observation by Vézin cited in the epigraph ("[...] that this founder be able to produce grey iron for the use of the forges [...]") should therefore be taken literally. In *Métallurgie pratique du fer* (1835), Walter de Saint-Ange writes that grey iron always gives better wrought iron than white iron made from the same ore and adds that the German method of refining, in other words, the *renardière* process used at the Forges applies mainly to very grey iron, in which the carbon cannot be reduced in a single operation. Walter de Saint-Ange, *Métallurgie pratique du fer, ou description méthodique des procédés de fabrication de la fonte et du fer, accompagnée de documents relatifs à l'établissement des usines, à la conduite et aux résultats des opérations; avec Atlas des machines, appareils et outils actuellement employés renfermant tous les détails nécessaires pour exécuter les constructions* (Paris: Librairie scientifique et industrielle de L. Mathias, 1835–38) (hereafter cited as Saint-Ange 1835–38), pp. 42 and 45. Woronoff comments about the Franche-Comté method that it "only suits grey (highly carburized) iron, which is slow to melt but remains in a liquid state longer. It is longer than the former method [Walloon method], since it requires two-step decarburization and produces less iron. But it has decisive advantages.[...] Another argument that is often made is that unprecipitated purified iron is much better. The superiority of this process became, at the end of the 18th century, a commonplace"; Woronoff 1984, p. 288.

70. Michel Fiset, A. Galibois and T. Vo Van, "Analyse métallurgique d'un groupe d'objets en métal provenant de contextes archéologiques aux Forges du Saint-Maurice," typescript (Quebec City: Parks Canada, 1982) cited in Michel Bédard, "Tarification, commercialisation et vente des produits des Forges du Saint-Maurice," typescript (Quebec City: Parks Canada, 1982a), p. 21.

71. With the signature in 1865 of a large contract, the McDougalls sold most of the iron produced as pig iron (between 1,500 and 2,000 tons) to a factory in Montreal, only reserving around 10% of the cast iron (roughly 150 tons) for their own use; Michel Bédard, "La privatisation des Forges du Saint-Maurice 1846-1883: adaptation, spécialisation et fermeture," manuscript on file (Quebec City: Parks Canada, 1986) (hereafter cited as Bédard 1986), pp. 167–69.

72. The first samples of iron produced in Francheville's bloomery forge were judged to be as good as Berry iron; Nish tells how the director of the Council of Commerce in France equipped his carriage with one wheel made of Berry iron and another of St Maurice iron after receiving samples from the Forges in 1737; NAC, MG 1, C¹¹A, vol. 110, fol. 285, Beauharnois and Hocquart to Minister Maurepas, 18 September 1734; Cameron Nish, *François-Étienne Cugnet. Entrepreneur et entreprises en Nouvelle-France* (Montreal: Fides, 1975) (hereafter cited as Nish 1975), p. 101. Subsequently, other observers confirmed the quality of the ore used at the Forges: Nordburgh (1761); Baddeley (1828); Logan (1853); Moussette 1978, p. 39.

73. Harrington 1874, p. 235. In 1828, Baddeley spoke of yields of 45%; Baddeley 1828, p. 10.

74. Franquet 1752, p. 20.

75. Casting the pig did not remove all the iron from the hearth. Since the taphole was higher than the bottom of the hearth, the equivalent of a half pig remained in the crucible after tapping, which explains the moulding technique described by Franquet. Monique Barriault, "Rapport préliminaire sur l'identification des techniques de moulage utilisées aux Forges du Saint-Maurice, étude faite à partir des déchets de moulage," Manuscript Report No. 330, (Ottawa: Parks Canada, 1978) (hereafter cited as Barriault 1978), p. 9. Here, Barriault cites Abbé Jaubert, "Fonte du fer," *Dictionnaire raisonné universel des Arts et Métiers*, vol. II (Lyon: Amable Leroy, 1801), p. 261.

76. NAC, MG 1, C¹¹A, vol. 110, fols. 144–49, "Dépenses annuelles de l'exploitation," Hocquart and Vézin, 17 October 1735.

77. The production figures we have, which cover the period from August 1742 to January 1746 (102.5 t of cast iron), show that this level was reached. Average yearly production is therefore 30.75 t; Réal Boissonnault, "Quelques notions sur l'orientation de la production et les types de produits fabriqués aux Forges du Saint-Maurice, 1729-1883," manuscript on file (Quebec City: Parks Canada, 1981) (hereafter cited as Boissonnault 1981), fol. 21 (n.p.).

78. NAC, MG 1, C¹¹A, vol. 112, fol. 141, "Mémoire concernant [...]."

79. Boissonnault 1981, fols. 22 and 33.

80. Selkirk 1804, p. 230.

81. Johanne Cloutier, "Répertoire des produits fabriqués aux Forges du Saint-Maurice," manuscript on file (Quebec City: Parks Canada, 1980) (hereafter cited as Cloutier 1980), p. 96.

82. The announcements by Monro & Bell (see Chapter 6, Tables 6.5 and 6.6) infer that the moulders themselves make the patterns.

83. In 1745, Étienne Cantenet *fils* was listed as a sand moulder, as well as a moulder like his father. See Chapter 8, note 109. See excerpt from the *Encyclopédie* on sand moulders in note 85.

84. See section in Chapter 2 on moulding sand.

85. "Four sand moulders can serve a furnace that produces two thousandweight in 24 hours. When they have prepared the number of moulds required for the iron being smelted, they line their ladles with clay mixed with horse dung so the iron does not stick to them, and heat them up. The iron handle of the ladle is covered with two hollowed out pieces of wood bound together with an iron ring. The worker holds the handle in his left hand and ladles molten iron from the furnace. He supports the ladle with his left hand and uses his right hand to pour the molten iron into the moulds. Since castings must be poured in a single motion, when they are large, one man pours while the others keep the flow going in the first man's ladle by pouring their own into his: all sand moulds are filled in the same way [...] a ladle may hold 40 to 50 pounds of metal." *Encyclopédie* 1757, vol. 7, pp. 155–56. For more information on the pouring techniques used and the ladle remains found, see Barriault 1978, pp. 52–60.

86. Barriault 1978, pp. 25–37.

87. See Chapters 1 and 6.

88. According to Barriault, the box moulding technique was first used to make stove plates in the years between 1820 and 1850. The year 1820 comes from a parallel source (Tyler 1973) and that of 1850, from the disappearance of moulding scraps. However, the degree of fluidity of the cast iron required according to Barriault (citing Tyler) would have been found only after the furnace was modified in 1854 until 1860, when stove plate production was halted at the Forges. Barriault 1978, p. 33.

89. Ibid., p. 42.

90. Ibid., p. 43; gun carriages and cannonballs were being manufactured as early as 1744; Cloutier 1980, pp. 17–23.

91. Barriault, citing the *Encyclopédie*, notes a difference between how cannon moulds were dried (over a wood fire on horses, which took the place of the bench) and how kettles were dried (in a kiln at a temperature of 180°C); Barriault 1978, pp. 47–48, and p. 22 of French-English glossary of moulding terms.

92. Ibid., p. 48.

93. Ibid., p. 48, 57.

94. Baddeley 1828, p. 12; Monique Barriault, "La capacité adaptative des forges de St. Maurice face aux changements économiques et technologiques, vue à travers l'évolution fonctionnelle des ateliers de moulage adjacents au haut-fourneau," Master's thesis, Archaeology, Université Laval, 1984 (hereafter cited as Barriault 1984), pp. 101–20.

95. Barriault 1984, pp. 49–63.

96. A 1748 reference confirms that 99 cannonballs of 24 pounds each were made. The numerous cannonballs unearthed during the excavations were 1, 2, 3, 4, 5, 8, 12, 16 and 24 pounds in size. However, the largest mould found was for the production of 6 pound balls. Simon Courcy, "L'artillerie fabriquée aux Forges du Saint-Maurice," document file and artifact inventory (Quebec City: Parks Canada, 1989).

97. Baddeley 1828, p. 10.

98. Barriault 1984, p. 64.

99. There were five moulds or more for grape shot (small balls). Simon Courcy, "L'artillerie fabriquée aux Forges du Saint-Maurice," document file and artifact inventory (Quebec City: Parks Canada, 1989).

100. Barriault 1984, p. 67, cites Nicole Casteran, "Fabrication d'armement aux Forges du Saint-Maurice" (Ottawa: Parks Canada, 1975), p. 8.

101. Ibid. Courcy cites the following references, among others: NAC, MG 1, C^{11}A, vol. 112, fols. 242–46, Hocquart to the Minister, 18 October 1744 (4,624 balls); vol. 110, "produit du fourneau des Forges du Saint-Maurice entre le 23 avril et le 31 août 1746" (4,844 balls); vol. 112, "compte du sieur Martel de Belleville du 1er janvier 1745 au 1er janvier 1746" (10,450 balls).

102. Barriault, like Bérubé, cannot specify whether it was a cupola or reverberatory air furnace. Barriault 1984, p. 132. Bérubé 1976, pp. 62–64.

103. Bérubé 1976, p. 63. The evidence that coke was being imported is an invoice of 1857, detailing the shipping costs for "20 chaldrons Coals on board Perseverance"; a "chaldron" was a dry measure of coal (36 bushels), equivalent to 1,308 litres or 1.3 m^3; ASTR, Forges Papers, St. Maurice Forges, December 1857, fol. 278.

104. JLAPC, 1852–53, vol. 11, app. CCC, report of William Hunter, engineer, St Maurice Forges, 24 August, 1852.

105. Barriault notes that loam moulding was abandoned around 1830, but indicates that there were moulding pits in the floor of the moulding shed next to where the cupola was. Barriault 1984, p. 117, 132–37.

106. Ibid., p. 156.

107. Bédard 1986, pp. 128, 167.

108. Stove plates were made for a little longer; Samson 1983, Appendix 11; Bédard 1986, p. 169: "in 1871, around 200 stoves and other unspecified items were manufactured. In 1872 or 1873, stove production was abandoned for good."

109. Barriault cites the pig mould in the 1786 inventory; Barriault 1978, pp. 12–19.

110. This was the fifth one constructed, and also the largest, 15.5 m by 10.6 m; Barriault 1984, p. 151.

111. One has only to visit a foundry that is using traditional methods, like the St Anselme foundry near Lévis, across the river from Quebec City, to see how dangerously crowded the moulding room floor is, and how the moulders have to manoeuvre between the multiplicity of box moulds covering the entire floor.

112. Metal moulds may also have been used. Claire-Andrée Fortin and Benoit Gauthier, "Description des techniques et analyse du déclin de la sidérurgie mauricienne, 1846-1910," research report submitted to the Regional Branch of the Quebec Department of Cultural Affairs, Centre de recherches en études québécoises (Trois-Rivières: Université du Québec à Trois-Rivières, February 1988), pp. 85–86.

113. Ibid.

114. Bédard 1986, p. 89.

115. Marshall M. Kirkman, *The Science of Railways: Train Service* (Chicago: The World Railway Publishing Company, 1903), vol. IV, p. 204; cited in Bédard 1986, p. 89.

116. The author of this document, who seemed to be fairly well acquainted with the workers whose "qualities" he was enumerating, attributed their differences in behaviour to their origins: "If workers from the Ardennes can be engaged they are easier to manage than those from Franche-Comté who, being raised in a province with abundant wine, are almost all drunkards and moreover naturally independent and difficult to control. The Ardennes workers may not know how to work in the *renardière* fashion, but they will learn to do so in very little time if they are mixed with Comtois workers and are under a good master." In the same document condemning the drunkenness of the men from Franche-Comté, a sum of 300 *livres* is set aside for bonuses for hammermen and finers "either in wine or in money"; NAC, MG 1, C^{11}A, vol. 112, fols. 150 and 147, "Mémoire concernant [...]."

117. In 1743, Lalouette, who was a finer at the upper forge, was said to be "the finer who made the least and the poorest-quality iron"; the workers' performance thus varied and paying the forgemen per thousandweight was an incentive to productivity; NAC, MG 1, C^{11}A, vol. 112, fol. 149, "Mémoire concernant [...]."

118. *Encyclopédie* 1757, vol. 7, p. 136.

119. "One of the four finers assigned [in two shifts] to the forge had, as hammerman, the task of making repairs to the hammer and the finery hearth. As one report noted, he was better paid than the others without being at a higher level. Originally, the hammerman worked at the hammer and the finer, at the finery hearth, hence the origin of their titles. Later, it became the custom in the Franche-Comté process for the hammerman to do both tasks. Hardach 1969, p. 36 (author's translation).

120. See Chapter 8.

121. NAC, MG 1, C^{11}A, vol. 112, fol. 144 "Mémoire concernant [...]."

122. "There is no risk at all of this happening to the St Maurice and St Étienne streams, which, six years in a row, have never dropped considerably during the low water season and have hardly risen at all during high water season [...]"; NAC, MG 1, C^{11}A, vol. 112, fol. 156 "Mémoire concernant [...]."

123. "A *renardière* working continually can manufacture 25 thousandweight, which amounts to 300 thousandweight a year if the forge runs all year [...]." In actual fact, there would be very few months when this type of production, or even anything close to it, could be expected. NAC, MG 1, C^{11}A, vol. 112, fol. 144 "Mémoire concernant [...]."

124. NAC, MG 1, C^{11}A, vol. 111, fol. 271v, memorial from Vézin and Simonet to Monsieur Delaporte Lalane, 10 June 1741.

125. Ibid.

126. NAC, MG 1, C^{11}A, vol. 112, fol. 128, "Extrait du produit des Forges de Saint-Maurice, depuis le premier octobre 1741 jusques au dit jour 1742," Quebec, Estèbe, 2 October 1742.

127. NAC, MG 1, C^{11}A, vol. 111, fol. 82v, "Mémoire des articles arrêtés [...]" between Cugnet and Jacques Simonet, Quebec, 18 March 1740.

128. Ibid. Lambert would write in 1808: "[...] a great advantage is therefore derived by carrying on any work in summer instead of winter [...]" (contrary to Europe); Lambert 1808, p. 488.

129. See reference to the 1746 fire at the lower forge in Chapter 4, "Alterations to the Forges."

130. "The forges are going day and night and the men are relieved every 6 hours," Lambert 1808, p. 487. In 1804, Lord Selkirk also spoke of the night shifts: "The two forges each 4 men & 2 boys - half day - half night." Selkirk 1804, p. 231.

131. "This short shift of six hours also was a peculiarity of traditional ironworks. It was seen most often in small establishments; in the early 1870s, it was still common at a few works in Champagne, but was in the process of disappearing." Hardach adds that, when six hour shifts were the norm, the short rest periods meant that the workers had to live at the job site. Elsewhere (p. 27), he speaks of 70-kg loops made in 1 or 1 1/2 hours. Bérubé 1976 (p. 60) cites one hundredweight an hour and one thousandweight a day (this no doubt included interruptions in work). Our data for 1739 to 1741 show an average weekly production of 7,349 pounds at the two forges. Based on an average daily production of 1,000 pounds, the average production per shift (1,000 pounds in 4 shifts) was 250 pounds (122 kg). Hardach 1969, p. 54 (author's translation).

132. André Lepage, "Étude du travail et de la production aux Forges du Saint-Maurice à deux moments de l'histoire de l'entreprise," study conducted for Historical Research, Quebec Region, Parks Canada, June 1984, pp. 73–76.

133. NAC, MG 1, C^{11}A, vol. 112, fols. 148v–50.

134. NAC, MG 1, C^{11}A, vol. 110, fol. 292, "Mémoire du Sieur Olivier de Vézain sur les Forges du Saint-Maurice, à Monseigneur le Comte de Maurepas Ministre et Secrétaire d'état de la Marine," Versailles, 28 December 1739.

135. According to *Le Petit Robert*, an *éclusée* is "the quantity of water that flows after a floodgate has been opened until it is closed again." We do not have any other information on the meaning of this term when used in the context of the operation of a waterwheel at an ironworks but believe the term designates here the quantity of water provided by the forge's millpond in a day. Since the water intake for the third forebay, which fed the second chafery was higher than for the other two forebays (see Chapter 4), this forebay could be continually supplied with water as long as the level in the millpond remained high. However, when the water level dropped below that of the intake, the third forebay could not be filled, while the other two forebays would still receive enough water to turn the waterwheels for the first chafery and the hammer. What probably happened is that, when the second chafery was shut down, the millpond would fill up slowly until the level was high enough to supply the third forebay again. In spring when the water levels were high, the third forebay could be supplied continuously since the millpond would be constantly filled due to the high flow rates in the creek. However, the operation of the second chafery also depended on the availability of labour, since it could run only when the finers were not busy replacing workers elsewhere.

136. The figure of eight forgemen is corroborated by later sources. In 1764, when the Forges were run on a reduced staff, management drew up a list of the workers required to operate the Forges "if the establishment were at full strength," which comprised two hammermen and six finers; the two extra finers assigned to the second chafery of the upper forge were not included. This chafery would be demolished in the 1770s. An employee roll of 1829 shows eight forgemen and two retired (relief?) workers. NAC, Haldimand Papers (1764), MG 21, B21-2, (21681) reel A-615, fols. 186-87; NAC, RG 4, A1, vol. 225 (1829), p. 84.

137. "In a well-tended hearth, four workers can make 12 to 15 hundredweight in 24 hours. A single hammer can serve two *renardières.*" *Encyclopédie* 175, p. 162. This estimate of daily production is close to the 25,000 pounds a month given in the memorial of 1743 (25,000 for 24 working days = 1,041.66 pounds/day). A 1739 document specifies a capacity of 1,000 pounds per day per fire (Bérubé 1976, p. 60). After the upper forge was opened in 1739, Chaussegros de Léry wrote that the forges (upper and lower) produced "two thousandweight of iron per day"; Chaussegros de Léry, NAC, MG 1, C¹¹A, vol. 72, fols. 239–43, October 1739.

138. See figures for December 1739 in Figure 5.3. The high level of production in 1739 cannot be explained by the operation of two *renardières* during the day in each forge, since, in 1739, the second *renardière* in the upper forge had not been constructed yet, and the one in the lower forge was not operational.

139. Selkirk 1804, p. 230–31; Lambert 1808, p. 487–88.

140. In his French translation of the last sentence of this quotation, Sulte gives a different meaning to it: "On y gagne ce grand avantage de compenser en été la perte de temps des nuits d'hiver" [The great advantage of this is to make up in summer the loss of time on winter nights]; Sulte 1920, p. 179.

141. The weight of the loop was variable, and depended on the size of the bar to be forged (small, medium or large). Hardach gives the case of a small forge (the Aisy forge in 1836) that produced 8 loops of 35–70 pounds in a 6-hour shift. Léon, citing Grignon, gives the average dimensions of a loop as 15–22 in. long and 4–5 in. wide. In an 1856 list of Forges wares, the following dimensions are given for wrought iron: 1 1/2 × 1/2, 2 × 3/4 and 1 inch. Hardach 1969, p. 27; Léon 1961, p. 99; Bédard 1986, p. 103 (Table 3).

142. The hammer used in 1739 probably weighed around 400 pounds (195 kg) according to Chaussegros de Léry. NAC, MG 1, C¹¹A, vol. 110, fol. 178, notes and calculations attributed to Chaussegros de Léry (1739).

143. *Encyclopédie, Recueil de planches* 1765, "Forges ou Art du fer," sect.4, plate V; Gabriel Pelletier, *Les Forges de Fraisans. La métallurgie comtoise à travers les siècles* (Dampierre: published by author, 1980), p. 45.

144. "At least a third of the iron is wasted in making wrought iron, with 15 hundred-weight of cast iron yielding one thousandweight of wrought iron. The weight diminishes proportionally with the number of heats and times under the hammer, and it is not unusual that the wastage is greater in merchant iron than in other types. A bloom to be made into merchant iron is heated four or five times to produce the bar, another three times to be slit and rolled, and another two times for the drawing of wire; *Encyclopédie* 1757, vol. 7, p. 163.

145. The forge interior illustrated in the *Encyclopédie* is of a *renardière* chafery since only one hearth can be seen, and Plate VII of section 4 shows one man using the hammer while another reheats his bloom in the hearth next to the hammer; Figure 3 in the same plate shows a perspective drawing of a *renardière.*

146. Saint-Ange 1835–38, p. 52. The German (or Franche-Comté) method used at the time (early 19th century) as described by Saint-Ange required the refining of 100 to 150 kg of cast iron (p. 46) to form a large loop. The loop was then chopped into 4, 5 or 6 pieces, which were then hammered into bars (p. 51).

147. Hardach 1969, p. 36.

148. Cameron Nish was the first to consider this record in his 1975 study (Nish 1975, Table 14, pp. 105–7). However, his figures contain several miscalculations and transcription errors. Other miscalculations, sometimes significant ones, have been found in other tables in this work. Therefore, we recompiled the data in the original document, being careful to recreate, using a spreadsheet, the weekly calendar for the period from 15 October 1739 to 1 October 1741. We found that the person who did the calculations in the original document also made some errors, due to overlapping dates. Our calculations, given in Appendix 10, show some discrepancies with Nish's results. For example, he had a total of 558 working days, while we obtained an average of 533 (523 for the upper forge and 543.5 for the lower forge), which amounts to a difference of more than 20 working days. We also found discrepancies, though less significant ones, in the production totals. The unknown 18th-century compiler came within 100 pounds of the correct figure for total production (749,504 pounds of iron instead of the actual 749,604). This is something of a feat, given the mess the original document is in and the fact that it is barely legible. NAC, MG 1, C¹¹A, vol. 110, fols. 157-62," État des fers fabriqués dans les forges de St. Maurice, depuis le quinze octobre 1739 (jusqu'au 1ᵉʳ octobre 1741)," n.s., n.d.

149. Total weight of iron made from January to March 1740 (in French pounds): upper forge, 47,126.5 (23 t); lower forge, 61,030.5 (30 t) (see Appendix 10).

150. 103 weeks from 15 October 1739 to 1 October 1741, therefore 103 × 6 = 618 working days.

CHAPTER 6

1. *Encyclopédie ou Dictionnaire raisonné des Sciences, des Arts et des Métiers*, vol. 7, 1757, s.v. "Forges, (Grosses-) by M. Bouchu, ironmaster at Veuxsaules near Château-Vilain (hereafter cited as *Encyclopédie* 1757), p. 154.

2. See note 11 of Chapter 5, in the section on "The Work Week."

3. In 1828, Lieutenant Baddeley established the daily capacity of the blast furnace at 2.5 t (64 t per month), and, according to a document entered as court evidence in 1861, in 1854 the monthly capacity of the blast furnace was 120 tons of pig iron (4 tons per day). In 1874, Harrington confirms an average daily capacity of 4 tons. "Lieutenant Baddeley's (Rl Engineers) report on the Saint Maurice Iron Works, Near Three Rivers, Lower Canada (1828)," in *APT*, vol. V, no. 3 (1973) (hereafter cited as Baddeley 1828), p. 10; ANQ-Q, Superior Court, District of Quebec, docket no. 2238, John Porter et al. v. Weston Hunt et al., plaintiff's exhibit no. 2, 1856; Dr B. J. Harrington, "Notes on the Iron Ores of Canada and Their Development," Geological Survey of Canada, *Report of Progress for 1873–74* (Montreal: Dawson Brothers, 1874) (hereafter cited as Harrington 1874), p. 247.

4. NAC, MG I, C¹¹A, vol. 57, fol. 112, Beauharnois and Hocquart to Maurepas, Quebec, 11 October 1732.

5. NAC, MG 1, C¹¹A, vol. 110, fols. 284–92, Beauharnois and Hocquart to the Minister, 28 September 1734.

6. On the subject of Intendant Hocquart's industrial strategy, see Chapter 1, as well as Jacques Mathieu, *La construction navale royale à Québec, 1739–59*, Cahiers d'histoire no. 23 (Quebec City: La Société Historique de Québec, 1971) (hereafter cited as Mathieu 1971), pp. 10–11.

7. NAC, MG1, C¹¹A, vol. 63, fol. 191, "Observations faites par moy […]", Olivier de Vézin, 17 October 1739.

8. NAC, MG 1, C¹¹A, vol. 110, fols. 299v–300, Maurepas to Beauharnois and Hocquart, 19 April 1735. In his reply, the Minister appears to assume the existence of a blast furnace producing "4,000 to 5,000ᵗ pounds of pig iron in 24 hours." The three iron bars sent to France in September 1734 were from La Brèche's dismal output in January and February 1734; NAC, MG 1, C¹¹A, vol. 110, fols. 284–92v, Beauharnois and Hocquart to the Minister, 18 September 1734.

9. "[…] The weight of iron used in the construction of a man-of-war depended, of course, on its type. It would appear that 210 thousandweight were needed for a ship of the line, 100 for a frigate, 50 for a pinnace. The ballast weighed at least one or two times that amount. Three-deckers would carry between five and six hundred thousandweight of pig iron in ship's cannon." Denis Woronoff, *L'industrie sidérurgique en France pendant la Révolution et l'Empire* (Paris: Éditions de l'École des hautes études en sciences sociales, 1984) (hereafter cited as Woronoff 1984), p. 371.

10. Peter Kalm, *Travels into North America*, trans. J. R. Forster (Warrington: William Eyres, 1771), vol. 3 (hereafter cited as Kalm 1771), p. 88.

11. Bertrand Gille, *Les origines de la grande industrie métallurgique en France* (Paris: Éditions Domat, 1947), p. 102; he mentions the La Chaussade Forges.

12. It must be borne in mind that New France was under the administration of the Ministry of Marine. In such a context, it is not surprising that the Minister intended to set up the Forges for the primary goal of meeting the particular needs of the Ministry, such as naval construction and manufacture of armaments.

13. NAC, MG 1, C¹¹A, vol. 100, fols. 304-8, Beauharnois and Hocquart to Maurepas, 26 October 1735. The first accounts of the Forges show that the King's advances were reimbursed in the form of bar iron supplied at reduced prices to the royal storehouses and arsenals at Quebec and in France.

14. NAC, MG1, C¹¹A, vol. 110, fol. 10, Vézin, 15 October 1738. A detailed statement of the bar iron ordered by Rochefort and of the difficulty of manufacturing round iron without a tilt hammer is also made in a subsequent document, which also includes a request for an exemption from weight duties for wrought iron exported to France. NAC, MG1, C¹¹A, vol. 74, fols. 132–47, "Les intéressés aux forges de fer, sur les forges de St. Maurice," signed Cugnet, Gamelin and Taschereau, 19 October 1740; see fol. 143 ff.

15. Maurepas would have liked the Forges to furnish all the necessary wrought iron for shipbuilding, but Intendant Hocquart reminded him that the company was unable to supply certain complex articles (knees); other wrought iron, as well as certain types of nails, were supplied by the Rochefort Arsenal; Mathieu 1971, p. 72.

16. NAC, MG 1, C¹¹A, vol. 70, fol. 144, Hocquart to Maurepas, 24 October 1738.

17. Cameron Nish, *François-Étienne Cugnet, Entrepreneur et entreprises en Nouvelle-France* (Montreal: Fides, 1975) (hereafter cited as Nish 1975), p. 102, Table 12.

18. Based on production (expressed in French pounds) from the start of operations at the lower forge, 20 August 1738 (date of the official blowing in of the blast furnace) to 15 October 1739. The figure of 42% was obtained as follows:

Wrought iron produced from start-up to
1 October 1741:   890,031.25 pounds

minus

Wrought iron produced between
15 October 1739 and 1 October 1741:
                749,604.00 pounds

Wrought iron produced from start-up to
15 October 1739:   140,427.25 pounds

58,513 = 42% of 140,427.25

It should be noted, however, that this start-up period (1738–41) was punctuated by multiple work stoppages due to equipment breakage and insufficient supplies of raw materials. NAC, MG 1, C¹¹A, vol. 110, fols. 157–62 and vol. 112, fols. 80–83. See Chapter 5, "The Art of Ironworking."

19. "No vessel, from a ship of the line to brig or shallop, could do without iron, in the form of nails, anchors, bars and hoops, which had to be replaced periodically during refittings. Ships also had to have cast iron ballast." Woronoff 1984, p. 370.

20. Without excusing the workers at the Forges, it should be noted that these criticisms were made at a time when everything was going wrong for the Compagnie des Forges; NAC, MG 6, C¹, series E, vol. 166, item 168, 6 December 1741; cited in Michel Bédard, "Tarification, commercialisation et vente des produits des Forges du Saint-Maurice," typescript (Quebec City: Parks Canada, 1982) (hereafter cited as Bédard 1982a), p. 18.

21. The industry faced problems of scarce manpower and the poor, low-quality wood supply also posed problems. Mathieu 1971, pp. 42–45 and 55–60; Alice Jean E. Lunn, *Développement économique de la Nouvelle-France, 1713–1760* (Montreal: Les Presses de l'Université de Montréal, 1986) (hereafter cited as Lunn 1986), pp. 161–77.

22. Mathieu states that 15 boats (including the *Canada*) were built by the royal shipyards, while Brisson counts only 13 (excluding the *Iroquoise* and the *Outaouaise*). Mathieu 1971, pp. 101–3; Mathieu also counts 38 ships built by private entrepreneurs between 1739 and 1749 (p. 77); Réal Brisson, *La charpenterie navale à Québec sous le régime français, les 100 premières années de la charpenterie navale à Québec: 1663-1763* (Quebec City: Institut québécois de recherche sur la culture, 1983) (hereafter cited as Brisson 1983), Appendix B.

23. NAC, MG 1, C¹¹A, vol. 88, fols. 85–94, Hocquart to the Minister, 15 October 1747.

24. It is surprising, to say the least, that Brisson, in his brief discussion of wrought iron supplied by Quebec blacksmiths, fails to make any mention of the St Maurice Forges in his work on shipbuilding during the French regime; Brisson 1983, pp. 146–47.

25. This was Claude-Joseph Courval-Cressé, director of the Forges from 1761 to 1764 (the Courval of the 1764 memorial); he was the son of Claude Courval-Cressé, clerk and director of the Forges under the French regime, and brother of Louis-Pierre Poulin, *dit* Courval-Cressé, apprenticed to Levasseur in 1745 and who directed the construction of the *Abénakise* (1753–56) at the royal shipyard at Quebec; his father Claude was a first cousin to Francheville, the original founder of the Forges. Mathieu 1971, p. 15, and Lunn 1986, pp. 168–69; AUL, FM-79, notes by Benjamin Sulte. Sulte confused the relationships among the Poulin de Courval family: see the Poulin family tree, Appendix 16.

26. NAC, MG 21, B²¹⁻², fols. 139–44, microfilm A-615, Haldimand Papers, memorial by Courval, 20 September 1764 (hereafter cited as Courval 1764).

27. "[…] the said Forges produced in that year [1756] 141,432.13 Livres equal to £5,893 sterling & that the whole charges thereon amounted for that year to 127,170.14 equal to £5,290 sterling, Exclusive of great deal of work done by People Impressed for that service (*corvées*) that the quantity of Cannon shots & other work delivered for the King's account at Quebec & Montreal was an Extraordinary produce of that year […]." NAC, MG 11, CO ⁴², vol. 1-1, pp. 159–65, 21 June 1764, "Memorial of John Marteilhe of Quebec to the Lords Commissioners of Trade & Plantation" (hereafter cited as Marteilhe 1764).

28. Courval relates the results of one of the preliminary tests carried out in preparation for using wrought iron from the Forges for shipbuilding: "[…] I can make to you a brief report of a test which Monsieur Hocquart, Intendant of Finance in Canada, caused to be carried out during the first of the constructions for the King in this country. He had two knees made, the first out of our wrought iron, and the other of the iron from Europe. These were then dashed onto rocks from a height of around 30 feet. Our iron broke into two pieces only, at the weld, while the European iron shattered into twenty pieces, which convinced this magistrate to have a quantity sent through the various ports of France each year […]." Courval 1764, fol. 139.

29. It is no doubt significant that only forgemen were retained, but no moulders. During the first seven years of the French regime, the Forges had produced on average upwards of 100 stoves per year. According to a 1756 statement of account, stoves were being produced at that time at the rate of 200 per year. Courval, in 1764, intended to earmark 150,000 pounds of pig iron for the manufacture of 300 stoves. Marcel Moussette, *Le chauffage domestique au Canada. Des origines à l'industrialisation* (Quebec City: Les Presses de l'Université Laval, 1983) (hereafter cited as Moussette 1983), p. 85. Courval 1764.

30. F. C. Wurtele, "Historical Record of the St. Maurice Forges, the Oldest Active Blast-Furnace on the Continent of America" *Proceedings and Transactions of the Royal Society of Canada for the year 1886,* vol. IV, sec.2, pp. 83–84; Marcel Trudel, "Les Forges Saint-Maurice sous le régime militaire (1760-1764)," *RHAF,* vol. V, no. 2 (September 1951) (hereafter cited as Trudel 1951), p. 170.

31. La Bruère of Boucherville, a lumber merchant, bought 37,000 pounds of wrought iron in 1762. Tonnancour of Trois-Rivières bought 15,786 pounds. Malcolm Fraser, an officer at Trois-Rivières, sold 120,000 pounds of wrought iron in six months (September 1764 to March 1765); Trudel 1951, pp. 174–80. On the subject of La Bruère, see Lunn 1986, p. 171, and Mathieu 1971, p. 98.

32. John Lambert, *Travels Through Canada and The United States Of North America, in the Years 1806, 1807 & 1808: To Which Are Added Biographical Notices and Anecdotes Of Some Leading Characters in the United States.* (London: printed for Baldwin, Cradock, and Joy; Edinburgh: W. Blackwood, and Dublin: J. Cumming, 1816) (hereafter cited as Lambert 1808); Réal Boissonnault, "Quelques notions sur l'orientation de la production et les types de produits fabriqués aux Forges du Saint-Maurice," typescript (Quebec City: Parks Canada, 1981) (hereafter cited as Boissonnault 1981), n.p.

33. Charles Robin appears to have been a disappointed customer. In this letter, he asks his buyer to order his iron from Europe, saying that the Forges iron, while resistant to heat, is cold-short: "The iron which is ordered is for our own use, if you can get it all of the Europeans I would prefer it, the Canadian works extremely well at the fire but when cold it gets so brittle that it cannot be drove, it proceeds from a defect in the Manufacture, it's a pity for the best Iron in the World might be had from Canada, but it's left in a too rough state. I am surprised they have not proper workmen at Three Rivers." NAC, MG 28, III, 18, microfiche 903, "Robin Jones and Whitman Limited, Paspebiac Letterbooks, 1790–1858," Charles Robin to Burns and Woolsey at Quebec, 29 August 1796 (as well as the sheet of paper included after this letter attesting "ordered from Quebec 13 July.") Another order of this type exists, from 21 August 1797 (placed before the letter of 20 September 1795 in the letterbook).

Lambert speaks of a slump in bar iron sales in the early days of the Davison, Monro and Bell administration (around 1793). There may well have been a quality problem then. See note 66. Lambert 1808, p. 485.

34. Woronoff 1984, p. 431

35. J. N. Fauteux, *Essai sur l'industrie au Canada sous le régime français* (Quebec City: Ls-A. Proulx, King's Printer, 1927), vol. 1, pp. 107–8; see also Lunn 1986, p. 217.

36. Woronoff 1984, p. 431.

37. "I am sending an order to Montreal for immediate delivery to Carillon of mill saws, anvils, anvil irons, and other articles requested by you, especially the files of good quality. I am also ordering the St Maurice Forges to make 24 sledgehammers, large and ready for the handle, and to send them promptly on to Chambli […]"; "I had ordered the bar iron from the Forges as described in your memorial, in which there were 3 thousandweight of rod iron. I am informed that they were sent to Chambli by Perault, haulier from Chambli to St-Jean. They may have become confused with that for the artillery. You will find them there, for your memorial has indeed been executed;" NAC, MG 18, K3, II, Chartier de Lotbinière Papers, Correspondence and other documents, Quebec, 23 August 1756, and Quebec, 30 June 1757.

38. NAC, MG 1, C¹¹A, vol. 91, fols. 228–32, Beauharnois to Maurepas, 1 October 1748. In 1748, six cannon and 10 mortars were sent to Forts Niagara, Frontenac and St Frédéric; "[…] two to arm the barque plying Lake Ontario"; Lunn 1986, p. 218; an excerpt of the document cited is reproduced.

39. In a "Memorial on Trade in Canada" dating from the early days of the Forges (around 1741), it was written that "there could also be cast cannon in these forges, which would serve to arm the ships to be built in this country, whether on the King's account, or that of the merchants themselves. The wrought iron from these same forges, which has been sent to France, is such that the mine is deemed proper for the making of artillery, but the Principals are in no wise able to undertake this. And they content themselves for the present with having pots, stoves, and other articles of that kind cast for sale in this country […]." NAC, MG 1, C¹¹A, vol. 76, fols. 171–72 (n. d.). In a letter written in 1744 to the Minister, Beauharnois and Hocquart toy with the idea of manufacturing ammunition and artillery pieces, and talk of attempting cannonballs at the Forges; they discuss hiring two "master founders with the necessary workers"; NAC, MG 1, C¹¹A, vol. 81-1, fols. 59–60, Beauharnois and Hocquart to the Minister, 17 October 1744.

40. For three-decker vessels, the weight of ship's cannon would be between 500 and 600 thousandweight of pig iron. Woronoff 1984, p. 371.

41. Mathieu 1971, p. 101. In a personal communication, Parks Canada historian Gilles Proulx states that a storeship (*flûte*) had a burden of between 500 to 800 tons; *flûtes* built at Rochefort during the 1750s were known to carry between 36 and 46 cannon, however, in times of war.

42. Lunn 1986, p. 217; "I have received, Monseigneur, a copy of the record of the test made at Brest of the three cannon cast at the St Maurice Forges. It is in no way surprising that these were found to be defective, the workers who cast them being moulders of bombs and pots, who are little able, if not completely incapable of founding cannon [...]"; NAC, MG 1, C¹¹A, vol. 91, fols. 228–32, Beauharnois to Maurepas, 1 October 1748; the writer is the Chevalier de Beauharnois, Artillery Commander, who spent six weeks at the Forges, and not Governor Beauharnois, who had left New France in October 1747.

43. The Metz, Douai, Parchemin (Planchêmenier) and Rancognes foundries; Lunn 1986, p. 218; NAC, MG 1, C¹¹A, vol. 112, fols. 321–22v°, Le Mercier to the Minister, Rancogne, 2 April 1750.

44. Ibid.

45. "The precautions taken when one makes cannon, not to mention as concerns the casting [...] are first—that the pig iron is refined with less mine, second—the furnaces must not run quickly, that is to say, one must not be obliged to charge the furnace more than eight to ten times at the most every 24 hours, so that the mine does not arrive in the receiver until it has melted [...]." NAC, MG 1, C¹¹A, vol. 112, fol. 336, "Mémoire dans lequel on a détaillé les précotions que l'on a prises pour fondre des Canons de fer dans la forge de Rancogne," Le Mercier, 2 April 1750.

46. "While I was at the Forges I had a casting pit dug for casting cannon, it is lined with masonry with a covered drain at the bottom to keep it dry and we could cast with this furnace cannon taking eight-pound balls. It would be necessary, Monseigneur, for the preparation of the loam for moulding, to have a good moulder, the iron is of such good quality that it deserves not to be neglected. If Mr. de La Jonquière were to send me there next spring I could attempt to cast some for you and send them to you to show you"; NAC, MG 1, C¹¹A, vol. 112, fols. 324–25v, Le Mercier to the Minister, Quebec, 4 November 1751. In this same document, Le Mercier informs the Minister that he has had some 420 iron wheels cast for Isle Royale. These apparently served to support gun carriages (for cannon and mortars), which Le Mercier sent at the same time. Artillery Officer Beauharnois had also had a cannon pit dug "near the furnace"; this was used to pour small cannon "upright." However, he says further on that the "swampy land allows no digging of casting pits [...]." This is no doubt why Le Mercier states two years later, concerning a

deeper pit that he has had dug, that he had built a "covered drain at the base to keep it dry." NAC, MG 1, C¹¹A, vol. 91, fols. 228–32, Beauharnois to Maurepas, 1 October 1748.

47. NAC, MG 11, C¹¹A, vol. 112, fols. 320–20v, Le Mercier to the Minister, Quebec, 17 October 1750. The items manufactured were iron tongs for cannon, mortars and perriers, and gun carriages.

48. Marteilhe 1764.

49. "Notary Badeaux of Trois-Rivières notes on 8 March 1776 that "Pélissier sent two thousandweight of wrought iron to the American commander to make picks to be used during the siege of Quebec [...]." Quoted in Benjamin Sulte, *Les Forges Saint-Maurice*, in *Mélanges historiques : études éparses et inédites* (Montreal: G. Ducharme, 1920), vol. 6 (hereafter cited as Sulte 1920), pp. 148–49. Sulte, who is citing Notary J.-B. Badeaux, *Journal de l'invasion du Canada par les américains en 1775*, writes in footnote 5, "the forgemen said that the bombs could not explode (there being a lack of the proper tools for their correct manufacture) and that they would not be ready for five more weeks"; this remark suggests that the workers had not cast bombs for some time.

50. According to Pierre de Sales Laterrière, quoted in Sulte 1920, p. 147; Pierre de Sales Laterrière, *Mémoires de Pierre de Sales Laterrière et de ses traverses* (Quebec City: Imprimerie de l'Événement, 1873) (hereafter cited as Laterrière 1873), p. 83.

51. Marie-France Fortier, "La structuration sociale du village industriel des Forges du Saint-Maurice : étude quantitative et qualitative," Manuscript Report No. 259 (Ottawa: Parks Canada, 1977), p. 88; Fortier cites an article that appeared in *American Catholic Historical Researches*, Pittsburgh, July 1909, p. 195.

52. Ibid.

53. Cited in Sulte 1920, p. 167.

54. Laterrière was detained for a month, in February and March 1776, on board the frigate *Triton*. He was arrested again in 1779 and held as a prisoner of the state for three and a half years. Laterrière 1873, pp. 97, 98, 103 and 109–23.

55. See note 49.

56. Réal Boissonnault, "La structure chronologique des Forges du Saint-Maurice des débuts à 1846," typescript (Quebec City: Parks Canada, 1980) (hereafter cited as Boissonnault 1980), p. 177. These are wooden patterns, not cast-iron chills such as those found in the digs. NAC, MG 21, B¹⁵⁴, fols. 106–10, Twiss to Haldimand, 9 November 1778 (on the manufacture of cannonballs).

57. "During the late war [1812] the Lake Service was supplied with tracks, (?) shot and other castings by this establishment. [...] This establishment can supply Gun Carriages, Shot etc., for the supply of Ordnance: the price would depend of the state of the trade." Lieutenant Baddeley's visit to the Forges was motivated by the need, expressed by the Board of Ordnance, to set up new forges in the country in order to have arms at a reasonable price. During the War of 1812, the Board had had to pay exorbitant prices for munitions and transport. This is explained in a letter referred to by Baddeley in his report. Baddeley 1828, pp. 10, 13 and 14; NAC, MG 13, WO⁵⁵, vol. 864, Engineer's papers, 1827, Canada, Nflnd. and Nova Scotia, letter signed R. Byham, 3 July 1827.

58. See the section on the upper forge in Chapter 4.

59. Bédard 1982a, pp. 14–15.

60. *Quebec Gazette*, 26 August 1784.

61. Michel Bédard, "Fluctuations des prix de certains produits manufacturés aux Forges du Saint-Maurice (1740–1858)," typescript (Quebec City: Parks Canada, 1982) (hereafter cited as Bédard 1982b), p. 8; Johanne Cloutier, "Répertoire des produits fabriqués aux Forges du Saint-Maurice," Manuscript Report No. 350 (Ottawa: Parks Canada, 1980) (hereafter cited as Cloutier 1980).

62. At the time Conrad Cugy, ex-secretary to Governor Haldimand and member of the Legislative Council, obtained a 16-year lease to the Forges starting in 1783. He had previously studied engineering. Sulte 1920, p. 170.

63. See section entitled "The Iron Trade" in this chapter.

64. The opening of the British imperial market to Canadian farm products would have a ripple effect on the colony, increasing production facilities, particularly new flour mills and sawmills, the machinery for which was manufactured at the Forges.

65. Bédard 1982a, pp. 23–24.

66. Lambert 1808, p. 485; he also suggests that the firm had already had problems selling its wrought iron: "I have heard that the present proprietors of the works (Monro & Bell), at the commencement of their taking them, in order to push the sale of their bar iron, which was at that time inconsiderable, purchased a large stock of very inferior British iron, and knowing that the Habitans regarded the price more than the quality, they sold it to them for a trifle less than the Three Rivers iron; but the British iron was so bad, that when they came to use it, ' sacre diable', they would have no more; and the next time bought the Three Rivers iron, which being really of a good quality, has continued in reputation among them ever since." Ibid., p. 487.

67. The master founder was also responsible for moulding objects in the casting house floor; Moussette 1983, pp. 84–85.

68. "More of the cast iron comes from England than from the continent; most of the mechanisms are made of cast iron; for example, one sees cast-iron window frames, so thin and light that they are barely perceptible; and everything from cast-iron nails of various types to flat-link chains. The methods used to cast these pieces are most ingenious [...]"; "Sur les défauts qu'on observe en général dans la construction et la conduite des Fourneaux de réduction directe, et dans les grosses forges," *Annales des Arts et des Manufactures*, le 30 thermidor an XI, [1803], pp. 122–23.

69. This list contains just the names of the first moulders and founders. The list of British employees, drawn up by Marie-France Fortier, contains 40 names of workers engaged in various trades; Fortier 1977, pp. 23–24, Table 4.

70. Moussette 1983, p. 105.

71. See Figure 6.5.

72. *The Montreal Herald*, 29 November 1823, cited in Cloutier 1980, p. 80.

73. Shortly after the Conquest, the Carron Iron Works in Scotland began to export stoves to Canada. In 1768, the firm's inventories contained references to "Canada stove plates" and later, in 1783, "Canada stove patterns," and "Canada stoves" (over 6 tons in stock), as well as, in 1788, "handles for Canada stoves with Castors." Moussette was not sure at first (p. 136) which firm (the Forges or Carron) should be credited with the design of the Canada stoves but specifies later on (p. 208), citing an American source, that "double Canada stoves were designed for the Scottish foundry Carron by the Haworth brothers around 1775." The name "Canada stove" suggests, however, that, at some point in time, someone imitated a stove made in Canada, and therefore at the St Maurice Forges. Edinburgh, Scottish Record Office, GD 58-18-14 and GD 58-16 3, p. 6, 22–23 and 160; Moussette 1983, p. 136 and 208.

74. "Anyone who wants to buy St Maurice stoves is consequently warned against the deception in which they are urged to buy an article that experience has shown to be incontestably less durable and consequently of less intrinsic value." *Le Canadien*, 21 September 1840.

75. "Quebec prices current, 31ᵗʰ december 1833," circular printed by Woolsey & Son, 31 December 1833, sent to Charles Robin & Co of Paspébiac; NAC, MG 28, III, 18, vol. 235 (vol. 39 according to volume number conversion table in finding aid no. 589, Robin, Jones and Whitman Ltd.). Anthropologist André Lepage kindly supplied this reference.

76. "The 1748 inventory indicates that at the Forges there were five wooden moulds for stoves, estimated to weigh 30 pounds each, and in the store "61 heating stoves, large and medium sized." This means that, at this time, two stove models of different sizes were made, but the exact dimensions are not known. Production must have continued throughout the 1750s, although there is no written evidence of this, only that four wooden moulds for stoves were included in the 1760 inventory." Moussette 1983, pp. 85–86.

77. A 1799 advertisement proposed 12 models, at the same prices as in 1794 (model L for £100 was not on the 1794 list and model N for £180 was not on the 1799 list); *Quebec Gazette*, 1 August 1799. It is assumed that a stove of a different size is a different model.

78. Cloutier 1980, pp. 65 and 80.

79. According to Michel Bédard, "La privatisation des Forges du Saint-Maurice 1846-1883: adaptation, spécialisation et fermeture," manuscript on file (Quebec City: Canadian Parks Service, 1986) (hereafter cited as Bédard 1986), pp. 101–8, Table 3. Cloutier lists 33 models; Cloutier 1980, p. 64, table.

80. According to Bédard 1986, pp. 114–15, Table 4.

81. Other firms had been making stoves for around 20 years; Marcel Moussette reports that "[...] between 1835 and 1845, there were no fewer than 69 makers or inventors of heaters in Upper Canada and 25 in Lower Canada"; Moussette 1983, p. 166.

82. See epigraph to section "The Forges and the Iron and Steel Industry;" ANQ-Q, Superior Court, docket no. 2238, John Porter et al. v. Weston Hunt et al., plaintiff's exhibit 3. Letter from William Henderson to Messrs. Stuart and Porter, 11 December 1854 (cited in Bédard 1986, p. 119).

83. Bédard 1986, p. 169.

84. NAC, MG 24, L³, Baby collection (M1396), A. Guy to Brother Baley (Quebec), 19 October 1772.

85. The manager of the Batiscan Iron Works, which also had to deal with British competition, explained how difficult it was to manufacture lightweight castings: "[...] our hollow ware is so much superior to that from England on convenience quality, and appearance, that I make no doubt it will ere long obtain a decided preference. The moulding of ware so very thin as this is being more apt to fail in pouring the metal, proves more expencive to the manufactury altho more economical to the user, our are now put as nigh as possible at the same rates as that last in the neughbouring states." N. Bayard to G. Platt, 12 July 1807. Batiscan Iron Works letterbook, 410 pages (several missing pages), Parks Canada, Quebec City.

86. From *Les épis*, 1914; cited in Robert-Lionel Séguin, *L'équipement aratoire et horticole du Québec ancien (XVIIᵉ, XVIIIᵉ et XIXᵉ siècle)* (Montreal: Guérin, Collection Culture populaire/Guérin littérature, 1989) (hereafter cited as Séguin 1989), vol. 2, p. 297.

87. "Once adopted by the *habitants*, the wheeled plough did not go out of favour for quite some time. After the Treaty of Paris, the swing plough was also used, but it did not gain ascendancy among farmers until three quarters of the way through the 19th century. According to Joseph-François Perrault, in 1831, "the wheeled plough, used in the country since settlement, was fairly well suited to heavy soil, while the English plough was better suited to light soil. Therefore, he writes in his *Traité d'agriculture*, Canadians, hold steadfast, do not change until a better one has been built." His advice was followed in many regions, including Charlevoix where the two-wheeled plough was used until the Second World War." Séguin 1989, p. 87.

88. "These parts are prepared at the forge, where there is a hearth for a tilt hammer or a chafery. Socks, in particular, were worked by hand first in the finery before being finished and mounted." Woronoff 1984, p. 429.

89. *Encyclopédie* 1757, vol. 7, p. 167. François Dornic, who provides a description of sock plate models in 19th century France, notes, among other things, that "The wrought iron for these items does not go through the slitting mill but is fashioned, on request, from bar iron made in the forge." François Dornic, *Le fer contre la forêt* (Ouest France, 1984), pp. 232–33.

90. *Quebec Gazette*, 1 June 1794 and 11 August 1817; PAO, Baldwin Papers, box 2, envelope 10, Monro and Bell to Quetton St. George, 10 March 1808.

91. The documentation available on the Forges does not contain any exact descriptions of the French term *plaque de soc*. According to a 19th-century English source, the term designates an iron plate, in the shape of a laurel leaf, that is used to produce two ploughshares: "The share is always formed from a plate forged for the express purpose at the ironmills, and known in the trade by the term sockplate. [...] each half being capable of forming a share." However, the estimated weight of the plates sold at the Forges suggests that they contained only a single share. The account books speak of *socs*, *plaques de socs*, soc plates, plough moulds, plough points and plough castings. Our analysis suggests that the first four terms refer to the same thing, the sock plate or share, while the term plough points designates the points of the share (welded to the share) and the term plough castings designates the pins attaching the sock to the bottom of the plough. James Slight and R. Scott Burn, engineers, *The Book of Farm Implements and Machines*, ed. Henry Stephens (Edinburgh and London: 1858), p. 157; APJQ, Superior Court, District of Quebec, docket no. 2238, J. Porter et al. v. Weston Hunt et al., exhibits 12, 13, 15, 16 and 19, January 1860. The terms plough moulds

and soc plates seem to refer to the same product, weighing 14 lbs. on average. In the 1852 account books, the Montreal retailer uses the term plough mould to designate a sock, while his Quebec City counterpart tends to use the term soc plate. In addition, speaking of a transaction on 31 October in Montreal, Bédard notes the sale of 237 plough moulds from the Quebec City retailer, who recorded in his own books, on 14 October, 237 items under the column "soc plates"; Bédard 1982b, p. 13.

92. The average number of sock plates (recorded as soc plates in the books) offered for sale at Quebec in 1852 (1,820), 1853 (3,740) and 1854 (2,477) was 2,679. In the books of the Montreal retailers in the same years, there were no references to soc plates but rather to plough moulds, their price and weight being equivalent to the sock plates sold at Quebec. The average number of sock plates offered for sale in Montreal in 1852 (1,021) and in 1,853 (41) was 531.

93. In the account books available from the French regime, sock plates were recorded in terms of the total weight produced or sold, without mentioning the individual weight or quantity. Based on the average weight of these items sold at Quebec in 1853 (14 lbs.), an estimated 781 plates were manufactured at the Forges in 1745 (10,140.5 French pounds or 10,941 imperial pounds in weight). Séguin states that over 600 sock plates were sold at Quebec in 1752, based on a document that indicates that 5,620 pounds of sock plates remained in stock. When this weight is converted into imperial pounds, we get 433 plates weighing 14 pounds each. To have 600 plates, the average weight would have to be 10 pounds but Séguin does not give information on weight. However, he does cite (vol. 2, p. 489) a 1716 document that refers to a sock plate weighing 18 pounds. NAC, MG 1, C¹¹A, vol. 112, fol. 267v–68, "Extrait des trois comptes de recette et Dépenses [...] (1741–46)"; Séguin 1989, p. 273; APJQ, Superior Court, District of Quebec, docket no. 2238, J. Porter et al. v. Weston Hunt et al., exhibits 12, 13, 15, 16 and 19, January 1860.

94. Woronoff 1984, p. 431.

95. Séguin 1989, pp. 192 and 489.

96. Boissonnault 1980, pp. 27–28. In Estèbe's inventory of 1741, there is mention of a "wrought-iron hammer" and a "cast-iron hammer" suitable for the tilt hammer that the late Sieur Francheville had had sent from France; also cited was "1 anvil suitable for a tilt hammer weighing 450 pounds, from the late Sieur Francheville." This equipment, which was lighter, suggests that the hammer in Francheville's forge was smaller than those used subsequently at the upper and lower forges. NAC, MG 1, C¹¹A, vol. 112, fols. 46v- 47v, "Inventaire des Forges 1741, Estèbe."

97. "The merchant Sieur Huguet notified me that he had just received the tools and other implements for a forge to be established in Canada sent by Sieur Le Blanc, ironmaster at the Clavières forges [...] these tools weigh 12 thousandweight [...]." The Clavières Forges were established in 1671 near Chateauroux. NAC, MG 2, B³, vol. 380, fols. 32–33, M. Dionis to "Colonies et autres lieux," Nantes, 15 March 1737; *Les forges du Pays de Châteaubriant*, Cahiers de l'Inventaire 3, Ministère de la Culture, Inventaire général des monuments et richesses artistiques de la France, Pays de Loire, Département de Loire-Atlantique, 1984, pp. 82–84.

98. In 1737, with the blast furnace not yet in operation, Vézin had to fashion a chafery hearth with stone hearth plates to conduct his first trial runs in making bar iron. See Chapter 4.

99. "The furnace is blowing well and it is an opportune time to cast a goodly number of anvils and hammers to replace the broken ones in the forges"; NAC, MG 1, C¹¹A, vol. 110, fols. 228–30v, Estèbe to Hocquart, 28 October 1741.

100. The moulds for these pieces were also inventoried. Other spare parts such as hammers, anvils and hursts were inventoried at the forges themselves. NAC, MG 1, C¹¹A, vol. 112, fols. 40v–42 and 46v– 47v, "Inventaire des Forges 1741, Estèbe."

101. ANQ-TR, Not. Rec. J.-B. Badeaux, indenture of Roc Baudry, joiner, 28 April 1787.

102. *Quebec Gazette*, 11 August 1817.

103. NAC, MG 28, III, 57, (AC) vol. 8, 26 August, 12 September, 5 October and 4 November 1809 and 11 April 1810; ibid., (NC) vol. 34-8, invoice from G. Platt, 28 September 1809; cited by Jean Bélisle and André Lépine, *Le site de l'épave d'un des premiers bateaux à vapeur de la Molson Line*, "Rapport préliminaire de la troisième campagne de fouilles (1986)," Comité d'histoire et d'archéologie subaquatique du Québec, December 1986.

104. Cloutier 1980, pp. 36–37; Boissonnault 1980, p. 255.

105. Boissonnault 1980, p. 257.

106. Cloutier 1980, p. 88.

107. AO, Baldwin Papers, box 1, envelope 5, and box 2, envelope 10, letter from Monro and Bell to Quetton St. George, 10 March 1808 (Price list); AUM, Baby collection, G2, 1819 (bilingual price list); a kettle delivered in 1829 weighed 1,376 lbs. (Qx 12.1.4), NAC, MG 23, G¹¹¹, vol. 25, document 819, receipt from Captain Pierre Cormier to M. Bell, 25 May 1829.

108. Lambert 1808, pp. 485–86.

109. Baddeley 1828, p. 12.

110. NAC, RG4, A1, vol. 5-225, fol. 84. List of people residing at the King's Iron Works of Saint-Maurice, August 1829.

111. Harrington 1874, p. 248.

112. In 1873, a world depression paralyzed the Quebec forest industry.

113. NAC, MG 13, WO[35], vol. 867, p. 155–56 (no. 21), Corps of Royal Engineers, Col. Durnford to Col. Mann, Quebec, 11 February 1830.

114. NAC, RG 8, C, vol. 50, pp. 53–56, Col. By to Rouths, 9 January 1830, and pp. 46–47, Edward Grieves to the officer in charge, 22 January 1830.

115. Patterns made by the engineers were sent to the Forges.

116. NAC, MG 13, WO[35], vol. 867, pp. 155–56 (no. 21), Corps of Royal Engineers, Col. Durnford to Col. Mann, Quebec, 11 February 1830.

117. The chairman of the Three Rivers Gas Company and the city's mayor was John McDougall, who bought the Forges 10 years later.

118. William Henderson described how the contract was obtained: "We have got the order for the Three Rivers Gas Works, in the teeth of Larue [Auguste Larue, one of the three owners of the Radnor Forges] and the 3 Rivers founders; it was most cleverly accomplished by Mr. Rickaby whom I sent in for the purpose to get the work at all hazards —he did so for £17 per ton — there will be it is [?] 16 to 20 tons all plain work." In another letter, he states: "In the 3 Rivers Gas Co. castings, it was only late one afternoon that I found out that tenders were required next day — I sent in Mr Rickaby immediately with full power to take them at anything [over?] £15— He obtained £17 for the contract and was just in time to get it — had he appeared sooner the Radnor Forges would have bid lower and we should have lost it." In this letter, Henderson had just explained that he could not expect to sell stoves for more than 12 shillings/cwt, and they were more expensive to produce, and that it would be better to produce plain castings for the same price. With this contract, he obtained 5 shillings more per cwt than he expected. APJQ, Superior Court, District of Quebec, docket no. 2238, J. Porter et al. v. W. Hunt et al., item 33, exhibit no. 3, Henderson to Porter, 11 December 1854, and exhibit 24, letter, 8 November 1854.

119. According to Nish 1975, p. 101. The bars were submitted to Sieur Lottin, master locksmith in Paris, for testing.

120. Cloutier 1980, pp. 32–38.

121. According to Bédard 1986, the railcar wheel operation was established in 1853 based on the presence of a cupola furnace at that time reported by the engineer Hunter in 1852. The cupola furnace could also have been located in a shed adjoining the blast furnace. Furthermore, there is no mention of the manufacture or sale of railcar wheels in 1853 or 1854 in the detailed records and accounts of John Porter & Company and the Forges inventories for August 1854 do not include any railcar wheels. At the end of the same year, in a letter dated 11 December, superintendent Henderson mentioned a contract for "Rail way wheels" that was to close in nine days. He says in another letter (8 November, item 24) that the cupola, built in August to last six months, already had to be rebuilt. It is very likely that production began after the installation of the cupola in August 1854; APJQ, Superior Court, District of Quebec, docket no. 2238, J. Porter et al. v. W. Hunt et al., items 24 and 33, exhibit no. 3, Henderson to Porter, 11 December 1854, and exhibits A and K, Inventories, August 1854. An 1857 account mentions the sale of 413 wheels, and an estimate of repairs between 1853 and 1857 mentions the presence of annealing pits used in wheel production. Furthermore, it would be more logical for this manufacture to have begun at the same time that the capacity of the blast furnace was increased.

122. Bédard writes that "the estimate for 4 September 1857 does not contain any reference to the construction of a new cupola furnace between 1853 and 1857." There is however a reference to a "complete furnace," which Bédard associates with a kiln (p. 91); but subsequently, there is reference to a 15-ft. chimney that we think should be associated with this furnace rather than a kiln. The document also mentions a "brick block containing 5 holes, each 8 ft. high and 3 ft. in diameter, to chill the wheels," which seem to have been built from scratch. Bédard 1986, p. 87; Quebec, Department of Lands and Forests, Service des terres, M30, B19, "Estimation de E. Normand," 4 September 1857.

123. In mid-19th century France, charcoal iron was also sought after for wire works, tires and railcar axles; Bédard 1986, p. 88; Serge Benoit, "La consommation de combustible végétal et l'évolution des systèmes techniques," in Denis Woronoff, *Forges et forêts : recherches sur la consommation proto-industrielle de bois*, Denis Woronoff, ed. (Paris: Éditions de l'École des hautes études en sciences sociales, 1990), p. 96.

124. André Bérubé, "Rapport préliminaire sur l'évolution des techniques sidérurgiques aux Forges du Saint-Maurice, 1729-1883," Manuscript Report No. 221 (Ottawa: Parks Canada, 1976) (hereafter cited as Bérubé 1976), pp. 61–62.

125. Kalm 1771, p. 89.

126. Bérubé 1976, p. 61.

127. Baddeley 1828, p. 10.

128. APJQ, Superior Court, District of Quebec, docket no. 2238, J. Porter et al. v. W. Hunt et al., item 33, exhibit no. 3, Henderson to Porter, 11 December 1854.

129. Ibid.

130. Fortin and Gauthier suggest that, because of the undeveloped state of mineral exploration and mining at the time, the Canadian iron industry did not have the requisite capacity to fuel the expansion of the wrought-iron and cast-iron markets. As a result, the railway industry had to resort to American iron. Claire-Andrée Fortin and Benoit Gauthier, "Description des techniques et analyse du déclin de la sidérurgie mauricienne, 1846-1910," research report submitted to the Regional Branch of the Department of Cultural Affairs, Centre de recherches en études québécoises (Trois-Rivières : Université du Québec à Trois-Rivières, February 1988) (hereafter cited as Fortin and Gauthier 1988), p. 120.

131. Ibid., p. 111–51.

132. However, it should be noted that prices rose suddenly by close to $20/ton owing to the economic crisis of 1872–73, sinking back close to their initial levels in 1876 and 1877. Bédard 1986, p. 180, Table 9, based on statistics by H. Mitchell, in C.A. Curtis et al., *Statistical Contribution to Canadian Economic History* (Toronto: The Macmillan Company of Canada Limited, 1931), p. 79. See the section of this chapter entitled "Overview of Pricing."

133. The L'Islet Forges were closed for good in 1878 and the Radnor Forges were put up for sale in 1883; ibid., pp. 126–27.

134. A letter from Henderson in 1854 confirms that several moulders who had been trained at the St Maurice Forges had already gone to work for Radnor by that time. The Clermont census (Radnor Forges) in 1861 contains several names of workers, particularly moulders, who had previously been employed at the Forges. McGill University Archives, Logan Papers, accession no. 1207/11, item no. 94, "John Porter & co. to W. E. Logan, Provincial Geologist, Montreal"; see Chapter 8.

135. The blast furnace at the Radnor Forges had a daily capacity of approximately 5 t of pig iron. The establishment also had three forges, a rolling mill, a foundry and a nail works; in 1860, production consisted of 300 stoves, 4,700 railcar wheels and 1,820 t of pig iron, worth $2,620; Fortin and Gauthier 1988, p. 57; Census of Canada, Champlain County, St Maurice Parish, 14 February 1861, Fermont (or Radnor) Forges.

136. The L'Islet Forges, which had a blast furnace with a 1,200-t annual capacity, was established by two former St Maurice workers (Dupuis, Robichon & cie), while many St Maurice moulders went to Radnor. The L'Islet Forges opened shortly before St Maurice shut down. In 1863, John McDougall bought the L'Islet Forges and the St Maurice Forges within a few weeks of each other. Therefore, L'Islet and St Maurice were in direct competition for only a short time; Claire-André Fortin and Benoit Gauthier, "Les entreprises sidérurgiques mauriciennes au XIXe siècle: approvisionnement en matières premières, biographies d'entrepreneurs, organisation et financement des entreprises," research report submitted to the Regional Branch of the Department of Cultural Affairs, Centre de recherches en études québécoises (Trois-Rivières : Université du Québec à Trois-Rivières, November 1986), pp. 113 and 116; Fortin and Gauthier 1988, p. 54.

137. Fortin and Gauthier wished to draw attention to the effects of this local competition. Fortin and Gauthier 1988, p. 149.

138. He also acquired, shortly before the St Maurice Forges, the nearby L'Islet Forges. The output of the latter, since its establishment in 1856, went mainly to the Louis Dupuis foundry in Trois-Rivières. Dupuis was a co-owner of L'Islet. Fortin and Gauthier 1988, pp. 113–15 and 160.

139. "Nos Mines," *Le Constitutionnel*, Trois-Rivières, 27 August 1869, p. 2. For a long time, the Forges had been selling pig iron in pigs of 84 to 168 lbs. (3/4 to 1 1/2 cwt). Pélissier bought back the shares of his partners in 1771 and 1772 against delivery at Quebec of 90 tons of "pig iron or pigs marked 3 Rivers" for each share (1/9) purchased. In 1771, Pierre de Sales Laterrière, then the Forges' agent at Quebec, shipped some pigs to London. A 1787 inventory mentions "1307 quintals of Pig iron" (65.35 t). In 1815 and 1832, Joseph Bouchette wrote that a large quantity of pig iron was exported. This pig iron was then remelted by foundries, which made parts and goods for industry and trade. The local foundries (including the 56 "inventors and manufacturers" of heaters listed by Marcel Moussette in Lower Canada between 1845 and 1855), which competed with the Forges in the manufacture of stoves, no doubt obtained their supplies of pig iron from the Forges. Moussette, who speaks of "the high cost of the pig iron" used to manufacture stoves, does not say where the pig iron was obtained. See Moussette 1983, p. 133 and pp. 173–230; ANQ-Q, Not. Rec. Saillant, 4 April 1771, No. 2152, and 26 June 1772, No. 2298; NAC, MG 11, Q[38], p. 125 (recorded by S. Courcy, "Gueuses et gueusets fabriqués aux Forges du Saint-Maurice," document file and inventory of archaeological artifacts prepared as part of the Grande Maison restoration project (Quebec City: Canadian Parks Service, 1989), pp. 9–11; Laterrière 1873, p. 65; Bouchette 1815, p. 314; Joseph Bouchette, *A Topographical Dictionary of the Province of Lower Canada* (London: Longman, Rees, Orme,

Brown, Green and Longman, 1832), s.v. "St-Étienne."

140. In 1880, George McDougall took over again (leased) the operation of the former Turcotte and Larue railcar wheel foundry in Trois Rivières. Between 1880 and 1883, most of the output of the Forges was pig iron to supply this foundry. Michel Bédard, "La structure chronologique des Forges du Saint-Maurice (1846-1883)," typescript (Quebec City: Parks Canada, 1979), p. 363; Bédard 1986, pp. 159, 168 and 179.

141. This amounts to around 6% of the total production of the St Maurice and L'Islet works, or 10% of the St Maurice Forges' output alone according to the production capacity (4 tons a day or 1,460 tons a year) indicated by Harrington in 1874. Harrington said the furnace was generally in blast for 10 to 13 months at a time; this suggests that the blast furnace operated year-round; Harrington 1874, p. 248.

142. John McDougall of Montreal was a distant relative of the Trois-Rivières John McDougall; the former owned the Caledonia Foundry, which then became the Montreal Car Wheel Works. A new two-year contract was subsequently negotiated. Although there is no trace of later contracts, according to the 1874 report by Dr B. J. Harrington, the firm was still supplying pig iron to railcar wheel manufacturers in Montreal. Bédard 1986, pp. 129–30 and 168–69; Harrington 1874, p. 248.

143. See Chapter 1, note 152, on the acquisition by John McDougall & Company, the main customer for the Forges' pig iron, of the St Pie de Guire Forges in 1874. Bédard 1986, pp. 169 and 184.

144. NAC, MG 1, CIIA, vol. 111, 51e, memorial from Vézin and Simonet to Monsieur Delaporte Lalane, 10 June 1741.

145. NAC, MG 1, CIIA, vol. 112, fols. 26–27, Hocquart to the Minister, Quebec, 20 October 1741.

146. Cugnet wrote that Vézin made a profit of at least 35% (see Chapter 8).

147. NAC, MG 1, CIIA, vol. 112, fols. 182–94, Hocquart to the Minister, Quebec, 25 October 1742.

148. NAC, MG 1, CIIA, vol. 112, fols. 173–74, "Mémoire concernant les Forges de St. Maurice," n.s., n.d. [1742–43].

149. Although based on Estèbe's accounting (1741–42), the price of wrought iron in France was lower than shown in his "Estat général de la dépense [...]." The adjustments (in italics in Table 6.11) resulting from the verification of the original source show that Estèbe did not sell at a loss and the profit margin was closer to 26%.

150. This document is found in four locations in the CIIA series: vol. 110, fols. 221v–27v, vol. 112, fols. 114–26, fols. 141–48 and fols. 172–75 (summary); the same numbers are found in vol. 112, fols. 182–94 (original), Hocquart to the Minister, Quebec, 25 October 1742. Of all these sources, the second is the most complete; it is included in the following memorial: NAC, MG 1, CIIA, vol. 112, fols. 135–79, "Mémoire concernant les Forges de St. Maurice," n.s., n.d. [1742–43]. In the summary of expenditures at the end of this memorial, the total annual expenditures of the forges are listed as 74,159.10 livres (fol. 174v); this amount includes 71,703.17.6 livres in total expenditures for the two forges, plus production costs for the 30 t of cast iron sold in the colony. The other sources fail to include this latter cost. Réal Boissonnault initially compiled this data — his work is quoted in Bédard 1982, (pp. 12–13). To determine overall production costs and distribute these costs among the blast furnace and the two forges, Boissonnault added the sum of 30,696.17.6 two times, which represents the production cost of the 375 t of pig iron converted into 250 t of wrought iron in the two forges. Including the production cost of the cast iron twice, once for the blast furnace and again for the two forges, artificially increases overall production costs and reduces the profit margin. The corrected prices in the table are taken from NAC, CIIA, vol. 111, fols. 278–305, "Estat général de la dépense faite pour l'exploitation des forges de St. Maurice depuis le 1er octobre 1741 jusqu'au premier aoust 1742," signed by Estèbe, Quebec, 2 October 1742. Boissonnault 1981, fols. 19-20.

151. NAC, MG 1, CIIA, vol. 88, fol. 88, Forges production from 1 January to 14 October 1747. A total of 689,464 French pounds (337.4 t) of cast iron was produced, 29% of which was used to make castings and 71% of which was converted into iron.

152. It is difficult, however, to interpret Laterrière's numbers, because he switches from *louis* (a twenty-franc piece) to pounds sterling in the same paragraph, without differentiating. Furthermore, his idea of profit is not clearly defined. If what he calls "profit" is actually how much production is worth, then his figures, if they are indeed in pounds sterling, show total revenues of £34,500 and a profit of £11,500. We still do not know whether the amount of £10,000 to £12,000 reported by Selkirk for the entire "produce" in 1804 is indeed the gross revenue or the profit. Laterrière 1873, pp. 84–85.

153. ANQ–Q, Superior Court, District of Quebec, docket no. 2238, John Porter et al. v. Weston Hunt et al., testimony of William Henderson, 14–17 December 1861, p. 8; quoted in Bédard 1986, p. 120.

154. ANQ–Q, Superior Court, District of Quebec, docket no. 2238, John Porter et al. v. Weston Hunt et al., exhibit no. 32.

155. Woronoff 1984, pp. 490-92.

156. Bédard 1986, pp. 180-81.

157. See the references in note 150.

158. These are only rough comparisons. We do not always know exactly what is included in the French figures, and more nuances must be brought to bear if the various cost categories are to be truly comparative. We must also consider the particular operating conditions in place at the French ironworks (land leasing, fees, etc.) that differed from those at the St Maurice Forges.

159. Lambert 1808, pp. 486–87.

160. From Bédard 1986, pp. 171–72, who cites Harrington 1874.

161. The company had a large depot and a store built in Trois-Rivières to store its products. The store was separate from the King's store, where wrought iron for export was kept. Iron was also stored in the King's stores at Quebec and Montreal, again in different premises from the company's. After the 1741 bankruptcy, iron was sold from the King's stores in each of the three cities. NAC, MG 1, CIIA, vol. 112, fols. 28–59, Inventaire des Forges 1741, Estèbe; Bédard 1982a, p. 26.

162. Bédard 1982a, pp. 25–27.

163. NAC, MG 1, CIIA, vol. 111, 53e, memorial from Vézin and Simonet to Monsieur Delaporte Lalane [...], 10 June 1741.

164. AO, Baldwin Papers, box 1, envelope 5, Monro and Bell to Quetton St. George, 21 July 1807; cited in Bédard 1982a, pp. 56-59.

165. Bédard 1982a, pp. 27–33.

166. NAC, MG 11, Q, vol. 109, pp. 38–39, Monro and Bell to Ryland, 31 December 1808; cited in Bédard 1982b.

167. Boissonnault 1980, pp. 192–96. They no doubt needed money with a view to buying back the shares of their partner, Davison, who had died in March; indeed, the shares were purchased two months later, in January 1800.

168. The Batiscan Iron Works Company was founded 18 September 1798. Claire-Andrée Fortin and Benoit Gauthier, "Aperçu de l'histoire des Forges Saint-Tite et Batiscan et préliminaires à une analyse de l'évolution du secteur sidérurgique mauricien, 1793-1910," research report submitted to the Regional Branch of the Department of Cultural Affairs, Centre de recherches en études québécoises (Trois-Rivières: Université du Québec à Trois-Rivières, December 1985), p. 4.

169. Ibid., pp. 52–54.

170. Ibid., pp. 46–47.

171. Ibid., pp. 48–50.

172. AJQ, Province of Canada, District of Three Rivers, Superior Court, docket no. 2238, John Porter et al v. Weston Hunt et al., exhibit I, "Account sales of St. Maurice Wares ... from 24th November 1853 to the 22nd August 1854."

173. Bédard 1982a, pp. 44–45.

174. During the War of the Austrian Succession (1740–48), which pitted France against Britain.

175. Fernand Ouellet, *Histoire économique et sociale du Québec, 1760-1850, Structures et Conjoncture* (Montreal and Paris: Fides, 1966), pp. 603–07, graphs.

176. Ibid., p. 614, graph.

177. See Fortin and Gauthier 1988, p. 142 ff. for more information on how the market changed from the 1850s on.

178. Bédard estimates that, at the rate of 120 axes per day, only 70 t of pig iron each year would have been needed as raw material; Bédard 1986, pp. 169–170 and 181.

179. Fortin and Gauthier 1988, pp. 128 and 131; Bédard 1986, p. 183.

CHAPTER 7

1. Well after the ironworks closed in 1883, the Forges formed part of the mission of St Michel Archange (St Michel des Forges), with a priest in residence from 1920 on. According to a recent monograph on the parish, it went back as far as 1740. The parishioners of St Michel des Forges who wanted to demonstrate their belonging to the Forges relied on a register of St Louis des Forges de St Maurice, kept in the Parish of L'Immaculée Conception de Trois-Rivières, in which the missionaries had recorded the entries relating to ironworks employees between 1740 and 1764. The municipality of St Michel des Forges (in the process of being incorporated in 1953) was not given parish status until 16 July 1959. Hormidas Magnan, *Dictionnaire historique et géographique des paroisses, missions et municipalités de la province de Québec* (Arthabaska: L'imprimerie d'Arthabaska, 1925), pp. 582–83; François De Lagrave and the Corporation communautaire de Saint-Michel-des-Forges, *Au pays des cyclopes: Saint-Michel-des-Forges, 1740–1990* (Trois-Rivières: Corporation communautaire de Saint-Michel-des-Forges, 1990); Hubert Charbonneau and Jacques Légaré, eds., "Saint-Louis-des-Forges de Saint-Maurice," *Gouvernement de Trois-Rivières, 1730–49*, vol. 23, and *1750–65*, vol. 36, *Répertoire des actes de baptême, mariage, sépulture et des recensements du Québec ancien*, Université de Montréal, Département de démographie (Montréal: Les Presses de l'Université de Montréal, 1984 and 1987).

2. Apart from the registers of L'Immaculée Conception de Trois-Rivières, there are those of the parishes of St Étienne des Grès, Pointe du Lac and Yamachiche, as well as that of the Anglican church in Trois-Rivières. Most of the records are at L'Immaculée Conception de Trois-Rivières. The Forges register (St Louis des Forges de St Maurice) for the years 1740 to 1764, which is kept in the Parish of L'Immaculée Conception de Trois-Rivières, does not seem to have entries for all the baptisms, marriages and deaths in the population of the Forges for those years, however, since other records associated with ironworks employees are kept in the general parish register. In the Forges register of St Louis des Forges de St Maurice, the first entry is a baptism on 12 June 1740, and the last entry is a death that occurred on 25 March 1764.

3. Regarding methods of extracting data, see Micheline Tremblay and Hubert Charbonneau, "La population des Forges St-Maurice (1729–1883)," a study done for Parks Canada by the University of Montreal Demography Department under the historical demography research program (Montreal: Université de Montréal, 1982) (hereafter cited as Tremblay and Charbonneau 1982), pp. 12–13. Our interpretation of the demographic trends of the population of the Forges is based on the findings of both this study and our own, Roch Samson, "Les ouvriers des Forges du Saint-Maurice: aspects démographiques, 1762–1851," Microfiche Report No. 119 (Ottawa: Parks Canada, 1983) (hereafter cited as Samson 1983).

4. Tremblay and Charbonneau 1982, pp. 12–13.

5. We are also aware that a selective method of extracting data could skew the results. To see whether this is really the case, a complete demographic study of the Trois-Rivières area for the same period would have to be done. In such a study, the demographic behaviour of the Forges ironworkers would be subsumed under that of a much larger population including people living in both urban and rural areas.

6. As the population figure obtained from the parish registers is much higher than that based on the censuses, it seems almost impossible to distinguish between natural growth (more or fewer births than deaths) and growth due to migration (immigration or emigration) in calculating total growth. Tremblay and Charbonneau 1982, pp. 20–25.

7. Ibid., p. 85.

8. The increase in deaths, which parallels that of births, is partly due to infant mortality, which was high at the time.

9. Tremblay and Charbonneau 1982, p. 36.

10. This table is taken from Tremblay and Charbonneau 1982, p. 41. It counts only instances in which the exact age of both spouses at the time of marriage was known.

11. Ibid., p. 43.

12. Ibid., p. 44.

13. Ibid., p. 47.

14. Ibid., p. 57.

15. Countries with some of the highest death rates in 1990 were:

Country	Infant mortality (per thousand)	Life expectancy (years)
Afghanistan	167	42.5
Mali	164	45
Sierra Leone	148.5	42
Guinea Bissau	145.5	42.5

Source: *L'état du monde 1992* (Montreal: Éditions la Découverte and Éditions du Boréal, 1991).

Canada now has a death rate of 7 per 1,000, with a life expectancy of 77 years.

16. This table is taken from Tremblay and Charbonneau 1982, p. 64.

17. Tremblay and Charbonneau 1982, p. 68.

18. The 3 or 4 moulders employed during the French regime increased to 9 during the years 1820–30 and to 26 in 1851. This increase did not correspond to an increase in the output of the blast furnace, however. Rather, it indicates that a greater share of the iron produced was used to make castings rather than refined into bar iron. The greater emphasis on casting had no effect on the number of finers and hammermen, which remained the same (10 forgemen) throughout the entire period.

19. The 1784 figure was probably a maximum, since a few British moulders had already been absorbed into the village population 20 years earlier.

20. "Workmen and families in number about 200." Monro and Bell, RG 4, A¹, S, vol. 86, L26692-26695 (1805).

21. If we consider that in Lower Canada in the 18th century, the high fertility of French Canadian women helped double the population every 30 years, it is hardly surprising that the same thing occurred at the Forges. We do not believe, however, that the doubling of the population living at the Forges is attributable to high birth rates of the original generation. If that had been the case, we would have seen, later in 1851, close to 600 inhabitants, rather than the 395 actually found in the census. In 1830, the population began to stabilize at about 400. Jacques Henripin, *La population canadienne au début du XVIIIᵉ siècle*, Institut national d'études démographiques, Travaux et documents, Cahier No. 22 (Paris: Presses universitaires de France, 1954), p. 74.

22. As early as 1769, the company had begun to recruit moulders and founders in England, Scotland and Ireland. Their arrival undoubtedly explains the increase in population that occurred between 1784 and 1805, when it went from 149 inhabitants to almost 200. Marie-France Fortier, "La structuration sociale du village industriel des Forges du Saint-Maurice: étude quantitative et qualitative," Manuscript Report No. 259 (Ottawa: Parks Canada, 1977), pp. 18, 23 and 24 (Table 4).

23. Here, of course, we are only talking about those of the new arrivals still at the Forges in 1829. Other workers already counted before 1829 were probably also part of the wave of immigration in the first quarter of the century.

24. ANQ-M, Not. Rec. N. B. Doucet, No. 714, 20 November 1805; No. 2738, 28 November 1810; and No. 3090, 27 November 1811. The indentures quoted stipulate that workers will be housed by the company. In other later indentures (1815–17) reported by Serge Saint-Pierre, one clause stipulates that "carters shall give notice three months before the end of the agreement if they intend to leave service, otherwise they will automatically be re-engaged for another year." This kind of provision shows that the carters lived year-round at the Forges and indicates the power the Forges had over them. The same provisions applied to the blacksmiths. See Serge Saint-Pierre, "Les artisans du fer aux Forges du St-Maurice; aspect technologique," Manuscript Report No. 307 (Ottawa: Parks Canada, 1977), pp. 20 and 44–45.

25. More than half of the other 22 horses belonged to skilled workers, while the other workers—including only two carters—shared the rest.

26. See chapters 2 and 9.

27. In 1804, Lord Selkirk wrote that the ore was brought to the ironworks from two leagues (9.7 km) away and charcoal from three leagues (16 km). Lieutenant Baddeley, visiting the Forges in 1827, reported that the nearest ore was found six to nine miles (9.7–14.5 km) away and charcoal carried seven to nine miles (11–14.5 km). The high cost of transporting raw materials over long distances was probably behind the company's decision to employ carters rather than pay them by the load. NAC, MG 19, E.1, 1 (Lord Selkirk 1804); "Lieutenant Baddeley's (Rl Engineers) report on the Saint Maurice iron works, near Three Rivers, Lower Canada, jany 24th 1828" in *APT*, vol. V, no. 3 (1973), pp. 9 and 10.

It is also possible that the decision to sign contracts with carters was related to new regulations concerning carters in Trois-Rivières, but this is merely a hypothesis. In 1802, a "police regulation at Trois-Rivières concerning hired hands and masters" was introduced for the entire judicial district. And some obligations arising from this regulation can be found in the contracts the Forges entered into after that. Another document, undated and issued by the court of Trois-Rivières, in response to a petition, orders that carters of the town henceforth obtain a licence, and that their number be limited to six. Rates are also set for each type of load. This regulation would have been a serious obstacle to the Forges, which used many carters at different times of year, and it might have prompted the Forges managers to stop using self-employed carters and to hire carters as employees instead. But this hypothesis cannot be verified until the document has been dated, and we have other evidence providing more details. Documents reproduced in Serge

Saint-Pierre, "Les charretiers aux Forges du Saint-Maurice," typescript (Quebec City: Parks Canada, 1977).

28. In his inventory of 9 November 1741, Estèbe lists 11 horses, while in an inventory dating from 1746, 28 horses are listed, and in 1748, 4 head of cattle and 24 horses were counted. These numbers are in line with the 1784 figure of 22 horses, but are a far cry from the 77 horses counted in 1831. The horses on site in the 18th century were likely draught horses used by the two or three carters permanently employed at the furnace and forges and by the household help. The other, more numerous, carters employed for shorter periods—two to five months, according to Estèbe's "Estat général de la dépense [...]" of 1742, or by the day—must have had to provide their own horses. André Lepage, "Étude du travail et de la production aux Forges du Saint-Maurice à deux moments de l'histoire de l'entreprise," typescript, study conducted for Historical Research, Quebec Region, Parks Canada (June 1984), tables 1.1 to 1.7; Saint-Pierre 1977, p. 8.

29. Joseph Bouchette, *A Topographical Description of the Province of Lower Canada, with Remarks upon Upper Canada, and on the Relative Connexion of Both Provinces with the United States of America* (London: printed for the author, and published by W. Faden, geographer to His Majesty and the Prince Regent, Charing-Cross, 1815), p. 304.

30. In 1829 and 1830, one of the main projects of the Forges was to provide the castings used in building the locks on the Rideau Canal (see Chapter 6). Passing through the Forges in 1808, John Lambert referred to these variations, speaking of all employees, both outside and inside: "Forty or fifty horses are employed and upwards of 300 men, more or less, according to the work in hand." John Lambert, *Travels through Canada and the United States of North America, in the Years 1806, 1807 & 1808: To Which Are Added Biographical Notices and Anecdotes of Some Leading Characters in the United States* (London: printed for Baldwin, Cradock, and Joy; Edinburgh: for W. Blackwood; Dublin: for J. Cumming, 1816), pp. 485–88.

31. The difference of about 30 workers, some of whom were skilled, seen between the 1829 figures and those of 1825 and 1831, shows that the ironworks had no trouble increasing its work force as needed from time to time. We will also see later that the population at the Forges actually produced more workers than could find employment there.

32. The male-to-female ratio, or sex ratio, is the number of men per 100 women in a population. It is used to measure the breakdown of the sexes in a given population, either overall or by age group. In human populations, the sex ratio at birth is 105, which means that 105 boys are born for every 100 girls. A ratio of less than 100 indicates an imbalance, with more women than men; this usually happens in older populations, because women generally live longer. A high ratio, of say 145, could indicate an immigration of men (workers) into a population. Several major factors—war, migration and epidemic, for example—may change the sex ratio. It should be pointed out, however, that in such a small population as at the Forges, sometimes the absence of just a few individuals of either sex is enough to change the ratio.

33. The low sex ratio seen in 1851 and 1861 is due to the Forges not operating at the time the censuses were taken. Hardy and Gauthier found a different ratio when they considered people 15 and over. René Hardy and Benoît Gauthier, "La sidérurgie en Mauricie au 19e siècle: les villages industriels et leurs populations," research report submitted to the Regional Branch of the Quebec Department of Cultural Affairs by the Centre de recherches en études québécoises (Trois-Rivières: Université du Québec à Trois-Rivières, May 1989) (hereafter cited as Hardy and Gauthier 1989), p. 87.

34. According to the censuses of 1829 and 1851. Samson 1983, pp. 49–50, tables 6 and 7.

35. The figures calculated by Hardy and Gauthier for all industrial villages between 1851 and 1891 are similar; they found means of 60–64% unmarried people for the six villages studied, including the St Maurice Forges. Hardy and Gauthier 1989, p. 92.

36. When Hardy and Gauthier compared the groups of 20–29-year-olds in all the industrial villages in the St Maurice Valley, they observed that this age group was smaller at the Forges (less than 14%) than in other villages (15–21% on average). Hardy and Gauthier 1989, p. 91.

37. Our assessment is based on the assumption that, in a normal situation where no immigration or emigration is taken into account, the number of 20–29-year-olds matches the number of 10–19-year-olds less the number of people who died in the intervening 10 years. When the deaths are subtracted, any significant drop in numbers from one age bracket to the next is probably due to the emigration of people in this bracket during those 10 years. We tested this assumption using the known numbers of these two age groups for each sex at different times (1829, 1831 and 1851) and concluded that deaths were only a small factor in the drop, which was due instead to the fact that half of the people in these groups emigrated during this decade of their lives. Samson 1983, p. 61, Table 10.

38. Legend mentions the names of some workers (such as Arthur Imbeau) who emigrated from the Forges, and marriage records show that a man born at the Forges who had been working at an American ironworks came home to marry a local girl. Inspired by this story, a sound and light show presented at the historic site tells of the return of a worker to the St Maurice Forges in 1845.

39. Analysis of vital data also confirms that young adults used to have a tendency to emigrate. Tremblay and Charbonneau came to that conclusion by calculating the ratio of marriages dissolved (through death) at the Forges to marriages entered into there, which came to just 32.6%. In other words, for every 100 couples married at the Forges, only 32 had all their children there. Tremblay and Charbonneau 1982, p. 15.

40. There were 4 two-family houses in 1831, 9 in 1851, 10 in 1861 (including one- and three-family houses), 6 in 1871 and 3 in 1881 (see Chapter 9).

41. According to the "number of families in the house" (1851) and the "number of families living in the house" (1861). The drop in the number of houses from 72 in 1851 to 42 in 1861 may have been due to the demolition of some tenement houses. The Henderson photograph shows that the McDougalls' new house replaced the old tenement house (in an L shape) shown in the engraving by Piggott (1845) (see Plate 9.2).

42. H. Clare Pentland has underscored the mutual dependency between the company and its workers. The labour market being what it was, the company did not have the luxury of being able to replace its skilled workers, and they themselves had no other job openings. The company therefore had to ensure that its employees remained loyal, and they in turn felt obligated towards the company. Pentland has described the labour relations that originated from this double bind as paternalistic and feudal, to differentiate them from those characterizing the Industrial Revolution, at a time when all the conditions were right to create a labour surplus in the job market. Although his interpretation may stand in need of review in light of recent research on the Forges as well as on the early iron industry specifically and on the pre-1850 labour market in general, Pentland is the only one, so far, to propose an interpretation of the social labour relations at the St Maurice Forges. It is worth noting, however, that his thesis (although published in 1981) dates from 1960 and was therefore written long before any detailed research had been conducted into the establishment. H. Clare Pentland, *Labour and Capital in Canada, 1650–1860* (Toronto: James Lorimer & Co., 1981). Pentland devotes 12 pages of his book (pp. 34–46) to the St Maurice Forges.

43. This rejuvenation of the working population, and thus of families, could indicate that the Forges assigned less importance to the passing on of technical know-how within families. The large number of workers (37) who declared themselves to be labourers in 1851 and the disappearance of a dozen or so stated occupations on that date in relation to 1831 would also seem to indicate that less importance was being given to the know-how a worker had to have to perform a job that would formerly have been given a special mention. In the second half of the 19th century, the company would hire more and more young workers, and fewer and fewer older ones, with the result that apprenticeship possibilities were no doubt reduced. This rejuvenation of the work force may have had an impact on the quality of the company's products and on the survival of the company itself. While these observations are essentially speculation, they allow us to underscore the fragility of companies that depend on the availability of families of skilled workers to ensure their survival.

CHAPTER 8

1. Hubert Charbonneau and Jacques Légaré, eds., *Répertoire des actes de baptême, mariage, sépulture et des recensements du Québec ancien*, Université de Montréal, Département de démographie (Montreal: Les Presses de l'Université de Montréal, 1984 and 1987), vol. 23, *Gouvernement de Trois-Rivières, 1730–49*, and vol. 36, *Gouvernement de Trois-Rivières, 1750–65*, s.v. "Saint-Louis-des-Forges-de-Saint-Maurice" (hereafter cited as Charbonneau and Légaré 1984 and 1987).

2. In a memorial, the different positions of ironmaster, clerk and foreman are described as follows:

   To a manager/ironmaster (expert, also a skilled founder, hammerman and even a finer, a true workman, able to oversee all the workmen at the furnace and the forges, able himself to know when and how they are lacking, rectify their errors both to correct their work and show them the correct proportions and the degree of fire required to produce good pig iron and well-made wrought iron) at 2000 *livres* per year, board not included [...]

   To a forge clerk to follow under the orders of the ironmaster the work of the furnace and forges, the production of charcoal and the cutting of wood at 1200 *livres*, board not included [...]

   To a foreman to serve under the orders of the ironmaster and clerk to watch over the work of the furnace and forges, the production of charcoal and the cutting of wood, his board included, at 700 [...]."

   Note that the latter job description corresponds to that given for the clerk, who was paid 500 *livres* more, but with his board not included. In fact, the clerk clearly looked after the books, while the foreman looked after the performance of the work.

   NAC, MG 1, C¹¹A, vol. 112, fol. 142, "Mémoire concernant les Forges de St. Maurice," [1743] (hereafter cited as NAC, MG 1, C¹¹A, vol. 112, "Mémoire concernant [...]").

3. Ibid., fol. 148.

4. APJQ, Superior Court, docket no. 2238, J. Porter et al. v. Weston Hunt et al., exhibit 5, deposition of Hamilton Rickaby, 27 January 1860.

5. ANQ-M, Not. Rec. N. B. Doucet, no. 3198, 19 March 1812, one-year deed of indenture of Joseph Peterson; ANQ, Not. Rec. Jacques Voyer, 8 July 1811, one-year deed of indenture of François Grenier.

6. In the Trois-Rivières region, several legends arose and stories were told, inspired by the men at the St Maurice Forges who lived and worked in the forest. It was believed that an entire sector of the forest bordering on the Forges was under the sway of the devil, who had apparently been left it by a woman descended from the Forges founder. This place, known as the "*vente-au-diable*" ("*vente*" meaning the clearing where the colliers made the charcoal), terrorized the *habitants*, who gave it a wide berth. In Europe, forest activity was so widespread that it was common to hear talk of serious risks for travellers' safety. Benjamin Sulte, Napoléon Caron et al., *Contes et légendes des Vieilles Forges* (Trois-Rivières: Éditions du Bien Public, 1954), pp. 19–21; Bertrand Gille, *Les origines de la grande industrie métallurgique en France* (Paris: Éditions Domat, 1947) (hereafter cited as Gille 1947), p. 150.

7. Lieutenant Baddeley noted in 1828, talking about the inside workmen: "the number of men employed vary at different seasons from 75 to 100"; he called the ironworkers "mechanics." In the final years of operation (1876–81), Alexander McDougall, the manager of the Forges, would mention "an average of about sixty men employed for the St. Maurice forges only." "Lieutenant Baddeley's (Rl. Engineers) Report of the Saint Maurice Iron Works, near Three Rivers, Lower Canada, jany 24th 1828," in *APT*, vol. V, no. 3 (1973) (hereafter cited as Baddeley 1828), p. 12; AJTR, Superior Court, docket no. 108, testimony of Alexander McDougall, 18 March 1881, exhibit 57, p. 8.

8. Selkirk had his information on the Forges from John Lees. *Lord Selkirk's Diary 1803–1804*, edited with an introduction by Patrick C. T. White (Toronto: The Champlain Society, 1958) (hereafter cited as Selkirk 1804), p. 231.

9. NAC, MG 1, C¹¹A, vol. 111, fols. 281v–87, "Estat général de la dépense faite pour l'exploitation des Forges de St. Maurice depuis le 1er octobre 1741 jusqu'au premier aoust 1742," signed Estèbe, Quebec, 2 October 1742 (hereafter cited as NAC, MG 1, C¹¹A, vol. 111, "Estat général de la dépense [...]").

10. Pierre de Sales Laterrière, *Mémoires de Pierre de Sales Laterrière et de ses traverses* (Quebec City: Imprimerie de l'Événement, 1873) (hereafter cited as Laterrière 1873), pp. 84–85; see also Serge Saint-Pierre, "Les artisans du fer aux Forges du Saint-Maurice: aspect technologique," Manuscript Report No. 307 (Ottawa: Parks Canada, 1977) (hereafter cited as Saint-Pierre 1977), p. 43.

11. Vézin had planned for 20 workmen to work "on the ore mines to gather the supplies for 1740"; NAC, MG 1, C¹¹A, vol. 110, fol. 393, memorial from Vézin to Maurepas, Versailles, 28 December 1739.

12. Réal Boissonnault, "La structure chronologique des Forges du Saint-Maurice des débuts à 1846," typescript (Quebec City: Parks Canada, 1980) (hereafter cited as Boissonnault 1980), pp. 134, 137 and 154.

13. Even in Europe, the early ironmaking industry was constantly grappling with this problem.

14. Selkirk 1804; Baddeley 1828, pp. 9–10. According to these two eye witnesses, the ore was mined 9.7 to 14.5 km from the Forges, while the charcoal was made some 11 to 16 km away.

15. Only the names of the ironworkers employed at the blast furnace and the two forges are sometimes mentioned. Some statements of expenditures, such as Estèbe's of 1742, list all the workmen who were employed during the operating period covered by the record. A rare document from the French regime provides a description of the personality, temperament and productivity of each of the main ironworkers. Reports on operating expenditures under the military regime (1760–64) also list the names of workmen employed. In his memoirs, inspector Laterrière (1775–79) also mentions the different categories of workmen employed during his 1775 campaign; he counted 125 inside workers "and the others employed at the charcoal pits, coaling, dressing, miners, colliers, road makers, *garde-feu*, ... etc. etc."; NAC, MG 1 C¹¹A, vol. 111, fols. 278–305, "Estat général de la dépense [...]"; NAC, MG 13, War Office 34, vol. 6, fols. 134–40, 1760–61; Laterrière 1873, pp. 84–85.

16. The 18th-century censuses we have (for 1762 and 1784) are really only enumerations. The population of the Forges is clearly defined in them, but only the sex and age group of the people listed there are mentioned, with no reference to their name or craft (see Appendix 13).

17. Note, however, that the individuals in this directory had to pay for their entry. Strangely, the alphabetical listing of workmen at the Forges published in 1871 stops at the letter R. A comparison with the list from the official census conducted the same year, as well as with the 1875 tally of parishioners, reveals that, on the one hand, a number of workmen do not have entries in the Lovell's directory and, on the other hand, some of them were beyond the letter R in alphabetical sequence. *Lovell's Province of Quebec Directory for 1871*, vol. 2, p. 1464.

18. Actually, eight colliers, two pit setters and one *garde-feu*.

19. Namely five moulders and two forgemen. NAC, RG 4, A 1, vol. S-225, p. 84, "List of People residing at the King's Iron works of St Maurice under the present Lessee Matthew Bell esquire, August 1829."

20. René Hardy and Benoît Gauthier, who totalled the figures for the population of the ironmaking villages in the St Maurice Valley from 1851 onwards, noted substantial underestimation at the Forges, particularly for the 1851 and 1861 censuses, but made no attempt to place a figure on it. Our review of the other censuses also revealed substantial underestimation before and after those dates. René Hardy and Benoît Gauthier, "La sidérurgie en Mauricie au 19e siècle: les villages industriels et leurs populations," research report submitted to the Regional Branch of the Quebec Department of Cultural Affairs, Centre de recherches en études québécoises (Trois-Rivières: Université du Québec à Trois-Rivières, May 1989) (hereafter cited as Hardy and Gauthier 1989), p. 105.

21. In 1851, one moulder was 12 years old, another 14, and yet another 16; moreover, one day labourer was aged 13, and another 14. In the 1851 and 1861 censuses, a good part of the underestimation is due to some ambiguous entries by the census takers. Under occupation, they put quotation marks (") or the abbreviation "do" (for "ditto") to indicate that the same occupation as written above applies to a name. But one cannot always be sure of the census taker's intention, because these notations were not always used very rigorously: sometimes all the members of a family, male and female, were marked "do"!

22. Men of working age were identified as follows:

    1784: Males 15 and over

    1825: Males 18 and over

    1831: Males 14 and over

    1842: Males 14 and over

    1851–81: Males 15 and over

    The figures for 1784 are based on a global count, not a nominal census; nor is there any mention of occupations. For comparison purposes, we took the 30 "married men" to be heads of household or workmen worthy of mention, and the other 12 "aged over 15" as workmen belonging to these families. Also, as the 1825 census does not give the occupations of the only heads of household counted, we attributed to those men the occupations reported in 1829; 9 men from 1825 would not appear in the 1829 census, so we had to add them to the 19 men aged 14 and over who were not on the list. The 1842 list does not specify occupations either, and includes only 97 men out of the 126 "aged over 14." The 1825, 1831 and 1842 censuses cover the age group "aged over 14"; including or excluding people aged 14 instead of 15 makes a difference of only a few individuals. If we go by the order of listing, the 11 unspecified workmen of 1829 could logically include 1 foreman, 3 colliers, 4 pit setters and 3 fillers. The 1829 figures exclude, on the one hand, 5 moulders and 2 forgemen, who were said to have been born at, and belong to, the Forges but were working at Trois-Rivières at that time and, on the other hand, 11 workmen (8 colliers, 2 pit setters and 1 *garde-feu*) who were said to be employed during the summer months.

(Adding these 11 men would give a total of 99 workmen, or close to the 1831 figure.) The census taker of 1829 included the 7 workmen working in Trois-Rivières, as well as their 13 family members, in his total figure (415) for the population of the Forges.

23. In the 1831 census, Forges employees are clearly identified as "non-property owners"; this lack of ownership was also clearly indicated in the 1851 agrarian census.

24. According to the order of listing, the 11 unspecified workmen of 1829 could logically be 1 foreman, 3 colliers, 4 pit setters and 3 fillers.

25. NAC, MG 1, C¹¹A, vol. 85, pp. 68–72, Hocquart to the Minister, 9 October 1746, and pp. 88–92, Beauharnois and Hocquart to the Minister, 16 October 1746.

26. Denis Woronoff, *L'industrie sidérurgique en France pendant la Révolution et l'Empire* (Paris: Éditions de l'École des hautes études en sciences sociales, 1984) (hereafter cited as Woronoff 1984), p. 140.

27. Women performed this type of work in France in the late 18th and early 19th centuries; Woronoff 1984, p. 141.

28. The small number of workmen (26 out of 53) for whom occupations are specified in the 1861 census may be explained by the fact that the Forges were closed when the census taker passed by; he also noted that there were 39 people absent.

29. There are in fact 14 specific references to moulders in the 1851 census; logically, however, 12 other names can be associated with this craft if the special method of entry used by the census takers is taken into account. We did not include a 13th moulder (by association), Jos Boisvert, owing to his advanced age of 87.

30. The presence of a cupola, reported in 1852 but likely installed earlier, is a notable sign of this intensification. Michel Bédard, "La privatisation des Forges du Saint-Maurice 1866–1883: adaptation, spécialisation et fermeture," manuscript on file (Quebec City: Parks Canada, 1986), pp. 47–48 and 87.

31. It is interesting to note that the crafts of founder and keeper would continue to be reported at the Radnor and L'Islet Forges in the 1861 and 1871 censuses, and until 1891 in the case of Radnor. In 1861, the Radnor founder, François Pellerin, was a former workman from the St Maurice Forges; the same was true of Isaac Boisvert, who was the founder at L'Islet in 1861 and 1871. Note that the L'Islet Forges belonged to the McDougalls, of the St Maurice Forges. See the list of workmen in Hardy and Gauthier 1989.

32. The workmen who had worked at the axe factory, in operation in 1872 and 1873, would, however, be distinguished by their specialization: strikers, sharpeners and hardeners. Marcelle Caron, "Analyse comparative des quatre versions de l'enquête de Dollard Dubé sur les Forges du Saint-Maurice," manuscript on file (Quebec City: Parks Canada, 1982) (hereafter cited as Caron 1982), pp. 136–42.

33. NAC, MG 1, C¹¹A, vol. 110, fol. 318, Minister Maurepas to Beauharnois and Hocquart, Versailles, 15 May 1736.

34. NAC, MG 1, C¹¹A, vol. 82, fols. 86–6v, Hocquart to the Minister, Quebec, 29 October 1744.

35. Denis Woronoff, "Le monde ouvrier de la sidérurgie ancienne: note sur l'exemple français." *Le Mouvement social*, no. 97 (1976) (hereafter cited as Woronoff 1976), pp. 113–4.

36. This is actually the commune of Labergement-Foigny, in the bailiwick of Dijon. According to his marriage certificate, Simonet was from "Danpierre, diocese of Langres," i.e., Dampierre-sur-Vingeanne in the bailiwick of Dijon, now Dampierre-et-Flée. "Carte de la province de Bourgogne au XVIIIᵉ siècle," *Annales de Bourgogne*, tome LV (1983), fascicule III (juillet–décembre).

37. NAC, MG 1, C¹¹A, vol. 110, fol. 385, memorial from Olivier de Vézin, 28 December 1739.

38. According to Marie-France Fortier, "La structuration sociale du village industriel des Forges du Saint-Maurice: étude quantitative et qualitative," Manuscript Report No. 259 (Ottawa: Parks Canada, 1977) (hereafter cited as Fortier 1977), pp. 178–82.

39. In 1740, Vézin had to seek the intervention of Sieur de La Porte to have Sieur de La Brisse (Intendant of Burgundy) ensure that the engaged workmen fulfilled their obligations: "[...] Sr Olivier de Vezin one of the undertakers of the forges being operated in Canada, has told you that he has engaged in Burgundy some workmen he requires for the operation of these forges, but they now refuse to fulfil their obligations. I shall, Sir, inform myself of the grounds of this refusal, and should I find it ill founded I shall grant this undertaker, as you order me, such assistance of my authority as he needs to force them to fulfil their obligations [...]" NAC, MG 2, B³, vol. 400, fol. 82, de la Brisse to de La Porte, Dijon, 7 April 1740.

40. Aside from the four workmen who came with Simonet, 55 people arrived in 1737 aboard the *Jason*; in 1739, Vézin recruited two German "mining experts" and, in 1740, a further 13 workmen. Among the latter was a founder, Cortillier, from the forge of Sieur Déscologne, ironmaster of Burgundy, who would die at Quebec barely one month after his arrival. Marie-France Fortier, "Une industrie et son village: Les Forges du Saint-Maurice, 1729–1764," Master's thesis (Quebec City: Université Laval, 1981) (hereafter cited as Fortier 1981), p. 50.

41. We found no trace of these Ardennes ironworkers whom it was wished to see mix with the workers from Franche-Comté, nor of the colliers of whom it was recommended that they be recruited around St Jean de Luz, in the Basque country of southwestern France. See the quotation in note 116, Chapter 5.

42. Cressé, no doubt; NAC, MG 1, C¹¹A, vol. 112, fol. 150, "Mémoire concernant [...]."

43. Ibid.

44. NAC, MG 1, C¹¹A, vol. 112, fols. 100–101v, memorial from Jacques Simonet, Paris, 17 March 1742. He was reproached, however, for being "too familiar and too easy-going with them." Cited in Fortier 1977, p. 71.

45. The War of the Austrian Succession (1740–48), and the Seven Years' War (1756–63), which actually began in the colonies in 1754.

46. Bigot would then call for the following workmen: "A good founder. A keeper. A moulder of cooking pots. A moulder's assistant. Four finers. Two helpers. Four colliers. A hammerman." NAC, MG 1, B, vol. 87, fol. 6769, Minister to Hocquart, 18 January 1748, and MG 1, C¹¹A, vol. 112, fols. 296–99, Bigot to the Minister, Quebec, 11 October 1748.

47. The Minister would follow this up in a letter to Mr de St-Comtest: "As early as 23 December I sent you a list of some workmen I am requested for the ironworks which have been established in Canada, asking you to have them engaged in the forges at Châtillon sur Seyne." NAC, MG 1, vol. 90, fol. 48, letter from the Minister of Marine, 19 February 1749.

48. NAC, MG 1, C¹¹A, vol. 112, fol. 300, note appended to "Mémoire [de Bigot] sur les forges de St. Maurice en Canada," 1748, accompanying a letter to the Minister, 11 October 1748.

49. NAC, MG 1, C¹¹A, vol. 96, pp. 58–62, Bigot to the Minister, 27 October 1750.

50. Bigot does not name him, but this was François Godard, whose death certificate, dated 23 January 1752, was found at Notre Dame de Montréal; he died at the age of 30, and only three priests attended his burial. We also note that no further deeds referring to this couple are recorded after 4 December 1751 in the register of St Louis des Forges. A baptism on 3 June 1754 (Pierre Thérault) confirms the widowhood of his spouse, Marie Blais. Charbonneau and Légaré 1987, vol. 36, s.v. "Saint-Louis-des-Forges-de-Saint-Maurice," and vol. 37, s.v. "Notre-Dame de Montréal."

51. NAC, MG 1, C¹¹A, vol. 112, fols. 326–27v, Bigot to the Minister, Quebec, 20 October 1752.

52. Parish registers show that other workmen, who were not targeted by this order, would also remain at the Forges or in the surrounding area. Among them are the colliers named Aubry, the master edge-tool maker Pierre Bouvet, who would die at the Forges in 1763 at the age of 48, and Jacques Tassé, whose sons and daughters would marry Aubry and Gilbert family members, as well as the Dupuis family, who would marry into the Michelin family in the 1770s. Bouvet also features on the list of workmen paid during the military administration of the Forges. NAC, MG 21, B²¹⁻² (21681), microfilm A-615, fols. 204–5, Haldimand Papers [1762].

53. *Report of the Public Archives for the Year 1918* (Ottawa: J. De Labroquerie Taché, 1920), pp. 85–6, "Order to M. Courval for the management of the Forges," signed J. Bruyère, 1 October 1760.

54. John Lambert, *Travels through Canada and the United States of North America, in the Years 1806, 1807 & 1808: To Which Are Added Biographical Notices and Anecdotes of Some Leading Characters in the United States* (London: printed for Baldwin, Cradock, and Joy; Edinburgh, for W. Blackwood, and Dublin, for J. Cumming, 1816), p. 488.

55. John Slicer senior had married Josephte Mailloux on 14 July 1771 in an Anglican ceremony; his son John would marry Véronique Élie, *dit* Breton, also in an Anglican ceremony, on 5 March 1803 and then again in a Catholic ceremony on 20 July 1811. Parish registers of L'Immaculée Conception de Trois-Rivières and St James Anglican Church of Three Rivers, 1767–1845.

56. It is the crafts specified in the parish registers that lead us to believe that both the descendants of the first generation of French workmen and the French-Canadian workmen who would make up the rest of the work force after the Conquest would adopt the moulder's craft only belatedly. The following list provides the first year in which the moulder's craft is mentioned in the parish registers for certain families from the Forges:

Gilbert: 1824 (Jean-Baptiste)

Imbleau: 1803 (Michel)

Mailloux: 1800 (Louis)

Tassé: 1811 (Antoine)

Terreau: 1823 (Éloi)

57. Cited by Peter Bischoff, "Des Forges du Saint-Maurice aux fonderies de Montréal: mobilité géographique, solidarité communautaire et action syndicale des mouleurs, 1829–1881," *RHAF*, vol. 43, no. 1 (Summer 1989) (hereafter cited as Bischoff 1989), p. 3. Excerpts from the petition addressed to Hon. D. Daly, Provincial Secretary, signed by six workmen from the St Maurice Forges on behalf of all the families in the industrial community upon the sale of the Forges to private enterprise. JLAPC, Montreal, Louis Perrault, 1846, vol. 5, p. 268.

58. Woronoff 1984, pp. 162–64.

59. Bischoff 1989, p. 11.

60. In 1829, two forgemen, Augustin and Antoine Gilbert, cousins of the Tassé and Terreau families, were grandsons of Augustin Gilbert. The latter had married Marguerite Parent at Quebec on 27 June 1757; we then find a trace of the couple in the register of St François Xavier de Batiscan, upon the baptism of their son Joseph, on 6 August 1764. The couple were living at that time in Ste Geneviève de Batiscan. We shall see that the marriages of founder Delorme and hammerman Marchand had already helped establish links with the people of Batiscan in the 1750s. Charbonneau and Légaré 1984 vol. 36, s.v. "Sainte-Geneviève de Batiscan."

61. According to Benjamin Sulte, there is no trace of the marriage of Chaillé and Marie-Anne Godard, although their union was subsequently confirmed by their children's baptisms. Family relationships are notably confirmed by the following baptismal records: 6 August 1745, Claude Chaillet, son of hammerman Michel, where Claude Godard, finer, is said to be the child's maternal uncle; 24 June 1747, Jean-François Godard, son of finer François, where Anne Godard, Dautel's wife, is said to be the child's maternal aunt. Marriages with members of families who practised other categories of crafts were also contracted during the French regime. Sulte 1920, pp. 64–65; Charbonneau and Légaré 1984, vol. 23, s.v. "Saint-Louis-des-Forges-de-Saint-Maurice."

62. Pierre Michelin is said to be a cousin of Pierre Marchand upon the latter's second marriage, 8 May 1750. Charbonneau and Légaré 1987, vol. 36, s.v. "Saint-François-Xavier-de-Batiscan."

63. Delorme and Marchand first married Charlotte and Marie Sauvage respectively (daughters of François and Françoise Moette; marriage of Delorme and Charlotte, at Trois-Rivières, 31 January 1739, and marriage of Marchand and Marie, 13 September 1740). Later, in 1750 and 1751, Marchand and Delorme respectively would wed, in second marriages, Gertrude and Marie-Louise Frigon, at St François Xavier de Batiscan. Master wheelwright François Caissé had also married a Sauvage (Louise), a sister to the other two, on 20 November 1740. Charbonneau and Légaré 1984 and 1987, vols. 23 and 36, s.v. "L'Immaculée-Conception-de-Trois-Rivières" and "Saint-François-Xavier-de-Batiscan."

64. Sulte does not, however, give any sources to justify this reputation. Furthermore, Marchand's daughter would marry Jean Sicard Carifel [sic], son of the seigneur of Carufel (near Maskinongé), on 6 August 1762. Sulte 1920, p. 67; Charbonneau and Légaré 1987, vol. 36, s.v. "Saint-Louis-des-Forges-de-Saint-Maurice."

65. See above the case of John Slicer, who married Josephte Mailloux in 1771; the latter could be a relative of the Mailloux moulders at the Forges, although her parents' names were not given at the wedding in the Anglican church at Trois-Rivières; Bischoff 1989, p. 11. Moulder Thomas Lewis had married Josette Delorme (daughter of founder Jean-Baptiste) in 1771. His son Jean Samuel, forgeman, would marry Thérèse Sulte dit Vadeboncoeur, and two of their sons, working at the Forges in 1829, one as a moulder and the other as a bateauman, married a Tassé (1819) and a Robichon (1837). Forgeman André Cook, who was also there in 1829, married a Moussette in 1824. Parish registers of L'Immaculée Conception de Trois-Rivières and St James Anglican Church of Three Rivers, 1767–1845.

66. Weddings performed in Trois-Rivières on 15 June 1840 (Louis Imbleau), and 5 August 1845 (Pierre Imbleau, second marriage). L'Immaculée Conception de Trois-Rivières parish register.

67. Selkirk 1804, p. 236. At the end of his 1828 report, Lieutenant Baddeley listed a number of ironworks in Canada (in particular Marmora in Upper Canada) and the United States, near the Canadian border. He specifically mentioned the ironworks at Vergennes, Vermont, on the eastern shore of Lake Champlain. Baddeley 1828, p. 13.

68. "The following are some of the localities of iron works and ore situated in those parts of the United States bordering on Canada communicated by Edward Grieves Esq. superintendent at Three Rivers." Ibid.

69. ANQ-TR, judicial records, 4 September 1806, complaint by the clerk of the Batiscan Iron Works against forgeman Charles Caul [?], who has left his employ. Concerning the hiring of American workmen: "I see myself absolutely forced to press you insistently to make this payment by return of mail. These individuals are almost all Americans who awaiting it so that they can return to their country, whom we are obliged to keep on the Company payroll until it arrives [...]." N. Bayard, December 1807, Batiscan Iron Works Letterbook, from 27 August 1807 to 14 July 1812, p. 70, Parks Canada, Quebec City (410 pages, several missing). The Batiscan Iron Works was founded in 1798 and closed down in 1814. Baddeley 1828, and E. Z. Massicotte, "Notes sur les Forges de Ste-Geneviève de Batiscan," *Bulletin de recherches historiques*, vol. XLI (1935), no. 10, pp. 708–11. Claire-Andrée Fortin and Benoît Gauthier, "Aperçu de l'Histoire des Forges Saint-Tite et Batiscan et préliminaires à une analyse de l'évolution du secteur sidérurgique mauricien, 1793–1910," research report submitted to the Regional Branch of the Quebec Department of Cultural Affairs, Centre de recherches en études québécoises (Trois-Rivières: Université du Québec à Trois-Rivières, December 1985), pp. 4–7.

70. Bischoff 1989, p. 5.

71. Even if, during this period, the Forges were substantially increasing the number of moulder positions, the rising population of the Forges during the first quarter of the 19th century also helped maintain the same surplus labour.

72. Numerous suitcases were listed in inventories of the dwellings of workmen at the Forges in the first half of the 19th century, and this could indicate that workmen travelled back and forth. See the section on physical conditions later in this chapter.

73. Bischoff 1989, p. 22, Table 1.

74. ANQ-TR, judicial records, 8 April 1807, concerning the engagement of a workman "at the cupola operated by Messrs Monro & Bell at Trois-Rivières." Baddeley, in 1828, described "two cupola furnaces" there; Baddeley 1828, p. 12.

75. Bischoff 1989, p. 22.

76. The Forges were also closed for one year, in 1849–50, owing to the failure to lay in a supply of raw materials the previous year. APJQ, Superior Court, docket no. 614, John Porter et al. v. James Ferrier, 3 May 1853.

77. Bischoff 1989, p. 16.

78. Ibid., pp. 24–28.

79. NAC, C¹¹A, vol. 111, fol. 31, Quebec, Hocquart to the Minister, 24 October 1740.

80. The contract specified that the workmen would not be paid if their apprentice failed through any fault of theirs. ANQ-Q, NF-25, no. 1300, 2 March 1740 and 19 March 1742. Vilard had already been at the Forges for at least six months since, in October 1739, he had witnessed the murder of Pierre Beaupré by one of his fellow soldiers, Jean Brissard; the same Vilard [St-Maixant] was employed for nine days as an ore breaker in 1742. ANQ-Q, NF-25, no. 1178, 1/2, Criminal records of the royal jurisdiction of Trois-Rivières, murder of Beaupré, 19 October 1739; NAC, MG 1, C¹¹A, vol. 111, fols. 278–305, "Estat général de la dépense [...]."

81. By that he meant that these workmen were merely blacksmiths and not forgemen. NAC, MG 1, C¹¹A, vol. 96, pp. 58–62, Bigot to the Minister, 27 October 1750.

82. Pierre Mayrand, "La culture et les souvenirs de voyage de l'ingénieur Louis Franquet," research notes, *RHAF*, vol. 25, no. 1 (June 1971), p. 91.

83. See especially ANQ-TR, Not. Rec. J.-B. Badeaux, 20 January 1795, no. 1882, three-year indenture of Modeste Antoine Cecile (aged 15) to John Anderson, master founder; 14 September 1790, five-year indenture of Samuel Lewis as moulder's apprentice. Not. Rec. E. Ranvoyzé, 11 November 1801, three-year indenture of François Dufresne (aged 15) as forgeman's apprentice; 11 June 1802, three-year indenture of Joseph Camirand (aged 17) as forgeman's apprentice. Not. Rec. Jos. Badeaux, 1 July 1814, discharge by Mathew Bell of a moulder's apprentice, James Cooper (from London), who had spent four years at the Forges under master moulder Guy Wauviel [?].

84. This was probably Pierre Terreau, son of Joseph, born at the Forges in 1754, and married to Marie-Louise Choret (1774), whom we find at the Forges as a "retired forgeman" in August 1829. He would die there on 28 September of the same year at the age of 75. Joseph, one of his sons, would become a forgeman like him and the other, Éloi, would become a moulder; they are recorded with him on the 1829 list of workmen at the St Maurice Forges. NAC, RG 4, A¹, vol. S-225, p. 84, "List of People residing at the King's Iron works of St Maurice under the present Lessee Matthew Bell esquire, August 1829"; ANQ-TR, Not. Rec. Jos Badeaux, 13 December 1801, indenture of P. Teraux to A. Craigie and T. Coffin; Maurice Terreau, another of Joseph's sons, would also be employed at Batiscan in 1808; ANQ-M, Not Rec. N. B. Doucet, no. 1721, 5 May 1808; L'Immaculée Conception de Trois-Rivières parish register.

85. "Ordonnance aux ouvriers, journaliers et autres employés d'obéir aux Sieurs Olivier, Simonet et autres chargés de leurs ordres [...]," 12 February 1739, cited in Pierre-Georges Roy, *Inventaire des ordonnances des Intendants de la Nouvelle-France* (Beauceville: L'Éclaireur, 1919) (hereafter cited as Roy 1919), vol. II, p. 263.

86. The ordinance also prohibited ship's captains from boarding workmen from the Forges without permission, as well as forbidding the *habitants* to "debauch" the workmen. The wording of the ordinance is reminiscent of a "Royal Declaration" of 22 May 1724 under which all were forbidden to enter the colonies under British control without permission. Roy 1919, vol. II, p. 228, 16 September 1737, cited in Cameron Nish, *François-Étienne Cugnet. Entrepreneur et entreprises en Nouvelle-France* (Montreal: Fides, 1975), p. 115, and NAC, MG 1, F³, vol. 12 (2), fols. 450–54, Beauharnois and Hocquart, Québec, 16 September 1737.

87. Woronoff 1984, p. 186.

88. Louis Franquet, *Voyages et mémoires sur le Canada (1752–1753)* (Montreal: Éditions Élysée, 1974), p. 21. English translation from F. C. Wurtele, "Historical Record of the St. Maurice Forges, the Oldest Active Blast-Furnace on the Continent of America," *Proceedings and Transactions of the Royal Society of Canada for the Year 1886*, vol. 4, Montreal (1887), section 2, p. 81.

89. Woronoff (citing F. Lassus) reports the words of ironmaster Rochet in Franche-Comté in 1788; Woronoff 1984, p. 186.

90. NAC, MG 1, C¹¹A, vol. 112, fol. 163 "Mémoire concernant [...]."

91. Luce Vermette, *Domestic Life at Les Forges du Saint-Maurice*, History and Archaeology, no. 58 (Ottawa: Parks Canada, 1982) (hereafter cited as Vermette 1982), p. 106.

92. NAC, MG 1, C¹¹A, vol. 111, fols. 111–16 (impr.) "Mémoire sur les comptes du S. Olivier et autres intéressés," 8 and 9 April 1741.

93. NAC, MG 1, C¹¹A, vol. 111, fol. 21, memorial from Cugnet, 25 September 1740.

94. Ibid., fols. 29–29v.

95. Ibid., cited in Vermette 1982, p, 107.

96. Ibid.

97. NAC, MG 23, G¹⁴, vol. 2, pp. 26–27, J. Bruyère to Courval, 22 October 1760.

98. Joseph Moussette (brother-in-law of hammerman Robichon) owed £140 to Trois-Rivières merchant Jean McBean. Cited by Luce Vermette, who found a similar case of a bateauman in 1847. Vermette 1982, p. 202.

99. NAC, RG 4, A¹, vol. S-191, nos. 17–17a, Mathew Bell to A. W. Cochran, Quebec, 4 December 1827.

100. Cited in Caron 1982, p. 77.

101. Reference is made in a 1748 memorial to the special circumstances of this period: "the expenditure on workmen who were brought over from France and whose wages had to be paid both during their voyage and during the long time they remained idle at St Maurice for lack of raw materials or because the Establishment was not yet in operation […]." NAC, MG 1, C¹¹A, vol. 112, fols. 309–10v, memorial, n.s., 1 December 1748.

102. NAC, MG 8, A⁶, vol. 115, fols. 19–21 and 21–8, "Ordonnance et règlement de police pour contenir dans la paix les ouvriers des Forges," 8 February 1739, and "Ordonnance de police aux Forges de Saint-Maurice au sujet de l'obéissance dû par les ouvriers aux commis chargés de la conduite," 12 February 1739.

103. ANQ-Q, NF-25, no. 1178, 1/2, Criminal records of the royal jurisdiction of Trois-Rivières. The death certificate registered on 20 October 1739 specifies that this was Pierre Beaupré, master locksmith, "struck yesterday by a mortal blow," Charbonneau and Légaré 1984, vol. 23, s.v. "L'Immaculée-Conception-de-Trois-Rivières," p. 411.

104. NAC, MG 1, C¹¹A, vol. 72, fols. 239–43, Chaussegros de Léry to Minister Maurepas, October 1739. In this letter, Chaussegros de Léry reveals that the second forge (upper forge) had been in operation since 10 October 1739. Vézin and Simonet would leave for France that fall, one week after the upper forge began operation. According to a weekly listing of work, the forges were shut down "on 21 October, six days after the departure of Mr Olivier, for lack of charcoal." So Vézin and Simonet left on 15 October. If they left one week after the forge started up, then the forge began operating on the 7th rather than the 10th. But the first production date recorded in the listing is 15 October. NAC, MG 1, C¹¹A, vol. 111, fols. 255v–56, memorial from Vézin and Simonet to Monsieur Delaporte Lalane, 10 June 1741; NAC, MG 1, C¹¹A, vol. 110, fols. 157–62.

105. Guyon was already at the Forges in 1743, where he had been engaged as a filler. NAC, MG 1, C¹¹A, vol. 110, fol. 174v, "Estat de ce qui est du par la Régie des forges de Saint Maurice tant aux ouvriers qu'à d'autres particuliers," 1743.

106. ANQ-Q, NF-25, 1406–1407–1419, pp. 1–147, Criminal records of the royal jurisdiction of Trois-Rivières. The death certificate of Pierre Dion [sic], aged 40, would be registered the following day, 20 September 1745, at the parish of L'Immaculée Conception de Trois-Rivières. Charbonneau and Légaré 1984, vol. 23, s.v. "L'Immaculée-Conception-de-Trois-Rivières," p. 415.

107. There were other soldiers among the fillers and ore breakers in the 1741–42 accounts. NAC, MG 1, C¹¹A, vol. 111, fols. 278–305, "Estat général de la dépense […]."

108. "Ordonnance qui porte règlement pour les ouvriers des Forges Saint-Maurice […]," 12 February 1745; Roy 1919, vol. III, p. 68.

109. Étienne Cantenet (also called Campenay or Campéné) the son, is described as a sand moulder at the inquiry, apparently corresponding to the task of a moulder's apprentice. In 1742, he and his father had worked as moulders between April and August, during the campaign; the previous winter, the son had been employed as a helper, doubtless in one of the forges, and his father as a sawyer of firewood and bolter of flour. NAC, MG 1, C¹¹A, vol. 111, fols. 278–305, "Estat général de la dépense […]."

110. Benoît Gauthier, "La criminalité aux Forges du Saint-Maurice," preliminary report, typescript (Quebec City: Parks Canada, 1982) (hereafter cited as Gauthier 1982). In 1796, it was reported that someone had hanged himself at the Forges and that an investigation would be carried out; ANQ-TR, judicial records, 21 January 1796.

111. Gauthier 1982, tables 41 and 45. Based on his declared age in 1871, he could be the son of Jean-Baptiste and Adélaïde Rivard, born on 24 January 1809. He is actually described as a forgeman upon the death of his wife Sophie Lanöete on 11 July 1867. Antoine was the great-grandson of Pierre Michelin and Claire Filet.

112. In 1752, Jacques Philippe Dolfin, a wood-cutter accused of stealing two axes (marked with *fleurs de lys*) belonging to the Forges, would first be sentenced to a beating and thrashing "at the crossroads of the King's Establishment at St Maurice," to be branded with a hot iron on the shoulder before the Grande Maison, and to be banished for three years from the jurisdiction of Trois-Rivières. He appealed this sentence, and was finally sentenced to "be placed in an iron collar" before the casting house with a sign reading as follows: "Thief of axes belonging to the King." The severity of the initial sentence was doubtless intended as an example. It highlights the fact that the King's property was inviolable and also no doubt shows that, at the time, iron objects or tools, which were not very widespread, were very valuable. ANQ-Q, NF-25, 1663, pp. 1–22.

113. NAC, MG 1, C¹¹A, vol. 110, fol. 173v, "Estat des divers ouvriers et autres particuliers […], 1742 à 1743."

114. Gauthier 1982.

115. ANQ-TR, judicial records, complaint by Étienne Durant against Pierre Rivard, 23 December 1839; 11 September 1793.

116. At her marriage to Jacques Simonet, on 17 November 1738, Geneviève Boucher was the widow of Charles Hertel of Chambly. She was the daughter of the late Lambert Boucher, Seigneur De Grandpré, esquire, major, of Trois-Rivières. At the time of this marriage, Jean-Baptiste's grandfather is described as counsellor, secretary to the King. Charbonneau and Légaré 1984, vol. 23.

117. NAC, MG 1, C¹¹A, vol. 77, pp. 337–41, Hocquart to the Minister, 28 June 1742.

118. Charbonneau and Légaré 1984 and 1987, vols. 23 and 36, s.v. "Saint-Louis-des-Forges."

119. The two children were almost one month old when they were baptized; this suggests that these Algonquins were passing through. The children were Louis, son of Joachim Hosetawa and Marie-Madeleine Tegenagis, and François, son of Jean Jeannot and Catherine Polichiche, who were baptized on 7 September 1744 and 8 November 1745 respectively. Charbonneau and Légaré 1984, vol. 23, s.v. "Saint-Louis-des-Forges."

120. Some are named. NAC, MG 1, C¹¹A, vol. 111, "Estat général de la dépense […]," fols. 290v, 292v (Joachim, savage) and 294v (Lolichiche, savage).

121. Laterrière 1873, pp. 86 and 92.

122. In his memoirs, Laterrière relates that the 66-year-old Pélissier had managed to marry by force this young "nymph of 14," who was nevertheless his "intended," in exchange for lending 300 *louis* to the girl's father. To console his young wife, Pélissier was generous enough to employ Laterrière at the Forges as an inspector! Ibid., pp. 70–73.

123. Ibid., pp. 83–115.

124. Living in an atmosphere of insecurity, since there were doubts as to his loyalty after the Americans had passed through, Laterrière hid his pregnant mate at the home of two English brothers, hammermen at the lower forge (no doubt the Slicer brothers), of whom he said that they were "good children, educated, full of good sentiments, discretion and delicacy." Their housekeeper, the widow Montour, would look after the young woman. Back to settle his affairs in the spring of 1778, Pélissier came to claim his wife. But, following a series of fantastic adventures, Laterrière managed to hide her until Pélissier, whom the authorities wanted out of the country, returned to France empty-handed. Ibid., p. 93.

125. Ibid., p. 122.

126. "Juillet 2ᵉ […] payé au R. P. Augustin [Quintal] pour une messe pour la réussite du fourneau […] 0.15.0." NAC, MG 1, C¹¹A, vol. 110, fol. 166v, "Estat des divers achats faits et autres payements pour l'exploitation des forges de Saint Maurice depuis le 27ᵉ d'Aoust 1742 jusqu'au 1ᵉʳ Octobre 1743 […]."

127. The date of this feast day is unknown. Probably the festival commemorating the transfer (or "translation") of the relics of St Éloi.

128. Cited in Fortier 1977, Appendix A, p. 247.

129. Vermette 1982, pp. 108–9.

130. "Nov. 30......payé aux forgerons par gratification pour la Saint Eloy.........12.0.0"; NAC, MG 1, C¹¹A, vol. 110, fol. 166v.

131. In the case of the upper forge, work was interrupted from 23 November 1739 to 14 December because of a breach in the dam. At the lower forge, work was interrupted for two days.

132. Fortier, 1981, p. 147; Vermette 1982, pp. 109–10.

133. In a 1760 inventory, Rouville refers to "a log church with laths inside and outside, whitewashed, 40 ft long by 30 ft wide," and immediately afterward, the church plate. NAC, MG 1, C¹¹A, vol. 112, fols. 340–2v, "Inventaire des fers, fontes, mines […] appartenant aux Forges Saint-Maurice," René-Ovide Hertel de Rouville, 8 September 1760.

134. ANQ-Q, AP-G, 3/3/1, Allsopp Papers, 21 April and 5 May 1772.

135. NAC, RG 4, A 1, vol. S-191, Nos. 17–17a, Mathew Bell to A. W. Cochran, Quebec, 4 December 1827.

136. "Lettre écrite à un prêtre de l'Archevêché par le rvd Père Bourassa, Bytown, 5 Janvier 1849," *Rapport sur les missions du diocèse de Québec et autres qui en ont ci-devant fait partie*, April 1849, no. 8 (Quebec: A. Côté & cie, 1849), pp. 75–76 and 81.

137. Trois-Rivières *Gazette*, 27 March 1847.

138. ANQ-TR, judicial records, deposition, 3 January 1848 (cited in Vermette 1982, pp. 222–23).

139. Cited in Woronoff 1984, p. 167.

140. Woronoff 1976, p. 118.

141. H. Clare Pentland says these payments to the workmen at the St Maurice Forges represented "heavy overhead costs of the Forges, untypical of industrial enterprises of their age," contrary to widespread practice in the early ironmaking world. H. Clare Pentland, *Labour and Capital in Canada, 1650–1860* (Toronto: James Lorimer & Co., 1981), p. 45. See Woronoff 1984, pp. 165–76.

142. NAC, RG 4, A¹, vol. S-191, Nos. 17–7a, Mathew Bell to A. W. Cochran, Quebec, 4 December 1827.

143. Dr. B. J. Harrington, "Notes on the Iron Ores of Canada and Their Development," Geological Survey of Canada, *Report of Progress for 1873–74* (Montreal: Dawson Brothers, 1874), p. 243.

144. NAC, MG 1, C¹¹A, vol. 111, 40th article, "Mémoire des représentations que les sieurs Olivier et Simonet [...]," 10 June 1741.

145. NAC, MG 1, C¹¹A, vol. 112, "Mémoire concernant [...]," fol. 164v.

146. NAC, MG 1, C¹¹A, vol. 111, fols. 117–32; "Le Sieur Cugnet répond ... au septième article" (in response to "sur les comptes du S. Olivier et autres intéressés", by Vézin).

147. NAC, MG 1, C¹¹A, vol. 111, fols. 117–32; "sur les comptes du S. Olivier et autres intéressés," "sur le premier article" by Vézin.

148. NAC, MG 1, C¹¹A, vol. 111, fols. 246–77, "Mémoire des représentations que les sieurs Olivier et Simonet [...]," 10 June 1741, from the 39th to the 43rd article.

149. NAC, MG 1, C¹¹A, vol. 111, fols. 20v–21, memorial from Cugnet, 25 September 1740. This criticism from Trois-Rivières merchants would be taken up again in Bell's time, when it was said that the presence of the Forges did not generate any trade in the region.

150. NAC, MG 1, C¹¹A, vol. 112, "Mémoire concernant [...]," fol. 164v.

151. Ibid., fol. 165v.

152. Peter Kalm, *Travels into North America*, trans. J. R. Forster (Warrington: William Eyres, 1771), vol. 3, p. 89; NAC, MG 4, C², fol. 210e, asst, "Des forges de St Maurice," excerpt from a memorial by engineer Franquet [1752].

153. NAC, MG 1, C¹¹A, vol. 111, fols. 246–77, "Mémoire des représentations que les sieurs Olivier et Simonet [...]," 10 June 1741, 40th article.

154. Haldimand to Amherst, 5 December 1762, cited in Marcel Trudel, "Les Forges Saint-Maurice sous le régime militaire (1760–1764)," *RHAF*, vol. V, no. 2 (September 1951), p. 183.

155. The "penchant for spending" was also widespread among inhabitants of the colony as a whole, as Fernand Ouellet points out, so it was not exclusive to the workmen at the Forges; Fernand Ouellet, *Histoire économique et sociale du Québec, 1760–1850. Structures et conjoncture* (Montreal and Paris: Fides, 1966), p. 561.

156. Laterrière 1873, pp. 84 and 86.

157. Vermette 1982, pp. 182–84.

158. APJQ, Superior Court, docket no. 2238, J. Porter et al. v. Weston Hunt et al., exhibit 4, deposition of Timothy Lamb, 28 January 1860.

159. Caron 1982, p. 48.

160. During the French period, the profit that could be made from the sale of victuals and goods in the Forges store was estimated at 20,000 *livres*. In "Mémoire concernant [...]," this figure includes the proceeds *and* the goods. NAC, MG 1, C¹¹A, vol. 112, "Mémoire concernant [...]," fols. 166–66v (see Chapter 6).

161. NAC, MG 1, C¹¹A, vol. 110, fols. 237–38, "Projet pour parvenir à faire une petite forge [...]," n.s, n.d. The date of 1738 is confirmed by cross-referencing with other documents.

162. Estèbe's trusteeship, which immediately followed the bankruptcy of the Compagnie des Forges, was not very productive. At the time when the establishment was taken over, in the fall of 1741, there was an inadequate supply of pig iron for fining, and of raw materials. The furnace, which was already out, would be put in blast only on the following 25 April. The Forges would produce only 97,000 pounds of iron between October 1741 and May 1742, whereas from May to late October 239,724 pounds would be produced. Estèbe's partial accounts (from October 1741 to 1 August 1742) show that, except in the case of three workmen in the fall of 1741, and of one workman for one week in June 1742, all the forgemen were paid wages whether they were unemployed or working. This statement of account, as well as subsequent memorials, appears to indicate that the hammermen responsible for each forge were paid only wages (75 *livres* a month). NAC, MG 1, C¹¹A, vol.112, fols. 114–26, "Mémoire sur les Forges de Saint-Maurice," article "Exploitation du Sr. Estèbe," n.s, n.d. [probably from 1746]; NAC, MG 1, C¹¹A, vol. 111, "Estat général de la dépense [...]," fols. 278–305.

163. When the finers were evaluated individually, it was specifically said of Lalouette that he was paid wages of 700 *livres* a year like the others, but that he was the finer who "makes the least iron." So the implication was that he received the same wages regardless. NAC, MG 1, C¹¹A, vol. 112, "Mémoire concernant [...]," fols. 148–50.

164. NAC, MG 4, C², fol. 210e, asst, "Des forges de St Maurice," Franquet [1752]; memorial from John Marteilhe on the St Maurice Forges in 1764, NAC, MG 11, Colonial Office 42, vol. 1, Part 1 (hereafter cited as Marteilhe 1764), pp. 159–65. It was Marteilhe who produced this statement, in which costs are expressed in English pounds; the price per thousandweight of iron was 18s 4d, or 22 *livres* (£1 = 24 *livres*). NAC, MG 13, vol. 6, (War Office 34), microfilm B-2640, fol. 134, "Memoire de la Depense qui reste a faire pour fabriquer en fer environ 100000 # de fonte qui nous restent qui pourront produire près de 70000 # de fer," signed Courval, 1761; NAC, MG 13, War Office 34, vol. 6, fol. 134, 1761. On the other hand, Courval's estimate produced in 1764 mentions only monthly wages for the forgemen. NAC, MG 21, B²¹⁻², fols. 139–44, microfilm A 615, Haldimand Papers, memorial from Courval, 20 September 1764 (hereafter cited as Courval 1764).

165. Selkirk 1804, p. 231; Lambert, p. 487.

166. The latter reported that the eight "forgemen," paid 3s per hundredweight, each made about £50 from May to December (8 months). This corresponds to an output of 333.3 cwt per man per month, or 83.3 cwt per week. Selkirk 1804, p. 231.

167. The list of account books from the administration of John Porter & Company (1851–61) includes two books, titled "Upper Forge" and "Lower Forge," which show that iron bars were still being produced. We saw in Chapter 1 that the Forges was closed from 1857 to 1862. See the list of books in Appendix 5.

168. NAC, MG 1, C¹¹A, vol. 112, "Mémoire concernant [...]," fol. 141v.

169. Marteilhe 1764; Courval 1764.

170. The deed of indenture of a founder's apprentice in the late 18th century, revealing that the apprentice would be paid and maintained by the master founder, implies that the latter may have enjoyed a degree of autonomy in his work conditions, and thus was possibly paid by the job; Not. Rec. J.-B. Badeaux, no. 1882, indenture of Modeste Antoine Cécile as apprentice to John Anderson, master founder at the St Maurice Forges, 20 January 1795.

171. ANQ-M, Not. Rec. N. B. Doucet, no. 3059, indenture of Robert Turnbull, 18 October 1811, "moulder" and "fireman" by Mathew Bell for one year "to a cupola furnace" at £78 per annum; Not. Rec. P. B. Dumoulin, indenture of Joseph Wright, "moulder and founder" by Mathew Bell for four years at the Forges and Trois-Rivières, at £160 per annum, 12 August 1825.

172. See note 83.

173. Saint-Pierre 1977, p. 43.

174. NAC, MG 1, C¹¹A, vol. 111, "Estat général de la dépense [...]," fols.291–92.

175. See note 83.

176. ANQ-TR, Not. Rec. J.-B. Badeaux, one-year indenture of Roc Baudry, joiner, 28 April 1787; ANQ-M, Not. Rec. N. B. Doucet, no. 3089, one-year indenture of Louis Pépin, joiner and carpenter, living at St Maurice Forges, 27 November 1811.

177. From André Lepage, "Étude du travail et de la production aux Forges du Saint-Maurice à deux moments de l'histoire de l'entreprise," study carried out for Parks Canada, Historical Research, Quebec Region, typescript, June 1984 (hereafter cited as Lepage 1984), tables 1.4, 1.6 and 3.5. As to the ore, it should, however, be noted that the accounts cover the period from November 1741 to August 1742. But we also know that when Estèbe arrived there were not enough raw materials following the bankruptcy and that the ore collected during the summer of 1742 was not hauled until the next winter.

178. From Lepage 1984, tables 1.4 and 1.7.

179. Ibid., pp. 8 and 31–32, Table 1.7.

180. NAC, MG 1, C¹¹A, vol. 96, fol. 59, Bigot to the Minister, 27 October 1750.

181. See note 24, Chapter 7.

182. Carters during the French period were also paid for other types of work, by the job or by the day.

183. There was also the "Batteau a/c"; APJQ, Superior Court, docket no. 2238, J. Porter et al. v. Weston Hunt et al., exhibit D (List of books kept), deposition of Hamilton Rickaby, 27 January 1860. See Appendix 5.

184. APJQ, Superior Court, docket no. 2238, J. Porter et al. v. Weston Hunt et al., "Cost of keeping horses."

185. APJQ, Superior Court, docket no. 2238, J. Porter et al. v. Weston Hunt et al., exhibit 5, deposition of Hamilton Rickaby, 27 January 1860.

186. APJQ, Superior Court, docket no. 2238, J. Porter et al. v. Weston Hunt et al., exhibit 4, deposition of Timothy Lamb, 28 January 1860.

187. Caron 1982, pp. 21 and 143.

188. From Lepage 1984, pp. 26–28, Table 1.4.

189. The 1760–61 accounts show that carters were paid 6 *livres* a day, including board, and the use of two horses with their fodder, to carry iron bars and stoves from the workshops to the foot of the rapids. Estèbe's accounts include no equivalent transport by cart, but they do mention "two trips with oats from the rapids to the house" at 10 *sols* per trip. NAC, MG 13, C¹¹A, vol. 111, "Estat général de la dépense [...]," fol. 287v; NAC, MG 13, War Office 34, vol. 6, fols. 134–40, 1760–61.

190. APJQ, Superior Court, docket no. 2238, J. Porter et al. v. Weston Hunt et al., exhibit 5, deposition of Hamilton R. Rickaby, 27 January 1860.

191. See the annual supply maintenance schedule, based on clerk Rickaby's account, given in Chapter 2. In 1855, Édouard Tassé, then foreman of the Forges, said: "The season of coaling has always, in my knowledge which extends over nearly 50 years, finished about 8 days after Toussaint. Season of coaling means that the work finishes 8 or 10 days after Toussaint." It is specified, in deeds of indenture for colliers in 1806, that the coaling campaign ran from early May to late October or November. AJTR, Criminal justice records, Stuart & Porter v. Thomas Boucher, 22 November 1855; ANQ-M, Not. Rec. N. B. Doucet, nos. 912, 913, 915 and 916, 10 and 14 May 1806.

192. From Lepage 1984, Table 2.6. Eliminating a few duplications, due mainly to the lack of first names for some individuals, the number of these woodcutters could be brought down to 209.

193. ANQ-TR, judicial records, Rex v. Michel Cyr, deposition of Henry Macaulay, 24 April 1833.

194. "For 16 cords green soft wood for fuel during past winter at 3/0 ...... 2,8,0"; this was firewood, and is the only case recorded in the accounts. ASTR, Forges Papers, a few pages taken from an account book dated July 1856 to April 1858, p. 279, April 1858. A shilling was worth 25¢ at the time; A. B. McCullough, *Money and Exchange in Canada to 1900* (Toronto and Charlottetown: Dundurn Press Limited, 1984), pp. 156–57.

195. But workmen employed sawing "pine logs" at the Forges sawmill were paid $16 a month. ASTR, Forges Papers, p. 164. Deeds of indenture from 1855 and 1856 show that monthly wages of $7–10 were being paid. René Hardy and Normand Séguin, *Forêt et société en Mauricie* (Montreal: Boréal Express, 1984) (hereafter cited as Hardy and Séguin 1984), p. 131, Table 5.

196. JLAPC, 1852–53, vol. II, app. CCC, p. 28, Timothy Lamb, 31 August 1852.

197. Ibid., p. 9, A. Stuart and J. Porter, 23 June 1852. Unlike Lamb, they claimed to cut 12,000 cords of wood a year; the difference could lie in the fact that Lamb included firewood in this figure.

198. Caron 1982, pp. 138 and 143.

199. Geological Survey of Canada, *Report of Progress for 1873–74* (Montreal: Dawson Brothers, 1874), pp. 242–46.

200. Hardy and Séguin 1984, p. 131, Table 5.

201. NAC, MG 1, C¹¹A, vol. 112, "Mémoire concernant [...]" cited in André Bérubé, "L'évolution des techniques sidérurgiques aux Forges du Saint-Maurice, 1: la préparation des matières premières," Manuscript Report No. 305 (Ottawa: Parks Canada, 1978), pp. 28–29.

202. Woronoff 1984, p. 243.

203. The engagement of Joseph Sévigny at 20s a month in 1811, for instance; ANQ-M, Not. Rec. N. B. Doucet, No. 2940, 11 May 1811.

204. Caron 1982, p. 143.

205. Lepage 1984, pp. 63–66, Table 3.1.

206. NAC, MG 13, War Office 34, vol. 6, fols. 134–40, 1760–61.

207. APJQ, Superior Court, docket no. 2238, J. Porter et al. v. Weston Hunt et al., exhibit 5, deposition of Hamilton Rickaby, 27 January 1860.

208. JLAPC, 1852–53, vol. 11, app. CCC, p. 28, Timothy Lamb, 31 August 1852.

209. Memorial from John Marteilhe concerning the St Maurice Forges in 1764, which was based on the accounts for 1756; Marteilhe 1764, pp. 159–65; Courval 1764.

210. In the 1854 inventory, item 18, 6758 ⁶⁸/₇₈₀ *barriques* of ore were counted (so a *barrique* would contain 780 pounds). These 780 pounds perhaps correspond, in fact, to 784 pounds, or 7 cwt (7 × 112), and the thousandweight to 1008 pounds, or 9 cwt (9 × 112). APJQ, Superior Court, docket no. 2238, J. Porter et al. v. Weston Hunt et al., exhibit 18, "Recapitulation of inventory," August 1854; ASTR, N3-H30, "St. Maurice Forges" (excerpt from an account book), 1856–58 (8 pages).

211. Caron 1982, pp. 93, 94 and 139.

212. NAC, MG 1, C¹¹A, vol. 111, fol. 8, memorial from Cugnet, 25 September 1740.

213. Dollard Dubé, *Les Forges il y a 60 ans* (Trois-Rivières: Les éditions du Bien Public, 1933), p. 39 (cited in Caron 1982, p. 143).

214. Eleven inventories were drawn up between 1745 and 1756, and one in 1765; seven concern workmen living at the Forges, and four cover workmen who belonged to the establishment but were no longer living there when they died. Twenty-three inventories were found for the period from 1793 to 1845, but only seven described the interiors of dwellings at that time. Vermette 1982, pp. 55–56, 164 and 252–53.

215. Woronoff 1984, p. 183.

216. Vermette 1982, pp. 70–71.

217. Vermette 1982, pp. 58–59. The 1760 inventory mentions four stoves in the Grande Maison, and 25 stoves for the six houses and 17 workmen's huts; NAC, MG 11, Colonial Office 5, vol. 59, fols. 307–13, Hertel de Rouville, 8 September 1760.

218. Vermette 1982, p. 123.

219. Cited in Vermette 1982, p. 160. Luce Vermette observed, however, that the workmen who moved to Trois-Rivières had houses with several rooms. The better-off workmen apparently did not have interiors at the Forges such as they allowed themselves in town. This could be explained by the fact that they did not own the houses they lived in at the Forges. By taking into account inventories of workmen living in Trois-Rivières or elsewhere, Luce Vermette tends, however, to produce a somewhat distorted picture of the actual physical conditions of the dwellings at the Forges. Ibid., p. 161.

220. Ibid., p. 163.

221. Ibid., pp. 160 and 187.

222. Ibid., pp. 165–69.

223. Ibid., p. 170–71.

224. Ibid., p. 70.

225. NAC, MG 1, C¹¹A, vol. 112, "Mémoire concernant [...]," fol. 143v.

226. Selkirk 1804, p. 231.

227. Woronoff 1984, pp. 185–86.

228. NAC, MG 1, C¹¹A, vol. 110, fols. 254–59, "Mémoire pour la régie des forges de Saint-Maurice en Canada [...]," cited in Vermette 1982, p. 75. See also NAC, MG 1, C¹¹A, vol. 112, "Mémoire concernant [...]," fol. 163.

229. NAC, MG 1, C¹¹A, vol. 111, "Estat général de la dépense [...]," fol. 281.

230. NAC, MG 1, C¹¹A, vol. 13, fol. 367, tally of troops in Canada, 1695. Reference provided by historian Marc Lafrance.

231. The first known was called Le Roy. NAC, MG 1, C¹¹A, vol. 111, "Estat général de la dépense [...]," fol. 281.

232. Vermette 1982, p. 183.

233. Founder Delorme has 26 *minots* of flour and day labourer Boisvert, 9. Vermette 1982, p. 75.

234. Vermette 1982, p. 76.

235. Ibid., p. 187.

236. Ibid., pp. 79 and 188–89.

237. Ibid., pp. 77–78.

238. Ibid., p. 186.

239. Ibid., pp. 78–79, 186 and 188.

240. APJQ, Superior Court, docket no. 2238, J. Porter et al. v. Weston Hunt et al., exhibit 5, deposition of Hamilton Rickaby, 27 January 1860.

241. Vermette 1982, p. 191.

242. APJQ, Superior Court, docket no. 2238, J. Porter et al. v. Weston Hunt et al., exhibit 5, deposition of Hamilton Rickaby, 27 January 1860.

243. Vermette 1982, p. 79.

244. Ibid., pp. 81 and 191.

245. NAC, MG 1, C¹¹A, vol. 112, "Mémoire concernant [...]," fols. 148–51. L'Immaculée-Conception de Trois-Rivières: Michelin, 27 March 1811; Imbleau, 29 May 1771; Delorme, 25 July 1785; Belu, 29 January 1779. Fortier 1977, pp. 193–99.

246. NAC, MG 1, C¹¹A, vol. 111, fol. 64, Vézin to the Minister, Quebec, 12 October 1740.

247. Vermette 1982, p. 107.

248. Ibid., p. 222.

249. Caron 1982, p. 132.

CHAPTER 9

1. NAC, RG 4, A 1, vol. S-191, fols. 17–17a, Mathew Bell to A. W. Cochran, Quebec, 4 December 1827.

2. JLAPC, 1844–45, vol. 4, app. O, "Copy of a Report of the Honourable the Executive Council, on the subject of the Forges of St. Maurice, dated the 15th of September, 1843," signed by Étienne Parent, p. 26.

3. NAC, MG 1, C¹¹A, vol. 110, fols. 304–8, Beauharnois and Hocquart to the Minister, Quebec, 26 October 1735.

4. Réal Boissonnault, "La structure chronologique des Forges du Saint-Maurice des débuts à 1846," typescript (Quebec City: Parks Canada, 1980) (hereafter cited as Boissonnault 1980), p. 193.

5. BM 21661, fol. 204v, Haldimand to Amherst, 22 June 1762, cited in Marcel Trudel, "Les Forges Saint-Maurice sous le régime militaire (1760–1764)," *RHAF*, vol. V, no. 2 (September 1951), p. 172.

6. John Lambert, *Travels through Canada and the United States of North America, in the Years 1806, 1807 & 1808: To Which are Added Biographical Notices and Anecdotes of Some Leading Characters in the United States* (London: printed for Baldwin, Cradock, and Joy; Edinburgh: for W. Blackwood; Dublin: for J. Cumming, 1816) (hereafter cited as Lambert 1808), vol. 1, p. 485.

7. The model is on display in the Grande Maison at the Les Forges du Saint-Maurice National Historic Site.

8. The work was particularly expensive because there was some difficulty enlisting the *habitants* to do it, and in the end skilled workers, too hastily sent down from Quebec, had to be employed. NAC, MG 1, C¹¹A, vol. 110, fols. 385–94, "Mémoire du sieur Olivier de Vézin sur les forges de Saint Maurice," 28 December 1739.

9. In relative terms, considering the limited technology available in the early 18th century, the setting up of the St Maurice Forges, with all the labour, know-how and capital that it involved, has something in common with the process of setting up a hydro-electric power plant today. Hydro-Québec engineers can trace the roots of their work back to the accomplishments of Vézin, Simonet and Léry at the St Maurice Forges.

10. Pierre Drouin and Alain Rainville, "L'organisation spatiale aux Forges du Saint-Maurice: évolution et principes," typescript (Quebec City: Parks Canada, 1980), Microfiche Report No. 6 (Ottawa: Parks Canada, 1983) (hereafter cited as Drouin and Rainville 1980), p. 20.

11. NAC, MG 1, C¹¹A, vol. 110, fols. 370–78v, Beauharnois and Hocquart to the Minister, Quebec, 30 October 1737.

12. On 30 October 1735, Vézin wrote of "a perfect establishment, of which he is sending you the *plan* as well as that of the creek," but it was actually the plan showing the location of the establishment and the mines reproduced in Plate 2.3. Beauharnois and Hocquart's letter to the Minister of 26 October specifies that Vézin had enclosed with his memorial "a plan of the situation of the mines." But the next year, in a letter accompanying a "report on the work done this year" by Vézin, Beauharnois and Hocquart wrote to the Minister that Vézin "has made considerable progress with the work on the establishment, as you will see, Sir, in the detailed report enclosed herewith signed by him *with the plans he has drawn* of it upon which Sieur Simonet can show you what has been done and what remains to be done." These plans have not been found, and they were never mentioned again. NAC, MG 1, C¹¹A, vol. 110, fols. 309–10v, Olivier de Vézin to the Minister of Marine, 30 October 1735; fols. 304–8, Beauharnois and Hocquart to the Minister, Quebec, 26 October 1735; and fols. 339–42v, Beauharnois and Hocquart to Minister Maurepas, 19 October 1736.

13. NAC, MG 1, C¹¹A, vol. 111, fols. 117–32, Sieur Cugnet's reply to "Sur les comptes du S. Olivier et autres intéressés" from Vézin, and fols. 246–77, "Mémoire des représentations que les sieurs Olivier et Simonet [...]," 10 June 1741, 33rd item.

14. Franquet undoubtedly was making a comparison with the more orderly French forges, but Woronoff points out that this lack of symmetry was also quite common in France at the time:

The arrangement of the dwellings does not obey a rule, but responds to the constraints of the site, and is determined by earlier installations. More often than not, an ironworks of the late 18th century is the result of a long maturing process, the gradual construction of industrial buildings among which the dwellings are set up, moved, extended. Where the establishment is built according to an overall architectural plan, as were the new works of Buffon, Vierzon or Tronçais, the organization of the space clearly separates workshops and lodgings. Lodgings (including at early plants) are generally built on a quadrangle, with one of the sides being reserved for the ironmaster's and clerk's houses.

Louis Franquet, *Voyages et mémoires sur le Canada (1752–1753)* (Montreal: Éditions Élysée, 1974), p. 20; Denis Woronoff, *L'industrie sidérurgique en France pendant la Révolution et l'Empire* (Paris: Éditions de l'École des hautes études en sciences sociales, 1984) p. 184 (hereafter cited as Woronoff 1984).

15. ANQ—TR, Not. Rec. Petrus Hubert, no. 4575, 27 April 1863, "Vente des Forges St-Maurice," Onésime Héroux to John McDougall.

16. Pierre Drouin, "Les chemins et bâtiments de service dans l'aire du stationnement aux Forges du Saint-Maurice: fouilles archéologiques, 1981–82," Microfiche Report No. 166, (Ottawa: Parks Canada, 1984), p. 33.

17. Other corduroying was also redone on the site. Ibid, p. 25.

18. "Everywhere else, the rule was that foundrymen and forgemen lived on the worksite, but in separate buildings. A few establishments offered accommodations to all permanent employees: the joiner, the carpeter, even the carters (when they were salaried employees) all benefited, through assimilation with the core group of direct producers." Woronoff 1984, p. 183.

19. Ibid.

20. NAC, MG 1, C¹¹A, vol. 112, fol. 40, "Inventaire des Forges, 1741," signed by Ignace Gamelin, Simonet *fils* and Estèbe, 22 November 1741.

21. Stewardship plan of 1740, cited in Drouin and Rainville 1980, pp. 130–31.

22. Luce Vermette, *Domestic Life at les Forges du Saint-Maurice*, History and Archaeology No. 58 (Ottawa: Parks Canada, 1982) (hereafter cited as Vermette 1982), p. 47 and appendices E and F, pp. 261–62.

23. Four out of seven in 1741, four out of six in 1746 and in 1748, and six out of eight in 1785. Drouin and Rainville 1980, pp. 40 and 194.

24. Drouin and Rainville 1980, p.39; Vermette 1982, pp. 51 and 160–64.

25. NAC, MG 1, C¹¹A, vol. 111, fols. 168–72, "Projet des dépenses à faire pour l'établissement et exploitation des forges de fer en Canada," signed by Hocquart and de Vézin, 17 October 1735.

26. Estimate of 1785 cited in Drouin and Rainville 1980, p. 194, n. 137; AJTR, Not. Rec. Jos Badeaux, "Inventaire du poste des Forges St Maurice," 1 April 1807.

27. Alain Rainville, "La maison 11.2," "Le logement ouvrier aux Forges du Saint-Maurice," Microfiche Report No. 12 (Ottawa: Parks Canada, 1983) (hereafter cited as Rainville 1983), pp. 42–73.

28. Woronoff 1984, p. 183.

29. Jean Bélisle, "Le domaine de l'habitation aux Forges du Saint-Maurice," Manuscript Report No. 307 (Ottawa: Parks Canada, 1976), pp. 53–72.

30. Rainville 1983, "La maison 12.3," pp. 74–103.

31. Drouin and Rainville 1980, p. 59.

32. Marcelle Caron, "Analyse comparative des quatre versions de l'enquête de Dollard Dubé sur les Forges Saint-Maurice," manuscript on file (Quebec City: Parks Canada, 1982) (hereafter cited as Caron 1982), p. 88.

33. Drouin and Rainville 1980, p. 37.

34. AJTR, Not. Rec. Jos Badeaux, "Inventaire du poste des Forges St Maurice," 1 April 1807.

35. Vermette 1982, pp. 48–49.

36. Ibid., p. 49.

37. Ibid, pp. 52–53.

38. NAC, MG 1, C¹¹A, vol. 110, fol. 77v, "Sur les Forges de St-Maurice administrées par le sieur Olivier," signed by Cugnet, 17 October 1741, and vol. 111, fol. 8v, memorial from Cugnet, 25 September 1740.

39. The Batiscan Iron Works also had a "main house." ASTR, N3, B1, "A Sketch of the Forges of Batiscan"; Jean Bélisle, "La Grande Maison des Forges du Saint-Maurice, témoin de l'intégration des fonctions, étude structurale," Manuscript Report No. 272 (Quebec City: Parks Canada, 1977) (hereafter cited as Bélisle 1977), Appendix B, p. 142, and p. 101; France, Ministère de la Culture, *Les forges du Pays de Châteaubriant*, Cahiers de l'Inventaire 3, Inventaire général des monuments et richesses artistiques de la France (Pays de Loire, Département de Loire-Atlantique, 1984), pp. 172–79.

40. NAC, MG 1, C¹¹A, vol. 111, fols. 168–72, "Projet des dépenses à faire pour l'établissement et exploitation des forges de fer en Canada," signed by Hocquart and de Vézin, 17 October 1735.

41. Cugnet himself had a very big house at Quebec, with 5.8 m of foundation and four storeys above ground, roofed in sheets of tin. To contest the cost of 80,000 *livres* proposed by Vézin for the construction of the Grande Maison, he compared his own house at Quebec and the Grande Maison at the Forges, saying that "it is no bigger than mine" and that his had cost no more than 33,000 *livres*. NAC, MG 1, C¹¹A, vol. 111, fols. 8v–9, memorial from Cugnet, 25 September 1740, cited in Bélisle 1977, Appendix B, p. 142.

42. Bélisle 1977, pp. 14–17 and 101–6.

43. Adapted from Table 11 in Vermette 1982, p. 147.

44. Vermette 1982, p. 146.

45. Bélisle 1977, pp. 35–36.

46. Vermette 1982, pp. 152–53.

47. Parks Canada has rebuilt the building, and the modern interior incorporates the remains that can be seen in the exhibition rooms, which were opened to the public in 1990.

48. Drouin and Rainville 1980, p. 42.

49. ASTR, "St Maurice Forges," April 1857, fol. 164.

50. Drouin and Rainville 1980, p. 88.

51. According to Estèbe's accounts of 1741–42, the blacksmiths made a large number of axes. "Étude du travail et de la production aux Forges du Saint-Maurice à deux moments de l'histoire de l'entreprise," study prepared by Historical Research, Quebec Region, Parks Canada, typescript, June 1984, p. 89, Table 3.11.

52. Drouin and Rainville 1980, pp. 43–44.

53. Ibid., p. 94.

54. Ibid., p. 95.

55. Cugnet estimated, in the memorial in which he put the blame on Vézin, that at least a quarter of this wood had been "wasted," that is, it was apparently not really used for the construction of buildings at the Forges; NAC, MG 1, C¹¹A, vol. 110, fol. 84, "Sur les Forges de St-Maurice administrées par le sieur Olivier," signed by Cugnet, 17 October 1741.

56. NAC, MG 11, Colonial Office 5, vol. 59, fols. 307–13, Hertel de Rouville, 8 September 1760; AJTR, Not. Rec. Jos Badeaux, "Inventaire du poste des Forges St Maurice," 1 April 1807.

57. Michel Bédard, "Les moulins à farine et à scie aux Forges du Saint-Maurice," Manuscript Report No. 301 (Ottawa: Parks Canada, 1978) (hereafter cited as Bédard 1978), pp. 39–79.

58. André Bérubé, "Rapport préliminaire sur l'évolution des techniques sidérurgiques aux Forges du Saint-Maurice, 1729–1883," Manuscript Report No. 221 (Ottawa: Parks Canada, 1976) (hereafter cited as Bérubé 1976), p. 29.

59. In 1738, flour was also purchased at Quebec. NAC, MG 1, C¹¹A, vol. 111, fols. 117–32, Sieur Cugnet's reply to "Sur les comptes du S. Olivier et autres intéressés" from Vézin.

60. NAC, MG 1, C¹¹A, vol. 112, fol. 158v, "Mémoire concernant les Forges de St. Maurice," n.d. [1743].

61. Bédard 1978, p. 4.

62. The first known miller was Joseph Comeau, who occupied this position from 1798 to 1800. His successor was Jean-Marie Bouchard, who came from Deschambault. The oldest official record of his existence dates from 1801. He is mentioned again, 30 years later, in the 1829 roll—when he was 64 years old—and in the 1831 census. Registers of the Parish of L'Immaculée Conception de Trois-Rivières. Vermette 1982, p. 183.

63. Michel Bédard, André Bérubé, Claire Mousseau, Marcel Moussette and Pierre Nadon, "Le ruisseau des Forges du Saint-Maurice," Manuscript Report No. 302 (Ottawa: Parks Canada, 1978), pp. 102–109; Claire Mousseau, "Reconnaissance archéologique, automne 1980, Forges du Saint-Maurice. Dossier 1: Opérations 25G14 et 25G15, fouilles des aménagements hydrauliques. Dossier 2: Opérations 25G15, fondations d'un bâtiment localisé à l'est du haut fourneau," typescript (Quebec City: Parks Canada, 1980). It is our deduction that there was an undershot wheel.

64. Drouin and Rainville 1980, p. 99; Bédard 1978, pp. 30 and 47–54.

65. Drouin and Rainville 1980, p. 43.

66. Ibid., p. 102.

67. Ibid., pp. 42–43 and 89–90.

68. NAC, MG 1, C¹¹A, vol. 111, fol. 81, "Mémoire des articles arrêtés entre nous soussignés [...] Cugnet [...] et [...] Simonet," 18 March 1740, and vol. 112, fols. 308–20, "Mémoire [...] dans les forges de Rancogne," Le Mercier, 2 April 1750; Pierre de Sales Laterrière, *Mémoires de Pierre de Sales Laterrière et de ses traverses* (Quebec City: Imprimerie de l'Événement, 1873) (hereafter cited as Laterrière 1873), pp. 84–85.

69. Bérubé 1976, p. 28.

70. Drouin and Rainville 1980, pp. 46–47 and 99–102.

71. Vermette 1982, p. 43.

72. Ibid., pp. 151–52; Drouin and Rainville 1980, pp. 47 and 102.

73. Vermette 1982, p. 86.

74. Vermette also notes that water jugs were found in the inventoried houses. Ibid., p. 191; Caron 1982, pp. 14 and 34.

75. Benjamin Sulte, *"Les Forges Saint-Maurice,"* in *Mélanges historiques: études éparses et inédites* (Montreal: G. Ducharme, 1920), vol. 6 (hereafter cited as Sulte 1920), p. 160.

76. Drouin and Rainville 1980, pp. 46 and 98–99.

77. Vermette 1982, pp. 42–43.

78. Sulte, who is Caron's source, used the same description in 1920. Napoléon Caron, *Deux voyages sur le Saint-Maurice* (Trois-Rivières: Librairie du Sacré-Coeur, 1889), p. 257; Sulte 1920, p. 93, n. 3.

79. Death certificates of Joseph Aubry, son of Joseph and of Marie Josephe Chèvrefils, 5 May 1744, aged 2, and of Joseph Aubry (his brother), on 17 May 1745, aged 2 years and 6 days. Hubert Charbonneau and Jacques Légaré (eds.), *Répertoire des actes de baptême, mariage, sépulture et des recensements du Québec ancien*, Département de démographie, Université de Montréal (Montreal: Les Presses de l'Université de Montréal, 1984), vol. 23, *Gouvernement de Trois-Rivières, 1730–49*, s.v. "Saint-Louis-des-Forges-de-Saint-Maurice" (hereafter cited as Charbonneau and Légaré 1984).

80. Death certificate of François Lemerle, son of Louis, carter, and of Marie-Anne Lagrave, 16 November 1747, aged 3; death certificate of Marie Ursule Dupuis, daughter of Antoine and Ursule Alary, 28 May 1748, aged 4 months. Charbonneau and Légaré 1984.

81. Between 1921 and 1932, long after the Forges had shut down, a cemetery was built for the mission of Saint-Michel-des-Forges, at the top of the big hill leading to the Forges. It was later moved westward to the current church. Michel Bédard, "Utilisation et commémoration du site des Forges du Saint-Maurice, 1883–1963," Manuscript Report No. 357 (Ottawa: Parks Canada, 1979), pp. 169–75.

82. NAC, RG 48, B 30, vol. 113, no. 79 (1838), cited in Vermette 1982, p. 225.

83. Vermette 1982, p. 226.

84. It was under the Union government (1841–67) that "the key components of the school system of Lower Canada were put in place, especially by the bills of 1845, 1846 and 1856." Louis-Philippe Audet and Armand Gauthier, *Le système scolaire du Québec* (Montreal: Librairie Beauchemin, 1969), p. 29.

85. The first school board at the Forges had five members, appointed in the 1860s. Vermette 1982, p. 226.

86. Census of 1871, families 227 to 278: 64 adults over age 20 are "not able to read" and 74 are "not able to write." Canada, Census of 1871, Subdistrict of St. Étienne, Division No. 1.

87. Caron 1982, pp. 32 and 128.

88. Dubé gives the names of eight school-mistresses between 1860 and 1890 but does not mention the names of those who were censused at the Forges in 1871 and 1881: Étudienne Blais, wife of Xavier, and Marie Mainville. Caron 1982, pp. 32 and 128; Canada, Census of 1871 (see note 86) and Census of 1881, Subdistrict of St. Étienne, Division No. 2.

89. Two other disadvantages were the lack of water in the creek and the difficulty in getting to the mines opened by Francheville on the other bank of the river, in the seigneury of Cap de la Madeleine. NAC, MG 1, C¹¹A, vol. 110, fols. 231-36, "Mémoire instructif des observations à faire par le sieur Olivier dans la visite des mines de fer de Saint-Maurice et lieux en dépendant," n.s., 1 September 1735.

90. NAC, MG 1, C¹¹A, vol. 63, fol. 136, "Observations faites par moy [...]," Vézin, 17 October 1735.

91. Vermette 1982, p. 73.

92. Marcel Moussette, "L'histoire écologique des Forges du Saint-Maurice," Manuscript Report No. 333 (Ottawa: Parks Canada, 1978) (hereafter cited as Moussette 1978), p. 90.

93. Ibid., pp. 91–92.

94. Moussette, referring to Napoléon Caron (1889), says that the Hooper farm was located on the seigneury of Cap de la Madeleine, but Hooper and the farm were censused at the "Forges St. Maurice" in 1825 and on the "Fief St. Maurice" in 1831. Moussette 1978, p. 92; Canada, Censuses of 1825 and 1831.

95. A marginal note, which is hard to make out, indicates instead 205[?] bushels for that year. Canada, Census of 1831, Fief St Maurice.

96. Michel Bédard, "La privatisation des Forges du Saint-Maurice 1846–1883: adaptation, spécialisation et fermeture," manuscript on file (Quebec City: Parks Canada, 1986), p. 123.

97. Bédard 1979, "La maison du docteur Beauchemin," p. 92.

98. René Hardy and Benoît Gauthier, "La sidérurgie en Mauricie au 19ᵉ siècle: les villages industriels et leurs populations," research report submitted to the Regional Branch of the Quebec Department of Cultural Affairs, Centre de recherches en études québécoises (Trois-Rivières: Université du Québec à Trois-Rivières, May 1989) (hereafter cited as Hardy and Gauthier 1989), p. 166.

99. Vermette 1982, p. 185; Hardy and Gauthier 1989, p. 164.

100. MER, Service de la concession des terres, St Maurice Township, general file no. 25203/1936, letter of H. R. Symmes to P. M. Vankoughnah, 30 September 1861.

101. Caron 1982, p. 124.

102. Moussette 1978, p. 94.

103. Laterrière 1873, pp. 84–85.

104. Lambert 1808, p. 485.

105. Moussette 1978, p. 95.

106. Ibid., p. 96, citing M. R. Watson, E. Parker and J. J. Stewart, *Les forges du St-Maurice, Trois-Rivières, Quebec, Landscape Feasibility Study* (Ottawa: Environmental Services Division, Engineering and Architecture Branch, Indian and Northern Affairs, 1977), pp. 7–8.

107. Ibid., pp. 95–96.

108. Alayn Larouche, "Analyse des macrorestes végétaux aux Forges du Saint-Maurice: les jardins potagers," typescript, Université de Montréal, Département de Géographie, January 1979, cited in Moussette 1978, pp. 96–97.

109. Vermette 1982, p. 76.

110. Moussette 1978, p. 92.

111. Quoted in Vermette 1982, p. 76 [new translation].

112. Based on Table 4 (modified) in Moussette 1978, p. 98.

113. APJQ, Superior Court, docket no. 2238, J. Porter et al. v. Weston Hunt et al., exhibit 5, deposition of Hamilton Rickaby, 27 January 1860.

114. Canada, Census of 1871, Subdistrict of St. Étienne, Division No. 1, "Beauchemin," p. 67, family no. 266.

115. Walter Henry (1843), quoted in Vermette 1982, p. 158.

116. JLAPC, 1844–45, vol. 4, app. O, p. 26, "Copy of a Report of the Honourable the Executive Council, on the subject of the Forges of St. Maurice, dated the 15th of September 1843," signed by Étienne Parent. Quoted in Vermette 1982, p. 158.

CONCLUSION

1. Mgr. Albert Tessier, *Les Forges Saint-Maurice*, (Montreal and Quebec City: Les Éditions du Boréal Express, 1974 [1952]), p. 121.

2. Alice Jean E. Lunn, *Développement économique de la Nouvelle-France, 1713–1769* (Montreal: Les Presses de l'Université de Montréal, 1986), p. 287.

Appendix 1   ATTEMPTS TO MINE IRON ORE IN CANADA, 1541–1729

Year	Persons involved	Circumstances	Areas explored	Action taken	Outcome
1541 (third voyage)	Jacques Cartier	Voyage of exploration	Canada; near Quebec: "a fine mine of the best iron ore [...] ready for the furnace"	Four ironworkers were brought along to prospect for iron deposits.	None
1604	Samuel de Champlain	Voyage of exploration	Acadia	Simon Le Maistre, a miner, was brought along.	Iron deposits discovered in Acadia
1617–18	Samuel de Champlain	Resource inventory	New France	Revenues of 1 million *livres* from the iron deposits were projected.	None
1663	Sieur Gaudais	Granted a royal commission to explore for minerals	New France	This special commissioner studied the possibility of operating an iron mine.	None
1664	Compagnie des Indes occidentales	Mining privilege granted by Louis XIV	New France	The company was granted the right to manufacture arms, cannons and cannonballs.	None
1666	Jacques de Cailhault, Sieur de la Tesserie, founder of the Compagnie des Indes occidentales	Mining prospecting expedition commissioned by Intendant Jean Talon	Baie St Paul	Visited a mine 4 leagues north of Baie St Paul, with a miner.	Report submitted, but no further action taken. Map produced showing the area devastated by the 1663 earthquake.
1670	Sieur de La Potardière, ironmaster	Mining prospecting expedition commissioned by Intendant Jean Talon	Champlain and Cap de la Madeleine seigneuries	Mining exploration and gathering of iron ore (1500 *pipes*) and black sand samples.	20 *barriques* of ore and black sand taken back to France by Sieur de La Potardière. No further action taken.
1672-73	Governor Louis de Buade, Comte de Frontenac and miners	Mining prospecting expedition	Champlain and Cap de la Madeleine seigneuries	Discovery of iron mines	Proposal to mine iron deposits along the Pépin River; no further action taken.
1677	Jean-Baptiste de Lagny, Sieur des Brigandières	Letters patent to work mines in Canada for 20 years	Canada	None	None
1679	Governor Louis de Buade, Comte de Frontenac and miners	Prospective ironworks	Canada	None	None
1682–83	Governor La Barre and Intendant De Meulles	Prospective mining	Trois–Rivières region	None	None, due to a 1684 directive from Minister Jean–Baptiste Colbert, Marquis de Seignelay, which put a stop to exploration.
1685–86	Governor Denonville	Mining exploration	Trois–Rivières region	Ore samples collected and sent to France for analysis	None
1687–88	Pierre Hameau, ironmaster from Brittany	Proposal to operate an ironworks by Pouriat, Boula and Hameau, ironmasters from Brittany	Acadia and Canada (Trois–Rivières, Baie St Paul and around Quebec)	Expedition led by Hameau	Two reports were submitted, one containing a proposal to set up an ironworks on the Etchemin River, on the south shore of the St Lawrence near Quebec; investment required estimated at over 200,000 *livres*; no further action taken.
1705–06	Crisafy, Governor of Trois-Rivières	Proposal to mine iron ore in the Trois-Rivières region	Trois-Rivières region	Proposal submitted to the Minister of Marine	Rejected by Minister Pontchartrain because France was at war at the time.
1708	Crisafy, Governor of Trois-Rivières	Proposal to mine iron ore in the Trois-Rivières region	Trois-Rivières region	Ore samples sent to Rochefort for analysis; attempt was made to get Ironmaster Hameau involved again in the project	None
1714–17	Intendant Bégon and Governor Rigaud de Vaudreuil	Proposal to establish an ironworks (to produce iron for shipbuilding)	Trois-Rivières and Baie St Paul	Prospecting activities by Sieur Joseph Godefroy de Tonnancour in 1715; samples of iron ore and smelted iron sent to France; request to have a mining expert sent over	The regent, the Duc d' Orléans, refused, saying "there is enough [iron] in France to supply all of Canada."
1724	M. de Ressous	Plan to operate ironworks in Canada	Canada	Made proposal to Minister; two experts sent to Canada	Although Minister Maurepas was in favour of project, no further action taken.
1727	Intendant Dupuy	Plan to work iron mines in the Trois-Rivières region	Trois-Rivières region	Intendant submitted proposal to Minister	Project viewed favourably by Minister Maurepas, who requested a memorial on the subject; no immediate follow-up.*
1729	François Poulin de Francheville	Plan to work iron mines in Trois-Rivières	Trois-Rivières	Proposal to invest in an operation with exclusive rights for 20 years	Exclusive rights for 20 years granted to Francheville

Sources: Fauteux 1927, chap. 1; E. Nish 1975, chap. III; Boissonnault 1980, pp.11–24; Vallières 1989, chap. 2.

* "[...] in 1727, the Minister [Maurepas] told Intendant Dupuy to wait until the King had the funds to open mines and set up an ironworks before envisaging building some of the king's ships in Canada." Boissonnault 1980, p. 20.

Appendix 2

THE FORGES LANDS

Date and tenure	Land grants or purchases				Additional land held in usufruct				Total area of usable land (ha)
	Tract	Dimensions breadth/ depth (leagues¹)	Area (ha)	Total area (ha)	Tract	Dimensions breadth/ depth (leagues)	Area (ha)	Total area (ha)	
1730 Francheville	St Maurice fief	1/2	4,822	4,822	Yamachiche fief ²	1.5/2	7,232	118,129	**122,951**
					Gatineau fief	0.75/1	1,808		
					Pointe du Lac fief	1.25/2	6,027		
					Ste Marguerite fief	0.75/1	1,808		
					St Étienne fief	1/2	4,822		
					Cap de la Madeleine fief	2/20	96,432		
1737 Cugnet et Compagnie	St Maurice fief	1/2	4,822	19,287	Yamachiche fief	1.5/2	7,232	113,307	**132,594**
	St Étienne fief and adjoining land³	3/2	14,465		Gatineau fief	0.75/1	1,808		
					Pointe du Lac fief	1.25/2	6,027		
					Ste Marguerite fief	0.75/1	1,808		
					Cap de la Madeleine fief	2/20	96,432		
1767 Pélissier syndicate	St Maurice fief	1/2	4,822	20,435	⁴				**20,435**
	St Étienne fief and adjoining land	3/2	14,465						
	Portion of Cap de la Madeleine	2/20 arpents	1,148						
1783 Conrad Gugy	St Maurice fief	1/2	4,822	19,287					**19,287**
	St Étienne fief and adjoining land	3/2	14,465						
1799 Davison, Monro and Bell	St Maurice fief	1/2	4,822	20,435	Crown land northeast of the Gatineau fief⁵		2,023	2,023	**22,458**
	St Étienne fief and adjoining land	3/2	14,465						
	Portion of Cap de la Madeleine	2/20 arpents	1,148						
1810 Monro and Bell	St Maurice fief ⁶	1/1.5	3,254	16,960	Crown land northeast of the Gatineau fief		2,023	16 009	**32,969**
	St Étienne fief and adjoining land	3/2	12,558		Western Tract⁷	3/2	13,986		
	Portion of Cap de la Madeleine	2/20 arpents	1,148						
1819 Mathew Bell	St Maurice fief	1/1.5	3,254	16,960	Crown land northeast of the Gatineau fief		2,023	31 741	**48,701**
	St Étienne fief and adjoining land	3/2	12,558		Western Tract	3/2	13,986		
	Portion of Cap de la Madeleine	2/20 arpents	1,148		Most of Cap de la Madeleine⁸		15,732		
1834 Mathew Bell	St Maurice fief	1/1.5	3,254	15,812	Crown land northeast of the Gatineau fief		2,023	24 880	**40,692**
	St Étienne fief and adjoining land	3/2	12,558		Western Tract	3/2	13,986		
					Most of Cap de la Madeleine		8,871		
1846 Henry Stuart	St Maurice fief		3,363	14,818	⁹				**14,818**
	St Étienne fief and adjoining land		11,455						
1851 John Porter & Company	Portion of St Maurice fief		397	9,530	¹⁰				**9,530**
	St Étienne fief and adjoining land		9,133						
1867 J. McDougall & Sons	Part of St Maurice Township (including the former fiefs of St Maurice and St Étienne) Part of the parishes of Mont Carmel and St Maurice		2,490	2,490					**2,490**
1876 - G. & A. McDougall	Same as above		4,251	4,251					**4,251**
1880 G. McDougall	Same as above		4,162	4,162					**4,162**
1883 G. McDougall	Same as above		3,823	3,823					**3,823**

1. The league used here is equivalent to 4.91 km. According to Trudel (1974), the league (*lieue*) used in New France equalled 84 arpents, an arpent being equivalent to 180 *pieds* (French feet). The 84-arpent league is equivalent to 4.91 km and not 4.99 km as shown in Trudel's table of measures.

2. The royal warrant issued to Francheville in 1730 stated that he had the right to exploit the lands "from and including the seigneury of Yamachiche as far as and including the seigneury of Madeleine du Cap." These lands thus include six seigneuries in addition to the St Maurice seigneury. NAC, MG1, F³, vol. 11-2, fols. 429-37, "Brevet qui permet au Sr Poulin de Francheville d'ouvrir, de fouiller et d'exploiter pendant vingt ans des mines de fer en Canada," 25 March 1730.

3. In 1737, the fief of St Étienne was reunited to the King's domain and then annexed to the fief of St Maurice along with the land above the fief of St Étienne, effectively adding 2 leagues in breadth to the original St Étienne fief. Although the two fiefs had been combined into a single one, the fief of St Maurice, they continued to be designated separately in subsequent boundary issues. Bouchette, describing the St Étienne fief in 1815, describes it as being 3 leagues in depth and 2 leagues in front. According to the survey by Bureau in 1845, however, it was 3 leagues wide along the St Maurice River and 2 leagues in depth. Greer 1975, p. 7; Bouchette 1815, p. 312.

4. From 1767 on, it appears that the huge area of land made up of the seigneuries bordering the St Maurice fief (augmented by the St Étienne fief) was no longer available for use as it had been under the French regime. NAC, RG4, A1, vol. 18, fols. 6425-27; Boissonnault 1980, p. 170.

5. According to the 1829 Felton Report, this Crown land consisted of approximately 10,000 acres. In a letter dated 31 December 1832, Mathew Bell wrote of his proposal to "relinquish all the lands on the south-west line, lying between the River Machiche and the Seigniory of Gatineau. This tract His Excellency, Sir James Kempt, accordingly caused to be surveyed and laid out in small lots, comprising altogether upwards of 11,000 acres." These lots were divided into four ranges, only three of which were put up for sale, according to Bell. He subsequently stated that he himself had purchased 4,500 acres. A decade later, in 1843, referring to the same land, Bell wrote that he had "purchased nearly the whole" of the 5,000 arpents (4,230 acres) he had agreed to cede. On 15 May 1830, the provincial secretary informed the Chief Justice that this land had been divided into four ranges, the fourth range and a part of the third included in the Forges lands. According to the 1830 data, prior to Bell's cession of the land, he thus used only around half of the 10,000 or 11,000 acres of this Crown land, which he bought back when it was put up for sale in 1831. This is why we have set the area of the Forges lands at 5,000 acres approximately (2,023 ha). Bell would sell this land in the 1840s. Bell, like Pélissier in 1767, had been granted by the Jesuits 1,148 ha at Cap de la Madeleine. JHALC 1836, App. X.X., William B Felton, 30 July 1829; NAC, RG 4, C 2, vol. 8, pp. 36-38, 15 May 1830, "Provincial secretary to the Chief Justice"; JLAPC, 1844-45, vol. 4, App. O, letters of M. Bell, 31 December 1832, and 25 February 1843; Greer 1975, p. 24.

6. The 1806 survey showed that a portion of the fief of St Maurice encroached on the seigneury of Pointe du Lac; therefore, the St Maurice fief was reduced to 1 ¹/₂ leagues in depth, instead of 2 leagues, in the 1810 lease. Here, we have given the exact area of the St Maurice and St Étienne fiefs and of the Western Tract, as measured in the 1806 survey. However, we did not take account of the islands in the St Maurice River that were incorporated into the Forges lands in 1806 and 1810. NAC, RG 1 L3L, vol. 144, p. 70766-69, "Return of Survey of the Seigniory of St. Maurice and the adjoining tracts thereto," Joseph Bouchette, 23 September 1806; Bouchette 1815, p. 311.

7. Bouchette writes in 1815: "on the north-west of St. Étienne is another tract of the same dimensions [2 leagues by 3 leagues], that has lately [the 1810 lease] been annexed to the above grants, as part of the lands belonging to the forges." In 1819, Bell probably obtained from the Duke of Richmond authorization to cut wood and mine ore on the ungranted lands in the Cap de la Madeleine seigneury, but we have not been able to find any supporting documentation for this hypothesis. In 1831, he officially leased a part of this land, the boundaries of which were carefully documented this time.

8. In 1831, after 12 years of Bell's tenure, the government officially set the land available to Bell in the seigneury of Cap de la Madeleine at some 46,000 arpents (15,732 ha). This is therefore taken here as the minimum core area. In April 1834, the government reduced this to 25,940 arpents (8,871.48 ha), which was henceforth leased to Bell for £75 a year. Greer 1975, pp. 25 and 29, Boissonnault 1980, pp. 236-38.

9. After1846, when the Forges were sold to private interests, the government no longer granted land to the Forges proprietors, who were forced to buy the land from which they obtained the resources they needed. The boundaries of the St Maurice and St Étienne fiefs would be redrawn again after the 1845 survey done by Bureau. In 1847, Henry Stuart sold almost all (2,966 ha) of the fief of St Maurice to Pierre-Benjamin Moulin.

10.In 1852, John Porter & Company had already sold 3,290 ha of the St Étienne fief.

PROBLEMS FACED BY CUGNET ET COMPAGNIE
IN BUILDING AND BLOWING IN THE FORGES, 1736–41.

Date	Time of year	Location	Problem	Reference
1736	Summer	Quarries	Because the St Maurice River was swollen from heavy rains, the quarry at La Gabelle could not be worked. Dressed sandstone and limestone were brought in from Ange Gardien and Beauport at great expense but in the end were not used.	NAC, MG1, C¹¹A, vol. 111, fol. 247v, 10 June 1741, Vézin and Simonet
	Summer	Blast furnace	The construction of a retaining wall caused the foundation to crack.	NAC, MG1, C¹¹A, vol. 110, fol. 71v, Cugnet, 17 Oct. 1741
1737	Summer	Lower forge	The foundation was raised by 2 feet because of spring flooding, and the diameter of the waterwheels had to be reduced from 10 to 8 feet, which meant less power.	NAC, MG 1, C¹¹A, vol. 110, fols. 385–94, Vézin to Maurepas, 28 December 1739.
	Nov.-Dec.	Blast furnace	Three attempts to blow in the furnace failed; bellows mechanisms had to be replaced.	Ibid., fols. 387v–89.
1738	1 April	Charcoal house at blast furnace	Collapsed from the weight of the snow.	Ibid., fol. 389.
	May-August	Blast furnace	Bellows mechanisms had to be replaced again; another six attempts were made to blow in the furnace but all failed.	Ibid., fols. 387v–89.
	August	Blast furnace	Base of hearth raised and bellows mechanisms adjusted accordingly.	Ibid., fols. 387v–89.
	Spring	Lower forge	Pine waterwheels replaced by oak ones; blowing in delayed until late May.	Ibid., fols. 387v–89.
	Summer	Lower forge	Frequent shutdowns; not enough stream flow for both chaferies, leaving a team of forgemen idle.	Ibid., fol. 391.
1739		Lower forge	Bellows rebuilt.	Ibid., fol. 391v.
	Summer-fall	Upper forge (and milldam)	Additional forge built although not included in original plans for project.	Ibid., fols. 391v–92.
1739-1740	Fall-summer		Vézin and Simonet in France.	NAC, MG 1, C¹¹A, vol. 111, fols. 255v–56
1739	Nov.	Upper forge	Operations shut down for a month due to failure of milldam.	NAC, MG 1,C¹¹A, vol. 111, fol.15, memorial from Cugnet, 25 September 1740.
	Nov.	Blast furnace	Furnace out of blast until following spring due to charcoal shortage.	Ibid., fol.15.
1740		Lower forge	Extensive repairs: forge roof timbers, both chaferies, hurst frame and hammer wheel.	NAC, MG 1, C¹¹A, vol. 110, fols. 71-72, Cugnet, 17 Oct. 1741; vol. 111, fols. 272v–73, 10 June 1741, Vézin and Simonet
1741	Summer	Blast furnace	Not blown in until 4 July (two months late) because of shortage of charcoal, due to the fact that the two forges were operated during the previous winter. Furnace out of blast for the entire month of August.	NAC, MG 1, C¹¹A, vol. 110, fols. 48–48v, Cugnet, 4 October 1741.
	October		Resignation of company's partners.	

Appendix 4

INVENTORY OF BUILDINGS AT THE ST MAURICE FORGES, 1807[a]
(in the same order as the original document)

Building	Occupant(s)	Materials	Dimensions		Condition	Origins (Tenure)
			Length	Width		
House	master carpenter		30'	20'	very poor, needs to be rebuilt	pre-1760
Houses (5) (all attached)	carters		80'	23'	very poor, need to be rebuilt	
Blast furnace: masonry and chimney; wheelrace 225'; Bellows shed, Waterwheel shed,					needs repairs half needs renovations	Monro & Bell
Casting house 100' x 30'	boards				good	
House	carters		50'	30'	poor, needs to be rebuilt	
Small house		stone	15'	15'	very poor[b]	
Stable			20'	20'	poor, not used	
House	forgemen (lower forge)		30'	40'	falling down, needs to be rebuilt	
Milldam and sluice of lower forge		stone	40'	H: 25'	very poor, needs to be rebuilt	
Milldam, sluice and forebay of upper forge					very poor, needs to be rebuilt	
House	2 blacksmith's shops		40'	15'	very poor	
Grande Maison		limestone	80'	40'	poor	
Wing			24'	22'	poor	
Shop	master carpenter	pièce sur pièce	30' H: 11' (to sandbank)	30'	very good	Monro & Bell
Attached houses (2)	2 moulders		40'	20'	good	Monro & Bell
Attached houses (2)	carpenter		40'	20'	fair, need repairs	Davison & Lees
House	foreman		30'	20'	good	Monro & Bell
Charcoal house		planches debout	60'	30'	good (10 years)	Monro & Bell
Grist mill		pair of millstones			very good	Monro & Bell recently
House	quarryman		25'	12'	fair	Monro & Bell
Moulding shop	near blast furnace		40'	30'	poor, needs renovation	
House	master moulder		30'	20'	needs repairs	Davison & Lees
Front of blast furnace	(casting house)		40'	20'	good	Monro & Bell
Shed	near furnace	boards	100'			Monro & Bell
Blacksmith's and joiner's shop	smith, joiner		35'	15'	needs repairs	Davison & Lees
Pattern shed		boards	50'	30'	good	Monro & Bell
Old shed	(stonecutting)		30'	25'		Davison & Lees
Building with several apartments	founder, keepers and moulders		100'	20'	poor, needs repairs	Davison & Lees
House	filler		20'	15'	fairly good	Monro & Bell
Charcoal shed			70'	30'	good	Monro & Bell
Shed			60'	30'	poor	Monro & Bell
House	carters and labourers		50'	20'	good	Monro & Bell
House	moulder		20'	20'		Davison & Lees
Barn		boards	50'	30'	very poor	Davison & Lees
Outbuilding		pièce sur pièce	60'	30'	poor, needs immediate repairs	Monro & Bell (1795)
Stable			50'	20'	poor, to be rebuilt	Davison & Lees
Stable			20'	20'	good	Monro & Bell
Granary		pièce sur pièce	40'	30'	good	Monro & Bell
Stable			35'	15'	poor, needs immediate repairs	Monro & Bell
Bakery			30'	20'	needs immediate repairs	Davison & Lees
Lime kiln						Monro & Bell
Shed (lower forge)			50'	25'	good	Monro & Bell
Lower forge 2 bellows and waterwheels			80'	30'	good, needs repairs	Monro & Bell
Iron store	(near lower forge)	boards	20'	20'	good	Monro & Bell
Sawmill (2 saws)		roof only	40'	20'	machinery needs repairs	Monro & Bell
Shed (upper forge)	(attached to forge)		35'	20'	poor	Monro & Bell
Upper forge: masonry (covered and repaired); bellows, waterwheels and mechanisms					very old need repairs	
(14) Small houses	labourers and workers					built by workers living in them

a. ANQ-TR. Not. Rec. Jos Badeaux, 1 April 1807, "Inventaire du poste des Forges du Saint-Maurice".

b. Benjamin Sulte, writing in 1860, quoting André Robichon (born in 1793) that around 1800, there was a 20 ft. by 20 ft. sacristy with thick stone walls that was attached to an old wooden church then being used to house carts; Sulte 1920, pp. 180–81.

Note: dimensions given in English measure; 1 foot = 30.5 cm

Appendix 5

ACCOUNT BOOKS KEPT AT THE ST MAURICE FORGES
UNDER JOHN PORTER & COMPANY, 1851–61[a]

Lime

Men's Ledger

Check

Moulders Work

Job Work

Lower forge

Upper forge

Blast Furnace

Goods received

Goods sent away

Hay, Straw & Oats bought

Hay, Straw & Oats sold

Cash

Bill

Wares & iron sold

Orders for wares

Orders for goods wanted

Stock of wares book

Bakery a/c

Store Journal

Tools lent

Hard wood cut

Hard wood feuilled

Hard wood dressed

Hard wood charred

Soft wood cut

Soft wood feuilled

Soft wood dressed

Soft wood charred

Logs brought home

Logers Book

(?) Book

men's a/c & check

do """""""& Cash

(?) recd at Forges

Carters a/c book

Journal JB[b]

Ledger JB

Calculation Book JB

Lavoir a/c

Farm a/c

Batteau a/c

Carters a/c (Little)

Work done at shops

Horse Book

Garde feu Book

Lime Stone a/c

Grai stone a/c

a. ANQ-Q, Superior Court, docket no. 2238, J. Porter et al. v. Weston Hunt, exhibit D "List of books kept," deposition of Hamilton Rickaby, 27 January 1860. A total of 48 different account books are listed.

b. JB: Jeffrey Brock, the Forges manager representing Weston Hunt & Company, partners of John Porter & Company.

EXCERPTS FROM THE NOTES AND CALCULATIONS ON HYDRAULIC MECHANISMS *
AT THE ST MAURICE FORGES ATTRIBUTED TO ENGINEER CHAUSSEGROS DE LÉRY, 1738–39

(Excerpt 1)

Estat des bois de chesne nécessaires pour les harnois d'une forge./.bois de chesne

Chantiers
3 pièces de bois droit de 8 pieds de longueur
chacune, de 18 pouces de Large et de 15 pouces
d'Epaisseur

Solles
3 id. de 12 pieds de Longueur sur 24 pouces de
Largeur et 20 pouces d'Epaisseur

Court Carreau
1 idem de 8 pieds de longueur sur 30 pouces
de largeur et autant d'épaisseur

grand Carreau
1 idem de 8 a 9 pieds de Longueur, sur 30 pouces
en quarré

jambes du Marteau
2 autres pieces de 11 a 12 pieds de Longueur,
16 pouces en quarré dont une doit couder
par le pied suivant le gabarit et le dessin cy joint

arbre de Marteau
1 piece ronde de 25 pieds de Longueur, sur
30 a 36 pouces de diamètre au gros bout et 20 pouces
au petit bout - a peler l'Ecorce seulement.

Stocque
1 piece ronde de 7 a 8 pieds de Longueur sur
36 pouces de diamètre

Drosme
1 piece de 40 pieds de Longueur sur 3 pieds
de Largeur et 2 pieds 1/2 d'Epaisseur -/ et au défaut de cette piece
si Elle ne peut pas se trouver /
substituer 2 pieces de 40 pieds de Longueur
sur 20 pouces de Largeur et de 20 pouces d'Epaisseur Si Elles avaient 2
pouces de moins de Largeur
autant de moins d'Epaisseur, Elles serviraient,
mais si on peut les trouver des premières proportions cela est a souhaiter;
Elles passeront encore a 36 pieds de longueur.

petit Carreau
1 pièce de 12 pieds de Longueur, sur 24 pouces,
en quarré -
sans qu'il soit nécessaire qu'Elle fut a
vive arreste

Solle du petit Carreau
1 pièce de 10 pieds de Longueur sur
20 pouces sur une face et 18 pouces sur L'autre

Arbres
arbres d'affinerie et de chaufferie
4 arbres en rondeur, pelés de 2 pieds de
diamètre et de 30 pieds de Longueur; 20 pouces
de diamètre au petit bout

Arbres du fourneau
2 arbres pelés en rond de 24 pouces de diamètre
au gros bout, de 18 pouces au petit. L'un de 20 pieds de Longueur
L'autre de 30 pieds de longueur

Les moises
2 pieces rondes pelées seulement, de 15 pieds de Longueur, sur
30 pouces de diamètre au gros bout, et le petit bout comme
il se trouvera.
1 piece de 10 pieds de Longueur, 2 pieds de
Large; 7 a 8 pouces d'Epaisseur
1 piece de 10 pieds de Long
sur 2 pouces en quarré
1 idem de 10 pieds de Longueur sur 12 a 13 pouces
en quarré
dix pieces de bois de chene
de 15 a 20 pieds de Longueur, sur 12 a 15 pouces
d'Ecarissage./. et 5 de plus

arbre d'un martinet de 20 pieds de longueur
sur 30 pouces de Diamètre

---

*   NAC, MG 1, C¹¹A, vol.110, fol. 175 (n.s., 1739); this transcription is faithful to the form and layout of the original.

(Excerpt 2)

Fol. 176

24 pieds de Chute dans le Chemin d'Eau. - La Chute sur la Roue
de la chaufferie sera de   16 pieds - 192 pouces {est actuellement} de ... 82
id. sur la Roue du ...   _____ - 192 pouces   {est actuellement} de ... 103
La vitesse de 82 pouces est de 9 pouces 1/20 un peu plus
La vitesse de 192 pouces est de 14 ......un peu moins

ouverture de la Chauferie   28 pouces
ouverture de la pelle du
Marteau 72

Effort a la Chaufferie avec la Vitesse d'aujourd'huy qui est de 9 1/20 a ......
                    une ouverture de 28 po cy   2296
Effort a la Chaufferie avec une ouverture de 12 et la vitesse de 14 - cy en donnant
                    192 pouces de chute   2296

Mais la dépense de l'Eau dans le premier cas, sera a la dépense de l'Eau
dans le second cas comme les vitesses et réciproquement
partant - 9 1/20 : sera a 14:: 12:: 18 2/3 qui sera la dépense
de l'Eau dans le second cas au lieu de 28 pouces -

Pour le Marteau -

L'Effort au Marteau avec la vitesse d'aujourd'huy qui est de 10 1/6 par une
ouverture de 72 pouces sera de 7416
L'Effort id. avec la vitesse de 14 par une ouverture de
39 pouces sera de 7488

Mais la dépense de l'Eau dans le premier cas sera
a la dépense de l'Eau dans le second cas comme les
vitesses partant -
10 1/6 : 14:: 39 pouces est a 54 pouces environ -
donc en donnant les chutes convenables vous ménagerez - sur
la chaufferie d'en haut   ....9 1/3 d'Eau
sur la Roue du Marteau   ....18 pouces
État du ménage de l'Eau 27 1/3

Toutes ces opérations ont été vérifiées
le 22ᵉ Mars 1739

(Excerpt 3)

Fol. 185

14 Janvier 1739

Observations sur l'épreuve de la
gueuse N 167 pesant ....
Convertie en fer a St. Maurice en gros fer
Plat et quarré au mois de Janvier 1739

a rendu ....1155 fer plat │
id ...   327 │1482

a Consommé en Charbon 62 Resses de Charbon Lesquelles
a raison de 3 Resses par barriques donc ___ 21 Barriques
ou _____ 10 pipes et demi ou bien 1848 £ pesant
de charbon a raison de 88 £ la barrique pesant

La grande Roue du feu d'en haut fait 24 tours
en six minutes _ a pleine d'Eau le bief plein._
celle du feu d'en bas a fait autant de tours, et
dans le même temps _ mais avec moins d'Eau la pelle
n'Etant pas Entièrement Levée _ mais J'ay
remarqué aussy que les soufflets du feu d'En bas
perdent bEaucoup plus de vent que ceux d'en haut

du 12 fevrier

hauteur
difference de niveau de l'extremité du Chemin d'Eau
de la Roue du fourneau Jusque a celui de l'abreuvoir
au-dessous du fourneau où l'on propose de faire une
chaussée _____ 19 pieds un pouce
N Comme les huches sont Encore plus bas que la Chaussée
et que le Terrain va toujours en descendant le Nouveau
Chemin d'Eau de la forge proposée pourra estre encore baissé
d'un pied ou deux _

La Chute d'Eau en Ligne perpendiculaire du fond de la huche
sur la circonférence de la Roue a la Chauferie d'en haut est de dix pouces
id celle du feu d'en bas est de quinze pouces _
id de la Roue du Marteau est de deux pieds sept pouces _
Largeur de la Roue du Marteau 2 pieds neuf pouces
de dedans en dedans
Largeur de chacune des deux Chauferies deux pieds de dedans
en dedans
toutes les trois Roues ont chacune 8 pieds
de diamètre de dehors en dehors

La profondeur
des godets en
Ligne perpendiculaire sur la
Courbure des Roues
est

│ de ___ 14 pouces aux chauferies
│ de ___ 18 pouces a la Roue du
                Marteau

Les courbes des Chauferies ont 14 pouces de
largeur
celle de la Roue du Marteau 18 pouces
L'ouverture de la pelle du Marteau a 4 pouces
sur 18 pouces
celles des pelles des chauferies sont de deux pouces
sur 14 pouces _
On a observé que quoy que l'ouverture des pelles des
deux chauferies sont égales, celle du feu d'en haut
consomme beaucoup plus d'Eau que celle d'en bas - ce qui
provient sans doute de la différence qu'il y a entre leurs
Chutes d'Eau /.

PROJECTED EXPENSES TO START UP AND OPERATE AN IRONWORKS IN CANADA, 1735[a]

Structure	Material	Unit	Unit price[b]	Quantity	Cost[b] l - s - d
Blast furnace					
Stack	Stone	Cubic toise	50-00-00	87.33	4,366-13-04
Outer shell	Stone	Running toise	16-00-00	24	391-02-03
Inner shell	Stone	Running toise	16-00-00	12.22	195-11-01
Lining	Brick	Thousand	40-00-00	10,000	400-00-00
Inwalls	Firestone				600-00-00
Vents					300-00-00
Total for masonry					6,253-06-08
Bellows and waterwheel shed	Boards, planks and shingles				600-00-00
Casting house					600-00-00
Bellows					800-00-00
Tuyere, nozzles and striker plates					300-00-00
Gearing (wheels, cogwheels, lantern wheels)					600-00-00
Ringers and other tools					400-00-00
Millpond, dam, spillways and races					1,500-00-00
Charcoal house	Posts, planks and shingles				1,200-00-00
Stamp mill and ore washery					600-00-00
Forge, made up of two fineries, 1 chafery, 1 plate mill, complete with gearing and hurst frames for the forge hammer and tilt hammer					
Sheds	Posts, planks and shingles				2,000-00-00
Foundations	Stone	Running toise	12-00-00	108.33	1,300-00-00
Shed with two sides (wheels)	Posts, planks and shingles				600-00-00
4 hearths (2 fineries, 1 chafery, 1 plate mill)					
Masonry for the 4 hearths	Stone	Running toise	12-00-00	42.66	512-00-00
Chimneys	Stone	Running toise	12-00-00	25.66	308-00-00
4 pairs of bellows and fittings					1,000-00-00
Tuyeres, nozzles and striker plates					400-00-00
Cast iron plates					100-00-00
Ringers, tongs and other tools					600-00-00
4 sets of gearings (camshafts and cogwheels)					1,200-00-00
Hurst frame for forge hammer					1,000-00-00
Cast iron head for forge hammer					500-00-00
Cast iron anvil					100-00-00
Ties and gudgeons for the axle-tree					200-00-00
Hurst frame for tilt hammer					500-00-00
Wrought iron head for tilt hammer					200-00-00
Cast iron anvil					30-00-00
Millpond, dam, spillway and races					2,000-00-00
Charcoal house	Posts, planks and shingles				1,200-00-00
Blacksmith's shop	Boards, planks and shingles				300-00-00
Bellows					200-00-00
Anvil					400-00-00
Forge and chimney					100-00-00
Tools					100-00-00
Moulds for kettles and other castings					1,000-00-00
Slitting mill					
Sheds	Boards, planks and shingles				2,400-00-00
2 sheds for wheels	Posts, planks and shingles				150-00-00
Foundation (for wheels)	Stone	Running toise	12-00-00	32.22	386-13-04
Wheels, cogwheels, lantern wheels, camshafts, ties and gudgeons					900-00-00
Cutters, moulds, related equipment and tools					500-00-00
Millpond, spillway and races					1,000-00-00
Reverberatory furnace	Stone	Running toise	12-00-00	20	240-00-00
Cast iron grills and openings					100-00-00
Blacksmith's shop (similar to one at forge)					1,100-00-00
Anvil block					100-00-00
Files and other tools					100-00-00
House for master founder and his four hands					1,000-00-00
16 carthorses and carts				16	150-00-00
Ironmaster's house	Boards, planks and shingles (lath and plaster inside and out)				4,000-00-00
Iron store	Boards, planks and shingles				550-00-00
Stable					250-00-00
Oven					100-00-00
Total					43,980-00-00
Total (not including slitting mill and shop)					37,003-06-08
Vézin's total					36,003-06-08
Total (not including slitting mill and kettle moulds)					36,003-06-08

Note: Vézin forgot to include the master founder's house (1,000 *livres*) in his total (36,003 *livres* 6 *sols* 8 *deniers*); the total not including the slitting mill and shop is therefore 37,003 *livres* 6 *sols* 8 *deniers*.

ANNUAL OPERATING AND PRODUCTION EXPENSES[c]

Campaign: 8 months

Output: 1 million pounds of cast iron

Expenses and wages	Wages[a] (livres)	Unit	Unit cost[d] (livres)	Quantity	Cost[d] (livres)
Ore (at the furnace)		pipe	3	2,000	6,000
Charcoal (at the furnace)		pipe	1	20,000	20,000
Limestone		pipe	1	1,000	1,000
Mud and clay flux		pipe	1	600	600
Tallow		pound	0.5	100	50
Subtotal					**27,650**
1 ironmaster	3,000				3,000
1 clerk	700				700
1 master founder	1,500				1,500
1 furnace keeper	400				400
4 fillers	300				1,200
Subtotal					**6,800**
Cast iron		1,000 pounds	34.45	900	31,005
Charcoal		pipe	1	14,400	14,400
1 hammerman	1,200				1,200
3 chaferymen	600				1,800
1 helper	300				300
1 finer	1,200				1,200
7 helpers	600				4,200
1 carpenter	500				500
1 blacksmith	500				500
Note: these workers can also work in the plate mill as needed.					
Subtotal					**55,105**
Cost of wrought iron per 1,000 pounds			91.84		
Castings					
Cost of cast iron per 1,000 pounds		thousandweight	34.45	100	3,445
1 moulder	1,500				1,500
2 hands	600				1,200
Subtotal					**6,145**
Cost of castings per 1,000 French pounds (not noted by Vézin)			61.45		
Total					**61,250**
Total (Vézin)					**61,250**
Expenses for slitting mill					
1 slitter	1,500				1,500
4 hands	300				1,200
Subtotal (labour)					
FORGES OUTPUT					
Bar iron (colony)		thousandweight	200	200	40,000
Bar iron (France)		thousandweight	140	400	56,000
Subtotal				**600**	**96,000**
Castings		thousandweight	200	100	20,000
Total					**116,000**

[a] NAC, MG 1, C¹¹A, vol. 110, fols. 144–49, October 1735, n.s. (Vézin)* and vol. 111, fols. 168–72, 17 October 1735, signed Hocquart and Vézin.
Two copies of this document appear consecutively in vol.110, fols. 144–46 and fols. 147–49; the same forecasts are found in vol. 111, but also include annual operating expenses.

[b] Prices and costs are in livres, sols, and deniers.

[c] NAC, C¹¹A, vol. 110, fols. 144–49, estimate signed Hocquart and Vézin, 17 October 1735.

[d] Salaries, prices and costs are in livres or fractions thereof.

**LETTER FROM WILLIAM HENDERSON, MANAGER OF THE ST MAURICE FORGES,
TO GEOLOGIST WILLIAM EDMUND LOGAN, DIRECTOR OF THE GEOLOGICAL SURVEY OF CANADA***

St Maurice Forges
11 January 1855

Sir,

By Mr Richardsons request, who called here on 8 Inst, I have forwarded to his directives two Boxes of samples intended for the Paris exibition marked SMF, A and B.

No A contains two large samples of conglomonited [?] Bog ore. The larger one from Nicolet on SW side of the St Lawrence and the smaller and lighter colored from Point du lac on this side the river. In the same box are two pieces of our wrought iron & a piece of Pig iron from the blast furnace, both of average quality. We did not think it necessary to send any castings, as we understand that some will be sent by our neighbours at Radnor Forges who use the same material as we do and have many moulders who formerly learned their business here.

In box B is a smaller box containing Bog ore in small grains brought from Machiche. We did not think it worthwhile to send any from Cap Madelaine in Champlain as it is quite similar to what is now forwarded and besides the Radnor Forges will probably furnish samples from that quarter.

In the same box is a piece of lime stone, such as we use in smelting the ore, found at a place called La Gabelle about 5 miles from this up the River St Maurice, and on its banks. Limestone is plentifull in many other localities in this neighbourhood all of the same dark colour. I have also in the same box packed up two pieces of what we here call Firestone, as it is used in building the furnace and stands the fire so well that a furnace usually lasts from 3 to 4 years. It is also an excellent building material, and is found only at one place viz the Gabelle above referred to. I am not aware of its existing elsewhere in this part of the Province. It forms the face of the steep & high bank of the River and in quantity appears inexaustable, the veins in the rock are at narrow widths, some only a few inches & others two to three feet thus affording blocks of stone for every purpose.

This locality also affords moulding sand, and Bog pits for Peats which will hereafter become very valuable after the woods is exausted. The frozen state of the ground, at this season, prevented me from sending samples.

Some time back we sent a box of samples of ore fire stone lime stone and moulding sand to Mess Anderson Evans Co. If you should prefer the sample of indunited [?] ore, which is buryer than that now forwarded, I have no doubt [. . .] that they would cheerfully exchange it, especially as I have written to them respecting it.

It may here be well to add, that although in the process of time, when the wood has all been cleared off, there may be some difficulty in obtaining fuel for the blast furnaces. The supply of ore appears to be inexaustable. It is found near the shores of the St Lawrence and mere 10 to 15 miles back in allmost every part of this district & on both sides of the River. That from Nicolet is however found to be less intermixed with clay & requires less washing than which we find on this side of the River. One with the other it requires about from 50 to 60 bushels of harder charcoal to melt one ton of metal. If any other explanation or information required I shall be most happy to furnish to the best of my ability

I am here

Your [. . . . . .]
for Mr John Porter & Co
W Henderson
manager

* McGill University Archives, Logan Papers, accession 1207/11, item 94, "John Porter & Co.
To W.E. Logan, Provincial Geologist, Montreal," St Maurice Forges, 11 January 1855.

Appendix 9

OVERVIEW OF FACILITIES AT THE UPPER AND LOWER FORGES, ACCORDING TO ESTÈBE, 1741, AND CHAUSSEGROS DE LÉRY, 1738-39[a]

Hydraulic and Hydromechanical Works

Description	Lower Forge	Upper Forge
Milldam	d: 95 *pi.* across mat.: 144 *toises* of masonry	d.: 130 *pi.* x 25[b] *pi.* x 20 *pi.* mat.: wood, 19 trusses
Headrace	d: 30 *pi.* x 11 *pi.* x 7 *pi.*[c] (high), supported on one side by a wall of 30 *pi.* x 17 *pi.* x 3 *pi.*[d] str.: framework surrounded by planks caulked on three sides	d.: 70 *pi.* x 5 *pi.* x 10 *pi.*[e] (high) str.: not mentioned
Spillway	d: 118 *pi.* x 5 *pi.* x 5.5 *pi.* str.: planked on three sides, covered with a lean-to roof of *planches debout* construction, 80 *pi.* long bridge: 110 *pi.* wide, of timbers squared on the face	d.: 25 *pi.* x 8 *pi.* x 12 *pi.* str.: not mentioned bridge: 120 *pi.* x 24 *pi.*
Forebay (see also headrace)	d: possibly 30 *pi.* x 11 *pi.* x 7 *pi.* (high) (dimensions of headrace) head (chafery): 82 *po.* head (hammer): 103 *po.*	"forebays" d.: possibly 70 *pi.* x 5 *pi.* x 10 *pi.* (high) (dimensions of headrace)
Forebay (according to Léry)	head (waterwheels): 15 *po.* apertures: 2 *po.* x 14 *po.*	head (waterwheels): 10 *po.* hammer: 31 *po.* apertures: 2 *po.* x 14 *po.* hammer: 4 *po.* x 18 *po.*
Waterwheels (upper forge according to Léry)	diameter: 8 *pi.* (outside) width: 2 *pi.* (inside)	**bellows wheel:** diameter: 10 *pi.* (outside) width: 2 *pi.* (inside) felloes: 14 *po.* depth of buckets: 14 *po.* **hammer wheel:** diameter: 8 *pi.* (outside) width: 2 *pi.* 9 *po.* (inside) felloes: 18 *po.* depth of buckets: 18 *po.*

Note:
The abbreviations *pi.* (*pied*) and *po.* (*pouce*) refer to French feet and inches respectively. A French foot corresponds to 32.484 cm.

Abbreviations
d.: dimensions   str.: structure
mat.: materials

a. NAC, MG 1, C¹¹A, vol. 112, Estèbe, Inventaire des Forges 1741, fols. 28–59v, 22 November 1741, signed Ignace Gamelin, Simonet fils, Estèbe; NAC, MG 1, C¹¹A, vol.110, fols. 175– 208 (n.s., 1739, Léry).
b. "Excavated to a width of 25 *pi.*".
c. "Forebay" [...] receiving the water at the entrance to the milldam."
d. In 1736, plans were made to build a wall the following year "at least 20 *pi.* long, 3 *pi.* thick,and forming an angle between the forge and the milldam."
e. "A race that takes the water in the milldam and channels it to the forebays [...]"

Buildings Housing the Shops

Description	Lower Forge	Upper Forge
Building	d: 80 *pi.* (w) x 36 *pi.* (l) x 15 *pi.*(h) mat.: tongue and groove planks r.: *planches debout* fo.: 80 *pi.* x 6 *pi.* (h) x 2.5 *pi.* shed side 80 *pi.* x 8 *pi.* (h) x 3 *pi.*, wheelrace side 52 *pi.* x 14 *pi.* (h) x 3 *pi.* gable end (including wheelrace and iron store) 42 *pi.* x 6 *pi.* (h) x 2 *pi.* gable end (including charcoal house) o.: 1 large double door, a door on each gable end and 1 dormer	d.: 70 *pi.* (w) x 30 *pi.* (l) x 17 *pi.* (h) mat.: mudwall and posts and tongue and groove planks r.: *planches debout* fo.: 70 *pi.* x 3 *pi.* (h) x 2 *pi.* 70 *pi.* x 8 *pi.* (h) x 3 *pi.* wheelrace side 30 *pi.* x 4 *pi.* (h) x 2 *pi.*, both gables 30 *pi.* x 4 *pi.* (h) x 2 *pi.*, both gables o.: 1 large double door, 1 small door and a door on each gable end
Sheds[a] (charcoal shed, iron store)	d.: 80 *pi.* x 12 *pi.* x 10 (h) *pi.* mat.: tongue and groove planks r.: overlapping boards fo.: 80 *pi.* x 6 *pi.* (h) x 2.5 *pi.*[b] w.: not mentioned o.: not mentioned	d.: 70 *pi.* x 15 *pi.* x 8 *pi.* mat.: pilings and tongue and groove planks for 35 *pi.* r.: overlapping boards fo.: not mentioned w.: 15 *pi.* between the shed and store o.: 1 double door, 2 doors
Wheelrace[c]	d.: : 80 *pi.* (w) x 12 *pi.* (l) x 15 *pi.* (h) *pi.*[d] mat.: tongue and groove planks f.: tongue and groove planks cl.: tongue and groove planks b.: tongue and groove planks fo.: 80 *pi.* (w) x 8 *pi.* (h) x 3 *pi.* (t) o. : 2 gates with iron fittings	d.: 70 *pi.* (w) x 14.5 *pi.* (l) x 17.5 *pi.* (h)[e] mat.: tongue and groove planks f.: tongue and groove planks cl.: tongue and groove planks b: tongue and groove planks fo.: 70 *pi.* (w) x 8 *pi.* (h) x 3 (t) *pi.* o.: 6 gates with iron fittings

Note: The abbreviations pi. (pied) and po. (pouce) refer to French feet and inches respectively. A French foot corresponds to 32.484 cm.

Abbreviations
w.: walls   h.: height   o.: openings
d.: dimensions   w.: width   f.: floor
t.: thickness   l.: length   fo.: foundation
b.: bottom of   mat.: materials   r.: roof
   wheelrace   cl.: cladding

a. Work in 1736: "[...] for this forge, approximately 3,000 toises of clay soil and sand had to be removed from the site."
b. Second foundation.
c. From 1736. The wheelrace was probably 10 *pi.* high.
d. Head: 15 *pi.* + 8 *pi* = 23 *pi.*
e. Head: 17.5 *pi.* + 8 *pi* = 25.5 *pi.*

Appendix 10

WROUGHT IRON PRODUCTION, ST MAURICE FORGES, 15 OCTOBER 1739 TO 1 OCTOBER 1741 (in French pounds)

Year	Month	Period[1]	Days worked Upper forge	Lower forge	Average no. of days	Production Upper forge	Lower forge	Total Period	Month worked	Total Month	No. of days worked in month	Total for year	Daily average	PIG IRON	Pig iron for month	Pig iron for year
1739	October	15–22	6	6	6	2,000	5,000	7,000	October 1739	7,000	6		1,166.67			
	October	22–31[2]	0	0	0			0								
	November	2–9	6	6	6	2,697	4,833	7,530					1,255.00			
	November	9–16	6	6	6	1,000	5,193	6,193					1,032.17			
	November	16–23	6	6	6	1,163	5,464	6,627					1,104.50			
	November	23–30[3]	0	6	3		5,300	5,300	November	25,650	21		1,766.67			
	November–December	30–7[4]	0	4	2		3,079	3,079					1,539.50			
	December	7–14	0	6	3		4,241	4,241					1,413.67			
	December	14–21	6	6	6	5,970	8,046	14,016					2,336.00			
	December	21–28[5]	3	3	3	2,266	3,977	6,243				**1739**	2,081.00			
1740	December–January	28–4[6]	5	5	5	4,223	5,200	9,423	December	37,002	19	69,652	1,884.60			
	January	4–11[7]	5	5	5	4,220	5,617	9,837					1,967.40			
	January	11–18	6	6	6	3,366	3,881	7,247					1,207.83			
	January	18–25[8]	5	6	5.5	3,600	6,004	9,604					1,746.18			
	January	25–1[9]	2.5	2	2.25	4,670	5,315	9,985	January 1740	36,673	18.75		4,437.78			
	February	1–7[10]	3	2.5	2.75	3,074	4,408	7,482					2,720.73			
	February	7–14[11]	3	6	4.5	2,074	7,232	9,306					2,068.00			
	February	14–21[12]	3.5	6	4.75	1,500	4,305	5,805					1,222.11			
	February	22–29[13]	1	6	3.5	540	3,397	3,937	February	26,530	15.50		1,124.86			
	February–March	29–7	6	6	6	3,160	4,489	7,649					1,274.83			
	March	7–14	6	6	6	3,780	5,425	9,205					1,534.17			
	March	14–21	6	6	6	5,955	3,070	9,025					1,504.17			
	March	21–27	6	6	6			8,175					1,362.50			
	March–April	28–2[14]	6	6	6	7,100	3,800	10,900	March	44,954	30		1,816.67			
	April	2–9	6	6	6			11,890					1,981.67			
	April	9–18	6	6	6			7,500					1,250.00			
	April	18–23	6	6	6	3,213	1,350	4,563					760.50			
	April	25–30	6	6	6	4,360	3,730	8,090	April	32,043	24		1,348.33			
	May	1–7	6	6	6	2,900	4,650	7,550					1,258.33			
	May	8–15	6	6	6			3,050					508.33			
	May	16–23	6	6	6	3,000	4,200	7,200					1,200.00			
	May	23–28[15]	6	6	6			3,000	May	20,800	24		500.00			
	May–June	28–4	6	6	6			5,900					983.33			
	June	4–10[16]	6	6	6	4,250	1,500	5,750					958.33			
	June	11–18	6	6	6	3,250	3,800	7,050					1,175.00			
	June	20–23	4	4	4	2,350	2,700	5,050					1,262.50			
	June–July	27–2[17]	6	6	6			8,388	June	32,138	28		1,398.00	192,200	192,200	
	July	2–9	6	6	6			12,168					2,028.00	50,775		
	July	11–16	6	6	6			7,250					1,208.33	44,275		
	July	16–23[18]	6	0	3	6,235		6,235					2,078.33	40,675		
	July	25 and 26[19]	0	0	0			0								
	July	27–30	4	4	4			5,464	July	31,117	19		1,366.00	34,825	170,550	
	August	1–6	6	6	6			8,890					1,481.67	20,300		
	August	7–14	6	6	6			6,250					1,041.67	43,475		
	August	14–22	6	6	6			5,500					916.67	47,825		
	August	22–28	6	6	6			3,100					516.67	31,125		
	August–September	28–3	6	6	6			8,842	August	32,582	30		1,473.67	32,375	175,100	
	September	4–10	6	6	6			5,732					955.33	26,175		
	September	10–17	6	6	6			10,500					1,750.00	39,500		
	September	17–25	6	6	6			7,400					1,233.33	41,150		**1740**
	September	25–30	5	5	5			4,950	September	28,582	23		990.00	29,500	136,325	674,175
	October	2–9	6	6	6			6,625					1,104.17			
	October	10–15	6	6	6			11,950					1,991.67			
	October	17–22	6	6	6			4,900					816.67			
	October	22–30	6	6	6			7,918	October	31,393	24		1,319.67			
	October–November	30–6	6	6	6			6,402					1,067.00			
	November	7–12	6	6	6			15,342					2,557.00			
	November	13–20	6	6	6			13,930					2,321.67			
	November	20–26	6	6	6			13,754					2,292.33			
	November–December	27–3[20]	3	3	3			5,750	November	55,178	27		1,916.67			
	December	4–11	6	6	6			13,388					2,231.33			
	December	11–18	6	6	6			17,725					2,954.17			
	December	18–25	6	6	6			11,766				**1740**	1,961.00			
	December	25–31[21]	4	4	4			6,172	December	49,051	22	421,041	1,543.00			

Year	Month	Period¹	Days worked			Production		Total	Month worked	Total			Daily average	PIG IRON	Pig iron for month	Pig iron for year
			Upper forge	Lower forge	Average no. of days	Upper forge	Lower forge	Period		Month	No. of days worked in month	Total for year				
1741	January	1–9	6	6	6	2,087	2,400	4,487					747.83			
	January	9–15	6	6	6	2,312	950	3,262					543.67			
	January	15–22²²	0	0	0			0								
	January	22–29	6	6	6	4,344	4,625	8,969					1,494.83			
	January–February	29–5	6	6	6	4,301	4,598	8,899	January 1741	25,617	24		1,483.17			
	February	5–12	6	6	6	1,440	2,776	4,216					702.67			
	February	12–19²³	0	0	0			0								
	February	20–26	6	6	6	3,195	4,200	7,395	February	11,611	12		1,232.50			
	February–March	27–5²⁴	6	6	6	8,155	6,039	14,194					2,365.67			
	March	5–12	6	6	6			13,812					2,302.00			
	March	12–19	6	6	6	8,249	4,610	12,859					2,143.17			
	March	19–25	6	6	6	5,057	810	5,867					977.83			
	March–April	25–2	6	6	6	7,990	3,685	11,675	March	58,407	30		1,945.83			
	April	2–8²⁵	0	0	0			0								
	April	9–16	6	6	6			5,477					912.83			
	April	16–22²⁶	0	0	0			0								
	April	23–30	0	0	0			0	April	5,477	6					
	May	2–8	4	4	4			1,550					387.50			
	May	8–13	6	6	6			1,825					304.17			
	May	14–21	6	6	6			11,170					1,861.67			
	May	21–29	6	6	6			7,960					1,326.67			
	May–June	29–5	6	6	6			6,337	May	28,842	28		1,056.17			
	June	5–12	6	6	6			7,880					1,313.33			
	June	12–19	6	6	6			15,034					2,505.67			
	June	19–26	6	6	6			7,395					1,232.50			
	June–July	26–3	6	6	6			7,979	June	38,288	24		1,329.83			
	July	3–9²⁷	6	6	6			5,708					951.33			
	July	9–16	6	6	6			3,220					536.67	45,250		
	July	16–23	6	6	6			3,831					638.50			
	July	23–30	6	6	6			1,000	July	13,759	24		166.67			
	July–August	30–6	6	6	6			1,350					225.00	27,675	72,925	
	August	6–13	6	6	6			6,255					1,042.50			
	August	13–20²⁸	6	6	6			10,955					1,825.83	14,875		
	August	20–26	6	6	6			6,700					1,116.67			
	August–September	26–3²⁹	6	6	6			10,680	August	35,940	30		1,780.00		14,875	
	September	3–10	6	6	6			9,875					1,645.83	13,000		
	September	10–17	6	6	6			11,750					1,958.33	30,250		
	September	17–24	6	6	6			8,750				1741	1,458.33			1741
	September–October	24–1	6	6	6			10,595	September	40,970	24	258,911	1,765.83	21,725	64,975	152,775
						subtotal	subtotal									
TOTAL	weeks	**103**	523	543.5	533.25	139,046	163,899	749,604		749,604	533.25	749,604	1,405.73	**826,950**		826,950
	days	618								monthly average						
Total of subtotals from original document³⁰			...			749,504				31,234						
Average no. of days worked			21.79	22.65	22.22											

* The period was taken from the original document exactly as it appears and was checked using the 1739 to 1741 calendars Although the periods recorded vary a great deal (Monday to Monday, Monday to Sunday, Monday to Saturday, Saturday to Saturday), they all cover six working days, from Monday to Saturday. The figures in the "days worked" column therefore reflect the working days, as verified in the calendar. The inconsistency in the notation of working days in the original document is no doubt due to the fact that figures for pig iron production in the blast furnace were also kept and the furnace operated seven and not six days a week.

1. This period, which is not stated as such in the original document, takes into account the following comment, which refers to both forges: "taken out of blast on 21 October, 6 days after M. Oliver left, for want of charcoal." Work would not begin again until 2 November.

2. Work was halted at the upper forge until 14 December (three weeks) due to the failure of the dam. In the column "days idle" for the upper forge appears: "shut down due to accident to the dam".

3. "2 days holiday, St Éloi".

4. "3 days holiday, Christmas".

5. "1 holiday", New Year's Day.

6. "1 holiday", 6 January.

7. In the column "days idle" for the upper forge appears the comment: "1 day to repair the helve of the hammer and the anvil".

8. In the column "days idle" for the upper forge appears: "3 ¹/₂ days to repair [...] forebay damaged by freezing, and illness of a finer" and for the lower forge, "1 ¹/₂ days to repair hammer and hammer post and replace cams of hammer shaft, 2 ¹/₂ days to make other adjustments and tighten the anvil".

9. "1 holiday" (the 2nd, Candlemas— la Chandeleur). In the column "days idle" for the upper forge appears: "1 day to repair the bellows shaft and 2 nights lost due to illness of finer"; and for the lower forge, "2 ¹/₂ days to repair and tighten anvil". We deducted half a day for each night of work lost.

10. In the column "days idle" for the upper forge: "shut down Monday, Tuesday and Wednesday to replace hammer helve, install anvil, adjust bellows; only worked on and off the other days due to freezing".

11. In the column "days idle" for the lower forge (but applies to the upper forge): "shut down Monday to replace hammer helve and three full nights to tighten the hammer shaft, install cams and arms". The note for this day referring to a shutdown for 1 day and 3 nights seems to apply to the upper rather than the lower forge although it appears under the "days idle" column at the lower forge (someone began to write it under the upper forge and then crossed it out); furthermore, this shutdown seemed to have a greater effect on production at the upper forge (1,500 pounds). We deducted half a day for each night of work lost.

12. In the column "days idle" for the upper forge: "only one day was worked at the upper forge, three days being spent installing a new gudgeon and sharpening the anvil, and the rest of the time being spent in making mocket heads".

13. The second chafery at the upper forge was built during the winter of 1740 when Vézin had left for France, which is confirmed in the memorial of Vézin and Simonet of 10 June 1741 (p. 246) and in Cugnet's memorial of 17 October 1741 (p. 152). The high production figures (7,100 pounds) for the upper forge in early April seem to be due to the fact that a second hearth was in use, thanks to the spring flood. See also the period February-March 1741 below.

14. "The furnace was blown in on 23 May and produced until 30 June ___ 192,200 and from 30 June to 9 July ___ 50,775". The figures are written at the ending date for each of these periods.

15. Three production figures were given for the two forges: 1,500, 2,000 and 2,250. We have taken the first two figures as applying to the upper forge.

16. The following note on the blast furnace appears at the bottom of the page: "Furnace was blown in on 23 May and produced ___ 192,200 between then and 30 June; from 30 June to 9 July ___ 50,775."

17. A total of 5,385 pounds of wrought iron were recorded for the upper forge and 850 pounds for the "second fire." We have taken this as the second hearth at the upper forge.

18. Monday and Tuesday (25 and 26 July) were not included in the days worked, with the records skipping from the 23rd to the 27th of the month. The 26th corresponds to the feast of St Anne.

19. "3 days holiday", St Éloi, the patron saint of ironworkers, 1 December.

20. We deducted 2 days holiday in addition to Christmas Day, as was done for 1739.

21. "The forges did not operate because of the freezing weather." In their memorial of 10 June 1741, Vézin and Simonet stated that they had been forced to operate the forges despite the extreme cold in the winter of 1741.

22. "The forges did not operate because of the intense cold".

23. These figures undoubtedly take into account the use of second hearth at the upper forge, made possible by the spring flood.

24. "Total shutdown." In their memorial of 10 June 1741, Vézin and Simonet state that, because of winter working despite the intense cold, "there was a shortage of charcoal at the forges because, during the period of intense cold, close to 800 cartloads were wasted due not only to the extraordinary amount consumed by the chaferies but to the amount burned at each forge to prevent the movements from freezing up [...]".

25. "Total shutdown," from 16 April to 1 May.

26. The following note appears for the blast furnace: "furnace blown in on the 4th" and "the furnace from 4 July to 16 ___ 45,260, to 6 August ___ 27,675". The delay in blowing in the furnace was most likely due to the shortage of charcoal caused by excessive consumption the previous winter. Vézin and Simonet warned about this in their memorial of 10 June 1741. Cugnet wrote in his memorial of 17 October 1741 that "the furnace barely operated at all last summer and had only produced 145 thousandweight by 16 September. It had given 900 in 1740."

27. The figure of 14,875 pounds of pig iron was not attributed to a specific period; it followed right after the period ending on the 6th.

28. Note for the blast furnace: "the furnace produced from the 4th to the 10th only 13,000 of pig iron, from the 10th to the 16th, 30,750 of pig iron, 21,725".

29. The compiler of the day only made one error on the last page (of 80 pounds), with the errors involving the other 20 missing pounds occurring on the preceding pages. Given the total jumble these pages were in, the calculations are remarkably accurate.

Appendix 11

EXCERPT FROM A DOCUMENT ON PRICES AND CONSUMPTION OF SELECTED METALS IN CANADA, 1742–43[1]

[…] suit le détail du bénéfice que peut procurer ce commerce exclusif à raison de la consommation ordinaire en Canada de toutes ces espèces de marchandises, de leur prix d'achat en France, et du prix qu'elles se vendent en Canada au cours de la vente en gros.

On ne comprend point dans ce détail les fers forgez en barres, plaques de soc ni les poêles à chauffer, plaques de poêles, marmittes, chaudieres et autres de fonte en fer coulé parce que ces articles sont déjà compris dans le produit du fourneau et des forges de St-Maurice où ils doivent être fabriqués.

Qualité des marchandises	Consommation ordinaire en Canada	Prix d'achat en France (L.s.d)	Total (L.s.d)	Prix de la vente en Canada (L.s.d)	Total (L.s.d)	Différence de prix France/Canada[2] (L.s.d)
Poêles à frire à la pièce	300	01.11.06	472.10.00	02.05.00	675.00.00	202.10.00
Acier à la livre	5,000	00.07.00	1,750.00.00	00.12.00	3,000.00.00	1,250.00.00
Taule à la livre	6,000	00.08.00	2,400.00.00	00.12.00	3,600.00.00	1,200.00.00
Clous à couvrir au milllier	275 milliers	12.10.00	3,437.10.00	17.00.00	4,675.00.00	1,237.10.00
Clous à plancher au millier	550 milliers	06.05.00	3,437.10.00	09.10.00	4,675.00.00	1,237.10.00
Clous à bardeau au millier	1,000 milliers	01.12.00	1,600.00.00	02.05.00	2,250.00.00	650.00.00
Pour différentes sortes de clous de construction, à boîtes, brequette, mandière etc.			2,000.00.00		3,000.00.00	1,000.00.00
Chaudières de cuivre de toute espèce à la livre	3,000	01.12.00	14,400.00.00	02.10.00	22,500.00.00	8,100.00.00
Plomb en grain au quintal	750	26.00.00	19,500.00.00	35.00.00	26,250.00.00	6,750.00.00
Étain à la livre	2,000	01.04.00	2,400.00.00	01.15.00	3,500.00.00	1,100.00.00
Poêlons à la livre	2,000	01.06.00	1,040.00.00	01.15.00	1,400.00.00	360.00.00
Total			**52,437.10.00**		**75,525.00.00**	**23,087.10.00**

Déduction faite sur le bénéfice cy dessus porté à la somme de   23,087.10.00
celle de   7,865.12.06

Pour les frais de transport de France à Québec des marchandises du commerce exclusif demandé à raison de quinze pour cent de la somme de 52,437.10.00 a laquelle on supose monter l'achat des dites marchandises en France.

Il reste en proffit net la somme de 15,221.17.06

Laquelle ajoutée au proffit net de l'Exploitation des forges en leur état actuel cy devant établi a la somme 19,840.10.00

Fait en total celle de   35,062.07.06

Qui donneroit l'intérêt[3] à environ neuf et deux tiers pour cent des avances nécessaires à faire par le concessionnaire montantes sçavoir.

Pour les dépenses des forges à   309,941.17.08

Et pour l'achat en France des marchandises du commerce exclusif à   52,437.10.00

Total   362,379.07.08

1. NAC, MG 1, C¹¹A, vol. 112, fols. 170–170v, "Mémoire concernant les Forges de St. Maurice," n.s., n.d. (1742–43).
2. "Différence du prix de la vente en Canada à celuy d'achat en France faisant le bénéfice à espérer du Commerce."
3. "Intérêt" here means return, since 35, 062 7. 6. represents 9.66% of 362, 379 7. 8. (total expenses).

Appendix 12

FORGES AGENTS AT MONTREAL, QUEBEC AND TROIS-RIVIÈRES, 1733–1876 [1]

Year	Type of agent	Montreal	Quebec	Trois-Rivières
1733	Shareholder	Poulin de Francheville	François-Étienne Cugnet	Jean-Baptiste Labrèche [2]
1737–41	Shareholder	Ignace Gamelin	François-Étienne Cugnet	Olivier de Vézin
1767	Lessee	Dumas Saint-Martin	Alexandre Dumas	Christophe Pélissier
1769	Lessee	Jacob Jordan [3]	Johnston and Purss	Christophe Pélissier
1771			Pierre de Sales Laterrière	
1784	Merchant	Uriah Judah	Libéral Dumas	A. Proust et fils
1784–87				J. L. Leproust [4]
1793				J. L. Leproust
1794	Merchant	James Laing	Thomas Naismith	J. L. Leproust
1811	Merchant		Roy and Baby	
1812	Merchant		Roy and Baby (Upper Town)	
	Merchant		François Langlois (Lower Town)	
1817	Merchant	John and Thomas Porteous	Bell and Stewart	
1818	Merchant	John and Thomas Porteous	Bell and Stewart	
1819	Merchant	John Porteous	Bell and Stewart	
1820	Merchant	John Porteous	Bell and Stewart	John Monro
	Merchant			Edward Grieves
1821	Merchant	John Porteous		
	Merchant	Bridge and Penn		
1823–24	Merchant	John Porteous		
1833	Merchant	John Porteous		
1838	Merchant	Forsyth, Richardson and Co.		
1839	Merchant	Forsyth, Richardson and Co.	J. M. Fraser and Co.	
1840			Forsyth and Bell	
1842	Merchant		Forsyth and Bell	
			J. M. Fraser and Co.	
1843	Merchant	Bryson and Ferrier	Charles Collet	
	Merchant	Forsyth, Richardson and Co.	Donald Fraser	
	Merchant		Dupont and Co.	
1844	Merchant	Cuvillier et fils		
	Merchant	Bartett and Hagar		
1845	Merchant	Cuvillier et fils	Dupont and Co.	
	Merchant		M. D. Fraser	
1846	Merchant	Cuvillier et fils		
1847	Merchant	Cuvillier et fils	Dupont and Co.	
1848	Merchant	Cuvillier et fils	A. Burns	
	Merchant	Bryson and Ferrier	C. and W. Wurtele	
	Merchant	Daniel McGie		
1849	Merchant	Bryson and Ferrier		John McDougall
1850–51	Merchant		C. and W. Wurtele	
1852	Merchant	James Ferrier	C. and W. Wurtele	
	Merchant	Gillespie and Moffatt and Co.	W. Hunt and Co. [5]	
1853	Merchant	Gillespie and Moffatt and Co.	W. Hunt and Co.	
	Merchant	Frothingham and Workman		
1854	Merchant	Frothingham and Workman	W. Hunt and Co.	
	Merchant	Anderson, Evans and Co.		
1856	Merchant	Ryan Brothers and Co.	J. W. Leaycraft	
1857–58	Merchant		J. W. Leaycraft	
1861	Merchant		Charles Huot	
1863–66	Shareholder			John McDougall
1872-73	Shareholder			John McDougall and Sons
1876	Shareholder			James McDougall

1. From Michel Bédard, "Tarification, commercialisation et vente des produits des Forges du Saint-Maurice," preliminary report (Quebec City: Parks Canada, 1982), Appendix B, pp. 65–66.
2. Jean-Baptiste Labrèche, although not a shareholder, as ironmaster was mandated to sell the company's wares in Trois-Rivières.
3. Jacob Jordan was not included among the lessees.
4. It is not known whether Leproust was actually a merchant.
5. Hunt & Company, merchants at Quebec, owned a third of the shares in John Porter & Company, proprietor of the St Maurice Forges.

CENSUSES OF THE FORGES POPULATION, 1762–1881 [a]

Year	Type of census	Area covered	Characteristics	Number of persons
1762, April	Total numbers	St Maurice Forges	Population censused according to sex and marital status.	72
1765[b]				
1784	Total numbers	St Maurice Forges	Breakdown of population by sex, marital status and broad age group.	149
1825, September	Names of heads of households	St Maurice Forges	Number of people in each household by sex, age group and marital status.	321
1829[c], August	Names of heads of households	St Maurice Forges	Heads of household shown by occupation, age, marital status, the presence of a wife and the number of sons and daughters.	415 (395)[d]
1831	Names of heads of households	Fief of St Maurice	Trade of head of household and number of people in household by sex, age group, marital status and several other variables dealing with agriculture and household possessions.	335
1838[e]	Total numbers	St Maurice Forges		393
1842[f], December	Names of heads of households	St Maurice Forges	Census of inhabitants, broken down by sex and two age groups.	425
1851	Nominal census: entire population	Fief of St Étienne	Name, trade, religion, age, birthplace, residence, sex and other variables related to possessions and crops.	395
1861	Nominal census: entire population	District no. 1 of St Étienne[g]	Identical to 1851 census.	215
1866	Names of heads of households	St Étienne des Grès parish: St Maurice Forges	Census of inhabitants, communicants and souls per family.	234
1871[h]	Nominal census: entire population	St Étienne subdistrict, division no. 1	Same as 1851 census.	265
1875	Nominal census: entire population	Les Vieilles Forges (St Étienne des Grès parish)	Census of inhabitants, broken down into communicants and noncommunicants per family, including name and age of each person.	274
1881	Nominal census: entire population	St Étienne subdistrict, division no. 2	Same as 1851 census.	248

a. From Micheline Tremblay and Hubert Charbonneau, *La population des Forges St-Maurice (1729-1883)*, study done for Parks Canada by the Department of Demography, as part of a historical demographics research program (Montreal: Université de Montréal, 1982), Table 2.1, p. 21.

b. The figures for the 1765 census (273 inhabitants) cited in Trudel 1951 (p. 183, notes 68 and 69) have not been included because we believe that they do not apply to the St. Maurice Forges. A comparison of the data from the St Maurice parish with other data that we believe are more reliable suggest that these data do not in fact involve the Forges. We checked the original document (ANQ-Q, microfilm 4 M00-4643 A), and, indeed, the parish of St Maurice is mentioned; however, in the list of parishes, St Maurice appears between St Pierre and La Chenaye, in the Montreal region. The data for this parish (the ironworks was never a parish in its own right and at that time was part of the parish of L'Immaculée Conception de Trois-Rivières) seem completely contrary to the situation at the Forges, which was not operating at full capacity at the time. For example, a few years earlier, in 1762, only 11 families were censused at the Forges. Furthermore, a comparison of the 1765 data with those from the 1760 Forges inventory also shows significant discrepancies. The St. Maurice parish had 55 houses and 3,205 arpents of land (1,094 ha), 809 of which were under cultivation, as well as 29 oxen, 107 cows, 67 heifers, 115 sheep, 89 horses and 184 pigs. All these figures seem highly unlikely for the Forges when compared with the 1760 inventory, which only lists 6 houses, 17 shanties and 6 horses, and no other farm animals or land under cultivation. Even the figures for the 1784 census, done when the Forges was operating at full capacity, are well below those for the 1765 census.

1765 census: see RAPQ 1936–37, p. 119 or ANQ-Q, microfilm 4 M00-4643 A; see Marcel Trudel "Les Forges Saint-Maurice sous le régime militaire (1760-1764)", *RHAF*, vol. V, no. 2 (September 1951). p. 183; NAC, MG 11, Colonial Office 5, vol. 59, fol. 307–13, "Inventaire des Forges," signed by Hertel de Rouville, 8 September 1760.

c. Census ordered by Mathew Bell and conducted by the Forges superintendent, Henry Macaulay.

d. The census listed 20 people "born at and belonging to the Forges" but who were working temporarily in Trois-Rivières. For the purposes of our study, we deducted this group from the total of 415.

e. NAC, RG 48, B 30, vol. 113, no. 79 (1838), cited in Luce Vermette, "Domestic Life at Les Forges du Saint-Maurice," History and Archeology No. 58 (Ottawa: Parks Canada, 1982), p. 225.

f. Census ordered by Mathew Bell and conducted by the Forges superintendent, Henry Macaulay.

g. The inhabitants of the Forges were counted according to their trade.

h. There is a discrepancy between our totals and those obtained by Hardy and Gauthier in 1871 (265 instead of the 299 reached by Hardy and Gauthier). This is due first to the fact that, after 1861, the Forges were no longer censused as a separate unit and the inhabitants were identified according to their stated craft. Based on these crafts and comparisons with other lists, we settled on a group of families, numbered from 227 to 278 in the census. However, Hardy and Gauthier arrived at their figure of 299 by including families 214-17, 221–22, 224–45, 247–65, 267–78 and 284–86. René Hardy and Benoît Gauthier, "La sidérurgie en Mauricie au 19ᵉ siècle: les villages industriels et leurs populations," research report presented to the Regional Branch of the Quebec Department of Cultural Affairs by the Centre de recherches en études québécoises (Trois-Rivières: Université du Québec à Trois-Rivières, May 1989).

Appendix 14

DEMOGRAPHIC DATA ON THE ST MAURICE FORGES, 1737–1881

Date	Number of persons cited	Category of persons cited	Type of document	Context	Author of document	Reference
1737, 8 November	100	"Workers and day labourers"	Letter	Requirements for subsistence	Beauhamois and Hocquart	NAC, MG 1, C¹¹A, vol. 67, fol. 39 (1737).
1742, 2 October	352	Employees, including 143 workers (23 of them full time) and 209 woodcutters	Expenses	Balance sheet after 10 months of operations	Estèbe	NAC, MG 1, C¹¹A, vol. 111, fols. 278–305v.
1750	400	"Men that work here every day "	Journal	Visit to the Trois Rivières area	George Clinton	NAC, MG 18, J' (1751).
1752, July	120	"employing upwards of 120 men"	Travel journal	Visit to the site	Louis Franquet	Voyages et mémoires sur le Canada.
1754	24	"a manager—a storekeeper— 2 employees—20 family heads—"	Correspondance	Poll tax	—	NAC, MG 1, C¹¹A, vol. 99, fol. 529 (microfilm F. 99).
1760, 1 October	7	"Workers"	Letter	Order to keep workers at the Forges	J. Bruyère, on behalf of Colonel Ralph Burton	NAC, MG 23, G¹⁴, vol. 2, pp. 5-6.
1762, April	72	Inhabitants, including 11 family heads	Census	Military control	—	RAPQ 1936-37, pp. 1-121.
1775 a	400 to 800	People (employees)	Memoirs	Personal reminiscences of the time when he managed the Forges	Pierre de Sales Laterrière (inspector for two years and manager for three years)	Mémoires de Pierre de Sales Laterrière et de ses traverses (Quebec City: Imprimerie de l'Évènement, 1873) pp.84-85.
1775 b	around 125	Workers (working on the post)	same as above	same as above	same as above	same as above
1784	149	Inhabitants (including 30 married couples)	Census	Census of district and city of Trois- Rivières	—	AUM, Baby collection, CC, box 48.
1804	24-25	"Hands" on the post and other employees recruited from the neighbourhood	Journal	Visit to the site	Lord Selkirk	NAC, MG 19, E.1, 1.
1805	around 200	"Workmen and families in number about 200"	Letter	Report by the Forges lessees	Monro and Bell	NAC, RG 4, A 1,S, vol. 86, L26692-95.
1808	Over 300	"Men" (employed) (according to the work in hand)	Travel journal	Visit to the site	John Lambert	Travels through Canada and the United States of North America, in the years 1806, 1807 & 1808: to which are added biographical notices and anecdotes of some leading characters in the United States, (London: printed for Baldwin, Cradock, and Joy, Edinburgh: for W. Blackwood, and Dublin: for J. Cumming, 1814), vol. 1, pp. 485–88.
1815	250 to 300	"Men employed"	Report	Topographic description of Lower Canada	Joseph Bouchette	A Topographical Description of the Province of Lower Canada, with Remarks upon Upper Canada, and on the Relative Connexion of both Provinces with the United States of America, (London: printed for the author and published by W. Faden, Geographer to His Majesty and the Prince Regent [Charing Cross], 1815), p. 305.
1825, 20 September	321	Inhabitants (including 55 family heads)	Census	Census of Lower Canada	J-M. Badeaux	NAC, RG 31, A 1, pp. 1574-75.
1828	75 to 100	"Men employed" ("The number of men employed vary at different seasons from 75 to 100")	Report	Visit	Lieutenant Baddeley	"(R1. Engineers) report of the Saint Maurice iron works, near Three Rivers, Lower Canada, jany. 24th 1828," in APT, vol. V, no.3 (1973), p. 12.
1829, 27 June	300 to 400	Persons	Report	—	Lieutenant Ingall	JHALC, appendix s (no 1).
1829, 3 August	Close to 400	"Souls" (including 74 family heads and 34 single men)	Letter	Renewal of Forges lease	M. Bell	NAC, RG 4 A1, series S, vol. 216, p. 21.
1829, August	415	Inhabitants (395 on the post, including 95 workers)	Census	List - census of workers and their families	H. Macaulay	NAC, RG 4, A1, vol. 225 p. 84.
1831, May to September	335	Inhabitants (including 60 family heads)	Census	Census of Lower Canada		NAC, RG 31, A1, p. 2295-98.
1832	400	"Souls" (including 80-90 regular workers and 100-150 seasonal workers)	Letter and testimony	Renewal of lease	M. Bell	NAC, RG 4, B 15, vol. 18 p. 8823 and JHALC, 1832-33, vol. 42. Testimony by Bell between January and March of 1833.
1835, 29 June	79	Militiamen (from 18 to 60 years old)	List of militiamen	"Review of Capt. H.Macaulay's company [...] "		ASTR, N 3 H 20 History of Militia1835-38.
1842, 27 December	425	Inhabitants (including 99 workers and family heads)	List-census	Renewal of lease	M. Macaulay	NAC, RG 4, B15, vol. 18, p. 8824.
1851-52 a) b)	395 350	Inhabitants Employees	Census	Census of Canada	S. Dumoulin	NAC, RG 31, A 1 microfilm C-1139.
1858	120	Workers	Year book	Canada Year Book	—	Canada Year Book 1858.
1861	215	Inhabitants	Census	Census of Canada	P. Ferron	NAC, RG 31, A 1, microfilm C-1322, pp. 96-123.
1866	234	"Souls" (including 44 family heads)	State of souls	Census of inhabitants, communicants and souls in St Étienne des Grès	Rev. H. Bouchard	St Étienne des Grès parish records.
1871	265	Inhabitants	Census	Census of Canada	Edmond Duchêne	NAC, RG 31, A 1 microfilm C-10076.
1875	274	Inhabitants	State of souls	Forges census (Les Vieilles Forges) done in 1875	-	AETR
1881	248	Inhabitants	Census	Census of Canada	Ferdinand Plourde	NAC, RG 31, A 1, microfilm C-13214, p. 1-59.
1876-81	60	"Men employed (average)"	Testimony	Superior Court case: A. McDougall v. Geo. McDougall	Al. McDougall Manager FSM	AJTR, Superior Court, docket no. 108, item 57, p.8, 18 March 1881.

Appendix 15

CHANGES IN THE FORGE'S PHYSICAL PLANT OVER THE YEARS

	1741	1746	1748	1760	1764	1767	1785	1786	1807	1845	1870
Departments											
Blast furnace	1	1	1	1	1	1	1	1	1	1	1
casting house	x						x		x		x
moulding shed	x	x	x	x			x		x	x	x
charcoal house	x			x			x		x		x
bellows shed	x						x		x		x
lodging	x										
Upper forge	1	1	1	1	1	1	1	1	1	1	1
iron store	x			x	x	x	x				
charcoal house	x			x	x	x	x				x
Lower forge	1	1	1	1	1	1	1	1	1	1	1
iron store	x			x	x	x	x				x
charcoal house	x			x	x	x	x				x
Tilt hammer			1	1	1	1	1				
Mills											
sawmill				1	1	1	1	1	1	1	1
grist mill								1	1	1	1
Furnaces											
kiln											6
lime kiln		1	1						1		1
brick works											1
Shops											
smithies	1			u	u	u	1	1	2	1	3
woodworking shop	1	u	u	u	u	u		1	1	2	
Sheds and outbuildings											
sheds	3			3	2	2	3		8	12	8
outbuildings	1	1	2	5	5	5	1	3	2	3	4
store											1
barns				1					2	1	
Stables											
company	1	1	2	1			1		4	2	1
private individuals	4	2	2	1						13	
Dwelling houses											
Grande Maison	1	1	1	1	1	1	1	1	1	1	1
shanties	8	7	7	17							
houses											
single family	1	2	2	6	14	14	1		7	22	23
multifamily	4	4	4				6		7		
unknown									14		
Service buildings											
chapel				1			1	1			1
bakery	1	1	1	1			1	1	1	1	1
ice house		1	1				1				1
bread oven	2	1	1								
Total	31	25	28	42	27	27	22	13	55	63	57

Source: Pierre Drouin and Alain Rainville, "L'organisation spatiale aux Forges du Saint-Maurice: évolution et principes", typescript
(Quebec City: Parks Canada 1980), appendix A, p. 126.

Abbreviations
u: shops that were not separate buildings
x: buildings attached to the main departments

Appendix 16   THE POULINS AND COURVAL-CRESSÉS OF THE ST MAURICE FORGES

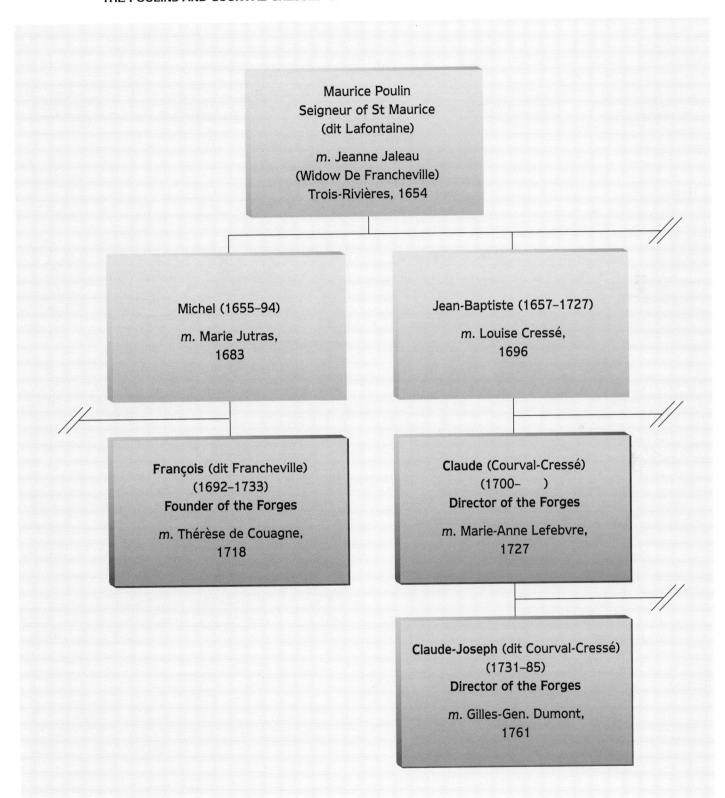

Maurice Poulin
Seigneur of St Maurice
(dit Lafontaine)

m. Jeanne Jaleau
(Widow De Francheville)
Trois-Rivières, 1654

Michel (1655–94)

m. Marie Jutras,
1683

Jean-Baptiste (1657–1727)

m. Louise Cressé,
1696

François (dit Francheville)
(1692–1733)
Founder of the Forges

m. Thérèse de Couagne,
1718

Claude (Courval-Cressé)
(1700–   )
Director of the Forges

m. Marie-Anne Lefebvre,
1727

Claude-Joseph (dit Courval-Cressé)
(1731–85)
Director of the Forges

m. Gilles-Gen. Dumont,
1761

Appendix 17

**INTERMARRIAGE BETWEEN FORGEMEN'S FAMILIES
AT THE ST MAURICE FORGES, 1829**

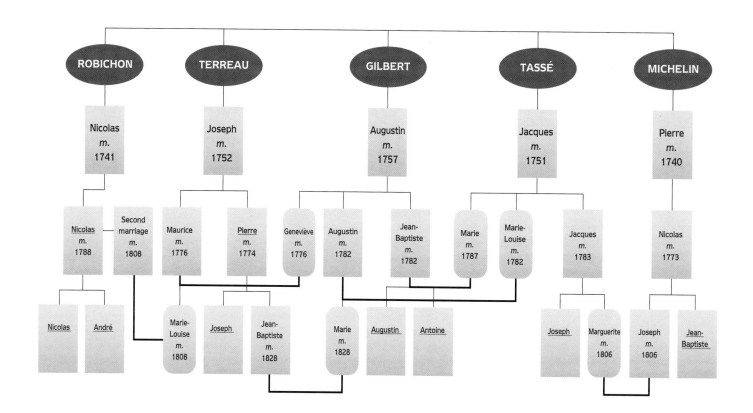

**MARRIAGE OF FORGEMAN JOHN ABBOTT
INTO THE CIRCLE OF ALLIANCES**

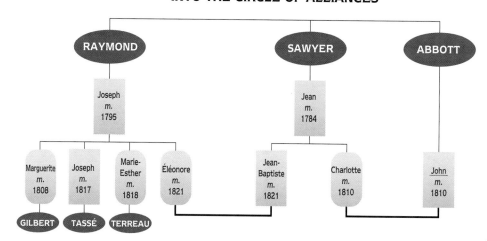

The names of forgemen are underlined.

List of Abbreviations

AETR	Archives de l'évêché de Trois-Rivières
AJM*	Archives judiciaires de Montréal
AJQ*	Archives judiciaires de Québec
AJTR*	Archives judiciaires de Trois-Rivières
ANQ-M	Archives nationales du Québec à Montréal
ANQ-Q	Archives nationales du Québec à Québec
ANQ-TR	Archives nationales du Québec à Trois-Rivières
AO	Archives of Ontario
APJQ	Archives du palais de justice de Québec
ASTR	Archives du Séminaire de Trois-Rivières
JHALC	Journals of the House of Assembly of Lower Canada (1791–1840)
JLAPC	Journals of the Legislative Assembly of the Province of Canada (1841–66)
MER	Ministère de l'Énergie et des Ressources du Québec (Quebec Department of Energy and Resources, formerly the Quebec Department of Lands and Forests)
MG	Manuscript Group (National Archives of Canada)
NAC	National Archives of Canada
Not. Rec.	Notarial Records (*greffe*)
RPA	Report of the Public Archives
RG	Record Group (National Archives of Canada)
RAPQ	Report of the Archivist for the Province of Quebec
APT	*Bulletin of the Association for Preservation Technology*
RHAF	*Revue d'histoire de l'Amérique française*

* These collections of court records are held by the local branch of the Quebec Archives in their respective cities.

Bibliography

MANUSCRIPT SOURCES

Archives de l'Évêché de Trois-Rivières
(Trois-Rivières Diocesan Archives)
Recensement des Vieilles Forges fait en
l'année 1875 (Census of the Old Forges,
1875)

Archives de l'Université de Montréal
(University of Montreal Archives)
Baby Collection

Archives de l'Université Laval
(Laval University Archives)
FM-79, Benjamin Sulte's Notes

Archives du Séminaire de Trois-Rivières
(Trois-Rivières Seminary Archives)
Notarial Records, Badeaux, J.-Bte
Forges Papers
Dollard Dubé Papers

Canada. Department of Canadian
Heritage, Parks Canada
Céline Marchand Papers
Batiscan Letterbook, 1807–12

Canada. National Archives
MG 1, Archives des Colonies, Paris
MG 2, Archives de la Marine, Paris
MG 4, Archives de la Guerre, Paris
MG 6, Archives départementales, muni-
cipales, maritimes et de bibliothèques
(France) [Departmental, municipal,
maritime and library archives (France)]
MG 8, Documents relatifs à
la Nouvelle-France et au Québec
(XVIIᵉ et XVIIIᵉ siècles) [Papers relating
to New France and Quebec
(17th and 18th centuries)]
MG 11, Colonial Office (London)
MG 12, Admiralty (London)
MG 13, War Office (London)
MG 18, Pre-Conquest Papers
MG 21, British Museum
(now British Library)
MG 23, Late 18th-century Papers
MG 28, Records of Post-Confederation
Corporate Bodies
RG 1, Executive Council, 1764–1867
RG 4, Provincial and Civil Secretaries'
Offices: Quebec, Lower Canada,
Canada East, 1760–1867

RG 8, British Military and Naval
Records, 1757–1903
RG 42, Marine Branch, 1762–1967
RG 68, Registrar General, 1763–1980
Nominal censuses of Canada
for the years 1825, 1831, 1842, 1851,
1861, 1871 and 1881

Canada Year Book
1858

McGill University Archives
Logan Papers

Ontario (Province) Archives
Baldwin Papers

Parish Records
L'Immaculée Conception
de Trois-Rivières
St Étienne des Grès
St James Anglican Church
of Three Rivers, civil records

Quebec (Province). Archives
nationales du Québec à Montréal
Notarial Records
Campbell, W. D.
Doucet, N. B.
Lighthall, W. F.

Quebec (Province). Archives
nationales du Québec à Québec
Notarial Records
Pinguet, Jacques
Saillant, Jean-Antoine
Superior Court
NF-25: Judicial and Notarial
Records, 1638–1759
Allsopp Papers

Quebec (Province). Archives
nationales du Québec à Trois-Rivières
Notarial Records
Badeaux, A.
Badeaux, J.-B.
Badeaux, Joseph
Hubert, Petrus
Superior Court
Judicial Records
Civil Records

Quebec (Province). Ministère de
l'énergie et des ressources du Québec
Service de la concession des terres

Scotland
Scottish Record Office, Edinburgh, GD
58-18-14, GD 58-16-3

PRINTED SOURCES

Annales des Arts et des Manufactures
"Sur les défauts qu'on observe en général
dans la construction et la conduite des
Fourneaux de réduction directe, et dans
les grosses forges," 30 thermidor an XI,
1803, pp. 122–23.

Baddeley, F. H.
"Lieutenant Baddeley's (Rl Engineers)
Report on the Saint Maurice Iron Works,
near Three Rivers, Lower Canada
(1828)." *APT* 5, no. 3 (1973): 9–20.

"An Essay on the Localities of Metallic
Minerals in the Canadas, with Some
Notices of Their Geological Associations
and Situation." *Transactions of the Literary
and Historical Society of Quebec* 2 (1831):
332–426.

Bouchette, Joseph
*A Topographical Description of the Province
of Lower Canada, with Remarks upon Upper
Canada, and on the Relative Connexion of
Both Provinces with the United States of
America.* London: printed for the author,
and published by W. Faden, geographer
to His Majesty and the Prince Regent,
Charing-Cross, 1815.

Bruyère, J.
"Order to M. Courval, for the manage-
ment of the Forges," 1 October 1760.
*Report of the Public Archives for the Year
1918.* Ottawa: J de Labroquerie Taché,
Printer to the King's Most Excellent
Majesty, 1920, p. 85.

Caron, Napoléon
Deux voyages sur le Saint-Maurice. Trois-
Rivières: Librairie du Sacré-Cœur, 1889.

Douglas, Thomas, Earl of Selkirk
*Lord Selkirk's Diary, 1803–1804: A Journal
of His Travels in British North America and
the Northeastern United States.* Edited with
an introduction by Patrick C. T. White.
Toronto: Champlain Society, 1958.

Franquet, Louis
*Voyages et Mémoires sur le Canada (1752–
1753).* Montreal: Éditions Élysée, 1974.

Harrington, Dr B. J.
"Notes on the Iron Ores of Canada and
Their Development." Geological Survey
of Canada, *Report of Progress for 1873–74.*
Montreal: Dawson Brothers, 1874.

Hunt, Dr T. Sterry
"Report of Dr. T. Sterry Hunt, II, Iron and Iron Ores." Geological Survey of Canada, *Report of Progress from 1866 to 1869.* Montreal: Dawson Brothers, 1870.

**Journals of the House
of Assembly of Lower Canada**
1829, app. S (no. 1).
1834, app. X.

**Journals of the Legislative Assembly
of the Province of Canada**
1844–45, vol. 4, app. O.
1852, app. C.C.C.

Kalm, Peter
Travels into North America. 3 vols. Trans. John Reinhold Forster. Warrington and London: William Eyres, 1770–71.

Lambert, John
Travels through Canada and the United States of North America, in the Years 1806, 1807 & 1808: To Which Are Added Biographical Notices and Anecdotes of Some Leading Characters in the United States. London: printed for Baldwin, Cradock, and Joy; Edinburgh: for W. Blackwood; Dublin: for J. Cumming, 1816.

Lovell's Province of Quebec Directory for 1871
Montreal: J. Lovell, 1871.

Newspapers
La Gazette de Québec/Quebec Gazette
Le Canadien
Le Constitutionnel
Le Journal des Trois-Rivières
L'Ère nouvelle

Sales Laterrière, Pierre de
Mémoires de Pierre de Sales Laterrière et de ses traverses. Quebec City: Imprimerie de l'Événement, 1873.

INVENTORIES

Beaulieu, Yves, and André Lavoie
Répertoire des parlementaires québécois 1867–1978. Quebec City: Bibliothèque de la législature, Service de documentation politique, 1980.

**Charbonneau, Hubert,
and Jacques Légaré, eds.**
Répertoire des actes de baptême, mariage, sépulture et des recensements du Québec ancien. Université de Montréal, Département de démographie. Montreal: Presses de l'Université de Montréal, 1980–88.

Magnan, Hormidas
Dictionnaire historique et géographique des paroisses, missions et municipalités de la province de Québec. Arthabaska: Imprimerie d'Arthabaska, 1925.

Roy, Pierre-Georges
Inventaire des ordonnances des intendants de la Nouvelle-France conservées aux Archives provinciales de Québec. Vol. 2. Beauceville: L'Éclaireur limitée, 1919.

DICTIONARIES AND TREATISES

Aubuisson de Voisins, J. F. d'
Traité d'hydraulique, à l'usage des ingénieurs. 2d ed. Paris: Langlois & Leclercq éditeurs, 1845.

Aviler, Augustin-Charles d'
Dictionnaire d'architecture civile et hydraulique, et des Arts qui en dépendent [...]. Paris: Charles-Antoine Jombert, 1755.

Béguillet, M.
Manuel du Meunier et du Charpentier de Moulins; ou abrégé classique du Traité de la mouture par économie [...]. Paris: Panckoucke & Delalain, 1775.

Bélidor, Bernard Forest de
Architecture hydraulique ou l'art de conduire, d'élever et de ménager les eaux pour les différents besoins de la vie [...]. Paris: C. A. Jombert, 1737–53.

**Courtivron, M. le Marquis de,
and M. Bouchu**
"Art des Forges et fourneaux à fer," *Descriptions des Arts et Métiers faites et approuvées par Messieurs de l'Académie Royale des Sciences.* Paris: Dessaint et Saillant, 1761.

Dictionary of Canadian Biography. Toronto: University of Toronto Press, 1966–.

Diderot, Denis et Jean d'Alembert
A Diderot Pictorial Encyclopedia of Trades and Industry: Manufacturing and the Technical Arts in Plates Selected from "L'Encyclopédie, ou Dictionnaire Raisonné des Sciences, des Arts et des Métiers" of Denis Diderot. 2 vols. Edited with introduction and notes by Charles Coulston Gillispie. New York: Dover, 1959.

Encyclopédie ou Dictionnaire raisonné des Sciences, des Arts et des Métiers. s.v. "Forges, (Grosses-)," by M. Bouchu, ironmaster at Veuxsaules, near Château-Vilain. Vol. 7, 1757.

Encyclopédie [...]. Recueil de planches, sur les sciences, les arts libéraux, et les arts méchaniques, avec leur explication. Pt. 3. 298 plates. Paris: Briasson, David, Le Breton, 1765.

Evans, Oliver
The Young Mill-wright and Miller's Guide. Philadelphia: Blanchard and Lea, 1853.

Fabre, M.
Essai sur la manière la plus avantageuse de construire les machines hydrauliques et en particulier les moulins à bled. Paris: Alexandre Jombert jeune, 1783.

Saint-Ange, Walter de
Métallurgie pratique du fer, ou description méthodique des procédés de fabrication de la fonte et du fer, accompagnée de documents relatifs à l'établissement des usines, à la conduite et aux résultats des opérations; avec Atlas des machines, appareils et outils actuellement employés renfermant tous les détails nécessaires pour exécuter les constructions. Drawings and engravings by Le Blanc. Paris: Librairie scientifique et industrielle de L. Mathias, 1835–38.

Slight, James, and R. Scott Burn
The Book of Farm Implements and Machines. Ed. Henry Stephens. Edinburgh and London, 1858.

Templeton, William
The Millwright and Engineer's Pocket Companion. London: Simpkin, Marshall and Co., 1871.

STUDIES AND PAPERS

Antonetti, Guy
"Recherches sur la propriété et l'exploitation des hauts fourneaux du Châtillonnais." Fascicule 13, *La Révolution en Côte-d'Or.* Dijon: Archives départementales de la Côte-d'Or, 1973.

Audet, Louis-Philippe,
and Armand Gauthier
Le système scolaire du Québec. Montreal: Librairie Beauchemin, 1969.

Barriault, Monique
"Rapport préliminaire sur l'identification des techniques de moulage utilisées aux Forges du Saint-Maurice. Étude faite à partir des déchets de moulage, accompagnée d'un lexique français-anglais des termes de fonderie." Manuscript Report No. 330. Ottawa: Parks Canada, 1978.

"La capacité adaptative des Forges de St-Maurice face aux changements économiques et technologiques, vue à travers l'évolution fonctionnelle des ateliers de moulage adjacents au haut-fourneau." Master's thesis, Archaeology, Université Laval, 1984.

Bartlett, James Herbert
The Manufacture, Consumption and Production of Iron, Steel, and Coal, in the Dominion of Canada. Montreal: Dawson Brothers, 1885.

Beaudet, Pierre
"Excavations to the South of the Blast Furnace at the Forges du Saint-Maurice, Quebec 1975: Vertical and Horizontal Displacements." Manuscript Report No. 209. Ottawa: Parks Canada, 1976.

"Vestiges des bâtiments et ouvrages à la forge basse, Forges du Saint-Maurice." Manuscript Report No. 315. Ottawa: Parks Canada, 1979.

"The Saint-Maurice Ironworks, Canada." *Journal of the Historical Metallurgy Society* 17, no. 1 (1983): 39–41.

Bédard, Michel
"La fontaine du diable." Typescript. Quebec City: Parks Canada, 1976.

"Le territoire des Forges du Saint-Maurice, 1863–1884." Manuscript Report No. 220. Ottawa: Parks Canada, 1976.

"Localisation d'emplacements mentionnés dans les légendes des Forges du Saint-Maurice." Typescript. Quebec City: Parks Canada, 1977.

"Les moulins à farine et à scie aux Forges du Saint-Maurice." Manuscript Report No. 301. Ottawa: Parks Canada, 1978.

"La structure chronologique des Forges du Saint-Maurice (1846–1883)." Typescript. Quebec City: Parks Canada, 1979.

"Utilisation et commémoration du site des Forges du Saint-Maurice, 1883–1963." Manuscript Report No. 357. Ottawa: Parks Canada, 1979.

"Le contexte de fermeture des Forges du Saint-Maurice, 1846–1883." Typescript. Quebec City: Parks Canada, 1980.

"Tarification, commercialisation et vente des produits des Forges du Saint-Maurice." Typescript. Quebec City: Parks Canada, 1982a.

"Fluctuations des prix de certains produits manufacturés aux Forges du Saint-Maurice (1740–1858)." Typescript. Quebec City: Parks Canada, 1982b.

"L'énigmatique fermeture des Forges du Saint-Maurice." *Image de la Mauricie* 7, no. 9 (June 1983): 6–7.

"La privatisation des Forges du Saint-Maurice, 1846–1883: adaptation, spécialisation et fermeture." Master's thesis, Université Laval, 1986. Manuscript on file. Quebec City: Environment Canada, Canadian Parks Service, 1986.

Bédard, Michel, André Bérubé,
Claire Mousseau, Marcel Moussette
and Pierre Nadon
"Le ruisseau des Forges du Saint-Maurice." Manuscript Report No. 302. Ottawa: Parks Canada, 1978.

Bélisle, Jean
"Le domaine de l'habitation aux Forges du Saint-Maurice." Manuscript Report No. 307. Ottawa: Parks Canada, 1976.

"La Grande Maison des Forges du Saint-Maurice, témoin de l'intégration des fonctions, étude structurale." Manuscript Report No. 272. Ottawa: Parks Canada, 1977.

"La maçonnerie d'époque aux Trois-Rivières." Manuscript Report No. 265. Ottawa: Parks Canada, 1977.

"Le phénomène du transport des bâtiments aux Forges du Saint-Maurice." Manuscript Report No. 265. Ottawa: Parks Canada, 1977.

Bélisle, Jean, and André Lépine
"Le site de l'épave d'un des premiers bateaux à vapeur de la Molson Line." Preliminary report on the third excavations (1986). Comité d'histoire et d'archéologie subaquatique du Québec, December 1986.

Bellemare, J.-E.
"Les Vieilles Forges Saint-Maurice et les Forges Radnor." *Bulletin des recherches historiques* 24, no. 9 (September 1918): 257–69.

Benoit, Serge
"La consommation de combustible végétal et l'évolution des systèmes techniques." In *Forges et forêts: recherches sur la consommation proto-industrielle de bois*, ed. Denis Woronoff, pp. 87–150. Paris: Éditions de l'École des hautes études en sciences sociales, 1990.

Bérubé, André
"Rapport préliminaire sur l'évolution des techniques sidérurgiques aux Forges du Saint-Maurice, 1729–1883." Manuscript Report No. 221. Ottawa: Parks Canada, 1976.

"L'évolution des techniques sidérurgiques aux Forges du Saint-Maurice." *Research Bulletin*, no. 49. Ottawa: Parks Canada, 1977.

"L'évolution des techniques sidérurgiques aux Forges du Saint-Maurice, 1: La préparation des matières premières." Manuscript Report No. 305. Ottawa: Parks Canada, 1978.

"Les changements à l'intérieur de la filière technique des Forges du Saint-Maurice entre 1729 et 1883." Typescript. Paper presented to the annual conference of the Institut d'histoire de l'Amérique française, 11 October 1980.

"The Evolution of Technology at Les Forges du Saint-Maurice, 1729–1883." In *The Industrial Heritage*. Transactions of the third international conference on the conservation of industrial monuments, ed. Marie Nisser, pp. 170–73. Vol. 3 (1981). Nordiska Museet, Stockholm.

"Technological Changes at Les Forges du Saint-Maurice, 1729–1883." *The Canadian Mining and Metallurgical Bulletin* 76, no. 853 (May 1983): 75–80.

"Vérification sur le terrain de la position des bâtiments 17 et 29 représentés sur la photographie des Forges du Saint-Maurice dite 'photo McDougall.'" Typescript. Quebec City: Parks Canada, 1983.

"The St. Maurice Industrial Community: Some Preliminary Thoughts." In *Industrial Heritage '84*. Proceedings of the fifth international conference on the conservation of the industrial heritage, ed. Helena Wright and Robert M. Vogel, pp. 122–27. Vol. 2. (1984).

"Les Forges du Saint-Maurice: un exemple de recherche historique appliquée au développement d'un parc historique national." Paper presented at the annual meeting of the Canadian Historical Association, Université de Montréal, 28–30 May 1985.

Bevan, Bruce
"A Magnetic Survey at Les Forges du Saint-Maurice." Typescript. Museum Applied Science Center for Archaeology, The University Museum, University of Pennsylvania, October 1975.

Bischoff, Peter
"Des Forges du Saint-Maurice aux fonderies de Montréal: mobilité géographique, solidarité communautaire et action syndicale des mouleurs, 1829–1881." *Revue d'histoire de l'Amérique française* 43, no. 1 (summer 1989): 3–29.

Boissonnault, Réal
"La structure chronologique des Forges du Saint-Maurice des débuts à 1846." Typescript. Quebec City: Parks Canada, 1980.

"Quelques notions sur l'orientation de la production et les types de produits fabriqués aux Forges du Saint-Maurice." Typescript. Quebec City: Parks Canada, 1981.

Les Forges du Saint-Maurice, 1729–1883, 150 Years of Occupation and Operation. Les Forges du Saint-Maurice National Historic Site, Booklet No. 1. Quebec City: Parks Canada, 1983.

Brisson, Réal
La charpenterie navale à Québec sous le régime français. Les 100 premières années de la charpenterie navale à Québec: 1663–1763. Quebec City: Institut québécois de recherche sur la culture, 1983.

Carlier, Michel
Hydraulique générale et appliquée. Collection de la Direction des études et recherches d'Électricité de France. N.p.: Eyrolles, 1980.

Caron, Marcelle
"Analyse comparative des quatre versions de l'enquête de Dollard Dubé sur les Forges Saint-Maurice." Manuscript on file. Quebec City: Parks Canada, 1982.

Casteran, Nicole
"Répertoire des produits des Forges du Saint-Maurice." Manuscript Report No. 132. Ottawa: Parks Canada, 1973.

"Fabrication d'armement aux Forges du Saint-Maurice." Ottawa: Parks Canada, 1975.

Charbonneau, André, Yvon Desloges and Marc Lafrance
Quebec, the Fortified City: From the 17th to the 19th Century. Ottawa: Parks Canada, 1982.

Charbonneau, Hubert, and Micheline Tremblay
"La population des Forges du St-Maurice." Typescript. Université de Montréal, Département de démographie, 1982.

Clavet, Alain
"Les premières tentatives de production de fer aux Forges du Saint-Maurice: le bas-fourneau de François-Poulin de Francheville (1729–1734)." Master's thesis, Université de Sherbrooke, 1983.

Cloutier, Johanne
"Répertoire des produits fabriqués aux Forges du Saint-Maurice." Manuscript Report No. 350. Ottawa: Parks Canada, 1980.

Cloutier-Nadeau, Céline
"Les abords de la maison Francheville aux Forges du Saint-Maurice, stratégie de fouille." Manuscript Report No. 360. Ottawa: Parks Canada, 1978.

Courcy, Simon
"Étude préliminaire du matériel céramique provenant des Forges." Typescript. Ottawa: Parks Canada, 1974.

"Les poêles des Forges du Saint-Maurice, 1742–1860." Microfiche Report No. 332. Ottawa: Parks Canada, 1986.

"Gueuses et gueusets fabriqués aux Forges du Saint-Maurice." Document file and inventory of archaeological artifacts prepared as part of the Grande Maison restoration project. Quebec City: Environment Canada, Canadian Parks Service, 1989.

"L'artillerie fabriquée aux Forges du Saint-Maurice." Document file and artifact inventory. Quebec City: Environment Canada, Canadian Parks Service, 1989.

"Les objets archéologiques exposés dans les caves de la grande Maison des Forges du Saint-Maurice." Quebec City: Environment Canada, Canadian Parks Service, 1991.

"Les objets archéologiques exposés au rez-de-chaussée de la grande Maison des Forges du Saint-Maurice." Quebec City: Environment Canada, Canadian Parks Service, 1991.

Courcy, Simon, and Marcel Tardif
"Essai de chronologie appliquée au secteur domestique 25G7G8 (1976)." Manuscript Report No. 448. Ottawa: Parks Canada, 1976.

Cox, Richard
"Maçonnerie de la salle des soufflets et emplacement des engrenages." Typescript. Quebec City: Parks Canada, 1976.

"Les Forges du Saint-Maurice." *Research Bulletin*, no. 51. Ottawa: Parks Canada, March 1977.

Daumas, Maurice
Histoire générale des techniques. Vol. 3, *L'expansion du machinisme.* Paris: Presses universitaires de France, 1968.

Donald, W. J. A.
The Canadian Iron and Steel Industry: A Study in the Economic History of a Protected Industry. Boston and New York: Houghton Mifflin, 1915.

Dorion, Jacques
"Le folklore oral des Forges du Saint-Maurice." Manuscript Report No. 255. Ottawa: Parks Canada, 1977.

Dornic, François
Le fer contre la forêt. N.p.: Ouest France, 1984.

Drouin, Pierre
"Un secteur d'habitation d'ouvriers (25G7–25G8) aux Forges du Saint-Maurice." Manuscript Report No. 254. Ottawa: Parks Canada, 1977.

"The Worker's Habitation Area at the St. Maurice Ironworks." Master's thesis, Arizona State University, 1977.

"La Maison des forgerons de la forge basse (structure 24.1)." Manuscript Report No. 313. Ottawa: Parks Canada, 1978.

"Reconnaissance archéologique aux Forges du Saint-Maurice, été 1979." Manuscript Report No. 448. Ottawa: Parks Canada, 1980.

"Report on Archaeological Research at the Master's House of the Forges du Saint-Maurice." *Research Bulletin*, no. 144. Ottawa: Parks Canada, 1980.

"La Grande Maison des Forges du Saint-Maurice: rapport archéologique 1977–1978." Typescript. Quebec City: Parks Canada, 1982.

"Les chemins et bâtiments de service dans l'aire du stationnement aux Forges du Saint-Maurice: fouilles archéologiques 1981–1982." Microfiche Report No. 166. Ottawa: Parks Canada, 1984.

"Interventions archéologiques aux Forges du Saint-Maurice en 1983." Microfiche Report No. 212. Ottawa: Parks Canada, 1985.

"Des charrons aux Forges du Saint-Maurice." In *Archéologies québécoises,* ed. Anne-Marie Balac, pp. 369–384. Paléo-Québec no. 23. Montreal: Recherches amérindiennes au Québec, 1995.

Drouin, Pierre, and Alain Rainville
"Données sur l'évolution de l'organisation spatiale aux Forges du Saint-Maurice. Étape 1." Preliminary report. Quebec City: Environment Canada, Canadian Parks Service, 1978.

"Dossiers sur l'organisation spatiale du site des Forges du Saint-Maurice." Typescript. Quebec City: Parks Canada, 1979.

"L'organisation spatiale aux Forges du Saint-Maurice: évolution et principes." Typescript. Quebec City: Parks Canada, 1980. Microfiche Report No. 6. Ottawa: Parks Canada, 1983.

Drouin, Pierre, Françoise Niellon and F. Sée
"Les Forges du Saint-Maurice (25G): rapport de fouille préliminaire." Manuscript Report No. 175. Ottawa: Parks Canada, 1973.

Dubé, Dollard
Les vieilles forges il y a 60 ans. Pages Trifluviennes, Série A, no. 4. Trois-Rivières: Les Éditions du Bien Public, 1933.

Espesset, Hélène, Jean-Pierre Hardy and Thierry Ruddell
"Le monde du travail au Québec au XVIIIᵉ et au XIXᵉ siècle: historiographie et état de la question." *Revue d'histoire de l'Amérique française* 25, no. 4 (March 1972): 499–539.

Évrard, René
Les artistes et les usines à fer. Liège: Éditions Solédi, 1955.

Faucher, Albert
Québec en Amérique au XIXᵉ siècle. Montreal: Fides, 1973.

Fauteux, Jean-Noël
Essai sur l'industrie au Canada sous le régime français. 2 vols. Quebec City: Ls-A. Proulx, King's Printer, 1927.

Filion, Maurice
La pensée et l'action coloniales de Maurepas vis-à-vis du Canada, 1723–1749, l'âge d'or de la colonie. Montreal: Leméac, 1972.

Fiset, Michel, A. Galibois and T. Vo Van
"Étude métallurgique d'un groupe d'objets en fer provenant d'un contexte archéologique sûr à la forge basse des Forges du Saint-Maurice." Typescript. Quebec City: Parks Canada, 1980.

"Analyse métallurgique d'un groupe d'objets en métal provenant de contextes archéologiques aux Forges du Saint-Maurice." Typescript. Quebec City: Parks Canada, 1982.

Fontaine, Achille
"Génie d'hier et d'aujourdhui aux Forges du Saint-Maurice." Information given to guides, summer 1978. Manuscript on file. Quebec City: Engineering and Architecture, Parks Canada, May 1978.

"Étude des mécanismes hydrauliques du haut fourneau. Forges du Saint-Maurice." Internal document. Quebec City: Engineering and Architecture, Parks Canada, December 1980.

Fortier, Marie-France

"Rapport d'étude sommaire des chiffres concernant la population des Forges du Saint-Maurice." Manuscript on file. Quebec City: Parks Canada, 1976.

"La structuration sociale du village industriel des Forges du Saint-Maurice: étude quantitative et qualitative." Manuscript Report No. 259. Ottawa: Parks Canada, 1977.

"Une industrie et son village: Les Forges du Saint-Maurice, 1729–1764." Master's thesis, Université Laval, 1981.

Fortin, Claire-Andrée, and Benoît Gauthier, under the direction of René Hardy

"Aperçu de l'histoire des Forges Saint-Tite et Batiscan et préliminaires à une analyse de l'évolution du secteur sidérurgique mauricien, 1793–1910." Research report submitted to the Regional Branch of the Quebec Department of Cultural Affairs. Centre de recherches en études québécoises, Université du Québec à Trois-Rivières, December 1985.

"Les entreprises sidérurgiques mauriciennes au XIXe siècle: approvisionnement en matières premières, biographies d'entrepreneurs, organisation et financement des entreprises." Research report submitted to the Regional Branch of the Quebec Department of Cultural Affairs. Centre de recherches en études québécoises, Université du Québec à Trois-Rivières, November 1986.

"Description des techniques et analyse du déclin de la sidérurgie mauricienne, 1846–1910." Research report submitted to the Regional Branch of the Quebec Department of Cultural Affairs. Centre de recherches en études québécoises, Université du Québec à Trois-Rivières, February 1988.

France. Ministère de la Culture

Les forges du Pays de Châteaubriant. Cahiers de l'Inventaire 3, Inventaire général des monuments et richesses artistiques de la France, Pays de Loire, Département de Loire-Atlantique, 1984.

Gale, W. K. V.

The British Iron & Steel Industry: A Technical History. Newton Abbot: David & Charles, 1967.

The Iron and Steel Industry: A Dictionary of Terms. Newton Abbot: David & Charles, 1971.

Gaumond, Michel

Les Forges du Saint-Maurice. Textes no. 2. Quebec City: Société historique de Québec, April 1969.

Gauthier, Benoît

"La criminalité aux Forges du Saint-Maurice, rapport préliminaire." Typescript. Quebec City: Parks Canada, 1982.

"Les sites sidérurgiques en Mauricie (Radnor, Saint-Tite, L'Islet)." Author's manuscript, April 1983.

Gauvin, Robert

"Archaeological Activities at the Master's House at the Forges du Saint-Maurice, Autumn 1987." *Research Bulletin,* no. 276 (March 1989).

Gille, Bertrand

Les origines de la grande industrie métallurgique en France. Paris: Éditions Domat, 1947.

"Le moyen âge en Occident (Ve siècle–1350)." In *Les origines de la civilisation technologique,* vol. 1, *Histoire générale des techniques,* ed. M. Daumas, pp. 431–598. Paris: Presses universitaires de France, 1962.

Greer, Allan

"Le territoire des Forges du Saint-Maurice, 1730–1862." Manuscript Report No. 220. Ottawa: Parks Canada, 1975.

Habashi, Fathi

"Chemistry and Metallurgy in New France." *Chemistry in Canada* (May 1975): 25–27.

Hamelin, Jean, and Jean Provencher

"La vie de relations sur le Saint-Laurent entre Québec et Montréal au milieu du XVIIIe siècle." *Cahiers de géographie de Québec* 11, no. 23 (1967).

Hamelin, Jean, and Yves Roby

Histoire économique du Québec, 1851–1896. Montreal: Fides, 1971.

Hardach, Dr Gerd H.

Der soziale Status des Arbeiters in der Frühindustrialisierung. Eine Untersuchung über die Arbeitnehmer in der französischen eisenschaffenden Industrie zwischen 1800 und 1870. Schriften zur Wirtschafts– und Sozialgeschichte. Berlin: Duncker & Humblot, 1969.

Hardy, René

"La sidérurgie de la Mauricie: matières premières et main-d'œuvre rurale." *Cahier des Annales de Normandie* (Caen: Musée de Normandie), no. 24 (1992): 287–97.

La sidérurgie dans le monde rural: les hauts fourneaux du Québec au XIXe siècle. Quebec City: Presses de l'Université Laval, 1995.

Hardy, René, and Benoît Gauthier

La sidérurgie en Mauricie au 19e siècle: les villages industriels et leurs populations. Research report submitted to the Regional Branch of the Quebec Department of Cultural Affairs. Centre de recherches en études québécoises, Université du Québec à Trois-Rivières, May 1989.

Hardy, René, and Normand Séguin

Forêt et société en Mauricie. Montreal: Boréal Express, 1984.

Hartley, E. N.

Ironworks on the Saugus. Norman: University of Oklahoma Press, (1957) 1971.

Henripin, Jacques

La population canadienne au début du XVIIIe siècle. Travaux et documents, cahier no. 22. Paris: Institut national d'études démographiques, 1954.

Inwood, Kris E.

The Canadian Charcoal Iron Industry, 1870–1914. New York: Garland, 1986.

"The Influence of Resource Quality on Technological Persistence: Charcoal Iron in Quebec." *Material History Review* 36 (autumn 1992): 49–56.

Kury, Theodore W.

"The Iron Plantation: Agent in the Formation of the Cultural Landscape." Paper presented at the symposium on the industrial archaeology of the American iron industry, 13th annual conference, Society for Industrial Archaeology, Boston, Mass., 16 June 1984.

Lagrave, François de, and la Corporation communautaire de Saint-Michel-des-Forges

Au pays des Cyclopes, Saint-Michel-des-Forges, 1740–1990. Trois-Rivières: Corporation communautaire de Saint-Michel-des-Forges, 1990.

Landes, David S.
The Unbound Prometheus: Technological Change and Industrial Development in Western Europe from 1750 to the Present. Cambridge: Cambridge University Press, 1969.

Lanthier, Pierre, and Alain Gamelin
L'industrialisation de la Mauricie: dossier statistique et chronologique, 1870–1975. Groupe de recherche sur la Mauricie, cahier no. 6. Trois-Rivières: Université du Québec à Trois-Rivières, November 1981.

Lapointe, Camille
"Étude d'un atelier de finition et d'assemblage de poêles et contenants de fonte aux Forges du Saint-Maurice." Manuscript Report No. 344. Ottawa: Parks Canada, 1979.

Larouche, Alayn
"Étude préliminaire des macrorestes de la localité des Forges du Saint-Maurice." Typescript. Quebec City: Parks Canada, 1975.

"Analyse des macrorestes végétaux d'une carotte prélevée dans la localité des Forges du Saint-Maurice et contenant un horizon tourbeux enfoui." Typescript. Quebec City, Université Laval, 1977.

"Histoire comparée de deux sites (25G3–25G7) des Forges du Saint-Maurice, telle que révélée par l'analyse des macrorestes." Typescript. March 1977.

"Analyse des macrorestes végétaux aux Forges du Saint-Maurice: les jardins potagers." Typescript. Université de Montréal, Département de géographie, January 1979.

Leboutte, René
La grosse forge wallonne (du XVᵉ au XVIIIᵉ siècle). Liège: Éditions du Musée de la vie wallonne, 1984.

Lee, David
"A Short History of the St. Maurice Forges." Manuscript Report No. 132. Ottawa: Parks Canada, 1965.

Léon, Pierre
Les techniques métallurgiques dauphinoises au dix-huitième siècle. Paris: Hermann, 1961.

Lepage, André
"Étude du travail et de la production aux Forges du Saint-Maurice à deux moments de l'histoire de l'entreprise." Study conducted for Historical Research, Quebec Region, Parks Canada, June 1984.

L'Heureux, Réjean
Vocabulaire du moulin traditionnel au Québec des origines à nos jours. Documents lexicaux et ethnographiques. Langue française au Québec, section 3. Quebec City: Presses de l'Université Laval, 1982.

Lunn, Alice Jean E.
Développement économique de la Nouvelle-France, 1713–1760. Montreal: Presses de l'Université de Montréal, 1986.

Mathieu, Jacques
La construction navale royale à Québec, 1739–1759. Cahiers d'histoire no. 23. Quebec City: Société historique de Québec, 1971.

Mayrand, Pierre
"La culture et les souvenirs de voyage de l'ingénieur Louis Franquet." Research notes. *Revue d'histoire de l'Amérique française* 25, no. 1 (June 1971), 91–94.

McCullough, A. B.
Money and Exchange in Canada to 1900. Toronto and Charlottetown: Dundurn Press, 1984. Published in co-operation with Parks Canada and the Canadian Government Publishing Centre.

McDougall, David J.
"The St. Francis Forges and the Grantham Iron Works." Author's manuscript. Concordia University, n.d.

McGain, Alison
"Fouilles archéologiques d'un bloc domestique aux Forges du Saint-Maurice en 1974 (25G51)." Manuscript Report No. 232. Ottawa: Parks Canada, 1977.

"La maison du marteleur aux Forges du Saint-Maurice, rapport de fouille 1974–1975." Manuscript Report No. 313. Ottawa: Parks Canada, 1977.

"Travaux d'hiver aux Forges du Saint-Maurice, 1977." *Research Bulletin*, no. 68. Ottawa: Parks Canada, December 1977.

Miville-Deschênes, François
"Les Forges du Saint-Maurice ou l'art de la fonderie: contenants domestiques et pièces d'artillerie." Microfiche Report No. 146. Ottawa: Parks Canada, 1983.

Morasse, Claire
"Recherches archéologiques dans la cuisine de la Grande Maison des Forges du Saint-Maurice, 1982." Typescript. Quebec City: Parks Canada, 1983.

Mousseau, Claire
"Bibliographie sélective de travaux imprimés sur la machinerie et l'équipement ancien." Typescript. Quebec City: Parks Canada, 1977.

"L'évolution fonctionnelle de la forge haute à travers la transformation des ouvrages, 1739–1883." Manuscript Report No. 398. Ottawa: Parks Canada, 1978.

"Sondage et forage: évaluation d'un outil de recherche en archéologie." Typescript. Quebec City: Parks Canada, April 1978.

"Reconnaissance archéologique, automne 1980, Forges du Saint-Maurice. Dossier 1: Opérations 25G14 et 25G15, fouilles des aménagements hydrauliques. Dossier 2: Opérations 25G15, fondations d'un bâtiment localisé à l'est du haut fourneau." Typescript. Quebec City: Parks Canada, 1980.

Moussette, Marcel
"Essais de typologie des poêles des Forges du Saint-Maurice." Manuscript Report No. 332. Ottawa: Parks Canada, 1975.

"L'histoire écologique des Forges du Saint-Maurice." Manuscript Report No. 333. Ottawa: Parks Canada, 1978.

Le chauffage domestique au Canada, des origines à l'industrialisation. Quebec City: Presses de l'Université Laval, 1983.

Nadon, Pierre
"Recherches archéologiques aux Forges." Typescript. Ottawa: Parks Canada, April 1975.

"La recherche archéologique aux Forges." *Research Bulletin*, no. 44. Ottawa: Parks Canada, 1977.

"Les Forges du Saint-Maurice." *Les dossiers de l'archéologie*, no. 27 (March-April 1978): 96–101.

"The St. Maurice System." Typescript. Paper presented to the conference of the Canadian Archaeological Association, 1980.

Nef, J.
"La civilisation industrielle." First part of "Industrie," *Encyclopaedia Universalis* (1968), 5th ed., 1973.

Nicol, Heather
"Faunal Analysis of Selected Contexts at Les Forges du Saint-Maurice." Typescript. Ottawa: Zooarchaeological Identification Centre, Museum of Natural Science, 1978.

"Faunal Analysis of Bones from the Grande Maison at St. Maurice Forges." Typescript. Ottawa: Zooarchaeological Identification Centre, Museum of Natural Science, 1979.

Niellon, Françoise
"La maison du contremaître aux Forges du Saint-Maurice (25G20). Éléments d'architecture: synthèse préliminaire, rapport préliminaire sur la fouille de 1974." Manuscript Report No. 152. Ottawa: Parks Canada, June 1975.

"La maison du contremaître aux Forges du Saint-Maurice (25G20): rapport préliminaire sur la fouille de 1974." Manuscript Report No. 152. Ottawa: Parks Canada, 1975.

Nish, Cameron
François-Étienne Cugnet. Entrepreneur et entreprises en Nouvelle-France. Montreal: Fides, 1975.

Ouellet, Fernand
Histoire économique et sociale du Québec, 1760–1850. Structures et conjoncture. Montreal and Paris: Fides, 1966.

Pacco-Picard, Maïté
Les manufactures de fer peintes par Léonard Defrance. Musées vivants de Wallonie et de Bruxelles, no. 3. Liège: Pierre Mardaga éditeur, 1982.

Pelletier, Gabriel
Les Forges de Fraisans. La métallurgie comtoise à travers les siècles. Dampierre: published by the author, 1980.

Pentland, H. Clare
Labour and Capital in Canada, 1650–1860. Edited and with an introduction by Paul Philips. Toronto: James Lorimer & Co., 1981.

Rainville, Alain
"Les bâtiments de service aux Forges du Saint-Maurice." Manuscript Report No. 301. Ottawa: Parks Canada, 1978.

"Grande Maison. Dossier architectural." Typescript. Quebec City: Parks Canada, 1982.

"Le logement ouvrier aux Forges du Saint-Maurice." Microfiche Report No. 12. Ottawa: Parks Canada, 1983.

Renaud, Roxanne
"Surveillance archéologique aux Forges du Saint-Maurice (1988). Preliminary report. Quebec City: Environment Canada, Canadian Parks Service, 1989.

Risi, Joseph
L'industrie de la carbonisation du bois dans la Province de Québec. Bulletin no. 3, nouvelle série. Province de Québec, Ministère des Terres et Forêts, Service Forestier, 1942.

Saint-Pierre, Serge
"La technologie artisanale aux Forges du Saint-Maurice, 1729–1883." Manuscript Report No. 307. Ottawa: Parks Canada, 1976.

"Les artisans du fer aux Forges du Saint-Maurice: aspect technologique." Manuscript Report No. 307. Ottawa: Parks Canada, 1977.

"Les charretiers aux Forges du Saint-Maurice." Typescript. Quebec City: Parks Canada, 1977.

"La technologie artisanale aux Forges du Saint-Maurice, 1729–1883. *Research Bulletin,* no. 48. Ottawa: Parks Canada, 1977.

Samson, Roch
"Les ouvriers des Forges du Saint-Maurice: aspects démographiques, 1762–1851." Microfiche Report No. 119. Ottawa: Parks Canada, 1983.

"Men of Iron," *Horizon Canada* 3, no. 25 (1985): 668–72.

"Une industrie avant l'industrialisation: le cas des Forges du Saint-Maurice." *Anthropologie et Sociétés* 10, no. 1 (1986): 85–107.

"Les Forges du Saint-Maurice: un site d'archéologie pré-industrielle." Paper presented to the 18th annual conference of the Society for Industrial Archaeology, Quebec City, 1–4 June 1989.

"Les maquettistes (La maquette des Forges du Saint-Maurice)." *Traces* (Journal of the history teachers' association) 28, no. 2 (March-April 1990).

"Setting Up the First Industrial Community in New France: Les Forges du Saint-Maurice (1730–1883)." Paper presented to the conference of the Society of Historians of the Early American Republic, York University, Toronto, 2–4 August 1990.

"Les Forges du Saint-Maurice dans le moule de l'histoire: le projet de la Grande Maison." Paper presented to the 70th annual conference of the Canadian Historical Association, at the Learned Societies' Conference, Queen's University, Kingston, 3–5 June 1991.

"L'histoire du génie hydraulique au Canada, Chapitre I: L'ingénieur Chaussegros de Léry aux Forges du Saint-Maurice (1738–1739)." In *Engineering and Society.* Proceedings of the eighth Canadian conference on engineering education, Université Laval, Quebec City, May 1992, pp. 519–29.

"La maquette des Forges du Saint-Maurice." *Machines* (Newsletter of the Canadian Society for Industrial Heritage), no. 4 (spring 1992): 3.

"Les Forges du Saint-Maurice." *Cap-aux-Diamants* (spring 1994): 28–32.

"Les Forges du Saint-Maurice. Au début était le fer… ." *Continuité,* no. 70 (fall 1996): 23–25.

Samson, Roch, and André Bérubé
"Les rebâtisseurs (La grande maison des Forges du Saint-Maurice)." *Traces* (Journal of the history teachers' association) 28, no. 1 (January-February 1990).

Samson, Roch, and Achille Fontaine
"La mise en opération des Forges du Saint-Maurice (1736–1741): une étude pluridisciplinaire." *APT* 18, no. 1–2 (1986): 15–31.

Séguin, Normand
La conquête du sol au 19e siècle. Sillery: Boréal Express, 1977.

Séguin, Robert-Lionel
L'équipement aratoire et horticole du Québec ancien (XVIIe, XVIIIe et XIXe siècle). Vol. 1. Culture populaire/Guérin littérature. Montreal: Guérin, 1989.

Sulte, Benjamin
Les Forges Saint-Maurice. Mélanges historiques, vol. 6. Montreal: G. Ducharme, 1920.

Sulte, Benjamin, Napoléon Caron et al.
Contes et légendes des Vieilles Forges. Trois-Rivières: Éditions du Bien Public, 1954.

Tessier, Monsignor Albert
Les Forges Saint-Maurice. Montreal and Quebec City: Boréal Express, (1952) 1974.

Thuillier, Guy

Georges Dufaud et les débuts du grand capitalisme dans la métallurgie, en Nivernais, au XIXᵉ siècle. Paris: SEVPEN, 1959.

Tremblay, Yves

"Étude de la maison dite du mouleur, secteur haut fourneau." Manuscript Report No. 366. Ottawa: Parks Canada, 1978.

Trottier, Louise

Les Forges du Saint-Maurice: Their Historiography. History and Archaeology No. 42. Ottawa: Parks Canada, 1980.

Trudel, Marcel

"Les Forges Saint-Maurice sous le régime militaire (1760–1764)." *Revue d'histoire de l'Amérique française* 5, no. 2 (September 1951).

Initiation à la Nouvelle-France: histoire et institutions. Montreal: Holt, Rinehart and Winston, 1968.

Les débuts du régime seigneurial au Canada. Montreal: Fides, 1974.

Tylecote, R. T.

A History of Metallurgy. London: Metal Society, 1976.

Unglik, Henry

"Examination of Cast Irons and Wrought Irons from les Forges du Saint-Maurice." Typescript. Ottawa: Parks Canada, 1977.

Cast Irons from les Forges du Saint-Maurice, Québec: Metallurgical Study. Ottawa: Environment Canada, Parks Service, 1990.

Vallières, Marc

Des mines et des hommes, histoire de l'industrie minérale québécoise, des origines au début des années 1980. Quebec City: Les Publications du Québec, 1988.

Vermette, Luce

"Monographie d'employés aux Forges du Saint-Maurice." Manuscript Report No. 292. Ottawa: Parks Canada, 1978.

Domestic Life at Les Forges du Saint-Maurice. History and Archaeology No. 58. Ottawa: Parks Canada, 1982.

Vial, J.

"L'ouvrier métallurgiste français." *Droit social* (February 1950): 58–68.

Villeneuve, Daniel

"La fabrication des haches aux Forges du Saint-Maurice." *La vie quotidienne au Québec: Histoire, métiers, techniques et traditions.* Quebec City: Presses de l'Université du Québec, 1983, pp. 361–79.

Walker, Joseph

Hopewell Village: The Dynamics of a Nineteenth Century Ironworking Community. Philadelphia: University of Pennsylvania Press, 1974.

Wayman, Michael L., ed.

All That Glitters: Readings in Historical Metallurgy. The Metallurgical Society of the Canadian Institute of Mining and Metallurgy, 1989.

Willis, John

"Note critique" on H. Clare Pentland's *Labour and Capital in Canada, 1650–1860. Revue d'histoire de l'Amérique française* 38, no. 2 (1984).

"Seigneurialism, Immigration and the Merchants: The Transition to Industrial Capitalism on the Lachine Canal in the 19th century." Paper presented to the 64th annual conference of the Canadian Historical Association, Université de Montréal, 28–30 May 1985.

Woronoff, Denis

"Le monde ouvrier de la sidérurgie ancienne: note sur l'exemple français." *Le Mouvement social,* no. 97 (1976): 109–19.

L'industrie sidérurgique en France pendant la Révolution et l'Empire. Paris: Éditions de l'École des hautes études en sciences sociales, 1984.

"Forges prédatrices, forges protectrices." *Revue géographique des Pyrénées et du Sud-Ouest* 55, no. 2 (April-June 1984): 213–18.

Forges et forêts: recherches sur la consommation proto-industrielle de bois, ed. Denis Woronoff. Paris: Éditions de l'École des hautes études en sciences sociales, 1990.

Wurtele, F. C.

"Historical Record of the St. Maurice Forges, The Oldest Active Blast-Furnace on the Continent of America." *Proceedings and Transactions of the Royal Society of Canada for the Year 1886,* vol. 4, section 2, Montreal, 1887.

Glossary

ancony: dumbbell-shaped **bloom** with a drawn-out centre that is the intermediate stage of the **shingling** process (the **mocket head** is the next step) (Plate 4.16)

anvil: large block of **cast iron** on which the piece of iron being hammered is placed (Plate 4.16)

back wall: the rear wall of the **hearth** of the **blast furnace**, facing the **tymp** and the **taphole**

bag: upper, movable part of the **bellows**; the lower, fixed part is called the **box**

bar iron: the long, narrow iron bars that were the final product of the **shingling** process (preceded by the intermediate products, the **ancony** and **mocket head**), which were then sent to the shops to be made into various objects

barrique: large measure of volume, equal to 6 bushels, French measure

bateau: from the French *bateau*, light, flat-bottomed river boat for transporting raw materials and ironwares

bellows: device supplying air for combustion to the **blast furnace** or **forge** (Plates 3.1 and 4.15)

belly: the widest part of the interior cavity of the **blast furnace**

binne: term for a cartload at the St Maurice Forges, derived from the French *benne*, the dump cart consisting of a box mounted on two or four wheels used to transport **charcoal** and emptied through a door in the bottom (Plates 2.18 and 2.19)

black cast iron: **cast iron** with the highest carbon content

blast: (1) the current of air supplied by an engine or **blower** to a furnace

(2) the period the **blast furnace** was in operation before it closed to rebuild the **hearth**; **in blast** was the time the furnace was making iron; **out of blast** was any period the furnace was not operating

blast furnace: tall shaft furnace used to reduce iron ore to **cast iron**, the first step in the **indirect reduction process**; the second step is **fining**, in which **wrought iron** is produced (Plates 4.3, 4.6 and 4.9)

block: collection of pieces of wood sunk into the ground and held by a masonry-covered frame, forming a support in which the **anvil** is mounted

bloom: initial product of the **shingling** process, resulting when the **loop** is reduced to a squarish mass approximately 10 cm thick

bloomery: a small **charcoal**-fired hearth for the production of **wrought iron** direct from the ore

blow in: the process of gradually putting a **blast furnace** into commission until it could carry full **blast** and burden

blow out: the process of taking the **blast furnace** out of commission

blower: air supply system for a **blast furnace** or **forge** using either **bellows** or a compressor (Plates 3.1 and 4.15)

bog iron (**bog ore**): iron oxide found in shallow deposits just under the humus in bogs, swamps or shallow lakes (Plate 2.6)

boshes: lower, funnel-shaped part of the **blast furnace**, located between the **belly** and the **hearth**, where carburization took place and the materials gradually passed from a pasty stage to a more liquid state (Plate 4.3)

box: lower, fixed part of the **bellows** (the upper movable part is known as the **bag**)

box moulding: technique whereby a sand mould was made in a box of metal or wood, open top and bottom

breastshot: used to describe a waterwheel driven by the weight of water falling between its highest point and the level of its axis (Plate 3.1)

breeze: dust and tiny fragments of charcoal left on the ground after the pit had been charred

bucket: troughs on the waterwheel that catch the water as it falls, thus turning the wheel (Plate 3.1)

buddle: trough with grating in the bottom used to wash the ore

cam: small wooden or metal knob, or shoe, set on a waterwheel-driven shaft used to raise a **hammer** or operate a **bellows** (Plate 4.16)

campaign: from the French *campagne*, period or season during which the **blast furnace** is in continuous operation; see **blast**

cast iron: iron with a high carbon content and significant silicon content, used to make **castings** and as the raw material for steel-making, **wrought iron** and malleable iron making; see **pig iron**

casting: (1) the running of molten metal into a mould prepared for that purpose

(2) piece of metal that takes on the form of the mould in which it is poured

casting house: the enclosed building in front of the mouth of the **blast furnace** in which the **pig bed** was laid out and **pigs** or ingots were cast

chafery: from the French *chaufferie*, hearth in the forge used to heat the blooms to be drawn out by hammering. At the St Maurice Forges, the **chafery** (which was the single-hearth *renardière* type) was used for both **fining** and heating (Plate 4.15)

charcoal: wood that has been distilled, leaving only carbon; formerly used as fuel in ironmaking

charcoal burner: see **collier**

charcoal iron: iron made with **charcoal** fuel

charcoal pit: pile of wood, usually laid in a cone shape and covered with sod and dirt, that was lit and burned to produce **charcoal** (Plate 2.11); also referred to the openings or clearings in the forest (*ventes*) where coaling took place

charge: raw materials used to feed the **blast furnace**: iron ore, **charcoal** (used as fuel and reducing agent) and **flux**

charger: worker who organized the charge for the **blast furnace** into a manageable size and who helped the **filler** load up the baskets of **charcoal**

charging platform: top of the **blast furnace** where the **charge** was loaded into the **throat** (Plates 4.3 and 4.9); also called the filling place

charring: see **coaling**

chill: an iron mould, or a piece of iron in a sand mould for making "chilled" **castings**; specifically at the St Maurice Forges, an iron insert used in wheel moulds to cool the edge and centre of the wheel

chill casting: a moulding technique by which part of the **casting** is chilled or cooled quicker than the rest of the casting by means of an iron block or **chill** moulded in the appropriate part of a sand mould; the chilled part is very hard

chimney plate: upright cast-iron slab or pillar forming the corner between the two open walls of the **chafery**

cinder: see **slag**

cinder notch: small sloping channel in front of the **fore plate** into which the **cinder** from the **finery hearth** runs

clay marl: clay used as **flux** in the **smelting** process, particularly in the 18th century

coaling: the carbonization of wood into **charcoal**, which is almost pure carbon; see **kiln** and **charcoal pit** (Plates 2.11 and 2.14)

cogwheel: in **double gearing**, a wheel with teeth, the latter engaging the **lantern pinion**. On a camshaft, the cogwheel is placed on the end opposite to the **cams** (Plate 3.1)

cold blast: **blowing engine** using cold air

cold-short: a condition of brittleness in iron when cold as a result of an excess of phosphorus

cold working: see **hot working**

collier: worker who made charcoal in **charcoal pits** or **kilns**

come to nature: when the carbon was burnt out of the iron the metal was said to have come to nature

connecting rod: rigid bar with a joint on each end that transmits motion from one moving part to another

core: in **loam moulding**, an internal mould filling the space intended to be left hollow in a hollow **casting**; making the core was the first stage in a three-part process

core casting: hollow ware made from cast iron

crucible: the lowest part or well of the **blast furnace**, between the **hearth** proper and the **boshes**, where the end products of fusion accumulate, consisting of the molten iron and slag; also refers to the lower part of the **finery** or *renardière*-**type chafery** hearth where the pasty mass of iron and **scoria** accumulate

cupola: metal furnace lined with refractory materials (brick) which, like the **blast furnace**, is loaded from the top; generally used for second fusion (Plates 4.20a and 4.20b)

dam: slanted stone or cast iron plate partly closing off the **hearth**, where the **slag** runs out of the **hearth** (Plate 4.3)

direct process: a single-stage reduction process of iron **smelting** directly from the ore

double gearing: double-axled spur-gear mechanism, one for the waterwheel and the other for the camshaft (Plate 3.1)

drome-beam: from the French *drosme*, main beam in the **hurst frame** of the **forge hammer** (Plate 4.16)

dross: mixture of **slag** and **breeze** accumulating at the bottom of the hearth during the **fining** of **cast iron** in a *renardière*-type chafery

estrique: as far as is known, a traction and compression mechanism that drove the **bellows** from below, used briefly at the St Maurice Forges

felloes: the curved boards forming the circular rim of the waterwheel

filler: worker who prepared the **charge** and emptied it into the **throat** of the **blast furnace** (Plate 4.9)

finer: worker who fined the iron **pig** by reducing its carbon content to convert it into **wrought iron**

finery: from the French *affinerie*, hearth in which the iron **pig** is melted down and refined to produce a pasty mass that is subsequently worked with the **forge hammer**; the product of this operation is called the **loop**; see also *renardière* (Plate 4.15)

fining: from the French *affinage*, the process of producing **wrought iron** from **cast iron** by reducing the carbon content

flask moulding: see **box moulding**

flume: wooden channel to direct the flow of water to the waterwheel

flux: material such as **limestone** or **clay** used to separate the ore from the **gangue** in the **blast furnace**

fore plate: front part of **finery** hearth, with a hole to remove the **cinder**

fore spirit plate: from the French *esprit* (a movement of air, a wind), **hearth plate** in the **finery** opposite the **tuyere** and used to keep the **charcoal** covering the **pig** in check

forebay: wooden cistern or reservoir placed above the waterwheel (Plate 4.13)

forge: facility where **cast iron** is refined and converted into **wrought iron**; the two **hearths** or fires in the forge are called the **finery** and the **chafery** (or *renardière*); also refers to the shop where the wrought iron was subsequently forged into various objects (Plates 4.13, 4.15 and 4.16); forges were also sometimes known as "hammer mills," and their ponds, "hammer ponds"

forge hammer: great lift hammer used to draw out the iron after subsequent heats in the finery; see **hammer** (Plate 4.16); the forge was sometimes called the "hammer mill"

forgeman: generic term for forge workers in charge of **shingling** the iron and exposing it to subsequent heats at the forge; more specifically, comprised the **finers** and the **hammermen** (Plates 4.15 and 4.16)

founder: highly skilled worker in charge of overseeing all activities and processes related to the **blast furnace** and the workers performing them (**keeper, fillers, chargers** and **helpers**)

founding: see **ironfounding**

foundry: establishment where **founding**, or **casting**, is carried on; see also **iron mill** and **ironworks**

furnace: see **kiln** and **charcoal pit**; see also **blast furnace**

furnace top: top part of the **belly** of the **blast furnace** that extends above the **stack** (Plate 4.9):

gage: flail-like instrument for measuring the descent of the **charge** in the **blast furnace** and the progress of fusion

gangue: the valueless organic, rock or mineral elements mixed with the ore when it is extracted from the ground

gearing: wheel and its mechanism; see also **double gearing** and **single gearing**

grate: part of the **seasoning** process, the heating of the **hearth** to a white heat before **blowing in** the furnace for **cast iron** production

grey cast iron: **cast iron** used for **moulding**, that is fluid and resistant, and with a higher carbon content than **white cast iron**, making it suitable for the lighter and finer sort of **castings** such as grates and ornamental work

gudgeon: cylindrical piece of metal serving as the axis of the shaft on one end; it rests on the **plummer block**

gutter: channel used to pour **cast iron** into the kettle moulds in the **loam moulding** process

gutterman: worker who prepared the moulds in the sand to make the **iron pigs**; see also **pig bed**

hammer (head): the hammer itself, and its iron head, attached to the end of a swinging shaft and driven by the camshaft, used to work the iron after subsequent heats in the **finery**. Depending on the point at which the **cams** act on the shaft, the hammer was termed a belly helve (action between head and axis of shaft), nose helve (action in front of the head) or **tilt hammer** (action at the end of the shaft) (Plate 4.16)

hammer post: vertical member at either end of the **hurst frame** of the **forge hammer** that supported the **drome-beam** (Plate 4.16); the great hammer post was solidly buttressed on the working side while the lesser hammer post stood by itself on the far side of the forge

hammerman: master craftsman in charge of the **forge** or hammer mill and the operation of the **forge hammer**; at the St Maurice Forges, the hammering and heats were done both by the **finer** and the hammerman (Plate 4.16)

hardener: worker who tempered the axe blades

hare plate: from the French *aire*, iron plate forming the back of the **finery** hearth; one end of the pig was rested on the plate, as it was fed gradually into the fire

head: end of the **bellows** where the **nozzle** is inserted

headrace: channel, most often of wood, directing a stream of water to the top or rear of a **bucket wheel** (Plate 3.1)

hearth: strictly speaking the bottom of the blast furnace but often applied to the **crucible** as well and encompassing the **tuyere** zone where the final stage of fusion occurs

hearth plates: the set of cast-iron plates forming the walls of the **finery** hearth; depending on their position, called the **tuyere plate**, **fore spirit plate**, **hare plate** and **fore plate**

heat: the step in the **shingling** process, which is repeated a number of times, during which the metal is reheated so that it can be worked with the **hammer**

helper: worker who performed ancillary tasks at the **forge** and **blast furnace**; helpers were often apprentices and hence called "boys"

hot blast: blowing engine using hot air; cf. **cold blast**

hot working: operating mode of the **blast furnace** involving the ratio of **charcoal** to ore; the ratio determines the type of **cast iron** produced; a higher ratio of charcoal (hot working furnace) gives **grey cast iron**, while a higher ore ratio (cold working furnace) gives **white cast iron**

hurst: wrought- or cast-iron ring in which the tail of the **hammer** helve is inserted

hurst frame: the **hammer**, frame and the mechanism for driving the hammer (Plate 4.16)

in blast: see **blast**

indirect process: a two-stage process of iron manufacture, where **pig iron** is made from ore by **smelting** and then purified by **fining** to **wrought iron**

inside worker: worker working and living on the Forges **post**, working at the plant and living in the village

inwalls: lining of the **blast furnace**

iron mill: establishment where **bar iron** is made

ironfounding: the melting and **casting** of iron in a mould

ironmaster: the master, or general manager, of an **ironworks**

ironworks: a complete plant having both a **blast furnace** producing **cast iron** and **forges** producing **wrought iron**

keeper: the master **founder**'s principal assistant, who sometimes replaced him at night or on holidays; he was responsible for monitoring and controlling the fusion process in the **blast furnace**

kiln: large oven, usually made of brick, for making **charcoal** (Plate 2.14)

King's domain: the *Domaine du roi*, or the Tadoussac trading concession, a vast tract of land lying north of the Lower St Lawrence and originally belonging to the French kings, in which most of the King's **posts** were found

knot: unit of measurement applied to the speed of a river current, equivalent to one nautical mile/hour or 1.8 km/hour

ladle: bucket-like vessel used to carry and pour the molten iron

lantern pinion: in **double gearing**, a pinion with parallel vertical bars that engage the teeth of the cogwheel (Plate 3.1)

limestone: mineral used as **flux** in smelting; it becomes fused with impurities in the ore and thus is used to separate unwanted materials (**gangue**) from the ore, transforming them into **scoria**

limestone breaker: worker who broke up the iron ore and **limestone** with an iron sledge

lintel: see **morris-bar**

loam moulding: moulding in a wet mixture of sand, clay, straw and horse manure or other binder pasted over a former and strickled by a shaped **strickle** or loam board to the shape required

loop: from the French *loupe*, shapeless pasty mass of molten iron that is the initial product of the **fining** process and which is subsequently worked with the **hammer** (Plate 4.15)

loop plate: cast-iron plate set in the forge floor where the **loop** was placed for the first hammering (Plate 4.15)

lump sum: method of remuneration by which the collier was contracted to oversee the entire **coaling** operation (preparing and setting the pit, leafing and charring) for a fixed price

mantle: in **loam moulding**, the outer mould, or the last step in making the mould; this outer envelope allows the molten **cast iron** to pass between it and the **shell**, or inner mould

manufactures: used in the 19th century to refer to items made by hand in the forge shops

Marine: the Navy department was responsible for the Colonies, fisheries, consulates and the Marseilles Chamber of Commerce, and the Minister was one of the most powerful officials in France

memorial: note, memorandum, report, petition

merchant furnace: furnace set up to make **castings** directly on **tapping**, rather than **pig iron**

merchant iron: **wrought iron** ready to be sold and delivered to shops for processing; see **bar iron**

milldam: dam built across a stream to check its flow and raise its level to make the water available for driving a waterwheel (Plate 4.13)

millpond: reservoir constructed above the **blast furnace** and **forges** to supply flow to the waterwheels (Plate 3.8); forge ponds were known as hammer ponds mine: (1) open-pit iron mine formed of **veins** of different sizes lying just beneath the soil (2) the ore itself (archaic)

miner's bar: iron bar with a bevelled end used to extract **bog iron**

miner's inch: unit of measurement of the flow of water, the amount that will pass in 24 hours through an opening 1 inch square under constant pressure of 6 inches; of Roman origin, equivalent to 14 pints/minute according to the hydraulic engineer Edmé Mariotte (1686); at the St Maurice Forges, the inches of water calculated by the ironmasters referred to the sluice apertures (see Chapter 3)

mocket head: from the French *maquette*, iron bar with a square mass on one end that is the penultimate stage in the **shingling** process (after the **ancony** and before the **bar**)

morris-bar: from the French *marâtre*, large horizontal piece of cast iron used as a structural element in the **blast furnace** and the **finery** and *renardière* hearths

mottled cast iron: mixture of **grey** and **white cast iron** suitable for larger **castings**, such as wheels, beams, pillars, where strength and hardness are desirable

moulder: worker who makes the moulds into which the molten iron is poured

moulding: the pouring of molten metal into a mould made of either sand (**open sand moulding** and **box/flask moulding**), metal (**chill casting**) or loam (**loam moulding**)

moulding shop: shop where **castings** are made in moulds

nozzle: conical tip of the **bellows** through which the air is expelled into the **tuyere**

open-sand moulding: casting direct into a depression formed in sand by pressing a **pattern** into it; only simple shapes could be cast

out of blast: see **blast**

outside worker: worker living and working, strictly speaking, off the Forges **post**; employed at various times of the year at the **charcoal pits**, **mines**, in the bush, on the roads or on the river to collect, prepare or transport raw materials or goods

oven: fire for drying the moulds at various stages in their manufacture

overshot: used to describe a waterwheel driven by the weight of water falling into buckets at its highest point (Plate 4.13)

pattern: a matrix, a mould

patternmaker: see **moulder**

penthouse: masonry structure that surrounded the **furnace top** on three sides, sheltering the **fillers** as they worked and acting as a windbreak for the furnace (Plate 5.1)

pig: see **pig iron**

pig bed: arrangement in the sand of the casting house floor of channels called **sows** and side channels called **pigs**, branching off from the main runner leading from the furnace

pig iron: the crude product of the blast furnace; so called because of the way in which the moulds were arranged, which resembled a sow suckling her piglets, hence the name for the larger mould, the **sow**, from which smaller moulds, called **pigs**, branched off (Plate 4.3); see **cast iron**

pillar of the furnace: corner wall at the foot of the **blast furnace** between the **tymp** (working) arch and the **tuyere** arch

pipe: former French unit of measure used for charcoal and ore; at the St Maurice Forges, a *pipe* of iron ore weighed around 540 kg and a *pipe* of charcoal around 85 kg

placket: from the French *placoire*, a tool designed as a kind of trowel for smoothing and shaping the clay into which the **tuyere** was fixed

plummer block: bearing supporting the camshaft that activates the **forge hammer** (Plate 4.16)

post: for most of their history, the Forges were held in leasehold from the Crown, both French and British, and were thus a King's post

rabbet: flexible but strong counterbeam used to lower the **hammer head** each time it is raised by the **cams** (Plate 4.16)

rabble : (1) [verb] to pry off the **slag** stuck to the walls of the **hearth** and mix it back in with the molten iron, or remove it altogether when there is too much or it is bound with charcoal;
(2) [noun] an iron bar, sharply bent at the end, used at the **blast furnace**

ram: heavy block of iron used to crush the **limestone** (Plate 2.10)

receiver: see **crucible**

renardière: single hearth used as both a **finery** and **chafery**, in the *méthode comtoise* practised in Franche-Comté (Plate 4.15)

ringer: long iron bar used to stir the materials being reduced in the **blast furnace** or **forge**; had other uses as well

rocker: movable lever on a pivot, one end of which was raised by lowering the other and which, attached to the **connecting rod**, was used to open the **sluice** to the **forebay**

rocker arm: rocking lever-type mechanism used to alternately raise the two **bellows** (Plate 3.1)

runner: two pieces of sloping **cast iron** laid side by side into which the **slag** flowed out of the furnace (see **dam**)

run-out: see **tapping**

scoria: the waste products of smelting; see **slag**, **cinder** and **scum**

scrip: a certificate, coupon or voucher of indebtedness issued as currency or in lieu of money

scum: solid **scoria** produced as a byproduct of fusion

seasoning: the process of heating the furnace to evaporate moisture before introducing the **blast** and feeding the **charge**

sharpener: worker responsible for sharpening axes

shell: in **loam moulding**, the inner mould or the second stage in making the mould

shingling: from the French *cingler*, process of using the **forge hammer** and **tilt hammer** to work the **loop** and draw it into **bar iron** (Plate 4.16)

shuttle: floodgate that opens to allow the flow and regulate the supply of water in a mill stream; see **sluice**

single gearing: waterwheel with a single shaft with **cams** on the end, acting directly on the tool (**hammer** or **bellows**) being driven

slag: **scoria** or non-metallic frothy **scum** resulting from the action of the **flux** on the ore and which floats on top of the metal during **smelting**; referred to as slag at the furnace and **cinder** at the forge

sluice: device that allows the flow of water to the waterwheel to be controlled by raising or lowering it (Plate 4.13)

sluicing: supply of head of water required to fill the **millpond** in a given period; a waterwheel worked by sluicing operated only when the water level in the millpond was high enough to power the wheel; when the water level fell below this point, the workers had to wait until a sufficient head of water built up again

smelting: the process of melting the iron ore or method used for doing so

sow: see **pig iron**

stack: masonry structure of the **blast furnace** (Plate 4.3)

stamp mill: device used to crush ore (Plates 2.8 and 2.9)

stoker: worker who loaded and emptied the charcoal **kilns**

strickle: in **loam moulding**, the shaped former used to make the mould

striker: assistant to the **hammerman**, who used a sledgehammer

striker plate: metal part attached to each **bellows** and struck by the **cams**, thus operating each pair of bellows alternately (Plate 3.1)

tafia: a spirituous liquor made from molasses

tailrace: channel or spillway directing the flow of water to the base of a paddle wheel

taphole: opening made at the base of the **hearth** of the **blast furnace**, from which the molten iron is drawn out (Plate 4.3)

tapping: the act of running out the molten iron from the **blast furnace** into a mould to allow it to harden

tenement house: long, multi-unit dwelling housing the Forges workers' families (Plates 9.2 and 9.6)

thousandweight: 1,000 French pounds (489,41 kg) used as a measure of **cast** or **wrought iron**

throat: opening at the top of the **blast furnace** where the **charge** was loaded in the **furnace top** (Plates 4.3 and 4.4)

tilt hammer: waterwheel-driven trip hammer which was smaller and operated at a faster rhythm than the **forge hammer**, used mainly to make round iron and rods

tongs: tool used in various **shingling** operations

turn: method of dividing up the work day in shifts

tuyere: from the French *tuyère*, opening in the lower part of the side of the **blast furnace** or **forge** through which the air from the bellows is directed; in the **blast furnace** the founder used this opening, called the "founder's eye," to see how **smelting** was proceeeding

tuyere muzzle: conical copper piece that extends into the **hearth**, where the **nozzles** of the two **bellows** are attached

tuyere plate: part of the **finery** hearth under the **tuyere**, on which the **tuyere nozzle** rests

tymp: the mouth of the **hearth** of a **blast furnace**, through which the molten metal descends, formed by an arch of masonry (tymp arch), or a block of stone or iron (tymp stone, tymp plate), or by the two of these together

undershot: used to describe a waterwheel driven by the flow of water striking the blades or **buckets** at its base

usufruct: the right of using another's property without injuring or destroying it

vein: a seam of iron ore

vent: conduit for expelling moisture from the **stack** of the **blast furnace** and to distribute the heat and pressure more evenly

Walloon process: a two-hearth version of the **finery** process, in which the **pig iron** was converted into **wrought iron** in the finery hearth and reheating for forging was done in the **chafery** hearth

Western domain: a number of customs and other taxes relating to the Colonies and comprised in what was called the *Domaine d'Occident*, the director of which was entrusted with the administration of certain territories being exploited for the benefit of the King; see **King's domain**

wheelrace: the enclosed shed-like extension to the forge and blast furnace housing a waterwheel (Plate 4.13)

white cast iron: **cast iron** that is very hard and brittle, with a white fracture and with the lowest carbon content, used for **fining** into **wrought iron** and therefore sometimes called forge pig

wrought iron: a form of mechanically shaped pure iron with threads of **slag** or **cinder**

References

Barriault, Monique. *Lexique français-anglais des termes de fonderie* in "Rapport préliminaire sur l'identification des techniques de moulage utilisées aux Forges du St-Maurice: étude faite à partir des déchets de moulage." Manuscript Report No. 330. Quebec City: Parks Canada, 1978, 127, 55 p.

Binning, Arthur C. *Pennsylvania Iron Manufacture in the Eighteenth Century*. New York: Augustus M. Kelley Publishers, 1970 [1938], 2xxp.

Cossons, Neil, ed. *Rees's Manufacturing Industry (1819–20). A Selection from the Cyclopaedia; or Universal Dictionary of Arts, Sciences and Literature by Abraham Rees*. 5 vols. Newton Abbot: David & Charles, 19xx.

Diderot, Denis and Jean D'Alembert. *Encyclopédie ou Dictionnaire raisonné des sciences, des arts et des métiers, par une société de gens de lettres*. Paris: 1776 s.v. "Forges, (Grosses-)"; *idem, Recueil de planches sur les sciences, les arts libéraux et les arts méchaniques, avec leur explication: forges ou l'art du fer*. Paris: Inter-livres, reprinting, 45 p., n.p.

Gale, W. K. V. *The Iron and Steel Industry: a Dictionary of Terms*. Newton Abbot: David & Charles, 1971. 238 p.

Gillispie, Charles Coulston. *A Diderot Pictorial Encyclopedia of Trades and Industry: Manufacturing and the Technical Arts in Plates Selected from "L'Encyclopédie, ou Dictionnaire Raisonné des Sciences, des Arts et des Métiers of Denis Diderot*. 2 vols. New York: Dover Publications, 1959.

Hartley, E. N. *Ironworks on the Saugus: The Lynn and Braintree Ventures of the Company of Undertakers of the Ironworks in New England*. Norman: University of Oklahoma Press.

L'Heureux, Réjean. *Vocabulaire du moulin traditionnel au Québec des origines à nos jours: documents lexicaux et ethnographiques*. Quebec City: Les Presses de l'Université Laval, 1982, 465 p.

Oxford English Dictionary (Second edition on compact disc).

Schubert, H.R. *History of the British Iron and Steel Industry from c. 450 B.C. to A.D. 1775*. London: Routledge & Kegan Paul, 1959, 437 p.

Straker, E. *Wealden Iron*. Newton Abbot: David & Charles, 1969, 487 p.

Unglik, Henry. *Cast Irons from Les Forges du Saint-Maurice, Quebec: a Metallurgical Study*. Ottawa: Environment Canada, Canadian Parks Service, 1990, 61 p.

Index